T0342171

PROBABILITY & STATISTICS FOR ECONOMISTS

PROBABILITY & STATISTICS FOR ECONOMISTS

BRUCE E. HANSEN

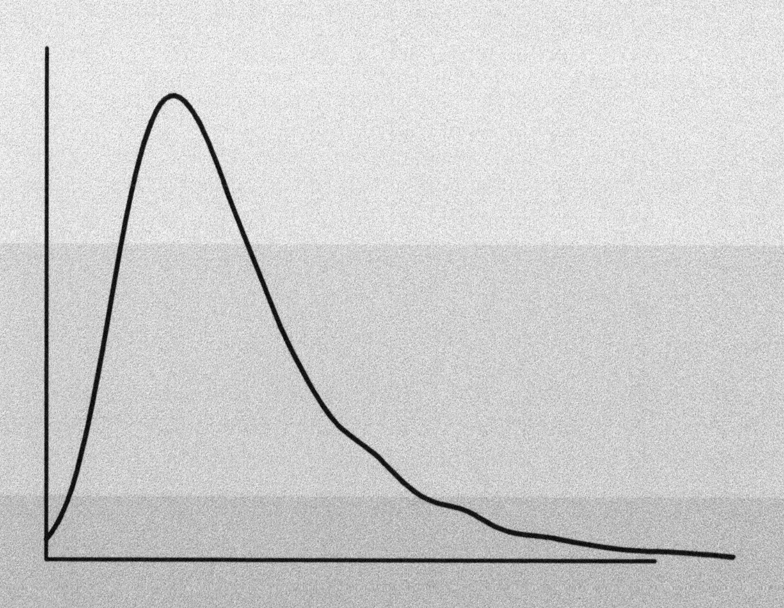

Princeton University Press
Princeton and Oxford

Published by Princeton University Press
41 William Street, Princeton, New Jersey 08540
99 Banbury Road, Oxford OX2 6JX

press.princeton.edu

Library of Congress Cataloging-in-Publication Data

Names: Hansen, Bruce E., 1962– author.
Title: Probability and statistics for economists / Bruce E. Hansen.
Description: Princeton : Princeton University Press, [2022] |
Includes bibliographical references and index.
Identifiers: LCCN 2021049277 (print) | LCCN 2021049278 (ebook) |
ISBN 9780691235943 (hardcover) | ISBN 9780691236148 (ebook)
Subjects: LCSH: Econometrics. | Probabilities. | Statistics.
Classification: LCC HB139 .H3638 2022 (print) | LCC HB139 (ebook) |
DDC 330.01/5195—dc23/eng/20220128
LC record available at https://lccn.loc.gov/2021049277
LC ebook record available at https://lccn.loc.gov/2021049278

British Library Cataloging-in-Publication Data is available

Editorial: Joe Jackson, Josh Drake
Production Editorial: Terri O'Prey
Text Design: Wanda España
Cover Design: Wanda España
Production: Jacqueline Poirier
Publicity: Kate Hensley, Charlotte Coyne
Copyeditor: Cyd Westmoreland

This book has been composed in Minion Pro

10 9 8 7 6 5 4 3 2 1

CONTENTS

18 EMPIRICAL PROCESS THEORY 352

APPENDIX: MATHEMATICS REFERENCE 367

PREFACE

This textbook is the first in a two-part series covering the core material typically taught in a one-year Ph.D. course in econometrics. The sequence is

1. *Probability and Statistics for Economists* (this volume)
2. *Econometrics* (the next volume).

The textbooks are written as an integrated series, but either can be used as a standalone course textbook.

This first volume covers intermediate-level mathematical statistics. It is a gentle yet a rigorous treatment using calculus but not measure theory. The level of detail and rigor is similar to that of Casella and Berger (2002) and Hogg and Craig (1995). The material is explained using examples at the level of Hogg and Tanis (1997) and is targeted to students of economics. The goal is to be accessible to students with a variety of backgrounds and yet maintain full mathematical rigor.

Readers who desire a gentler treatment may try Hogg and Tanis (1997). Readers who desire more detail are urged to read Casella and Berger (2002) or Shao (2003). Readers wanting a measure-theoretic foundation in probability should read Ash (1972) or Billingsley (1995). For advanced statistical theory, see van der Vaart (1998), Lehmann and Casella (1998), and Lehmann and Romano (2005), each of which has a different emphasis. Mathematical statistics textbooks with similar goals as this textbook include Ramanathan (1993), Amemiya (1994), Gallant (1997), and Linton (2017).

Technical material that is not essential for the main concepts is presented in the starred (*) sections. This material is intended for students interested in the mathematical details. Others may skip these sections with no loss of concepts.

Chapters 1–5 cover probability theory. Chapters 6–18 cover statistical theory.

The end-of-chapter exercises are an important part of the text and are central for learning the material presented in the book.

This textbook could be used for a one-semester course. It can also be used for a one-quarter course (as done at the University of Wisconsin) if some topics are skipped. For example, the material in Chapter 3 could be viewed as reference rather than being taught; Chapter 9 is for advanced students; Chapter 11 can be covered in brief; Chapter 12 can be left for reference; and Chapters 15–18 are optional, depending on the goals and interests of the instructor.

ACKNOWLEDGMENTS

This book and its companion, *Econometrics*, would not have been possible if not for the amazing flow of volunteered advice, corrections, comments, and questions I have received from students, faculty, and other readers over the 20 years I have worked on this project. I have received emailed corrections and comments from so many individuals that I have completely lost track of the list. So rather than publish an incomplete list, I simply give a grateful and sincere *Thank You* to every single one of them.

Special thanks go to Xiaoxia Shi, who typed up my handwritten notes for my Econ 709 course a few years ago, creating a preliminary draft for this manuscript.

My most heartfelt thanks go to my family: Korinna, Zoe, and Nicholas. Without their love and support over these years, this project would not have been possible.

All of the author's royalties will be gifted to charitable causes.

MATHEMATICAL PREPARATION

Students should be familiar with integral, differential, and multivariate calculus, as well as linear matrix algebra. This is the material typically taught in a four-course undergraduate mathematics sequence at a U.S. university. No prior coursework in probability, statistics, or econometrics is assumed, but it would be helpful.

It is also highly recommended, but not necessary, to have studied mathematical analysis and/or a "prove-it" mathematics course. The language of probability and statistics is mathematics. To understand the concepts, you need to derive the methods from their principles. This is different from introductory statistics, which unfortunately often emphasizes memorization. By taking a mathematical approach, little memorization is needed. Instead, such an approach requires a facility with detailed mathematical derivations and proofs. The reason it is recommended to have studied mathematical analysis is not that we will be using such results. The reason is that the method of thinking and proof structures are similar. We start with the axioms of probability and build the structure of probability theory from these axioms. Once probability theory is built, we construct statistical theory on that base. A timeless introduction to mathematical analysis is Rudin (1976). For those wanting more, Rudin (1987) is recommended.

The appendix to the textbook contains a brief summary of important mathematical results and is included for reference.

NOTATION

Real numbers (elements of the real line $\mathbb{R}$, also called **scalars**) are written using lowercase italics, such as x.

Vectors (elements of $\mathbb{R}^k$) are typically written using lowercase italics, such as x, and sometimes using lowercase bold italics, such as $\boldsymbol{x}$ (for matrix algebra expressions). For example, we write

$$x = \begin{pmatrix} x_1 \\ x_2 \\ \vdots \\ x_k \end{pmatrix}.$$

Vectors by default are written as column vectors. The **transpose** of x is the row vector

$$x' = \begin{pmatrix} x_1 & x_2 & \cdots & x_m \end{pmatrix}.$$

There is diversity across fields concerning the choice of notation for the transpose. The notation x' is the most common in econometrics. In statistics and mathematics, the notation $x^\top$ is typically used, or occasionally x^t.

Matrices are written using uppercase bold italics, for example,

$$\boldsymbol{A} = \begin{bmatrix} a_{11} & a_{12} \\ a_{21} & a_{22} \end{bmatrix}.$$

Random variables and vectors are written using uppercase italics, such as X.

Typically, Greek letters, such as β, θ, and σ^2, are used to denote parameters of a probability model. Estimators are typically denoted by putting a hat "^", tilde "~" or bar "-" over the corresponding letter, for example, $\widehat{\beta}$ and $\widetilde{\beta}$ are estimators of β.

COMMON SYMBOLS

a	scalar
a or $\boldsymbol{a}$	vector
$\boldsymbol{A}$	matrix
X	random variable or vector
$\mathbb{R}$	real line
$\mathbb{R}_+$	positive real line
$\mathbb{R}^k$	Euclidean k space
$\mathbb{P}[A]$	probability

$\mathbb{P}[A \mid B]$	conditional probability
$F(x)$	cumulative distribution function
$\pi(x)$	probability mass function
$f(x)$	probability density function
$\mathbb{E}[X]$	mathematical expectation
$\mathbb{E}[Y \mid X=x]$, $\mathbb{E}[Y \mid X]$	conditional expectation
$\text{var}[X]$	variance, or covariance matrix
$\text{var}[Y \mid X]$	conditional variance
$\text{cov}(X, Y)$	covariance
$\text{corr}(X, Y)$	correlation
$\overline{X}_n$	sample mean
$\widehat{\sigma}^2$	sample variance
s^2	biased-corrected sample variance
$\widehat{\theta}$	estimator
$s(\widehat{\theta})$	standard error of estimator
$\lim_{n\to\infty}$	limit
$\text{plim}_{n\to\infty}$	probability limit
$\to$	convergence
$\underset{p}{\longrightarrow}$	convergence in probability
$\underset{d}{\longrightarrow}$	convergence in distribution
$L_n(\theta)$	likelihood function
$\ell_n(\theta)$	log-likelihood function
$\mathscr{I}_\theta$	information matrix
$\text{N}(0, 1)$	standard normal distribution
$\text{N}(\mu, \sigma^2)$	normal distribution with mean μ and variance σ^2
χ_k^2	chi-square distribution with k degrees of freedom
$\boldsymbol{I}_n$	$n \times n$ identity matrix
$\text{tr}\,\boldsymbol{A}$	trace of matrix $\boldsymbol{A}$
$\boldsymbol{A}'$	vector or matrix transpose
$\boldsymbol{A}^{-1}$	matrix inverse
$\boldsymbol{A} > 0$	positive definite
$\boldsymbol{A} \geq 0$	positive semi-definite
$\|a\|$	Euclidean norm

$\mathbb{1}\{A\}$	indicator function (1 if A is true, else 0)
$\simeq$	approximate equality
$\sim$	is distributed as
$\log(x)$	natural logarithm
$\exp(x)$	exponential function
$\sum_{i=1}^{n}$	summation from $i=1$ to $i=n$

GREEK ALPHABET

It is common in economics and econometrics to use Greek characters to augment the Latin alphabet. The following table lists the various Greek characters and their pronunciations in English. The second character, when listed, is the upper case character (except for ϵ, which is an alternative script for ε.)

Greek Character	**Name**	**Latin Keyboard Equivalent**
α	alpha	a
β	beta	b
γ, Γ	gamma	g
δ, Δ	delta	d
ε, ϵ	epsilon	e
ζ	zeta	z
η	eta	h
θ, Θ	theta	y
ι	iota	i
κ	kappa	k
λ, Λ	lambda	l
μ	mu	m
ν	nu	n
ξ, Ξ	xi	x
π, Π	pi	p
ρ	rho	r
σ, Σ	sigma	s
τ	tau	t
υ	upsilon	u
ϕ, Φ	phi	f
χ	chi	x
ψ, Ψ	psi	c
ω, Ω	omega	w

PROBABILITY & STATISTICS FOR ECONOMISTS

CHAPTER 1

BASIC PROBABILITY THEORY

1.1 INTRODUCTION

Probability theory is foundational for economics and econometrics. Probability is the mathematical language used to handle uncertainty, which is central for modern economic theory. Probability theory is also the foundation of mathematical statistics, which is the foundation of econometric theory.

Probability is used to model uncertainty, variability, and randomness. When we say that something is "uncertain", we mean that the outcome is unknown. For example, how many students will there be in next year's Ph.D. entering class at your university? "Variability" means that the outcome is not the same across all occurrences. For example, the number of Ph.D. students fluctuates from year to year. "Randomness" means that the variability has some sort of pattern. For example, the number of Ph.D. students may fluctuate between 20 and 30, with 25 more likely than either 20 or 30. Probability gives us a mathematical language to describe uncertainty, variability, and randomness.

1.2 OUTCOMES AND EVENTS

Suppose you take a coin, flip it in the air, and let it land on the ground. What will happen? Will the result be "heads" (H) or "tails" (T)? We do not know the result in advance, so we describe the outcome as **random**.

Suppose you record the change in the value of a stock index over a period of time. Will the value increase or decrease? Again, we do not know the result in advance, so we describe the outcome as random.

Suppose you select an individual at random and survey them about their economic situation. What is their hourly wage? We do not know in advance. The lack of foreknowledge leads us to describe the outcome as random.

We will use the following terms.

An **outcome** is a specific result. For example, in a coin flip, an outcome is either H or T. If two coins are flipped in sequence, we can write an outcome as HT for a head and then a tails. A roll of a six-sided die has the six outcomes $\{1, 2, 3, 4, 5, 6\}$.

The **sample space** S is the set of all possible outcomes. In a coin flip, the sample space is $S = \{H, T\}$. If two coins are flipped, the sample space is $S = \{HH, HT, TH, TT\}$.

An **event** A is a subset of outcomes in S. An example event from the roll of a die is $A = \{1, 2\}$.

The one-coin and two-coin sample spaces are illustrated in Figure 1.1. The event $\{HH, HT\}$ is illustrated by the ellipse in Figure 1.1(b).

Set theoretic manipulations are helpful in describing events. We will use the following concepts.

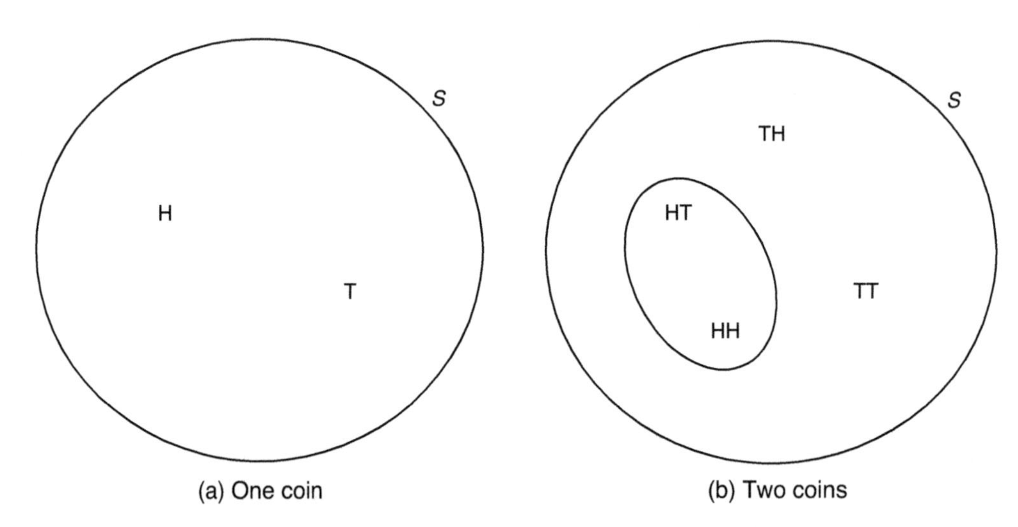

FIGURE 1.1 Sample space

Definition 1.1 For events A and B:

1. A is a **subset** of B, written $A \subset B$, if every element of A is an element of B.
2. The event with no outcomes $\varnothing = \{\}$ is called the **null** or **empty set**.
3. The **union** $A \cup B$ is the collection of all outcomes that are in either A **or** B (or both).
4. The **intersection** $A \cap B$ is the collection of elements that are in both A **and** B.
5. The **complement** A^c of A are all outcomes in S which are not in A.
6. The events A and B are **disjoint** if they have no outcomes in common: $A \cap B = \varnothing$.
7. The events $A_1, A_2, \ldots$ are a **partition** of S if they are mutually disjoint and their union is S.

Events satisfy the rules of set operations, including the commutative, associative, and distributive laws. The following theorem is useful.

Theorem 1.1 **Partitioning Theorem**. If $\{B_1, B_2, \cdots\}$ is a partition of S, then for any event A,

$$A = \bigcup_{i=1}^{\infty} (A \cap B_i) .$$

The sets $(A \cap B_i)$ are mutually disjoint.

A proof is provided in Section 1.15.

1.3 PROBABILITY FUNCTION

Definition 1.2 A function $\mathbb{P}$ which assigns a numerical value to events[1] is called a **probability function** if it satisfies the following **axioms of probability**:

1. $\mathbb{P}[A] \geq 0$.

[1] For events in a sigma field. See Section 1.14.

2. $\mathbb{P}[S] = 1$.

3. If $A_1, A_2, \ldots$ are disjoint, then $\mathbb{P}\left[\bigcup_{j=1}^{\infty} A_j\right] = \sum_{j=1}^{\infty} \mathbb{P}\left[A_j\right]$.

This textbook uses the notation $\mathbb{P}[A]$ for the probability of an event A. Other common notations include $P(A)$ and $\Pr(A)$.

Let us examine this definition. The phrase "a function $\mathbb{P}$ which assigns a numerical value to events" means that $\mathbb{P}$ is a function from the space of events to the real line. Thus probabilities are numbers. Now consider the axioms. The first axiom states that probabilities are nonnegative. The second axiom is essentially a normalization: the probability that "something happens" is 1.

The third axiom imposes considerable structure. It states that probabilities are additive on disjoint events. That is, if A and B are disjoint, then

$$\mathbb{P}[A \cup B] = \mathbb{P}[A] + \mathbb{P}[B].$$

Take, for example, the roll of a six-sided die which has the possible outcomes $\{1, 2, 3, 4, 5, 6\}$. Since the outcomes are mutually disjoint, the third axiom states that $\mathbb{P}[1 \text{ or } 2] = \mathbb{P}[1] + \mathbb{P}[2]$.

When using the third axiom, it is important to be careful that it is applied only to disjoint events. Take, for example, the roll of a pair of dice. Let A be the event "1 on the first roll" and B the event "1 on the second roll". It is tempting to write $\mathbb{P}[\text{"1 on either roll"}] = \mathbb{P}[A \cup B] = \mathbb{P}[A] + \mathbb{P}[B]$, but the second equality is incorrect, since A and B are not disjoint. The outcome "1 on both rolls" is an element of both A and B.

Any function $\mathbb{P}$ which satisfies the axioms is a valid probability function. Take the coin flip example. One valid probability function sets $\mathbb{P}[H] = 0.5$ and $\mathbb{P}[T] = 0.5$. (This is typically called a **fair coin**.) A second valid probability function sets $\mathbb{P}[H] = 0.6$ and $\mathbb{P}[T] = 0.4$. However, a function which sets $\mathbb{P}[H] = -0.6$ is not valid (it violates the first axiom), and a function which sets $\mathbb{P}[H] = 0.6$ and $\mathbb{P}[T] = 0.6$ is not valid (it violates the second axiom).

While the definition states that a probability function must satisfy certain rules, it does not describe the *meaning* of probability. The reason is because there are multiple interpretations. One view is that probabilities are the relative frequency of outcomes, as in a controlled experiment. The probability that the stock market will increase is the frequency of increases. The probability that an unemployment duration will exceed one month is the frequency of unemployment durations exceeding one month. The probability that a basketball player will make a free throw shot is the frequency with which the player makes free throw shots. The probability that a recession will occur is the relative frequency of recessions. In some examples, this definition is conceptually straightforward, as the experiment repeats or has multiple occurances. In other cases, a situation occurs exactly once and will never be repeated. As I write this paragraph, questions of uncertainty of general interest include "Will global warming exceed 2 degrees?" and "When will the COVID-19 epidemic end?" In these cases, it is difficult to interpret a probability as a relative frequency, as the outcome can only occur once. The interpretation can be salvaged by viewing "relative frequency" abstractly by imagining many alternative universes which start from the same initial conditions but evolve randomly. While this solution works (technically), it is not completely satisfactory.

Another view is that probability is subjective. This view holds that probabilities can be interpreted as degrees of belief. If I say "The probability of rain tomorrow is 80%", I mean that this is my personal subjective assessment of the likelihood based on the information available to me. This view may seem too broad, as it allows for arbitrary beliefs, but the subjective interpretation requires subjective probability to follow the axioms and rules of probability. A major disadvantage associated with this approach is that it is not necessarily appropriate for scientific discourse.

What is common between the two definitions is that the probability function follows the same axioms—otherwise, the label "probability" should not be used.

This concept can be illustrated with two real-world examples. The first is from finance. Let U be the event that the S&P stock index increases in a given week, and let D be the event that the index decreases. This is similar to a coin flip. The sample space is $\{U, D\}$. We compute[2] that $\mathbb{P}[U] = 0.57$ and $\mathbb{P}[D] = 0.43$. The probability 57% of an increase is somewhat higher than a fair coin. The probability interpretation is that the index will increase in value in 57% of randomly selected weeks.

The second example concerns wage rates in the United States. Take a randomly selected wage earner. Let H be the event that their wage rate exceeds \$25/hour, and L be the event that their wage rate is less than \$25/hour. Again the structure is similar to a coin flip. We calculate[3] that $\mathbb{P}[H] = 0.31$ and $\mathbb{P}[L] = 0.69$. To interpret this as a probability, we can imagine surveying a random individual. Before the survey, we know nothing about the individual. Their wage rate is uncertain and random.

1.4 PROPERTIES OF THE PROBABILITY FUNCTION

The following properties of probability functions can be derived from the axioms of probability.

Theorem 1.2 For events A and B, the following properties hold:

1. $\mathbb{P}[A^c] = 1 - \mathbb{P}[A]$.
2. $\mathbb{P}[\varnothing] = 0$.
3. $\mathbb{P}[A] \leq 1$.
4. **Monotone Probability Inequality**: If $A \subset B$, then $\mathbb{P}[A] \leq \mathbb{P}[B]$.
5. **Inclusion-Exclusion Principle**: $\mathbb{P}[A \cup B] = \mathbb{P}[A] + \mathbb{P}[B] - \mathbb{P}[A \cap B]$.
6. **Boole's Inequality**: $\mathbb{P}[A \cup B] \leq \mathbb{P}[A] + \mathbb{P}[B]$.
7. **Bonferroni's Inequality**: $\mathbb{P}[A \cap B] \geq \mathbb{P}[A] + \mathbb{P}[B] - 1$.

Proofs are provided in Section 1.15.

Property 1 states that the probability that an event does not occur equals 1 minus the probability that the event occurs.

Property 2 states that "nothing happens" occurs with 0 probability. (Remember this when asked "What happened today in class?")

Property 3 states that probabilities cannot exceed 1.

Property 4 shows that larger sets necessarily have larger probability.

Property 5 is a useful decomposition of the probability of the union of two events.

Properties 6 and 7 are implications of the inclusion-exclusion principle and are frequently used in probability calculations. Boole's inequality shows that the probability of a union is bounded by the sum of the individual probabilities. Bonferroni's inequality shows that the probability of an intersection is bounded below by an expression involving the individual probabilities. A useful feature of these inequalities is that the right-hand sides only depend on the individual probabilities.

---

[2]Calculated from a sample of 3,584 weekly prices of the S&P Index between 1950 and 2017.

[3]Calculated from a sample of 50,742 U.S. wage earners in 2009.

A further comment related to property 2 is that any event which occurs with probability 0 or 1 is called **trivial**. Such events are essentially nonrandom. In the coin flip example, we could define the sample space as $S = \{H, T, \text{Edge}, \text{Disappear}\}$, where "Edge" means the coin lands on its edge and "Disappear" means the coin disappears into the air. If $\mathbb{P}\left[\text{Edge}\right] = 0$ and $\mathbb{P}\left[\text{Disappear}\right] = 0$, then these events are trivial.

1.5 EQUALLY LIKELY OUTCOMES

When we build probability calculations from foundations, it is often useful to consider settings where symmetry implies that a set of outcomes is equally likely. Standard examples are a coin flip and the toss of a die. We describe a coin as **fair** if the event of a head is as equally likely as the event of a tail. We describe a die as **fair** if the event of each face is equally likely. Applying the axioms, we deduce the following.

Theorem 1.3 Principle of Equally Likely Outcomes: If an experiment has N outcomes $a_1, \dots, a_N$ which are symmetric in the sense that each outcome is equally likely, then $\mathbb{P}[a_i] = \frac{1}{N}$.

For example, a fair coin satisfies $\mathbb{P}[H] = \mathbb{P}[T] = 1/2$, and a fair die satisfies $\mathbb{P}[1] = \cdots = \mathbb{P}[6] = 1/6$.

In some contexts, deciding which outcomes are symmetric and equally likely can be confusing. Take the two-coin example. We could define the sample space as {HH,TT,HT}, where HT means "one head and one tail". If we guess that all outcomes are equally likely, we would set $\mathbb{P}[HH] = 1/3$, etc. However, if we define the sample space as {HH,TT,HT,TH} and guess that all outcomes are equally likely, we would find $\mathbb{P}[HH] = 1/4$. Both answers (1/3 and 1/4) cannot be correct. The implication is that we should not apply the principle of equally likely outcomes simply because there is a list of outcomes. Instead, there should be a justifiable reason for the outcomes to be equally likely. In this two-coin example, there is no principled reason for symmetry without further analysis, so the property should not be applied. We return to this issue in Section 1.8.

1.6 JOINT EVENTS

Take two events H and C. For concreteness, let H be the event that an individual's wage exceeds $25/hour, and let C be the event that the individual has a college degree. We are interested in the probability of the joint event $H \cap C$. This is the event "H and C", or in words, that the individual's wage exceeds $25/hour and they have a college degree. Previously it was noted that $\mathbb{P}[H] = 0.31$. We can similarly calculate that $\mathbb{P}[C] = 0.36$. What about the joint event $H \cap C$?

From Theorem 1.2, we can deduce that $0 \le \mathbb{P}[H \cap C] \le 0.31$. (The upper bound is Bonferroni's inequality.) Thus from the knowledge of $\mathbb{P}[H]$ and $\mathbb{P}[C]$ alone, we can bound the joint probability but not determine its value. It turns out that the actual[4] probability is $\mathbb{P}[H \cap C] = 0.19$.

From the three known probabilities and the properties of Theorem 1.2, we can calculate the probabilities of the various intersections. The results are displayed in the following chart. The four numbers in the central box are the probabilities of the joint events; for example, 0.19 is the probability of both a high wage and a college degree. The largest of the four probabilities is 0.52: the joint event of a low wage and no college degree. The four probabilities sum to 1, because the events are a partition of the sample space. The sums of the probabilities in

[4] Calculated from the same sample of 50,742 U.S. wage earners in 2009.

each column are reported in the bottom row: the probabilities of a college degree and no degree, respectively. The sums by row are reported in the rightmost column: the probabilities of a high and low wage, respectively.

Joint Probabilities: Wages and Education

	C	N	Any Education
H	0.19	0.12	0.31
L	0.17	0.52	0.69
Any Wage	0.36	0.64	1.00

As another illustration, let us examine stock price changes. We reported before that the probability of an increase in the S&P stock index in a given week is 57%. Now consider the change in the stock index over 2 sequential weeks. What is the joint probability? The results are displayed in the following chart. U_t means that the index increases, D_t means that the index decreases, U_{t-1} means that the index increases in the previous week, and D_{t-1} means that the index decreases in the previous week.

Joint Probabilities: Stock Returns

	U_{t-1}	D_{t-1}	Any Past Return
U_t	0.322	0.245	0.567
D_t	0.245	0.188	0.433
Any Return	0.567	0.433	1.000

The four numbers in the central box sum to 1, since they are a partition of the sample space. We can see that the probability that the stock price increases for 2 weeks in a row is 32% and that it decreases for 2 weeks in a row is 19%. The probability is 25% for an increase followed by a decrease, and also 25% for a decrease followed by an increase.

1.7 CONDITIONAL PROBABILITY

Take two events A and B. For example, let A be the event "Receive a grade of A on the econometrics exam", and let B be the event "Study econometrics 12 hours a day". We might be interested in the question: Does B affect the likelihood of A? Alternatively, we may be interested in questions such as: Does attending college affect the likelihood of obtaining a high wage? Or: Do tariffs affect the likelihood of price increases? These are questions of **conditional probability**.

Abstractly, consider two events A and B. Suppose that we know that B has occurred. Then the only way for A to occur is if the outcome is in the intersection $A \cap B$. So we are asking: "What is the probability that $A \cap B$ occurs, given that B occurs?" The answer is not simply $\mathbb{P}[A \cap B]$. Instead, we can think of the "new" sample space as B. To do so, we normalize all probabilities by $\mathbb{P}[B]$. We arrive at the following definition.

Definition 1.3 If $\mathbb{P}[B] > 0$, the **conditional probability** of A given B is

$$\mathbb{P}[A \mid B] = \frac{\mathbb{P}[A \cap B]}{\mathbb{P}[B]}.$$

The notation "$A \mid B$" means "A given B" or "A assuming that B is true". To add clarity, we will sometimes refer to $\mathbb{P}[A]$ as the **unconditional probability** to distinguish it from $\mathbb{P}[A \mid B]$.

For example, take the roll of a fair die. Let $A = \{1, 2, 3, 4\}$ and $B = \{4, 5, 6\}$. The intersection is $A \cap B = \{4\}$, which has probability $\mathbb{P}[A \cap B] = 1/6$. The probability of B is $\mathbb{P}[B] = 1/2$. Thus $\mathbb{P}[A \mid B] = (1/6)/(1/2) = 1/3$. This can also be calculated by observing that conditional on B, the events $\{4\}$, $\{5\}$, and $\{6\}$ each have probability 1/3. Event A only occurs given B if $\{4\}$ occurs. Thus $\mathbb{P}[A \mid B] = \mathbb{P}[4 \mid B] = 1/3$.

Consider our example of wages and college education. From the probabilities reported in Section 1.6, we can calculate that

$$\mathbb{P}[H \mid C] = \frac{\mathbb{P}[H \cap C]}{\mathbb{P}[C]} = \frac{0.19}{0.36} = 0.53$$

and

$$\mathbb{P}[H \mid N] = \frac{\mathbb{P}[H \cap N]}{\mathbb{P}[N]} = \frac{0.12}{0.64} = 0.19.$$

There is a considerable difference in the conditional probability of receiving a high wage conditional on a college degree: 53% versus 19%.

As another illustration, let us examine stock price changes. We calculate that

$$\mathbb{P}[U_t \mid U_{t-1}] = \frac{\mathbb{P}[U_t \cap U_{t-1}]}{\mathbb{P}[U_{t-1}]} = \frac{0.322}{0.567} = 0.568$$

and

$$\mathbb{P}[U_t \mid D_{t-1}] = \frac{\mathbb{P}[U_t \cap D_{t-1}]}{\mathbb{P}[D_{t-1}]} = \frac{0.245}{0.433} = 0.566.$$

In this case, the two conditional probabilities are essentially identical. Thus the probability of a price increase in a given week is unaffected by the previous week's result. This is an important special case and is explored further in the next section.

1.8 INDEPENDENCE

We say that events are **independent** if their occurrence is unrelated, or equivalently, that the knowledge of one event does not affect the conditional probability of the other event. Take two coin flips. If there is no mechanism connecting the two flips, we would typically expect that neither flip is affected by the outcome of the other. Similarly, if we take two die throws, we typically expect there is no mechanism connecting the throws and thus no reason to expect that one is affected by the outcome of the other. As a third example, consider the crime rate in London and the price of tea in Shanghai. There is no reason to expect one of these two events to affect the other event.[5] In each of these cases, we describe the events as independent.

This discussion implies that two unrelated (independent) events A and B will satisfy the properties $\mathbb{P}[A \mid B] = \mathbb{P}[A]$ and $\mathbb{P}[B \mid A] = \mathbb{P}[B]$. In words, the probability that a coin is H is unaffected by the outcome (H or T) of another coin. From the definition of conditional probability, this implies $\mathbb{P}[A \cap B] = \mathbb{P}[A]\,\mathbb{P}[B]$. Let us use this as the formal definition.

Definition 1.4 The events A and B are **statistically independent** if $\mathbb{P}[A \cap B] = \mathbb{P}[A]\,\mathbb{P}[B]$.

We typically use the simpler label **independent** for brevity. As an immediate consequence of the derivation, we obtain the following equivalence.

[5] Except in a James Bond movie.

Theorem 1.4 If A and B are independent with $\mathbb{P}[A] > 0$ and $\mathbb{P}[B] > 0$, then

$$\mathbb{P}[A \mid B] = \mathbb{P}[A]$$
$$\mathbb{P}[B \mid A] = \mathbb{P}[B].$$

Consider the stock index illustration in Section 1.6. We found that $\mathbb{P}[U_t \mid U_{t-1}] = 0.57$ and $\mathbb{P}[U_t \mid D_{t-1}] = 0.57$. This means that the probability of an increase is unaffected by the outcome from the previous week, which satisfies the definition of independence. It follows that the events U_t and U_{t-1} are independent.

When events are independent, then joint probabilities can be calculated by multiplying individual probabilities. Take two independent coin flips. Write the possible results of the first coin as $\{H_1, T_1\}$ and the possible results of the second coin as $\{H_2, T_2\}$. Let $p = \mathbb{P}[H_1]$ and $q = \mathbb{P}[H_2]$. We obtain the following chart for the joint probabilities.

Joint Probabilities: Independent Events

	H_1	T_1	
H_2	pq	$(1-p)q$	q
T_2	$p(1-q)$	$(1-p)(1-q)$	$1-q$
	p	$1-p$	1

The chart shows that the four joint probabilities are determined by p and q, the probabilities of the individual coins. The entries in each column sum to p and $1-p$, and the entries in each row sum to q and $1-q$.

If two events are not independent, we say that they are **dependent**. In this case, the joint event $A \cap B$ occurs at a different rate than predicted if the events were independent.

For example, consider wage rates and college degrees. We have already shown that the conditional probability of a high wage is affected by a college degree, which demonstrates that the two events are dependent. What we now do is see what happens when we calculate the joint probabilities from the individual probabilities under the (false) assumption of independence. The results are shown in the following chart.

Joint Probabilities: Wages and Education

	C	N	Any Education
H	0.11	0.20	0.31
L	0.25	0.44	0.69
Any Wage	0.36	0.64	1.00

The entries in the central box are obtained by multiplication of the individual probabilities (e.g., $\mathbb{P}[H \cap C] = 0.31 \times 0.36 = 0.11$). What we see is that the diagonal entries are much smaller, and the off-diagonal entries are much larger, than the corresponding correct joint probabilities. In this example, the joint events $H \cap C$ and $L \cap N$ occur more frequently than that predicted if wages and education were independent.

We can use independence to make probability calculations. Take the two-coin example. If two sequential fair coin flips are independent, then the probability that both are heads is

$$\mathbb{P}[H_1 \cap H_2] = \mathbb{P}[H_1] \times \mathbb{P}[H_2] = \frac{1}{2} \times \frac{1}{2} = \frac{1}{4}.$$

This addresses the issue raised in Section 1.5. The probability of HH is 1/4, not 1/3. The key is the assumption of independence, not how the outcomes are listed.

As another example, consider throwing a pair of fair dice. If the two dice are independent, then the probability of two 1's is $\mathbb{P}[1] \times \mathbb{P}[1] = 1/36$.

Naïvely, one might think that independence relates to disjoint events, but the converse is true. If A and B are disjoint, then they cannot be independent. That is, disjointness means $A \cap B = \varnothing$, and by property 2 of Theorem 1.2,

$$\mathbb{P}[A \cap B] = \mathbb{P}[\varnothing] = 0 \neq \mathbb{P}[A]\,\mathbb{P}[B]$$

and the right side is nonzero by the definition of independence.

Independence lies at the core of many probability calculations. If you can break an event into the joint occurance of several independent events, then the probability of the event is the product of the individual probabilities.

Take, for example, the two-coin example and the event $\{HH, HT\}$. This equals {First coin is H, Second coin is either H or T}. If the two coins are independent, this has probability

$$\mathbb{P}[H] \times \mathbb{P}[H \text{ or } T] = \frac{1}{2} \times 1 = \frac{1}{2}.$$

As a bit more complicated example, what is the probability of "rolling a seven" from a pair of dice, meaning that the two faces add to seven? We can calculate this as follows. Let (x, y) denote the outcomes from the two (ordered) dice. The following outcomes yield a seven: $\{(1,6), (2,5), (3,4), (4,3), (5,2), (6,1)\}$. The outcomes are disjoint. Thus by the third axiom, the probability of a seven is the sum

$$\mathbb{P}[7] = \mathbb{P}[1,6] + \mathbb{P}[2,5] + \mathbb{P}[3,4] + \mathbb{P}[4,3] + \mathbb{P}[5,2] + \mathbb{P}[6,1].$$

Assume that the two dice are independent of one another, so the probabilities are products. For fair dice, the above expression equals

$$\begin{aligned}
&\mathbb{P}[1] \times \mathbb{P}[6] + \mathbb{P}[2] \times \mathbb{P}[5] + \mathbb{P}[3] \times \mathbb{P}[4] + \mathbb{P}[4] \times \mathbb{P}[3] + \mathbb{P}[5] \times \mathbb{P}[2] + \mathbb{P}[6] \times \mathbb{P}[1] \\
&= \frac{1}{6} \times \frac{1}{6} + \frac{1}{6} \times \frac{1}{6} + \frac{1}{6} \times \frac{1}{6} + \frac{1}{6} \times \frac{1}{6} + \frac{1}{6} \times \frac{1}{6} + \frac{1}{6} \times \frac{1}{6} \\
&= 6 \times \frac{1}{6^2} \\
&= \frac{1}{6}.
\end{aligned}$$

Now suppose that the dice are not fair. Suppose they are independent, but each is weighted so that the probability of a "1" is 2/6 and the probability of a "6" is 0. We revise the calculation to find

$$\begin{aligned}
&\mathbb{P}[1] \times \mathbb{P}[6] + \mathbb{P}[2] \times \mathbb{P}[5] + \mathbb{P}[3] \times \mathbb{P}[4] + \mathbb{P}[4] \times \mathbb{P}[3] + \mathbb{P}[5] \times \mathbb{P}[2] + \mathbb{P}[6] \times \mathbb{P}[1] \\
&= \frac{2}{6} \times \frac{0}{6} + \frac{1}{6} \times \frac{1}{6} + \frac{1}{6} \times \frac{1}{6} + \frac{1}{6} \times \frac{1}{6} + \frac{1}{6} \times \frac{1}{6} + \frac{0}{6} \times \frac{2}{6} \\
&= \frac{1}{9}.
\end{aligned}$$

1.9 LAW OF TOTAL PROBABILITY

An important relationship can be derived from the partitioning theorem (Theorem 1.1) which states that if $\{B_i\}$ is a partition of the sample space S, then

$$A = \bigcup_{i=1}^{\infty} (A \cap B_i).$$

Since the events $(A \cap B_i)$ are disjoint, an application of the third axiom and the definition of conditional probability implies

$$\mathbb{P}[A] = \sum_{i=1}^{\infty} \mathbb{P}[A \cap B_i] = \sum_{i=1}^{\infty} \mathbb{P}[A \mid B_i]\,\mathbb{P}[B_i].$$

This is called the Law of Total Probability.

Theorem 1.5 **Law of Total Probability**. If $\{B_1, B_2, \ldots\}$ is a partition of S, and $\mathbb{P}[B_i] > 0$ for all i, then

$$\mathbb{P}[A] = \sum_{i=1}^{\infty} \mathbb{P}[A \mid B_i]\,\mathbb{P}[B_i].$$

For example, take the roll of a fair die and the events $A = \{1, 3, 5\}$ and $B_j = \{j\}$. We calculate that

$$\sum_{i=1}^{6} \mathbb{P}[A \mid B_i]\,\mathbb{P}[B_i] = 1 \times \frac{1}{6} + 0 \times \frac{1}{6} + 1 \times \frac{1}{6} + 0 \times \frac{1}{6} + 1 \times \frac{1}{6} + 0 \times \frac{1}{6} = \frac{1}{2},$$

which equals $\mathbb{P}[A] = 1/2$, as claimed.

1.10 BAYES RULE

A famous result is credited to Reverend Thomas Bayes.

Theorem 1.6 **Bayes Rule**. If $\mathbb{P}[A] > 0$ and $\mathbb{P}[B] > 0$, then

$$\mathbb{P}[A \mid B] = \frac{\mathbb{P}[B \mid A]\,\mathbb{P}[A]}{\mathbb{P}[B \mid A]\,\mathbb{P}[A] + \mathbb{P}[B \mid A^c]\,\mathbb{P}[A^c]}.$$

Proof. The definition of conditional probability (applied twice) implies

$$\mathbb{P}[A \cap B] = \mathbb{P}[A \mid B]\,\mathbb{P}[B] = \mathbb{P}[B \mid A]\,\mathbb{P}[A].$$

Solving, we find

$$\mathbb{P}[A \mid B] = \frac{\mathbb{P}[B \mid A]\,\mathbb{P}[A]}{\mathbb{P}[B]}.$$

Applying the law of total probability to $\mathbb{P}[B]$ using the partition $\{A, A_c\}$, we obtain the stated result. ■

Bayes Rule is terrifically useful in many contexts.

As one example, suppose you walk by a sports bar where you see a group of people watching a sports match which involves a popular local team. Suppose you suddenly hear a roar of excitement from the bar. Did the local team just score? To investigate this by Bayes Rule, let $A = \{\text{score}\}$ and $B = \{\text{crowd roars}\}$. Assume that $\mathbb{P}[A] = 1/10$, $\mathbb{P}[B \mid A] = 1$, and $\mathbb{P}[B \mid A^c] = 1/10$ (there are other events which can cause a roar). Then

$$\mathbb{P}[A \mid B] = \frac{1 \times \frac{1}{10}}{1 \times \frac{1}{10} + \frac{1}{10} \times \frac{9}{10}} = \frac{10}{19} \simeq 53\%.$$

This is slightly over one-half. Under these assumptions, the roar of the crowd is informative though not definitive.[6]

As another example, suppose there are two types of workers: hard workers (H) and lazy workers (L). Suppose that we know from previous experience that $\mathbb{P}[H] = 1/4$ and $\mathbb{P}[L] = 3/4$. Suppose we can administer a screening test to determine whether an applicant is a hard worker. Let T be the event that an applicant has a high score on the test. Suppose that $\mathbb{P}[T \mid H] = 3/4$ and $\mathbb{P}[T \mid L] = 1/4$. That is, the test has some signal but is not perfect. We are interested in calculating $\mathbb{P}[H \mid T]$, the conditional probability that an applicant is a hard worker, given that they have a high test score. Bayes Rule tells us

$$\mathbb{P}[H \mid T] = \frac{\mathbb{P}[T \mid H]\,\mathbb{P}[H]}{\mathbb{P}[T \mid H]\,\mathbb{P}[H] + \mathbb{P}[T \mid L]\,\mathbb{P}[L]} = \frac{\frac{3}{4} \times \frac{1}{4}}{\frac{3}{4} \times \frac{1}{4} + \frac{1}{4} \times \frac{3}{4}} = \frac{1}{2}.$$

The probability the applicant is a hard worker is only 50%! Does this mean the test is useless? Consider the question: What is the probability an applicant is a hard worker, given that they had a poor (P) test score? We find

$$\mathbb{P}[H \mid P] = \frac{\mathbb{P}[P \mid H]\,\mathbb{P}[H]}{\mathbb{P}[P \mid H]\,\mathbb{P}[H] + \mathbb{P}[P \mid L]\,\mathbb{P}[L]} = \frac{\frac{1}{4} \times \frac{1}{4}}{\frac{1}{4} \times \frac{1}{4} + \frac{3}{4} \times \frac{3}{4}} = \frac{1}{10}.$$

This is only 10%. Thus what the test tells us is that if an applicant scores high, we are uncertain about that applicant's work habits; but if an applicant scores low, it is unlikely that they are a hard worker.

To revisit our real-world example of education and wages, recall that we calculated that the probability of a high wage (H) given a college degree (C) is $\mathbb{P}[H \mid C] = 0.53$. Applying Bayes Rule, we can find the probability that an individual has a college degree given that they have a high wage is

$$\mathbb{P}[C \mid H] = \frac{\mathbb{P}[H \mid C]\,\mathbb{P}[C]}{\mathbb{P}[H]} = \frac{0.53 \times 0.36}{0.31} = 0.62.$$

The probability of a college degree given that they have a low wage (L) is

$$\mathbb{P}[C \mid L] = \frac{\mathbb{P}[L \mid C]\,\mathbb{P}[C]}{\mathbb{P}[L]} = \frac{0.47 \times 0.36}{0.69} = 0.25.$$

Thus given this one piece of information (if the wage is above or below \$25), we have probabilistic information about whether the individual has a college degree.

1.11 PERMUTATIONS AND COMBINATIONS

For some calculations, it is useful to count the number of individual outcomes. For some of these calculations, the concepts of counting rules, permutations, and combinations are useful.

The first definition we explore is the counting rule, which shows how to count options when we combine tasks. For example, suppose you own ten shirts, three pairs of jeans, five pairs of socks, four coats and two

[6]Consequently, it is reasonable to enter the sports bar to learn the truth!

hats. How many clothing outfits can you create, assuming you use one of each category? The answer is $10 \times 3 \times 5 \times 4 \times 2 = 1200$ distinct outfits.[7]

Theorem 1.7 **Counting Rule.** If a job consists of K separate tasks, the kth of which can be done in n_k ways, then the entire job can be done in $n_1 n_2 \cdots n_K$ ways.

The counting rule is intuitively simple but is useful in a variety of modeling situations.

The second definition we explore is that of a permutation. A **permutation** is a rearrangement of the order. Suppose you take a classroom of 30 students. How many ways can you arrange their order? Each arrangement is called a "permutation." To calculate the number of permutations, observe that there are 30 students who can be placed first. Given this choice, there are 29 students who can be placed second. Given these two choices, there are 28 students for the third position, and so on. The total number of permutations is

$$30 \times 29 \times \cdots \times 1 = 30!$$

Here, the symbol ! denotes the factorial. (See Section A.3.)

The general solution is as follows.

Theorem 1.8 The number of **permutations** of a group of N objects is $N!$

Suppose we are trying to select an ordered five-student team from a 30-student class for a competition. How many ordered groups of five are there? The calculation is much the same as above, but we stop once the fifth position is filled. Thus the number is

$$30 \times 29 \times 28 \times 27 \times 26 = \frac{30!}{25!}.$$

The general solution is as follows.

Theorem 1.9 The number of **permutations** of a group of N objects taken K at a time is

$$P(N, K) = \frac{N!}{(N-K)!}.$$

The third definition we explore is that of a combination. A **combination** is an unordered group of objects. For example, revisit the idea of selecting a five-student team for a competition, but now assume that the team is unordered. Then the question is: How many five-member teams can we construct from a class of 30 students? In general, how many groups of K objects can be extracted from a group of N objects? We call this the "number of combinations".

The extreme cases are easy. If $K = 1$, then there are N combinations (each individual student). If $K = N$, then there is one combination (the entire class). The general answer can be found by noting that the number of ordered groups is the number of permutations $P(N, K)$. Each group of K can be ordered $K!$ ways (since this is the number of permutations of a group of K). Thus the number of unordered groups is $P(N, K)/K!$. We have found the following theorem.

Theorem 1.10 The number of **combinations** of a group of N objects taken K at a time is

$$\binom{N}{K} = \frac{N!}{K!\,(N-K)!}.$$

[7] Remember this when you (or a friend) asserts "I have nothing to wear!"

The symbol $\binom{N}{K}$, in words "N choose K", is a commonly used notation for combinations. They are also known as the **binomial coefficients**. The latter name is used because they are the coefficients from the binomial expansion.

Theorem 1.11 Binomial Theorem. For any integer $N \geq 0$,

$$(a+b)^N = \sum_{K=0}^{N} \binom{N}{K} a^K b^{N-K}.$$

The proof of the binomial theorem is given in Section 1.15.

The permutation and combination rules introduced in this section are useful in certain counting applications but may not be necessary for a general understanding of probability. My view is that the tools should be understood but not memorized. Instead, these tools can be looked up when needed.

1.12 SAMPLING WITH AND WITHOUT REPLACEMENT

Consider the problem of sampling from a finite set. For example, consider a $2 Powerball lottery ticket which consists of five integers each between 1 and 69. If all five numbers match the winning numbers, the player wins[8] $1 million!

To calculate the probability of winning the lottery, we need to count the number of potential tickets. The answer depends on two factors: (1) Can the numbers repeat? (2) Does the order matter? The number of tickets could have four distinct values, depending on the two choices just described.

The first question, of whether a number can repeat or not, is called "sampling with replacement" versus "sampling without replacement". In the actual Powerball game, 69 ping-pong balls are numbered and put in a rotating air machine with a small exit. As the balls bounce around, some of them find the exit. The first five to exit are the winning numbers. In this setting, we have "sampling without replacement", as once a ball exits, it is no longer among the remaining balls. A consequence for the lottery is that a winning ticket cannot have duplicate numbers. However, an alternative way to play the game would be to extract the first ball, replace it in the chamber, and repeat. This would be "sampling with replacement". In this game, a winning ticket could have repeated numbers.

The second question, of whether the order matters, is the same as the distinction between permutations and combinations as discussed in the previous section. In the case of the Powerball game, the balls emerge in a specific order. However, this order is ignored for the purpose of determining a winning ticket. This is the case of unordered sets. If the rules of the game were different, the order could matter. If so, we would use the tools of ordered sets.

We now describe the four sampling problems. We want to find the number of groups of size K which can be taken from N items, for example, the number of five integers taken from the set $\{1, \ldots, 69\}$.

Ordered, with replacement. Consider selecting the items in sequence. The first item can be any of the N, the second can be any of the N, the third can be any of the N, etc. So by the counting rule, the total number of possible groups is

$$N \times N \times \cdots \times N = N^K.$$

[8]There are also other prizes for other combinations.

In the Powerball example, this is

$$69^5 = 1{,}564{,}031{,}359.$$

This is a very large number of potential tickets!

Ordered, without replacement. This is the number of permutations $P(N, K) = N!/(N-K)!$ In the powerball example, this number is

$$\frac{69!}{(69-5)!} = \frac{69!}{64!} = 69 \times 68 \times 67 \times 66 \times 65 = 1{,}348{,}621{,}560.$$

This is nearly as large as the case with replacement.

Unordered, without replacement. This is the number of combinations $N!/(K!(N-K)!)$. In the powerball example, this number is

$$\frac{69!}{5!\,(69-5)!} = 11{,}238{,}513.$$

This is a large number but considerably smaller than the cases of ordered sampling.

Unordered, with replacement. This computation is tricky. It is not N^K (ordered with replacement) divided by $K!$, because the number of orderings per group depends on whether there are repeats. The trick is to recast the question as a different problem. It turns out that the number we are looking for is the same as the number of N-tuples of nonnegative integers $\{x_1, \ldots, x_N\}$ whose sum is K. To see this, a lottery ticket (unordered with replacement) can be represented by the number of "1's" x_1, the number of "2's" x_2, the number of "3's" x_3, and so forth, and we know that the sum of these numbers $(x_1 + \cdots + x_N)$ must equal K. The solution has a clever name based on the original proof notation.

Theorem 1.12 Stars and Bars Theorem. The number of N-tuples of nonnegative integers whose sum is K is equal to $\binom{N+K-1}{K}$.

The proof of the stars and bars theorem is omitted, as it is rather tedious. It does give us the answer to the question we started to address, namely, the number of unordered sets taken with replacement. In the Powerball example, this is

$$\binom{69+5-1}{5} = \frac{73!}{5!68!} = 15{,}020{,}334.$$

Table 1.1 summarizes the four sampling results.

Table 1.1
Number of possible arrangments of size K from N items

	Without Replacement	With Replacement
Ordered	$\frac{N!}{(N-K)!}$	N^K
Unordered	$\binom{N}{K}$	$\binom{N+K-1}{K}$

The actual Powerball game uses sampling that is unordered without replacement. Thus there are about 11 million potential tickets. As each ticket has an equal chance of occurring (if the random process is fair), this means the probability of winning is about 1/11,000,000. Since a player wins $1 million once for every 11 million tickets sold, the expected payout (ignoring the other payouts) is about $0.09. This is a low payout (considerably below a "fair" bet, given that a ticket costs $2) but is sufficiently high to attract meaningful interest from players.

1.13 POKER HANDS

A fun application of probability theory is to the game of poker. Similar types of calculations can be useful in economic examples involving multiple choices.

A standard game of poker is played with a 52-card deck containing 13 denominations {2, 3, 4, 5, 6, 7, 8, 9, 10, Jack, Queen, King, Ace} in each of four suits {club, diamond, heart, spade}. The deck is shuffled (so the order is random) and a player is dealt[9] five cards called a "hand". Hands are ranked based on whether there are multiple cards (pair, two pair, three-of-a-kind, full house, or four-of-a-kind), all five cards in sequence (called a "straight"), or all five cards of the same suit (called a "flush"). Players win if they have the best hand.

We are interested in calculating the probability of receiving a winning hand.

The structure is unordered sampling without replacement. The number of possible poker hands is

$$\binom{52}{5} = \frac{52!}{47!5!} = \frac{48 \times 49 \times 50 \times 51 \times 52}{2 \times 3 \times 4 \times 5} = 48 \times 49 \times 5 \times 17 \times 13 = 2,598,560.$$

Since the draws are symmetric and random, all hands have the same probability of receipt, implying that the probability of receiving any specific hand is $1/2,598,560$, an infinitesimally small number.

Another way of calculating this probability is as follows. Imagine picking a specific five-card hand. The probability of receiving one of the five cards on the first draw is 5/52, the probability of receiving one of the remaining four on the second draw is 4/51, the probability of receiving one of the remaining three on the third draw is 3/50, etc., so the probability of receiving the five-card hand is

$$\frac{5 \times 4 \times 3 \times 2 \times 1}{52 \times 51 \times 50 \times 49 \times 48} = \frac{1}{13 \times 17 \times 5 \times 49 \times 48} = \frac{1}{2,598,960}.$$

One way to calculate the probability of a winning hand is to enumerate and count the number of winning hands in each category and then divide by the total number of hands, $2,598,560$. Let us consider a few examples.

Four of a kind. Consider the number of hands with four of a specific denomination (such as Kings). The hand contains all four Kings plus an additional card, which can be any of the remaining 48. Thus there are exactly 48 five-card hands with all four Kings. There are 13 denominations, so there are $13 \times 48 = 624$ hands with four-of-a-kind. Thus the probability of drawing a four-of-a-kind is

$$\frac{13 \times 48}{13 \times 17 \times 5 \times 49 \times 48} = \frac{1}{17 \times 5 \times 49} = \frac{1}{4165} \simeq 0.0\%.$$

[9] A typical game involves additional complications, which we ignore.

Three of a kind. Consider the number of hands with three of a specific denomination (such as Aces). There are $\binom{4}{3} = 4$ groups of three Aces. There are 48 cards from which to choose the remaining two. The number of such arrangements is $\binom{48}{2} = \frac{48!}{46!2!} = 47 \times 24$. However, this includes pairs. There are twelve denominations each of which has $\binom{4}{2} = 6$ pairs, so there are $12 \times 6 = 72$ pairs. Thus the number of two-card arrangements excluding pairs is $47 \times 24 - 72 = 44 \times 24$. Hence the number of hands with three Aces and no pair is $4 \times 44 \times 24$. As there are 13 possible denominations, the number of hands with a three of a kind is $13 \times 4 \times 44 \times 24$. Thus the probability of drawing a three-of-a-kind is

$$\frac{13 \times 4 \times 44 \times 24}{13 \times 17 \times 5 \times 49 \times 48} = \frac{88}{17 \times 5 \times 49} \simeq 2.1\%.$$

One pair. Consider the number of hands with two of a specific denomination (such as a "7"). There are $\binom{4}{2} = 6$ pairs of 7's. From the 48 remaining cards, the number of three-card arrangements is $\binom{48}{3} = \frac{48!}{45!3!} = 23 \times 47 \times 16$. However, this includes three-card groups and two-card pairs. There are twelve denominations. Each has $\binom{4}{3} = 4$ three-card groups. Each also has $\binom{4}{2} = 6$ pairs and 44 remaining cards from which to select the third card. Thus there are $12 \times (4 + 6 \times 44)$ three-card arrangements with either a three-card group or a pair. Subtracting, we find that the number of hands with two 7's and no other pairs is

$$6 \times (23 \times 47 \times 16 - 12 \times (4 + 6 \times 44)).$$

Multiplying by 13, the probability of drawing one pair of any denomination is

$$13 \times \frac{6 \times (23 \times 47 \times 16 - 12 \times (4 + 6 \times 44))}{13 \times 17 \times 5 \times 49 \times 48} = \frac{23 \times 47 \times 2 - 3 \times (2 + 3 \times 44)}{17 \times 5 \times 49} \simeq 42\%.$$

From these simple calculations, you can see that if you receive a random hand of five cards, you have a good chance of receiving one pair, a small chance of receiving a three-of-a-kind, and a negligible chance of receiving a four-of-a-kind.

1.14 SIGMA FIELDS*

Definition 1.2 is incomplete as stated. When there are an uncountable infinity of events, it is necessary to restrict the set of allowable events to exclude pathological cases. This is a technicality which has little impact on practical econometrics. However, the terminology is used frequently, so it is prudent to be aware of the following definitions. The correct definition of probability is as follows.

Definition 1.5 A **probability function** $\mathbb{P}$ is a function from a sigma field $\mathscr{B}$ to the real line which satisfies the axioms of probability.

The difference is that Definition 1.5 restricts the domain to a sigma field $\mathscr{B}$. The latter is a collection of sets which is closed under set operations. The restriction means that there are some events for which probability is not defined.

A sigma field is defined as follows.

Definition 1.6 A collection $\mathscr{B}$ of sets is called a **sigma field** if it satisfies the following three properties:

1. $\varnothing \in \mathscr{B}$.
2. If $A \in \mathscr{B}$, then $A^c \in \mathscr{B}$.
3. If $A_1, A_2, \ldots \in \mathscr{B}$, then $\bigcup_{i=1}^{\infty} A_i \in \mathscr{B}$.

The infinite union in part 3 includes all elements which are an element of A_i for some i. An example is $\bigcup_{i=1}^{\infty} [0, 1 - 1/i] = [0, 1)$.

An alternative label for a sigma field is "sigma algebra". The following is a leading example of a sigma field.

Definition 1.7 The **Borel sigma field** is the smallest sigma field on $\mathbb{R}$ containing all open intervals (a, b). It contains all open intervals and closed intervals, and their countable unions, intersections, and complements.

A sigma field can be **generated** from a finite collection of events by taking all unions, intersections, and complements. Take the coin-flip example and start with the event $\{H\}$. Its complement is $\{T\}$, their union is $S = \{H, T\}$, and the union's complement is $\{\varnothing\}$. No further events can be generated. Thus the collection $\{\{\varnothing\}, \{H\}, \{T\}, S\}$ is a sigma field.

For an example on the positive real line, take the sets $[0, 1]$ and $(1, 2]$. Their intersection is $\{\varnothing\}$, their union is $[0, 2]$, and their complements are $(1, \infty)$, $[0, 1] \cup (2, \infty)$, and $(2, \infty)$. A further union is $[0, \infty)$. This collection is a sigma field, as no further events can be generated.

When there are an infinite number of events, then it may not be possible to generate a sigma field through set operations, as pathological counterexamples exist. These counterexamples are difficult to characterize, are nonintuitive, and seem to have no practical implications for econometric practice. Therefore the issue is generally ignored in econometrics.

If the concept of a sigma field seems technical, it is! The concept is not used further in this textbook.

1.15 TECHNICAL PROOFS*

Proof of Theorem 1.1 Take an outcome ω in A. Since $\{B_1, B_2, \cdots\}$ is a partition of S, it follows that $\omega \in B_i$ for some i. Set $A_i = (A \cap B_i)$. Thus $\omega \in A_i \subset \bigcup_{i=1}^{\infty} A_i$. This shows that every element in A is an element of $\bigcup_{i=1}^{\infty} A_i$.

Now take an outcome ω in $\bigcup_{i=1}^{\infty} A_i$. Thus $\omega \in A_i$ for some i. This implies $\omega \in A$. This shows that every element in $\bigcup_{i=1}^{\infty} A_i$ is an element of A.

For $i \neq j$, $A_i \cap A_j = (A \cap B_i) \cap (A \cap B_j) = A \cap (B_i \cap B_j) = \varnothing$ since B_i are mutually disjoint. Thus A_i are mutually disjoint. ■

Proof of Theorem 1.2 property 1 A and A^c are disjoint and $A \cup A^c = S$. The second and third axioms imply

$$1 = \mathbb{P}[S] = \mathbb{P}[A] + \mathbb{P}[A^c]. \tag{1.1}$$

Rearranging, we find $\mathbb{P}[A^c] = 1 - \mathbb{P}[A]$ as claimed. ■

Proof of Theorem 1.2 property 2 We have that $\varnothing = S^c$. By Theorem 1.2 and the second axiom of probability, $\mathbb{P}[\varnothing] = 1 - \mathbb{P}[S] = 0$, as claimed. ■

Proof of Theorem 1.2 property 3 The first axiom implies $\mathbb{P}[A^c] \geq 0$. This and equation (1.1) imply

$$\mathbb{P}[A] = 1 - \mathbb{P}\left[A^c\right] \leq 1$$

as claimed. ■

Proof of Theorem 1.2 property 4 The assumption $A \subset B$ implies $A \cap B = A$. By the partitioning theorem (Theorem 1.1) $B = (B \cap A) \cup (B \cap A^c) = A \cup (B \cap A^c)$ where A and $B \cap A^c$ are disjoint. The third axiom implies

$$\mathbb{P}[B] = \mathbb{P}[A] + \mathbb{P}\left[B \cap A^c\right] \geq \mathbb{P}[A]$$

where the inequality is $\mathbb{P}[B \cap A^c] \geq 0$ which holds by the first axiom. Thus, $\mathbb{P}[B] \geq \mathbb{P}[A]$, as claimed. ■

Proof of Theorem 1.2 property 5 $\{A \cup B\} = A \cup \{B \cap A^c\}$ where A and $\{B \cap A^c\}$ are disjoint. Also $B = \{B \cap A\} \cup \{B \cap A^c\}$ where $\{B \cap A\}$ and $\{B \cap A^c\}$ are disjoint. These two relationships and the third axiom imply

$$\mathbb{P}[A \cup B] = \mathbb{P}[A] + \mathbb{P}\left[B \cap A^c\right]$$

$$\mathbb{P}[B] = \mathbb{P}[B \cap A] + \mathbb{P}\left[B \cap A^c\right].$$

Subtracting,

$$\mathbb{P}[A \cup B] - \mathbb{P}[B] = \mathbb{P}[A] - \mathbb{P}[B \cap A].$$

Rearranging, we obtain the result. ■

Proof of Theorem 1.2 property 6 From the Inclusion-Exclusion Principle and $\mathbb{P}[A \cap B] \geq 0$ (the first axiom)

$$\mathbb{P}[A \cup B] = \mathbb{P}[A] + \mathbb{P}[B] - \mathbb{P}[A \cap B] \leq \mathbb{P}[A] + \mathbb{P}[B]$$

as claimed. ■

Proof of Theorem 1.2 property 7 Rearranging the Inclusion-Exclusion Principle and using $\mathbb{P}[A \cup B] \leq 1$ (Theorem 1.2 property 3), we have

$$\mathbb{P}[A \cap B] = \mathbb{P}[A] + \mathbb{P}[B] - \mathbb{P}[A \cup B] \geq \mathbb{P}[A] + \mathbb{P}[B] - 1$$

which is the stated result. ■

Proof of Theorem 1.11 (Binomial Theorem) Multiplying out, the expression

$$(a+b)^N = (a+b) \times \cdots \times (a+b) \tag{1.2}$$

is a polynomial in a and b with 2^N terms. Each term takes the form of the product of K of the a and $N-K$ of the b, thus is of the form $a^K b^{N-K}$. The number of terms of this form is equal to the number of combinations of the a's, which is $\binom{N}{K}$. Consequently, expression (1.2) equals $\sum_{K=0}^{N} \binom{N}{K} a^K b^{N-K}$, as stated. ■

1.16 EXERCISES

Exercise 1.1 Let $A = \{a, b, c, d\}$ and $B = \{a, c, e, f\}$.

(a) Find $A \cap B$.

(b) Find $A \cup B$.

Exercise 1.2 Describe the sample space S for the following experiments.

(a) Flip a coin.
(b) Roll a six-sided die.
(c) Roll two six-sided dice.
(d) Shoot six free throws (in basketball).

Exercise 1.3 From a 52-card deck of playing cards, draw five cards to make a hand.

(a) Let A be the event "The hand has two Kings". Describe A^c.
(b) A **straight** is five cards in sequence, for example, $\{5, 6, 7, 8, 9\}$. A **flush** is five cards of the same suit. Let A be the event "The hand is a straight" and B be the event "The hand is 3-of-a-kind". Are A and B disjoint or not disjoint?
(c) Let A be the event "The hand is a straight" and B be the event "The hand is flush". Are A and B disjoint or not disjoint?

Exercise 1.4 For events A and B, express the probability of "either A or B but not both" as a formula in terms of $\mathbb{P}[A]$, $\mathbb{P}[B]$, and $\mathbb{P}[A \cap B]$.

Exercise 1.5 If $\mathbb{P}[A] = 1/2$ and $\mathbb{P}[B] = 2/3$, can A and B be disjoint? Explain.

Exercise 1.6 Prove that $\mathbb{P}[A \cup B] = \mathbb{P}[A] + \mathbb{P}[B] - \mathbb{P}[A \cap B]$.

Exercise 1.7 Show that $\mathbb{P}[A \cap B] \leq \mathbb{P}[A] \leq \mathbb{P}[A \cup B] \leq \mathbb{P}[A] + \mathbb{P}[B]$.

Exercise 1.8 Suppose $A \cap B = A$. Can A and B be independent? If so, give the appropriate condition.

Exercise 1.9 Prove that

$$\mathbb{P}[A \cap B \cap C] = \mathbb{P}[A \mid B \cap C]\, \mathbb{P}[B \mid C]\, \mathbb{P}[C].$$

Assume $\mathbb{P}[C] > 0$ and $\mathbb{P}[B \cap C] > 0$.

Exercise 1.10 Is $\mathbb{P}[A \mid B] \leq \mathbb{P}[A]$, $\mathbb{P}[A \mid B] \geq \mathbb{P}[A]$, or is neither necessarily true?

Exercise 1.11 Give an example where $\mathbb{P}[A] > 0$, yet $\mathbb{P}[A \mid B] = 0$.

Exercise 1.12 Calculate the following probabilities concerning a standard 52-card playing deck.

(a) Drawing a King with one card.
(b) Drawing a King on the second card, conditional on a King on the first card.
(c) Drawing two Kings with two cards.
(d) Drawing a King on the second card, conditional on the first card is not a King.
(e) Drawing a King on the second card, when the first card is placed face down (so is unknown).

Exercise 1.13 You are on a game show, and the host shows you five doors marked A, B, C, D, and E. The host says that a prize is behind one of the doors, and you win the prize if you select the correct door. Given the stated information, what probability distribution would you use for modeling the distribution of the correct door?

Exercise 1.14 Calculate the following probabilities, assuming fair coins and dice.

(a) Getting three heads in a row from three coin flips.
(b) Getting a heads given that the previous coin was a tails.
(c) From two coin flips getting two heads given that at least one coin is a heads.
(d) Rolling a six from a pair of dice.
(e) Rolling "snakes eyes" from a pair of dice. (Getting a pair of ones.)

Exercise 1.15 If four random cards are dealt from a deck of playing cards, what is the probability that all four are Aces?

Exercise 1.16 Suppose that the unconditional probability of a disease is 0.0025. A screening test for this disease has a detection rate of 0.9, and has a false positive rate of 0.01. Given that the screening test returns positive, what is the conditional probability of having the disease?

Exercise 1.17 Suppose that 1% of athletes use banned steroids. Suppose that a drug test has a detection rate of 40% and a false positive rate of 1%. If an athlete tests positive, what is the conditional probability that the athlete has taken banned steroids?

Exercise 1.18 Sometimes we use the concept of **conditional independence**. The definition is as follows. Let A, B, C be three events with positive probabilities. Then A and B are conditionally independent given C if $\mathbb{P}[A \cap B \mid C] = \mathbb{P}[A \mid C]\,\mathbb{P}[B \mid C]$. Consider the experiment of tossing two dice. Let $A = \{\text{First die is 6}\}$, $B = \{\text{Second die is 6}\}$, and $C = \{\text{Both dice are the same}\}$. Show that A and B are independent (unconditionally), but A and B are dependent given C.

Exercise 1.19 **Monte Hall.** This is a famous (and surprisingly difficult) problem based on an old U.S. television game show "Let's Make a Deal" hosted by Monte Hall. A standard part of the show ran as follows: A contestant was asked to select from one of three identical doors: A, B, and C. Behind one of the three doors was a prize. If the contestant selected the correct door, they would receive the prize. The contestant picked one door (say, A) but it is not immediately opened. To increase the drama, the host opened one of the two remaining doors (say, door B) revealing that that door does not have the prize. The host then made the offer: "You have the option to switch your choice" (e.g., to switch to door C). You can imagine that the contestant may have made one of reasonings (a)–(c) below. Comment on each of these three reasonings. Are they correct?

(a) "When I selected door A, the probability that it has the prize was 1/3. No information was revealed. So the probability that Door A has the prize remains 1/3."
(b) "The original probability was 1/3 on each door. Now that door B is eliminated, doors A and C each have each probability of 1/2. It does not matter whether I stay with A or switch to C."

(c) "The host inadvertently revealed information. If door C had the prize, he was forced to open door B. If door B had the prize, he would have been forced to open door C. Thus it is quite likely that door C has the prize."

(d) Assume a prior probability for each door of 1/3. Calculate the posterior probabilities that door A and door C have the prize, respectively. What choice do you recommend for the contestant?

Exercise 1.20 In the game of blackjack, you are dealt two cards from a standard playing deck. Your score is the sum of the value of the two cards, where numbered cards have the value given by their number, face cards (Jack, Queen, King) each receive 10 points, and an Ace either 1 or 11 (player can choose). A **blackjack** is receiving a score of 21 from two cards, thus an Ace and any card worth 10 points.

(a) What is the probability of receiving a blackjack?

(b) The dealer is dealt one of their cards face down and one face up. Suppose the "show" card is an Ace. What is the probability that the dealer has a blackjack? (For simplicity, assume you have not seen any other cards.)

Exercise 1.21 Consider drawing five cards at random from a standard deck of playing cards. Calculate the following probabilities.

(a) A straight (five cards in sequence, suit not relevant).

(b) A flush (five cards of the same suit, order not relevant).

(c) A full house (3-of-a-kind and a pair, e.g., three Kings and two "3's").

Exercise 1.22 In the poker game "Five Card Draw", a player first receives five cards drawn at random. The player decides to discard some of their cards and then receives replacement cards. Assume a player is dealt a hand with one pair and three unrelated cards and decides to discard the three unrelated cards to obtain replacements. Calculate the following conditional probabilities for the resulting hand after the replacements are made.

(a) Obtaining a four-of-a-kind.

(b) Obtaining a three-of-a-kind.

(c) Obtaining two pairs.

(d) Obtaining a straight or a flush.

(e) Ending with one pair.

CHAPTER 2

RANDOM VARIABLES

2.1 INTRODUCTION

In practice, it is convenient to represent random outcomes numerically. If the outcome is numerical and one-dimensional, we call the outcome a "random variable". If the outcome is multidimensional, we call it a "random vector".

Random variables are one of the most important and core concepts in probability theory. It is so central that most of the time, we don't think about the foundations of this concept.

As an example, consider a coin flip which has the possible outcome H or T. We can write the outcome numerically by setting the result as $X = 1$ if the coin is heads and $X = 0$ if the coin is tails. The object X is random, since its value depends on the outcome of the coin flip.

2.2 RANDOM VARIABLES

Definition 2.1 A **random variable** is a real-valued outcome: a function from the sample space S to the real line $\mathbb{R}$.

A random variable is typically represented by an uppercase Latin character; common choices are X and Y. In the coin-flip example, the function is

$$X = \begin{cases} 1 & \text{if } H \\ 0 & \text{if } T. \end{cases}$$

To illustrate, Figure 2.1 illustrates a mapping from the coin-flip sample space to the real line, with T mapped to 0 and H mapped to 1. A coin flip may seem overly simple, but the structure is identical for any two-outcome application.

Notationally it is useful to distinguish between random variables and their realizations. In probability theory and statistics, the convention is to use uppercase X to indicate a random variable and use lowercase x to indicate a realization or specific value. This may seem a bit abstract. Think of X as the random object whose value is unknown and x as a specific number or outcome.

2.3 DISCRETE RANDOM VARIABLES

We have a defined a random variable X as a real-valued outcome. In most cases, X only takes values on a subset of the real line. Take, for example, a coin flip coded as 1 for heads and 0 for tails. This only takes the values 0 and 1. It is an example of what we call a "discrete distribution".

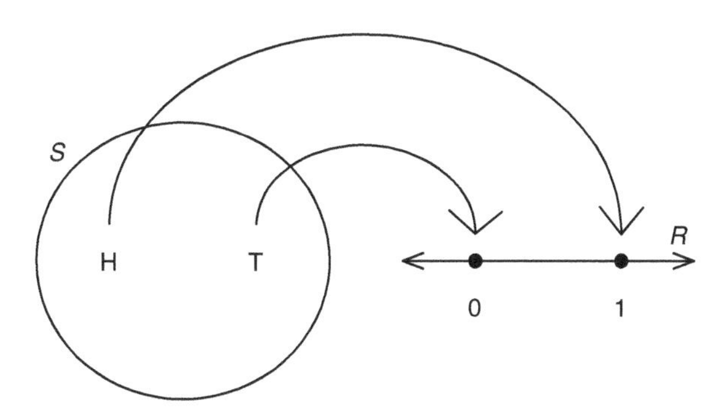

FIGURE 2.1 A random variable is a function

Definition 2.2 The set $\mathscr{X}$ is **discrete** if it has a finite or countably infinite number of elements.

Most discrete sets in applications are nonnegative integers. For example, in a coin flip, $\mathscr{X} = \{0, 1\}$ and in a roll of a die, $\mathscr{X} = \{1, 2, 3, 4, 5, 6\}$.

Definition 2.3 If there is a discrete set $\mathscr{X}$ such that $\mathbb{P}[X \in \mathscr{X}] = 1$, then X is a **discrete random variable**. The smallest set $\mathscr{X}$ with this property is the **support** of X.

The support is the set of values which receive positive probability of occurance. We sometimes write the support as $\mathscr{X} = \{\tau_1, \tau_2, \ldots, \tau_n\}$, $\mathscr{X} = \{\tau_1, \tau_2, \ldots\}$, or $\mathscr{X} = \{\tau_0, \tau_1, \tau_2, \ldots\}$ when we need an explicit description of the support. We call the values τ_j the **support points**.

The following definition is useful.

Definition 2.4 The **probability mass function** of a random variable is $\pi(x) = \mathbb{P}[X = x]$, the probability that X equals the value x. When evaluated at the support points τ_j, we write $\pi_j = \pi(\tau_j)$.

Take, for example, a coin flip with probability p of heads. The support is $\mathscr{X} = \{0, 1\} = \{\tau_0, \tau_1\}$. The probability mass function takes the values $\pi_0 = 1 - p$ and $\pi_1 = p$.

Take a fair die. The support is $\mathscr{X} = \{1, 2, 3, 4, 5, 6\} = \{\tau_j : j = 1, \ldots, 6\}$ with probability mass function $\pi_j = 1/6$ for $j = 1, \ldots, 6$.

An example of a countably infinite discrete random variable is

$$\mathbb{P}[X = k] = \frac{e^{-1}}{k!}, \qquad k = 0, 1, 2, \ldots. \tag{2.1}$$

This is a valid probability function, since $e = \sum_{k=0}^{\infty} 1/k!$. The support is $\mathscr{X} = \{0, 1, 2, \ldots\}$ with probability mass function $\pi_j = e^{-1}/(j!)$ for $j \geq 0$. (This is a special case of the Poisson distribution, which is defined in Section 3.6.)

It can be useful to plot a probability mass function as a bar graph to visualize the relative frequency of occurrence. Figure 2.2(a) displays the probability mass function for the distribution (2.1). The height of each bar is the the probability π_j at the support point. While the distribution is countably infinite, the probabilities are negligible for $k \geq 6$, so we have plotted the probability mass function for $k \leq 5$. You can see that the probabilities for $k = 0$ and $k = 1$ are about 0.37, that for $k = 2$ is about 0.18, for $k = 3$ is 0.06, and for $k = 4$ the probability is 0.015.

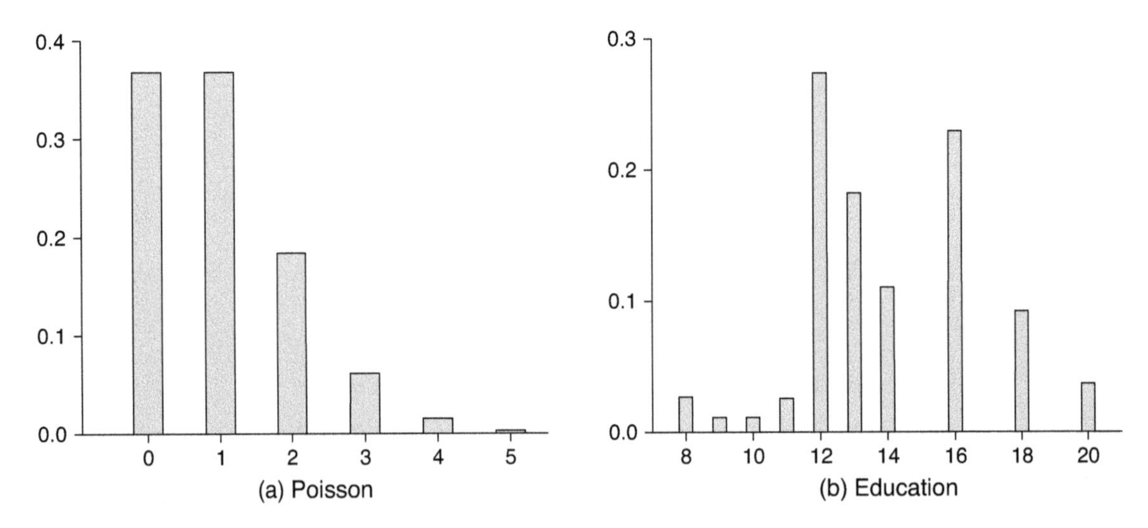

FIGURE 2.2 Probability mass functions

To illustrate using a real-world example, Figure 2.2(b) displays the probability mass function for the years of education[1] among U.S. wage earners in 2009. You can see that the highest probability occurs at 12 years of education (about 27%) and second highest at 16 years (about 23%).

2.4 TRANSFORMATIONS

If X is a random variable and $Y = g(X)$ for some function $g: \mathscr{X} \to \mathscr{Y} \subset \mathbb{R}$, then Y is also a random variable. To see this formally, note that since X is a mapping from the sample space S to $\mathscr{X}$, and g maps $\mathscr{X}$ to $\mathscr{Y} \subset \mathbb{R}$, then Y is also a mapping from S to $\mathbb{R}$.

We are interested in describing the probability mass function for Y. Write the support of X as $\mathscr{X} = \{\tau_1, \tau_2, \dots\}$ and its probability mass function as $\pi_X(\tau_j)$.

If we apply the transformation to each of X's support points, we obtain $\mu_j = g(\tau_j)$. If the μ_j are unique (there are no redundancies), then Y has support $\mathscr{Y} = \{\mu_1, \mu_2, \dots\}$ with probability mass function $\pi_Y(\mu_j) = \pi_X(\tau_j)$. The impact of the transformation $X \to Y$ is to move the support points from τ_j to μ_j, and the probabilities are maintained.

If the μ_j are not unique, then some probabilities are combined. Essentially the transformation reduces the number of support points. As an example, suppose that the support of X is $\{-1, 0, 1\}$ and $Y = X^2$. Then the support for Y is $\{0, 1\}$. Since both -1 and 1 are mapped to 1, the probability mass function for Y inherits the sum of these two probabilities. That is, the probability mass function for Y is

$$\pi_Y(0) = \pi_X(0)$$
$$\pi_Y(1) = \pi_X(-1) + \pi_X(1).$$

[1]Here, *education* is defined as years of schooling beyond kindergarten. A high school graduate has *education* = 12, a college graduate has *education* = 16, a Master's degree has *education* = 18, and a professional degree (medical, law, or PhD) has *education* = 20.

In general,

$$\pi_Y(\mu_i) = \sum_{j:g(\tau_j)=g(\tau_i)} \pi_X\left(\tau_j\right).$$

The sum looks complicated, but it simply states that the probabilites are summed over all indices for which there is equality among the transformed values.

2.5 EXPECTATION

The **expectation** of a random variable X, denoted by $\mathbb{E}[X]$, is a useful measure of the central tendency of the distribution. It is the average value with probability-weighted averaging. The expectation is also called the **expected value**, **average**, or **mean** of the distribution. I prefer the labels "expectation" or "expected value", as they are the least ambiguous. It is typical to write the expectation of X as $\mathbb{E}[X]$, $\mathbb{E}(X)$, or $\mathbb{E}X$. Some authors use the notation $E[X]$ or $\mathrm{E}[X]$.

Definition 2.5 For a discrete random variable X with support $\{\tau_j\}$, the **expectation** of X is

$$\mathbb{E}[X] = \sum_{j=1}^{\infty} \tau_j \pi_j$$

if the series is convergent. (For the definition of convergence, see Section A.1 in the Appendix.)

It is important to understand that while X is random, the expectation $\mathbb{E}[X]$ is nonrandom. It is a fixed feature of the distribution.

Example: $X = 1$ with probability p and $X = 0$ with probability $1-p$. Its expected value is

$$\mathbb{E}[X] = 0 \times (1-p) + 1 \times p = p.$$

Example: Fair die throw. The expected value is

$$\mathbb{E}[X] = 1 \times \frac{1}{6} + 2 \times \frac{1}{6} + 3 \times \frac{1}{6} + 4 \times \frac{1}{6} + 5 \times \frac{1}{6} + 6 \times \frac{1}{6} = \frac{7}{2}.$$

Example: $\mathbb{P}[X = k] = \frac{e^{-1}}{k!}$ for nonnegative integer k. For this probability distribution, the expected value is

$$\mathbb{E}[X] = \sum_{k=0}^{\infty} k \frac{e^{-1}}{k!} = 0 + \sum_{k=1}^{\infty} k \frac{e^{-1}}{k!} = \sum_{k=1}^{\infty} \frac{e^{-1}}{(k-1)!} = \sum_{k=0}^{\infty} \frac{e^{-1}}{k!} = 1.$$

Example: Years of education. For the probability distribution displayed in Figure 2.2(b), the expected value is

$$\begin{aligned}\mathbb{E}[X] &= 8 \times 0.027 + 9 \times 0.011 + 10 \times 0.011 + 11 \times 0.026 + 12 \times 0.274 \\ &\quad + 13 \times 0.182 + 14 \times 0.111 + 16 \times 0.229 + 18 \times 0.092 + 20 \times 0.037 = 13.9.\end{aligned}$$

Thus the average number of years of education is about 14.

One property of the expectation is that it is the **center of mass** of the distribution. Imagine the probability mass functions of Figure 2.2 as a set of weights on a board placed on top of a fulcrum. For the board to balance, the fulcrum needs to be placed at the expectation $\mathbb{E}[X]$. It is instructive to review Figure 2.2 with the knowledge that the center of mass for the Poisson distribution is 1, and that for the years of education is 14.

The expectation of transformations is similarly defined.

Definition 2.6 For a discrete random variable X with support $\{\tau_j\}$, the **expectation** of $g(X)$ is

$$\mathbb{E}\left[g(X)\right] = \sum_{j=1}^{\infty} g(\tau_j)\pi_j$$

if the series is convergent.

When applied to transformations, we may use simplified notation when it leads to less clutter. For example, we may write $\mathbb{E}|X|$ rather than $\mathbb{E}[|X|]$, and $\mathbb{E}|X|^r$ rather than $\mathbb{E}\left[|X|^r\right]$.

Expectation is an linear operator.

Theorem 2.1 **Linearity of Expectation**. For any constants a and b,

$$\mathbb{E}[a + bX] = a + b\mathbb{E}[X].$$

Proof: Using the definition of expectation, we have

$$\begin{aligned}\mathbb{E}[a + bX] &= \sum_{j=1}^{\infty} \left(a + b\tau_j\right)\pi_j \\ &= a\sum_{j=1}^{\infty} \pi_j + b\sum_{j=1}^{\infty} \tau_j\pi_j \\ &= a + b\mathbb{E}[X]\end{aligned}$$

since $\sum_{j=1}^{\infty} \pi_j = 1$ and $\sum_{j=1}^{\infty} \pi_j\tau_j = \mathbb{E}[X]$. ■

2.6 FINITENESS OF EXPECTATIONS

The definition of expectation includes the phrase "if the series is convergent". This caveat is included since some series are nonconvergent. In the latter case, the expectation is either infinite or not defined.

As an example, suppose that X has support points 2^k for $k = 1, 2, \ldots$ and has probability mass function $\pi_k = 2^{-k}$. This is a valid probability function, since

$$\sum_{k=1}^{\infty} \pi_k = \sum_{k=1}^{\infty} \frac{1}{2^k} = \frac{1}{2} + \frac{1}{4} + \frac{1}{8} + \cdots = 1.$$

The expected value is

$$\mathbb{E}[X] = \sum_{k=1}^{\infty} 2^k\pi_k = \sum_{k=1}^{\infty} 2^k 2^{-k} = \infty.$$

This example is known as the **St. Petersburg Paradox** and corresponds to a bet. Toss a fair coin K times until a heads appears. Set the payout as $X = 2^K$, which means that a player is paid \$2 if $K = 1$, \$4 if $K = 2$, \$8 if $K = 3$, and so forth. The payout has infinite expectation, as shown above. This game has been called a

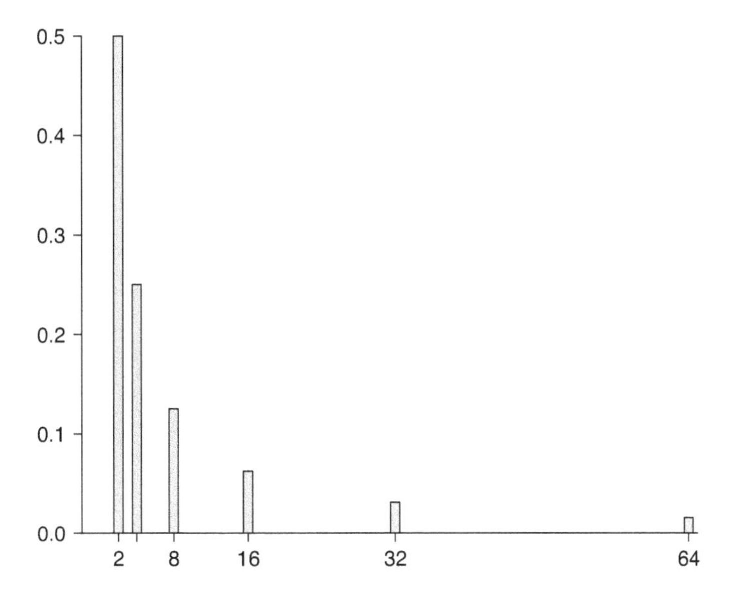

FIGURE 2.3 St. Petersburg paradox

"paradox" because few individuals are willing to pay a high price to receive this random payout, even though the expected value is infinite.[2]

The probability mass function for this payout is displayed in Figure 2.3. You can see how the probability decays slowly in the right tail. Recalling the property of the expected value as equal to the center of mass, this means that if we imagine an infinitely long board with the probability mass function of Figure 2.3 as weights, there would be no position where you could put a fulcrum such that the board balances. Regardless of where the fulcrum is placed, the board will tilt to the right. The small but increasingly distantly placed probability mass points dominate.

The above example is a nonconvergent case where the expectation is infinite. In some nonconvergent cases, the expectation is undefined. Suppose that we modify the payout of the St. Petersburg bet so that the support points are 2^k and -2^k, each with probability $\pi_k = 2^{-k-1}$. Then the expected value is

$$\mathbb{E}[X] = \sum_{k=1}^{\infty} 2^k \pi_k - \sum_{k=1}^{\infty} 2^k \pi_k = \sum_{k=1}^{\infty} \frac{1}{2} - \sum_{k=1}^{\infty} \frac{1}{2} = \infty - \infty.$$

This series is not convergent but is also neither $+\infty$ nor $-\infty$. (It is tempting but incorrect to guess that the two infinite sums cancel.) In this case, we say that the expectation is **not defined** or **does not exist**.

The lack of finiteness of the expected value can make economic transactions difficult. Suppose that X is loss due to an unexpected severe event, such as fire, tornado, or earthquake. With high probability, $X = 0$. With with low probability, X is positive and quite large. In these contexts, risk-adverse economic agents seek insurance. An ideal insurance contract compensates fully for a random loss X. In a market with no asymmetries or frictions, insurance companies offer insurance contracts with the premium set to equal the expected loss $\mathbb{E}[X]$. However, when the loss is not finite, this is impossible, so such contracts are infeasible.

[2]Economists should realize that there is no paradox once you introduce concave utility. If utility is $u(x) = x^{1/2}$, then the expected utility of the bet is $\sum_{k=1}^{\infty} 2^{-k/2} = 2^{-1/2} / \left(1 - 2^{-1/2}\right) \simeq 2.41$. The value of the bet (certainty equivalence) is \$5.83, because this also yields a utility of 2.41.

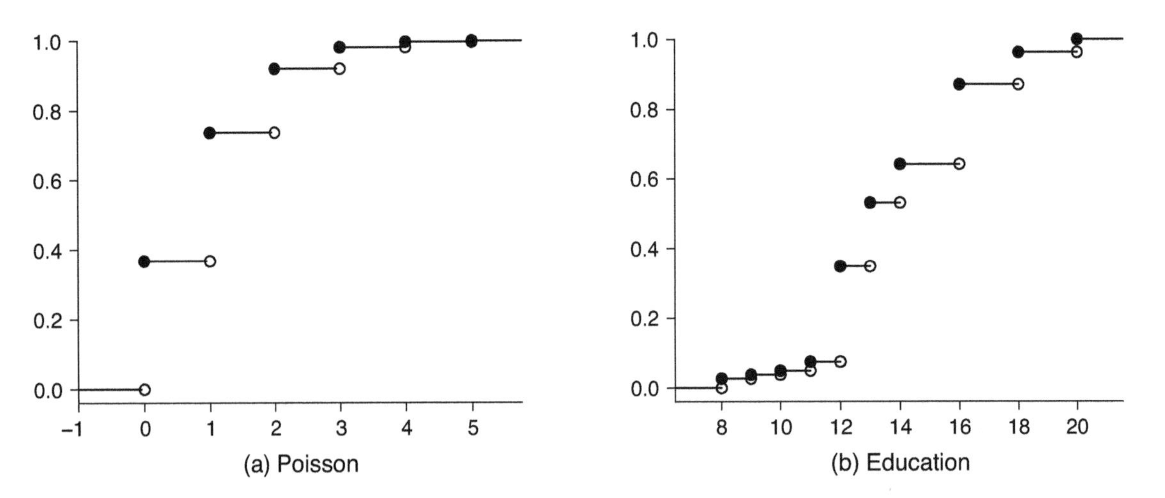

FIGURE 2.4 Discrete distribution functions

2.7 DISTRIBUTION FUNCTION

A random variable can be represented by its distribution function.

Definition 2.7 The **distribution function** is $F(x) = \mathbb{P}[X \leq x]$, the probability of the event $\{X \leq x\}$.

$F(x)$ is also known as the **cumulative distribution function (CDF)**. A common shorthand is to write "$X \sim F$" to mean "the random variable X has distribution function F" or "the random variable X is distributed as F". We use the symbol "$\sim$" to means that the variable on the left has the distribution indicated on the right.

It is standard notation to use uppercase letters to denote a distribution function. The most common choice is F, though any symbol can be used. When we need to be clear about the random variable, we add a subscript, writing the distribution function as $F_X(x)$. The subscript X indicates that F_X is the distribution of X. The argument "x" does not signify anything. We could equivalently write the distribution as $F_X(t)$ or $F_X(s)$. When only one random variable is being discussed, we simplify the notation and write the distribution function as $F(x)$.

For a discrete random variable with support points τ_j, the CDF at the support points equals the cumulative sum of the probabilities less than j:

$$F(\tau_j) = \sum_{k=1}^{j} \pi(\tau_k).$$

The CDF is constant between the support points. Therefore the CDF of a discrete random variable is a step function with jumps of magnitude $\pi(\tau_j)$ at each support point.

Figure 2.4 displays the distribution functions for the two examples displayed in Figure 2.2.

You can see how each distribution function is a step function with steps at the support points. The size of the jumps are varied, since the probabilities of the support points are unequal.

In general (not just for discrete random variables), the CDF has the following properties.

Theorem 2.2 **Properties of a CDF.** If $F(x)$ is a distribution function then

1. $F(x)$ is nondecreasing.

2. $\lim_{x \to -\infty} F(x) = 0$.
3. $\lim_{x \to \infty} F(x) = 1$.
4. $F(x)$ is right-continuous, meaning $\lim_{x \downarrow x_0} F(x) = F(x_0)$.

Properties 1 and 2 are consequences of the first axiom (probabilities are nonnegative). Property 3 is the second axiom. Property 4 states that at points where $F(x)$ has a step, $F(x)$ is discontinuous to the left but continuous to the right. This property is due to the definition of the distribution function as $\mathbb{P}[X \leq x]$. If the definition were $\mathbb{P}[X < x]$, then $F(x)$ would be left-continuous.

2.8 CONTINUOUS RANDOM VARIABLES

If a random variable X takes a continuum of values, it is not discretely distributed. Formally, we define a random variable to be continuous if the distribution function is continuous.

Definition 2.8 If $X \sim F(x)$ and $F(x)$ is continuous, then X is a **continuous** random variable.

Example: Uniform distribution:

$$F(x) = \begin{cases} 0 & x < 0 \\ x & 0 \leq x \leq 1 \\ 1 & x > 1. \end{cases}$$

The function $F(x)$ is globally continuous, limits to 0 as $x \to -\infty$, and limits to 1 as $x \to \infty$. Therefore, it satisfies the properties of a CDF.

Example: Exponential distribution:

$$F(x) = \begin{cases} 0 & x < 0 \\ 1 - \exp(-x) & x \geq 0. \end{cases}$$

The function $F(x)$ is globally continuous, limits to 0 as $x \to -\infty$, and limits to 1 as $x \to \infty$. Therefore, it satisfies the properties of a CDF.

Example: Hourly wages. As a real-world example, Figure 2.5 displays the distribution function for hourly wages in the United States in 2009, plotted over the range [\$0, \$60]. The function is continuous and everywhere increasing. This is because wage rates are dispersed. Marked with arrows are the values of the distribution function at \$10 increments from \$10 to \$50. This is read as follows. The distribution function at \$10 is 0.14. Thus 14% of wages are less than or equal to \$10. The distribution function at \$20 is 0.54. Thus 54% of the wages are less than or equal to \$20. Similarly, the distribution at \$30, \$40, and \$50 is 0.78, 0.89, and 0.94, respectively.

One way to think about the distribution function is in terms of differences. Take an interval $(a, b]$. The probability that $X \in (a, b]$ is $\mathbb{P}[a < X \leq b] = F(b) - F(a)$, the difference in the distribution function. Thus the difference between two points of the distribution function is the probability that X lies in the interval. For example, the probability that a random person's wage is between \$10 and \$20 is $0.54 - 0.14 = 0.30$. Similarly, the probability that their wage is in the interval [\$40, \$50] is $94\% - 89\% = 5\%$.

One property of continuous random variables is that the probability that they equal any specific value is 0. To see this, consider any number x. We can find the probability that X equals x by taking the limit of the

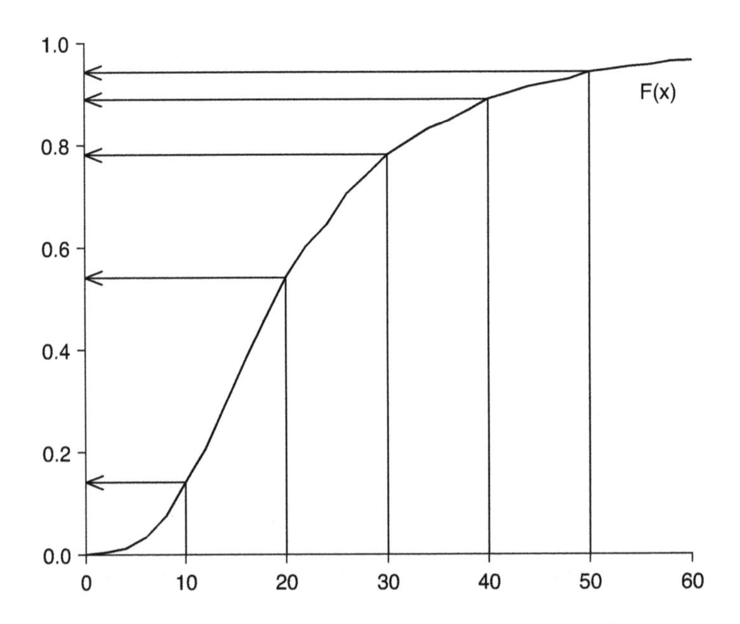

FIGURE 2.5 Distribution function—U.S. wages

sequence of probabilities that X is in the interval $[x, x+\epsilon]$ as ϵ decreases to 0. This is

$$\mathbb{P}[X=x] = \lim_{\epsilon\to 0} \mathbb{P}[x \le X \le x+\epsilon] = \lim_{\epsilon\to 0} F(x+\epsilon) - F(x) = 0$$

when $F(x)$ is continuous. This is a bit of a paradox. The probability that X equals any specific value x is 0, but the probability that X equals some value is 1. The paradox is due to the magic of the real line and the richness of uncountable infinity.

An implication is that for continuous random variables, we have the equalities

$$\mathbb{P}[X < x] = \mathbb{P}[X \le x] = F(x)$$
$$\mathbb{P}[X \ge x] = \mathbb{P}[X > x] = 1 - F(x).$$

2.9 QUANTILES

For a continuous distribution $F(x)$, the **quantiles** $q(\alpha)$ are defined as the solutions to the function

$$\alpha = F(q(\alpha)).$$

Effectively they are the inverse of $F(x)$, thus

$$q(\alpha) = F^{-1}(\alpha).$$

The quantile function $q(\alpha)$ is a function from $[0, 1]$ to the range of X.

Expressed as percentages, $100 \times q(\alpha)$ are called the **percentiles** of the distribution. For example, the 95th percentile equals the 0.95 quantile.

Some quantiles have special names. The **median** of the distribution is the 0.5 quantile. The **quartiles** are the 0.25, 0.50, and 0.75 quantiles. The later are called the "quartiles" as they divide the population into four equal groups. The **quintiles** are the 0.2, 0.4, 0.6, and 0.8 quantiles. The **deciles** are the 0.1, 0.2, . . . , 0.9 quantiles.

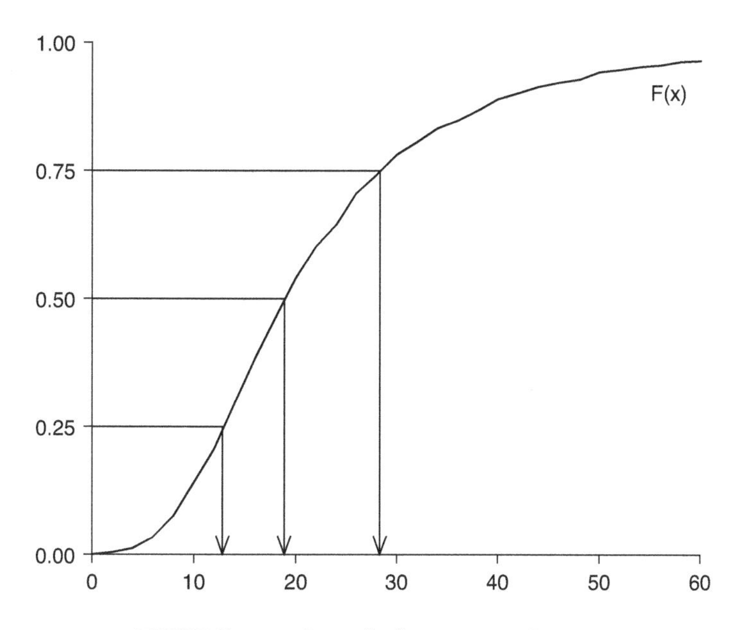

FIGURE 2.6 Quantile function—U.S. wages

Quantiles are useful summaries of the spread of the distribution.

Example: Exponential distribution. $F(x) = 1 - \exp(-x)$ for $x \geq 0$. To find a quantile $q(\alpha)$, set $\alpha = 1 - \exp(-x)$ and solve for x; thus, $x = -\log(1 - \alpha)$. For example, the 0.9 quantile is $-\log(1 - 0.9) \simeq 2.3$, the 0.5 quantile is $-\log(1 - 0.5) \simeq 0.7$.

Example: Hourly wages. Figure 2.6 displays the distribution function for hourly wages. From the points 0.25, 0.50, and 0.75 on the y-axis, lines are drawn to the distribution function and then to the x-axis with arrows. These are the quartiles of the wage distribution, and they are \$12.82, \$18.88, and \$28.35. The interpretation is that 25% of wages are less than or equal to \$12.82, 50% of wages are less than or equal to \$18.88, and 75% are less than or equal to \$28.35.

2.10 DENSITY FUNCTIONS

Continuous random variables do not have a probability mass function. An analog is the derivative of the distribution function, which is called the "density".

Definition 2.9 When $F(x)$ is differentiable, its **density** is $f(x) = \frac{d}{dx} F(x)$.

A density function is also called the **probability density function (PDF)**. It is standard notation to use lowercase letters to denote a density function; the most common choice is f. As for distribution functions, when we want to be clear about the random variable, we write the density function as $f_X(x)$, where the subscript indicates that this is the density function of X. A common shorthand is to write "$X \sim f$" to mean "the random variable X has density function f".

Theorem 2.3 Properties of a PDF. A function $f(x)$ is a density function if and only if

1. $f(x) \geq 0$ for all x.
2. $\int_{-\infty}^{\infty} f(x)dx = 1.$

A density is a nonnegative function which is integrable and integrates to 1. By the fundamental theorem of calculus, we have the relationship

$$\mathbb{P}[a \leq X \leq b] = \int_a^b f(x)dx.$$

This states that the probability that X is in the interval $[a, b]$ is the integral of the density over $[a, b]$, which shows that the area underneath the density function defines a probability.

Example: Uniform Distribution. $F(x) = x$ for $0 \leq x \leq 1$. The density is

$$f(x) = \frac{d}{dx}F(x) = \frac{d}{dx}x = 1$$

for $0 \leq x \leq 1$, 0 elsewhere. This density function is nonnegative and satisfies

$$\int_{-\infty}^{\infty} f(x)dx = \int_0^1 1dx = 1$$

and so satisfies the properties of a density function.

Example: Exponential distribution. $F(x) = 1 - \exp(-x)$ for $x \geq 0$. The density is

$$f(x) = \frac{d}{dx}F(x) = \frac{d}{dx}\left(1 - \exp(-x)\right) = \exp(-x)$$

on $x \geq 0$, 0 elsewhere. This density function is nonnegative and satisfies

$$\int_{-\infty}^{\infty} f(x)dx = \int_0^{\infty} \exp(-x)dx = 1$$

and so satisfies the properties of a density function. We can use it for probability calculations. For example,

$$\mathbb{P}[1 \leq X \leq 2] = \int_1^2 \exp(-x)dx = \exp(-1) - \exp(-2) \simeq 0.23.$$

Example: Hourly wages. Figure 2.7 plots the density of U.S. hourly wages.

The way to interpret the density function is as follows. The regions where the density $f(x)$ is relatively high are the regions where X has a relatively high likelihood of occurrence. The regions where the density is relatively small are the regions where X has a relatively low likelihood. The density declines to 0 in the tails—a necessary consequence of the property that the density function is integrable. Areas underneath the density are probabilities. For example, in Figure 2.7, the shaded region is for $20 < X < 30$. This region has area of 0.24, which is the probability that a wage is between \$20 and \$30. The density has a single peak around \$15. This is the **mode** of the distribution.

The wage density has an asymmetric shape. The left tail has a steeper slope than the right tail, which drops off more slowly. This asymmetry is called **skewness**. This is commonly observed in earnings and wealth

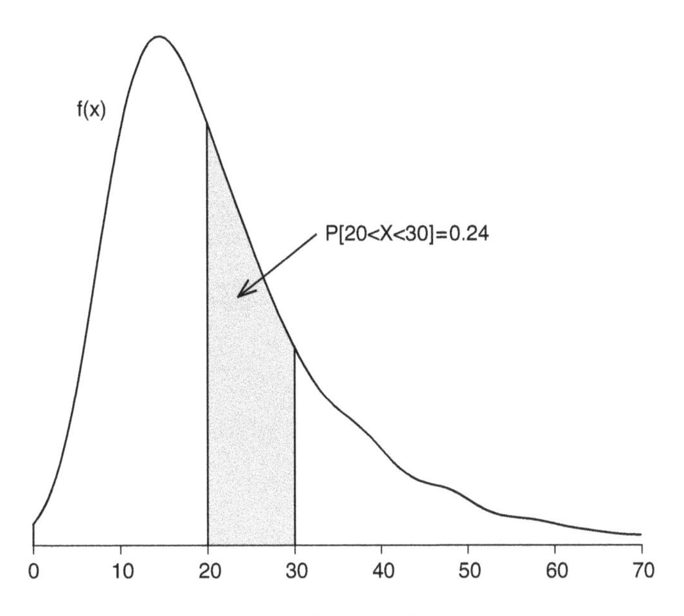

FIGURE 2.7 Density function for wage distribution

distributions. It reflects the fact that there is a small but meaningful probability of a very high wage relative to the general population.

For continuous random variables, the support is defined as the set of values for which the density is positive.

Definition 2.10 The **support** $\mathscr{X}$ of a continuous random variable is the smallest closed set containing $\{x : f(x) > 0\}$.

2.11 TRANSFORMATIONS OF CONTINUOUS RANDOM VARIABLES

If X is a random variable with continuous distribution function F, then for any function $g(x)$, $Y = g(X)$ is a random variable. What is the distribution of Y?

First consider the support. If $\mathscr{X}$ is the support of X, and $g : \mathscr{X} \to \mathscr{Y}$, then $\mathscr{Y}$ is the support of Y. For example, if X has support $[0, 1]$ and $g(x) = 1 + 2x$, the support for $Y = g(X)$ is $[1, 3]$. If X has support $\mathbb{R}_+$ and $g(x) = \log x$, then $Y = g(X)$ has support $\mathbb{R}$.

The probability distribution of Y is $F_Y(y) = \mathbb{P}\left[Y \le y\right] = \mathbb{P}\left[g(X) \le y\right]$. Let $B(y)$ be the set of $x \in \mathbb{R}$ such that $g(x) \le y$. The events $\{g(X) \le y\}$ and $\{X \in B(y)\}$ are identical. So the distribution function for Y is

$$F_Y(y) = \mathbb{P}\left[X \in B(y)\right].$$

Thus, the distribution of Y is determined by the probability function of X.

When $g(x)$ is strictly monotonically increasing, then $g(x)$ has an inverse function

$$h(y) = g^{-1}(y)$$

which implies $X = h(Y)$ and $B(y) = (-\infty, h(y)]$. The distribution function of Y is

$$F_Y(y) = \mathbb{P}\left[X \le h(y)\right] = F_X(h(y)).$$

The density function is its derivative. By the chain rule, we find

$$f_Y(y) = \frac{d}{dy}F_X(h(y)) = f_X(h(y))\frac{d}{dy}h(y) = f_X(h(y))\left|\frac{d}{dy}h(y)\right|.$$

The last equality holds since $h(y)$ has a positive derivative.

Now suppose that $g(x)$ is monotonically decreasing with inverse function $h(y)$. Then $B(y) = [h(y), \infty)$, so

$$F_Y(y) = \mathbb{P}\left[X \geq h(y)\right] = 1 - F_X(h(y)).$$

The density function is the derivative

$$f_Y(y) = -\frac{d}{dy}F_X(h(y)) = -f_X(h(y))\frac{d}{dy}h(y) = f_X(h(y))\left|\frac{d}{dy}h(y)\right|.$$

The last equality holds since $h(y)$ has a negative derivative.

We have found that when $g(x)$ is strictly monotonic, the density for Y is

$$f_Y(y) = f_X(g^{-1}(y))J(y)$$

where

$$J(y) = \left|\frac{d}{dy}h(y)\right| = \left|\frac{d}{dy}g^{-1}(y)\right|$$

is called the **Jacobian** of the transformation. It should be familiar from calculus.

We have shown the following.

Theorem 2.4 If $X \sim f_X$, $f_X(x)$ is continuous on $\mathscr{X}$, $g(x)$ is strictly monotone, and $g^{-1}(y)$ is continuously differentiable on $\mathscr{Y}$, then for $y \in \mathscr{Y}$

$$f_Y(y) = f_X(g^{-1}(y))J(y)$$

where $J(y) = \left|\frac{d}{dy}g^{-1}(y)\right|$.

Theorem 2.4 gives an explicit expression for the density function of the transformation Y. The following four examples illustrate Theorem 2.4.

Example: $f_X(x) = \exp(-x)$ for $x \geq 0$. Set $Y = \lambda X$ for some $\lambda > 0$. This means $g(x) = \lambda x$. Y has support $\mathscr{Y} = [0, \infty)$. The function $g(x)$ is monotonically increasing with inverse function $h(y) = y/\lambda$. The Jacobian is the derivative

$$J(y) = \left|\frac{d}{dy}h(y)\right| = \frac{1}{\lambda}.$$

The density of Y is

$$f_Y(y) = f_X(g^{-1}(y))J(y) = \exp\left(-\frac{y}{\lambda}\right)\frac{1}{\lambda}$$

for $y \geq 0$. This density is valid, since

$$\int_0^\infty f_Y(y)dy = \int_0^\infty \exp\left(-\frac{y}{\lambda}\right)\frac{1}{\lambda}dy = \int_0^\infty \exp(-x)\,dx = 1$$

where the second equality makes the change of variables $x = y/\lambda$.

Example: $f_X(x) = 1$ for $0 \le x \le 1$. Set $Y = g(X)$, where $g(x) = -\log(x)$. Since X has support $[0, 1]$, Y has support $\mathscr{Y} = (0, \infty)$. The function $g(x)$ is monotonically decreasing with inverse function

$$h(y) = g^{-1}(y) = \exp(-y).$$

Take the derivative to obtain the Jacobian:

$$J(y) = \left| \frac{d}{dy} h(y) \right| = \left| -\exp(-y) \right| = \exp(-y).$$

Notice that $f_X(g^{-1}(y)) = 1$ for $y \ge 0$. We find that the density of Y is

$$f_Y(y) = f_X(g^{-1}(y)) J(y) = \exp(-y)$$

for $y \ge 0$. This is the exponential density. We have shown that if X is uniformly distributed, then $Y = -\log(X)$ has an exponential distribution.

Example: Let X have any continuous and invertible (strictly increasing) CDF $F_X(x)$. Define the random variable $Y = F_X(X)$. Y has support $\mathscr{Y} = [0, 1]$. The CDF of Y is

$$\begin{aligned} F_Y(y) &= \mathbb{P}\left[Y \le y\right] \\ &= \mathbb{P}\left[F_X(X) \le y\right] \\ &= \mathbb{P}\left[X \le F_X^{-1}(y)\right] \\ &= F_X(F_X^{-1}(y)) \\ &= y \end{aligned}$$

on $[0, 1]$. Taking the derivative, we find the PDF:

$$f_Y(y) = \frac{d}{dy} y = 1.$$

This is the density function of a $U[0, 1]$ random variable. Thus $Y \sim U[0, 1]$.

The transformation $Y = F_X(X)$ is known as the **probability integral transformation**. The fact that this transformation renders Y to be uniformly distributed regardless of the initial distribution F_X is quite wonderful.

Example: Let $f_X(x)$ be the density function of wages from Figure 2.7. Let $Y = \log(X)$. If X has support on $\mathbb{R}_+$, then Y has support on $\mathbb{R}$. The inverse function is $h(y) = \exp(y)$, and the Jacobian is $\exp(y)$. The density of Y is $f_Y(y) = f_X(\exp(y)) \exp(y)$. It is displayed in Figure 2.8. This density is more symmetric and less skewed than the density of wages in levels.

2.12 NON-MONOTONIC TRANSFORMATIONS

If $Y = g(X)$, where $g(x)$ is not monotonic, we can (in some cases) derive the distribution of Y by direct manipulations. To illustrate, let us focus on the case where $g(x) = x^2$, $\mathscr{X} = \mathbb{R}$, and X has density $f_X(x)$.

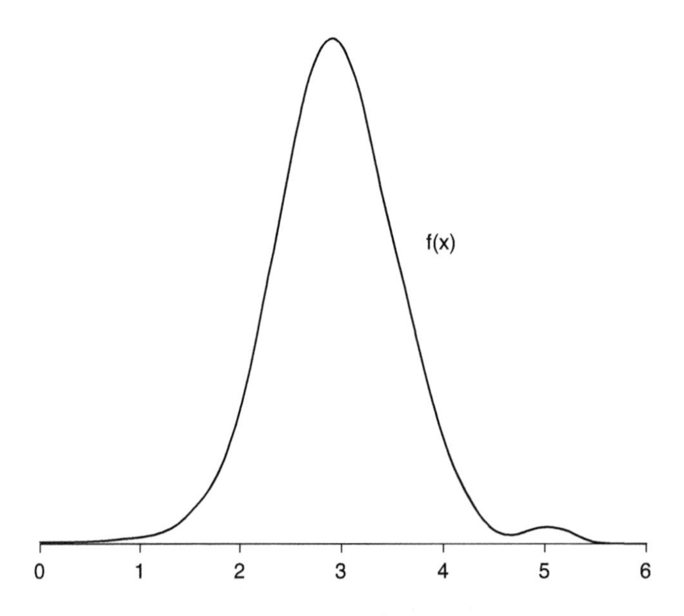

FIGURE 2.8 Density function for log wage distribution

Y has support $\mathscr{Y} \subset [0, \infty)$. For $y \geq 0$,

$$\begin{aligned} F_Y(y) &= \mathbb{P}\left[Y \leq y\right] \\ &= \mathbb{P}\left[X^2 \leq y\right] \\ &= \mathbb{P}\left[|X| \leq \sqrt{y}\right] \\ &= \mathbb{P}\left[-\sqrt{y} \leq X \leq \sqrt{y}\right] \\ &= \mathbb{P}\left[X \leq \sqrt{y}\right] - \mathbb{P}\left[X < -\sqrt{y}\right] \\ &= F_X\left(\sqrt{y}\right) - F_X\left(-\sqrt{y}\right). \end{aligned}$$

We find the density by taking the derivative and applying the chain rule:

$$\begin{aligned} f_Y(y) &= \frac{f_X\left(\sqrt{y}\right)}{2\sqrt{y}} + \frac{f_X\left(-\sqrt{y}\right)}{2\sqrt{y}} \\ &= \frac{f_X\left(\sqrt{y}\right) + f_X\left(-\sqrt{y}\right)}{2\sqrt{y}}. \end{aligned}$$

To further the example, suppose that $f_X(x) = \frac{1}{\sqrt{2\pi}} \exp(-x^2/2)$ (the standard normal density). It follows that the density of $Y = X^2$ is

$$f_Y(y) = \frac{1}{\sqrt{2\pi y}} \exp(-y/2)$$

for $y \geq 0$. This is known as the chi-square density with 1 degree of freedom, written χ_1^2. We have shown that if X is standard normal, then $Y = X^2$ is χ_1^2.

2.13 EXPECTATION OF CONTINUOUS RANDOM VARIABLES

In Section 2.5, we discussed the expectation for discrete random variables. In this section, we consider the continuous case.

Definition 2.11 If X is continuously distributed with density $f(x)$, its **expectation** is defined as

$$\mathbb{E}[X] = \int_{-\infty}^{\infty} xf(x)dx$$

when the integral is convergent.

The expectation is a weighted average of x using the continuous weight function $f(x)$. Just as in the discrete case, the expectation equals the center of mass of the distribution. To visualize, take any density function and imagine placing it on a board on top of a fulcrum. The board will be balanced when the fulcrum is placed at the expected value.

Example: $f(x) = 1$ on $0 \le x \le 1$.

$$\mathbb{E}[X] = \int_0^1 xdx = \frac{1}{2}.$$

Example: $f(x) = \exp(-x)$ on $x \ge 0$. We can show that $\mathbb{E}[X] = 1$. Calculation takes two steps. First,

$$\mathbb{E}[X] = \int_0^{\infty} x \exp(-x)\, dx.$$

Apply integration by parts with $u = x$ and $v = \exp(-x)$. We find

$$\mathbb{E}[X] = \int_0^{\infty} \exp(-x)\, dx = 1.$$

Thus $\mathbb{E}[X] = 1$ as stated.

Example: Hourly wage distribution (Figure 2.5). The expected value is \$23.92. Examine the density plot in Figure 2.7. The expected value is approximately in the middle of the grey shaded region. This is the center of mass, balancing the high mode to the left and the thick tail to the right.

Expectations of transformations are similarly defined.

Definition 2.12 If X has density $f(x)$ then the expected value of $g(X)$ is

$$\mathbb{E}\left[g(X)\right] = \int_{-\infty}^{\infty} g(x)f(x)dx.$$

Example: $X \sim f(x) = 1$ on $0 \le x \le 1$. Then $\mathbb{E}\left[X^2\right] = \int_0^1 x^2 dx = 1/3$.

Example: Log wage distribution (Figure 2.8). The expected value is $\mathbb{E}\left[\log\left(\text{wage}\right)\right] = 2.95$. Examine the density plot in Figure 2.8. The expected value is the approximate midpoint of the density, since the curve is approximately symmetric.

Just as for discrete random variables, the expectation is a linear operator.

Theorem 2.5 Linearity of Expectation. For any constants a and b,

$$\mathbb{E}[a+bX]=a+b\mathbb{E}[X].$$

Proof: Suppose X is continuously distributed. Then

$$\begin{aligned}\mathbb{E}[a+bX]&=\int_{-\infty}^{\infty}(a+bx)f(x)dx\\&=a\int_{-\infty}^{\infty}f(x)dx+b\int_{-\infty}^{\infty}xf(x)dx\\&=a+b\mathbb{E}[X]\end{aligned}$$

since $\int_{-\infty}^{\infty}f(x)dx=1$ and $\int_{-\infty}^{\infty}xf(x)dx=\mathbb{E}[X]$. ■

Example: $f(x)=\exp(-x)$ on $x\geq 0$. Make the transformation $Y=\lambda X$. By the linearity of expectations and our previous calculation $\mathbb{E}[X]=1$,

$$\mathbb{E}[Y]=\mathbb{E}[\lambda X]=\lambda\mathbb{E}[X]=\lambda.$$

Alternatively, by transformation of variables, in Section 2.11 we showed that Y has density $\exp(-y/\lambda)/\lambda$. By direct calculation, we can show that

$$\mathbb{E}[Y]=\int_{0}^{\infty}y\exp\left(-\frac{y}{\lambda}\right)\frac{1}{\lambda}dy=\lambda.$$

Either calculation shows that the expected value of Y is λ.

2.14 FINITENESS OF EXPECTATIONS

In our discussion of the St. Petersburg Paradox, we found that there are discrete distributions which do not have convergent expectations. The same issue applies in the continuous case. It is possible for expectations to be infinite or to be undefined.

Example: $f(x)=x^{-2}$ for $x>1$. This is a valid density, since $\int_{1}^{\infty}f(x)dx=\int_{1}^{\infty}x^{-2}dx=-x^{-1}\big|_{1}^{\infty}=1$. However, the expectation is

$$\mathbb{E}[X]=\int_{1}^{\infty}xf(x)dx=\int_{1}^{\infty}x^{-1}dx=\log(x)\big|_{1}^{\infty}=\infty.$$

Thus the expectation is infinite, because the integral $\int_{1}^{\infty}x^{-1}dx$ is not convergent. The density $f(x)=x^{-2}$ is a special case of the Pareto distribution, which is used to model heavy-tailed distributions.

Example: $f(x)=\dfrac{1}{\pi\left(1+x^{2}\right)}$ for $x\in\mathbb{R}$. The expected value is

$$\begin{aligned}\mathbb{E}[X]&=\int_{-\infty}^{\infty}xf(x)dx\\&=\int_{0}^{\infty}\frac{x}{\pi\left(1+x^{2}\right)}dx+\int_{-\infty}^{0}\frac{x}{\pi\left(1+x^{2}\right)}dx\end{aligned}$$

$$= \left. \frac{\log(1+x^2)}{2\pi} \right|_0^\infty - \left. \frac{\log(1+x^2)}{2\pi} \right|_0^\infty$$

$$= \log(\infty) - \log(\infty)$$

which is undefined. This is called the "Cauchy distribution".

2.15 UNIFYING NOTATION

An annoying feature of intermediate probability theory is that expectations (and other objects) are defined separately for discrete and continuous random variables. Thus all proofs have to be done twice, yet the same steps are used in both. In advanced probability theory, it is typical to instead define expectation using the Riemann-Stieltijes integral (see Section A.8 in the Appendix) which combines these cases. It is useful to be familiar with the notation, even if you are not familiar with the mathematical details.

Definition 2.13 For any random variable X with distribution $F(x)$, the **expectation** is

$$\mathbb{E}[X] = \int_{-\infty}^{\infty} x dF(x)$$

if the integral is convergent.

For the remainder of this chapter, we will not make a distinction between discrete and continuous random variables. For simplicity, we will typically use the notation for continuous random variables (using densities and integration), but the arguments apply to the general case by using Riemann-Stieltijes integration.

2.16 MEAN AND VARIANCE

Two of the most important features of a distribution are its mean and variance, typically denoted by the Greek letters μ and σ^2.

Definition 2.14 The **mean** of X is $\mu = \mathbb{E}[X]$.

The mean is either finite, infinite, or undefined.

Definition 2.15 The **variance** of X is $\sigma^2 = \text{var}[X] = \mathbb{E}\left[(X - \mathbb{E}[X])^2\right]$.

The variance is necessarily nonnegative: $\sigma^2 \geq 0$. It is either finite or infinite. It is 0 only for degenerate random variables.

Definition 2.16 A random variable X is **degenerate** if for some c, $\mathbb{P}[X = c] = 1$.

A degenerate random variable is essentially nonrandom and has a variance of 0.

The variance is measured in square units. To put the variance in the same units as X, we take its square root and give it an entirely new name.

Definition 2.17 The **standard deviation (sd)** of X is the positive square root of the variance, $\sigma = \sqrt{\sigma^2}$.

It is typical to use the mean and standard deviation to summarize the center and spread of a distribution. The following two theorems establish two useful calculations.

Theorem 2.6 $\operatorname{var}[X] = \mathbb{E}\left[X^2\right] - (\mathbb{E}[X])^2$.

To see this, expand the quadratic

$$(X - \mathbb{E}[X])^2 = X^2 - 2X\mathbb{E}[X] + (\mathbb{E}[X])^2$$

and then take expectations to find

$$\begin{aligned}
\operatorname{var}[X] &= \mathbb{E}\left[(X - \mathbb{E}[X])^2\right] \\
&= \mathbb{E}\left[X^2\right] - 2\mathbb{E}[X\mathbb{E}[X]] + \mathbb{E}\left[(\mathbb{E}[X])^2\right] \\
&= \mathbb{E}\left[X^2\right] - 2\mathbb{E}[X]\mathbb{E}[X] + (\mathbb{E}[X])^2 \\
&= \mathbb{E}\left[X^2\right] - (\mathbb{E}[X])^2 .
\end{aligned}$$

The third equality uses the fact that $\mathbb{E}[X]$ is a constant. The fourth combines terms.

Theorem 2.7 $\operatorname{var}[a + bX] = b^2 \operatorname{var}[X]$.

To see this, notice that by linearity,

$$\mathbb{E}[a + bX] = a + b\mathbb{E}[X] .$$

Thus

$$\begin{aligned}
(a + bX) - \mathbb{E}[a + bX] &= a + bX - (a + b\mathbb{E}[X]) \\
&= b(X - \mathbb{E}[X]) .
\end{aligned}$$

Hence

$$\begin{aligned}
\operatorname{var}[a + bX] &= \mathbb{E}\left[((a + bX) - \mathbb{E}[a + bX])^2\right] \\
&= \mathbb{E}\left[(b(X - \mathbb{E}[X]))^2\right] \\
&= b^2\mathbb{E}\left[(X - \mathbb{E}[X])^2\right] \\
&= b^2 \operatorname{var}[X] .
\end{aligned}$$

An important implication of Theorem 2.7 is that the variance is invariant to additive shifts: X and $a + X$ have the same variance.

Example: A Bernoulli random variable takes the value 1 with probability p and 0 with probability $1 - p$:

$$X = \begin{cases} 1 & \text{with probability} \quad p \\ 0 & \text{with probability} \quad 1 - p. \end{cases}$$

This has mean and variance

$$\mu = p$$
$$\sigma^2 = p(1-p).$$

See Exercise 2.4.

Example: An exponential random variable has density

$$f(x) = \frac{1}{\lambda} \exp\left(-\frac{x}{\lambda}\right)$$

for $x \geq 0$. This has mean and variance

$$\mu = \lambda$$
$$\sigma^2 = \lambda^2.$$

See Exercise 2.5.

Example: Hourly wages. (See Figure 2.5). The mean, variance, and standard deviation are

$$\mu = 24$$
$$\sigma^2 = 429$$
$$\sigma = 20.7.$$

Example: Years of education. (Figure 2.2(b)). The mean, variance, and standard deviation are

$$\mu = 13.9$$
$$\sigma^2 = 7.5$$
$$\sigma = 2.7.$$

2.17 MOMENTS

The moments of a distribution are the expected values of the powers of X. We define both uncentered and central moments.

Definition 2.18 The mth **moment** of X is $\mu'_m = \mathbb{E}[X^m]$.

Definition 2.19 For $m > 1$, the mth **central moment** of X is $\mu_m = \mathbb{E}[(X - \mathbb{E}[X])^m]$.

The moments are the expected values of the variables X^m. The central moments are those of $(X - \mathbb{E}[X])^m$. Odd moments may be finite, infinite, or undefined. Even moments are either finite or infinite. For nonnegative X, m can be real-valued.

For ease of reference, let us define the first central moment as $\mu_1 = \mathbb{E}[X]$.

Theorem 2.8 For $m > 1$, the central moments are invariant to additive shifts, that is, $\mu_m(a+X) = \mu_m(X)$.

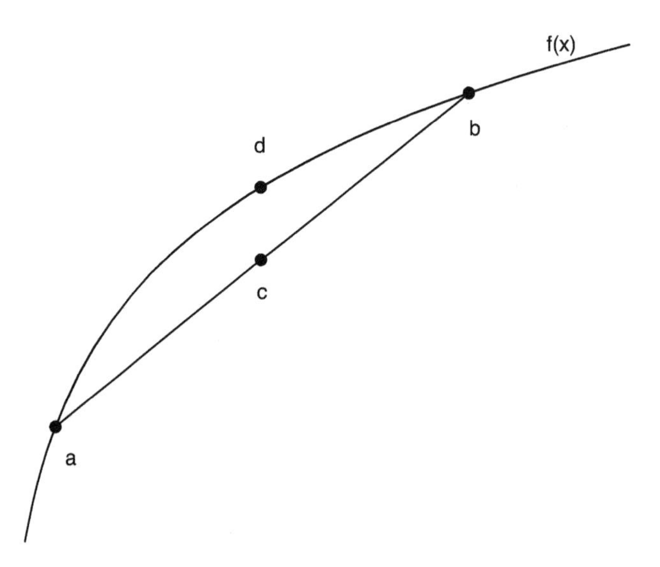

FIGURE 2.9 Concavity

2.18 JENSEN'S INEQUALITY

Expectation is a linear operator: $\mathbb{E}[a+bX]=a+b\mathbb{E}[X]$. It is tempting to apply the same reasoning to nonlinear functions, but this is not valid. We can say more for convex and concave functions.

Definition 2.20 The function $g(x)$ is **convex** if for any $\lambda \in [0,1]$ and all x and y

$$g\left(\lambda x+(1-\lambda)y\right) \leq \lambda g(x)+(1-\lambda)g(y).$$

The function is **concave** if

$$\lambda g(x)+(1-\lambda)g(y) \leq g\left(\lambda x+(1-\lambda)y\right).$$

Examples of convex functions are the exponential $g(x)=\exp(x)$ and quadratic $g(x)=x^2$. Examples of concave functions are the logarithm $g(x)=\log(x)$ and square root $g(x)=x^{1/2}$ for $x \geq 0$.

Concavity is illustrated in Figure 2.9. A concave function $f(x)$ is displayed, and a chord between the points a and b is drawn. The chord lies below the function. The point c lies on the chord, and is less than the point d, which lies on the function.

Theorem 2.9 Jensen's Inequality. For any random variable X, if $g(x)$ is a convex function, then

$$g(\mathbb{E}[X]) \leq \mathbb{E}\left[g(X)\right].$$

If $g(x)$ is a concave function, then

$$\mathbb{E}\left[g(X)\right] \leq g(\mathbb{E}[X]).$$

Proof: We focus on the convex case. Let $a+bx$ be the tangent line to $g(x)$ at $x=\mathbb{E}[X]$. Since $g(x)$ is convex, $g(x) \geq a+bx$. Evaluating at $x=X$ and taking expectations, we find

$$\mathbb{E}\left[g(X)\right] \geq a+b\mathbb{E}[X]=g(\mathbb{E}[X])$$

as claimed. ∎

Jensen's equality states that a convex function of an expectation is less than the expectation of the transformation. Conversely, the expectation of a concave transformation is less than the function of the expectation.

Examples of Jensen's inequality are:

1. $\exp(\mathbb{E}[X]) \leq \mathbb{E}[\exp(X)]$
2. $(\mathbb{E}[X])^2 \leq \mathbb{E}[X^2]$
3. $\mathbb{E}[\log(X)] \leq \log(\mathbb{E}[X])$
4. $\mathbb{E}[X^{1/2}] \leq (\mathbb{E}[X])^{1/2}$.

2.19 APPLICATIONS OF JENSEN'S INEQUALITY*

Jensen's inequality can be used to establish other useful results.

Theorem 2.10 Expectation Inequality. For any random variable X,

$$|\mathbb{E}[X]| \leq \mathbb{E}|X|.$$

Proof: The function $g(x) = |x|$ is convex. An application of Jensen's inequality with $g(x)$ yields the result. ■

Theorem 2.11 Lyapunov's Inequality. For any random variable X and any $0 < r \leq p$,

$$\left(\mathbb{E}|X|^r\right)^{1/r} \leq \left(\mathbb{E}|X|^p\right)^{1/p}.$$

Proof: The function $g(x) = x^{p/r}$ is convex for $x > 0$, since $p \geq r$. Let $Y = |X|^r$. By Jensen's inequality

$$g(\mathbb{E}[Y]) \leq \mathbb{E}[g(Y)]$$

or

$$\left(\mathbb{E}|X|^r\right)^{p/r} \leq \mathbb{E}|X|^p.$$

Raising both sides to the power $1/p$ completes the proof. ■

Theorem 2.12 Discrete Jensen's Inequality. If $g(x) : \mathbb{R} \to \mathbb{R}$ is convex, then for any nonnegative weights a_j such that $\sum_{j=1}^m a_j = 1$ and any real numbers x_j,

$$g\left(\sum_{j=1}^m a_j x_j\right) \leq \sum_{j=1}^m a_j g(x_j). \tag{2.2}$$

Proof: Let X be a discrete random variable with distribution $\mathbb{P}[X = x_j] = a_j$. Jensen's inequality implies equation (2.2). ■

Theorem 2.13 Geometric Mean Inequality. For any nonnegative real weights a_j such that $\sum_{j=1}^m a_j = 1$, and any nonnegative real numbers x_j,

$$x_1^{a_1} x_2^{a_2} \cdots x_m^{a_m} \leq \sum_{j=1}^m a_j x_j. \tag{2.3}$$

Proof: Since the logarithm is strictly concave, by the discrete Jensen inequality,

$$\log\left(x_1^{a_1} x_2^{a_2} \cdots x_m^{a_m}\right) = \sum_{j=1}^{m} a_j \log x_j \leq \log\left(\sum_{j=1}^{m} a_j x_j\right).$$

Applying the exponential yields (2.3). ■

Theorem 2.14 Loève's c_r Inequality. For any real numbers x_j, if $0 < r \leq 1$,

$$\left|\sum_{j=1}^{m} x_j\right|^r \leq \sum_{j=1}^{m} |x_j|^r \tag{2.4}$$

and if $r \geq 1$,

$$\left|\sum_{j=1}^{m} x_j\right|^r \leq m^{r-1} \sum_{j=1}^{m} |x_j|^r. \tag{2.5}$$

For the important special case $m = 2$, we can combine these two inequalities as

$$|a+b|^r \leq C_r \left(|a|^r + |b|^r\right) \tag{2.6}$$

where $C_r = \max\left[1, 2^{r-1}\right]$.

Proof: For $r \geq 1$, (2.5) is a rewriting of Jensen's inequality (2.2) with $g(u) = u^r$ and $a_j = 1/m$. For $r < 1$, define $b_j = |x_j| / \left(\sum_{j=1}^{m} |x_j|\right)$. The facts that $0 \leq b_j \leq 1$ and $r < 1$ imply $b_j \leq b_j^r$ and thus

$$1 = \sum_{j=1}^{m} b_j \leq \sum_{j=1}^{m} b_j^r.$$

This implies

$$\left(\sum_{j=1}^{m} x_j\right)^r \leq \left(\sum_{j=1}^{m} |x_j|\right)^r \leq \sum_{j=1}^{m} |x_j|^r.$$ ■

Theorem 2.15 Norm Monotonicity. If $0 < t \leq s$, for any real numbers x_j

$$\left|\sum_{j=1}^{m} |x_j|^s\right|^{1/s} \leq \left|\sum_{j=1}^{m} |x_j|^t\right|^{1/t}. \tag{2.7}$$

Proof: Set $y_j = |x_j|^s$ and $r = t/s \leq 1$. The c_r inequality (2.4) implies $\left|\sum_{j=1}^{m} y_j\right|^r \leq \sum_{j=1}^{m} |y_j|^r$ or

$$\left|\sum_{j=1}^{m} |x_j|^s\right|^{t/s} \leq \sum_{j=1}^{m} |x_j|^t.$$

Raising both sides to the power $1/t$ yields (2.7). ■

2.20 SYMMETRIC DISTRIBUTIONS

We say that a distribution of a random variable is **symmetric** about 0 if the distribution function satisfies

$$F(x) = 1 - F(-x).$$

If X has a density $f(x)$, X is symmetric about 0 if

$$f(x) = f(-x).$$

For example, the standard normal density $\phi(x) = (2\pi)^{-1/2}\exp(-x^2/2)$ is symmetric about 0.

If a distribution is symmetric about 0, then all finite odd moments are 0 (if the moment is finite). To see this, let m be odd. Then

$$\mathbb{E}\left[X^m\right] = \int_{-\infty}^{\infty} x^m f(x)dx = \int_0^{\infty} x^m f(x)dx + \int_{-\infty}^{0} x^m f(x)dx.$$

For the integral from $-\infty$ to 0, make the change of variables $x = -t$. Then the right-hand side equals

$$\int_0^{\infty} x^m f(x)dx + \int_0^{\infty} (-t)^m f(-t)dt = \int_0^{\infty} x^m f(x)dx - \int_0^{\infty} t^m f(t)dt = 0$$

where the first equality uses the symmetry property. The second equality holds if the moment is finite. This last step is subtle and easy to miss. If $\mathbb{E}\left|X\right|^m = \infty$, then the last step is

$$\mathbb{E}\left[X^m\right] = \infty - \infty \neq 0.$$

In this case, $\mathbb{E}\left[X^m\right]$ is undefined.

More generally, if a distribution is symmetric about 0, then the expectation of any odd function (if finite) is 0. An odd function satisfies $g(-x) = -g(x)$. For example, $g(x) = x^3$ and $g(x) = \sin(x)$ are odd functions. To see this, let $g(x)$ be an odd function, and without loss of generality, assume $\mathbb{E}\left[X\right] = 0$. Then making a change of variables $x = -t$ gives

$$\int_{-\infty}^{0} g(x)f(x)dx = \int_0^{\infty} g(-t)f(-t)dt = -\int_0^{\infty} g(t)f(t)dt$$

where the second equality uses the assumptions that $g(x)$ is odd and $f(x)$ is symmetric about 0. Then

$$\mathbb{E}\left[g(X)\right] = \int_0^{\infty} g(x)f(x)dx + \int_{-\infty}^{0} g(x)f(x)dx = \int_0^{\infty} g(x)f(x)dx - \int_0^{\infty} g(t)f(t)dt = 0.$$

Theorem 2.16 If $f(x)$ is symmetric about 0, $g(x)$ is odd, and $\int_0^{\infty}\left|g(x)\right|f(x)dx < \infty$, then $\mathbb{E}\left[g(X)\right] = 0$.

2.21 TRUNCATED DISTRIBUTIONS

Sometimes we only observe part of a distribution. For example, consider a sealed-bid auction for a work of art with a minimum bid requirement. Assuming that participants make a bid based on their personal valuation, they will not make bids if their personal valuation is lower than the required minimum. Thus we do not observe "bids" from these participants. This is an example of truncation from below. Truncation is a specific transformation of a random variable.

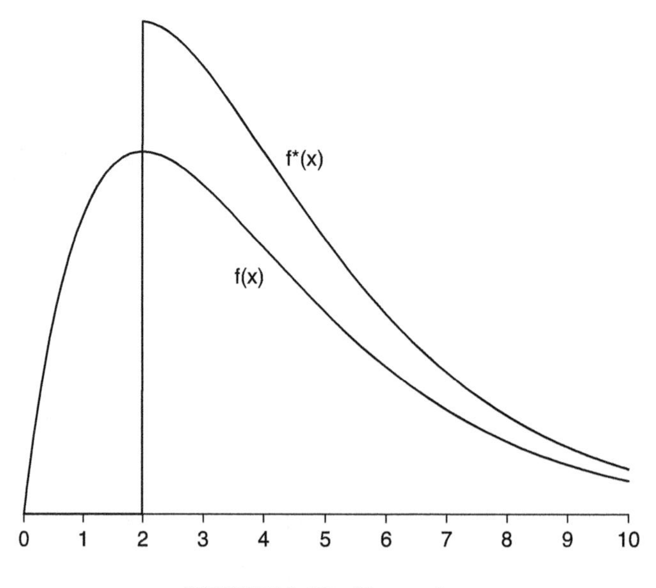

FIGURE 2.10 Truncation

If X is a random variable with distribution $F(x)$, and X is truncated to satisfy $X \leq c$ (truncated from above), then the truncated distribution function is

$$F^*(x) = \mathbb{P}[X \leq x \mid X \leq c] = \begin{cases} \dfrac{F(x)}{F(c)}, & x < c \\ 1, & x \geq c. \end{cases}$$

If $F(x)$ is continuous, then the density of the truncated distribution is

$$f^*(x) = f(x \mid X \leq c) = \frac{f(x)}{F(c)}$$

for $x \leq c$. The mean of the truncated distribution is

$$\mathbb{E}[X \mid X \leq c] = \frac{\int_{-\infty}^{c} x f(x) dx}{F(c)}.$$

If X is truncated to satisfy $X \geq c$ (truncated from below), then the truncated distribution and density functions are

$$F^*(x) = \mathbb{P}[X \leq x \mid X \geq c] = \begin{cases} 0, & x < c \\ \dfrac{F(x) - F(c)}{1 - F(c)}, & x \geq c \end{cases}$$

and

$$f^*(x) = f(x \mid X \geq c) = \frac{f(x)}{1 - F(c)}$$

for $x \geq c$. The mean of the truncated distribution is

$$\mathbb{E}[X \mid X \geq c] = \frac{\int_{c}^{\infty} x f(x) dx}{1 - F(c)}.$$

Trucation from below is illustrated in Figure 2.10. The untruncated density function is marked as $f(x)$. The truncated density is marked as $f^*(x)$. The portion of $f(x)$ below 2 is eliminated, and the density above 2 is shifted up to compensate.

An interesting example is the exponential $F(x) = 1 - e^{-x/\lambda}$ with density $f(x) = \lambda^{-1}e^{-x/\lambda}$ and mean λ. The truncated density is

$$f(x \mid X \geq c) = \frac{\lambda^{-1}e^{-x/\lambda}}{e^{-c/\lambda}} = \lambda^{-1}e^{-(x-c)/\lambda}.$$

This is also the exponential distribution, but it is shifted by c. The mean of this distribution is $c + \lambda$, which is the same as the original exponential distribution shifted by c. This is a "memoryless" property of the exponential distribution.

2.22 CENSORED DISTRIBUTIONS

Sometimes a boundary constraint is forced on a random variable. For example, let X be a desired level of consumption, but X^* is constrained to satisfy $X^* \leq c$ (censored from above). In this case, if $X > c$, the constrained consumption will satisfy $X^* = c$. We can write this as

$$X^* = \begin{cases} X, & X \leq c \\ c, & X > c. \end{cases}$$

Similarly if X^* is constrained to satisfy $X^* \geq c$, then when $X < c$, the constrained version will satisfy $X^* = c$. We can write this as

$$X^* = \begin{cases} X, & X \geq c \\ c, & X < c. \end{cases}$$

Censoring is related to truncation but is different. Under truncation, the random variables exceeding the boundary are excluded. Under censoring, they are transformed to satisfy the constraint.

When the original random variable X is continuously distributed, then the censored random variable X^* will have a mixed distribution, with a continuous component over the unconstrained set and a discrete mass at the constrained boundary.

Censoring is common in economic applications. A standard example is consumer purchases of individual items. In this case, one constraint is $X^* \geq 0$ and consequently, we typically observe a discrete mass at 0. Another standard example is "top-coding", where a continuous variable such as income is recorded either in categories or is continuous up to a top category "income above \$Y". All incomes above this threshold are recorded at the threshold \$Y.

The expected value of a censored random variable is

$$X^* \leq c: \qquad \mathbb{E}\left[X^*\right] = \int_{-\infty}^{c} xf(x)dx + c(1 - F(c))$$

$$X^* \geq c: \qquad \mathbb{E}\left[X^*\right] = \int_{c}^{\infty} xf(x)dx + cF(c).$$

2.23 MOMENT GENERATING FUNCTION

The following is a technical tool used to facilitate some proofs. It is not particularly intuitive.

Definition 2.21 The **moment generating function (MGF)** of X is $M(t) = \mathbb{E}\left[\exp(tX)\right]$.

Since the exponential is nonnegative, the MGF is either finite or infinite. For the MGF to be finite, the density of X must have thin tails. When we use the MGF, we are implicitly assuming that it is finite.

Example: $U[0, 1]$. The density is $f(x) = 1$ for $0 \le x \le 1$. The MGF is

$$M(t) = \int_{-\infty}^{\infty} \exp(tx) f(x) dx = \int_0^1 \exp(tx)\, dx = \frac{\exp(t) - 1}{t}.$$

Example: Exponential distribution. The density is $f(x) = \lambda^{-1} \exp(x/\lambda)$ for $x \ge 0$. The MGF is

$$M(t) = \int_{-\infty}^{\infty} \exp(tx) f(x) dx = \frac{1}{\lambda} \int_0^{\infty} \exp(tx) \exp\left(-\frac{x}{\lambda}\right) dx = \frac{1}{\lambda} \int_0^{\infty} \exp\left(\left(t - \frac{1}{\lambda}\right) x\right) dx.$$

This integral is only convergent if $t < 1/\lambda$. Assuming this holds, make the change of variables $y = \left(t - \frac{1}{\lambda}\right) x$, and we find that the above integral equals

$$M(t) = -\frac{1}{\lambda\left(t - \frac{1}{\lambda}\right)} = \frac{1}{1 - \lambda t}.$$

In this example, the MGF is finite only on the region $t < 1/\lambda$.

Example: $f(x) = x^{-2}$ for $x > 1$. The MGF is

$$M(t) = \int_1^{\infty} \exp(tx)\, x^{-2} dx = \infty$$

for $t > 0$. This nonconvergence means that in this example, the MGF cannot be succesfully used for calculations.

The MGF has the important property that it completely characterizes the distribution of X. It also has the following properties.

Theorem 2.17 Moments and the MGF. If $M(t)$ is finite for t in a neighborhood of 0, then $M(0) = 1$,

$$\left.\frac{d}{dt} M(t)\right|_{t=0} = \mathbb{E}[X]$$

$$\left.\frac{d^2}{dt^2} M(t)\right|_{t=0} = \mathbb{E}\left[X^2\right]$$

and

$$\left.\frac{d^m}{dt^m} M(t)\right|_{t=0} = \mathbb{E}\left[X^m\right]$$

for any moment which is finite.

This property is why it is called the "moment generating" function. The curvature of $M(t)$ at $t = 0$ encodes all moments of the distribution of X.

Example: $U[0,1]$. The MGF is $M(t) = t^{-1}\left(\exp(t) - 1\right)$. Using L'Hôpital's rule (Theorem A.12) results in

$$M(0) = \exp(0) = 1.$$

Using the derivative rule of differentiation and L'Hôpital's rule gives

$$\mathbb{E}[X] = \frac{d}{dt}\frac{\exp(t)-1}{t}\bigg|_{t=0} = \frac{t\exp(t) - \left(\exp(t)-1\right)}{t^2}\Bigg|_{t=0} = \frac{\exp(0)}{2} = \frac{1}{2}.$$

Example: Exponential distribution. The MGF is $M(t) = (1-\lambda t)^{-1}$. The first moment is

$$\mathbb{E}[X] = \frac{d}{dt}\frac{1}{1-\lambda t}\bigg|_{t=0} = \frac{\lambda}{(1-\lambda t)^2}\bigg|_{t=0} = \lambda.$$

The second moment is

$$\mathbb{E}\left[X^2\right] = \frac{d^2}{dt^2}\frac{1}{1-\lambda t}\bigg|_{t=0} = \frac{2\lambda^2}{(1-\lambda t)^3}\bigg|_{t=0} = 2\lambda^2.$$

Proof of Theorem 2.17 We use the assumption that X is continuously distributed. Then

$$M(t) = \int_{-\infty}^{\infty} \exp(tx) f(x)dx$$

so

$$M(0) = \int_{-\infty}^{\infty} \exp(0x) f(x)dx = \int_{-\infty}^{\infty} f(x)dx = 1.$$

The first derivative is

$$\begin{aligned}\frac{d}{dt}M(t) &= \frac{d}{dt}\int_{-\infty}^{\infty} \exp(tx) f(x)dx \\ &= \int_{-\infty}^{\infty} \frac{d}{dt}\exp(tx) f(x)dx \\ &= \int_{-\infty}^{\infty} \exp(tx)\, x f(x)dx.\end{aligned}$$

Evaluated at $t=0$, we find

$$\frac{d}{dt}M(t)\bigg|_{t=0} = \int_{-\infty}^{\infty} \exp(0x)\, x f(x)dx = \int_{-\infty}^{\infty} x f(x)dx = \mathbb{E}[X]$$

as claimed. Similarly

$$\begin{aligned}\frac{d^m}{dt^m}M(t) &= \int_{-\infty}^{\infty} \frac{d^m}{dt^m}\exp(tx) f(x)dx \\ &= \int_{-\infty}^{\infty} \exp(tx)\, x^m f(x)dx\end{aligned}$$

so

$$\frac{d^m}{dt^m}M(t)\bigg|_{t=0} = \int_{-\infty}^{\infty} \exp(0x)\, x^m f(x)dx = \int_{-\infty}^{\infty} x^m f(x)dx = \mathbb{E}\left[X^m\right]$$

as claimed. ∎

2.24 CUMULANTS

The **cumulant generating function** is the natural log of the MGF:

$$K(t) = \log M(t).$$

Since $M(0) = 1$, we see that $K(0) = 0$. Expanding as a power series, we obtain

$$K(t) = \sum_{r=1}^{\infty} \kappa_r \frac{t^r}{r!}$$

where

$$\kappa_r = K^{(r)}(0)$$

is the rth derivative of $K(t)$, evaluated at $t = 0$. The constants κ_r are known as the **cumulants** of the distribution. Note that $\kappa_0 = K(0) = 0$ since $M(0) = 1$.

The cumulants are related to the central moments. We can calculate that

$$K^{(1)}(t) = \frac{M^{(1)}(t)}{M(t)}$$

$$K^{(2)}(t) = \frac{M^{(2)}(t)}{M(t)} - \left(\frac{M^{(1)}(t)}{M(t)}\right)^2$$

so $\kappa_1 = \mu_1$ and $\kappa_2 = \mu_2' - \mu_1^2 = \mu_2$. The first six cumulants are as follows:

$$\begin{aligned}
\kappa_1 &= \mu_1 \\
\kappa_2 &= \mu_2 \\
\kappa_3 &= \mu_3 \\
\kappa_4 &= \mu_4 - 3\mu_2^2 \\
\kappa_5 &= \mu_5 - 10\mu_3\mu_2 \\
\kappa_6 &= \mu_6 - 15\mu_4\mu_2 - 10\mu_3^2 + 30\mu_2^3.
\end{aligned}$$

The first three cumulants correspond to the central moments, but higher cumulants are polynomial functions of the central moments.

Inverting, we can also express the central moments in terms of the cumulants, for example, the fourth through sixth are as follows:

$$\begin{aligned}
\mu_4 &= \kappa_4 + 3\kappa_2^2 \\
\mu_5 &= \kappa_5 + 10\kappa_3\kappa_2 \\
\mu_6 &= \kappa_6 + 15\kappa_4\kappa_2 + 10\kappa_3^2 + 15\kappa_2^3.
\end{aligned}$$

Example: Exponential distribution. The MGF is $M(t) = (1 - \lambda t)^{-1}$. The cumulant generating function is $K(t) = -\log(1 - \lambda t)$. The first four derivatives are

$$K^{(1)}(t) = \frac{\lambda}{1 - \lambda t}$$

$$K^{(2)}(t) = \frac{\lambda^2}{(1-\lambda t)^2}$$

$$K^{(3)}(t) = \frac{2\lambda^3}{(1-\lambda t)^3}$$

$$K^{(4)}(t) = \frac{6\lambda^4}{(1-\lambda t)^4}.$$

Thus the first four cumulants of the distribution are λ, λ^2, $2\lambda^3$, and $6\lambda^4$.

2.25 CHARACTERISTIC FUNCTION

Because the MGF is not necessarily finite, the characteristic function is used for advanced formal proofs.

Definition 2.22 The **characteristic function** of X is $C(t) = \mathbb{E}\left[\exp(itX)\right]$, where $i = \sqrt{-1}$.

Since $\exp(iu) = \cos(u) + i\sin(u)$ is bounded, the characteristic function exists for all random variables.

If X is symmetrically distributed about 0, then since the sine function is odd and bounded, $\mathbb{E}[\sin(X)] = 0$. It follows that for symmetrically distributed random variables, the characteristic function can be written in terms of the cosine function only.

Theorem 2.18 If X is symmetrically distributed about 0, its characteristic function equals $C(t) = \mathbb{E}[\cos(tX)]$.

The characteristic function has similar properties as the MGF, but it is a bit more tricky to deal with since it involves complex numbers.

Example: Exponential distribution. The density is $f(x) = \lambda^{-1}\exp(-x/\lambda)$ for $x \geq 0$. The characteristic function is

$$C(t) = \int_0^\infty \exp(itx)\frac{1}{\lambda}\exp\left(-\frac{x}{\lambda}\right)dx = \frac{1}{\lambda}\int_0^\infty \exp\left(\left(it - \frac{1}{\lambda}\right)x\right)dx.$$

Making the change of variables $y = \left(it - \frac{1}{\lambda}\right)x$, we find that the above integral equals

$$M(t) = -\frac{1}{\lambda\left(it - \frac{1}{\lambda}\right)} = \frac{1}{1-\lambda it}.$$

This is finite for all t.

2.26 EXPECTATION: MATHEMATICAL DETAILS*

In this section, we discuss a rigorous definition of expectation. Define the Riemann-Stieltijes integrals

$$I_1 = \int_0^\infty x dF(x) \tag{2.8}$$

$$I_2 = \int_{-\infty}^0 x dF(x). \tag{2.9}$$

The integral I_1 is the integral over the positive real line, and the integral I_2 is the integral over the negative real line. The number I_1 can be 0, positive, or positive infinity. The number I_2 can be 0, negative, or negative infinity.

Definition 2.23 The **expectation** $\mathbb{E}[X]$ of a random variable X is

$$\mathbb{E}[X] = \begin{cases} I_1 + I_2 & \text{if both } I_1 < \infty \text{ and } I_2 > -\infty \\ \infty & \text{if } I_1 = \infty \text{ and } I_2 > -\infty \\ -\infty & \text{if } I_1 < \infty \text{ and } I_2 = -\infty \\ \text{undefined} & \text{if both } I_1 = \infty \text{ and } I_2 = -\infty. \end{cases}$$

This definition allows for an expectation to be finite, infinite, or undefined. The expectation $\mathbb{E}[X]$ is finite if and only if

$$\mathbb{E}|X| = \int_{-\infty}^{\infty} |x|\, dF(x) < \infty.$$

In this case, it is common to say that $\mathbb{E}[X]$ is **well defined**.

More generally, X has a finite rth moment if

$$\mathbb{E}|X|^r < \infty. \tag{2.10}$$

By Lyapunov's Inequality (Theorem 2.11), the inequality (2.10) implies $\mathbb{E}|X|^s < \infty$ for all $0 \leq s \leq r$. Thus, for example, if the fourth moment is finite, then the first, second, and third moments are also finite, and so is the 3.9th absolute moment.

2.27 EXERCISES

Exercise 2.1 Let $X \sim U[0, 1]$. Find the PDF of $Y = X^2$.

Exercise 2.2 Let $X \sim U[0, 1]$. Find the distribution function of $Y = \log\left(\frac{X}{1-X}\right)$.

Exercise 2.3 Define $F(x) = \begin{cases} 0 & \text{if } x < 0 \\ 1 - \exp(-x) & \text{if } x \geq 0. \end{cases}$

(a) Show that $F(x)$ is a CDF.
(b) Find the PDF $f(x)$.
(c) Find $\mathbb{E}[X]$.
(d) Find the PDF of $Y = X^{1/2}$.

Exercise 2.4 A Bernoulli random variable takes the value 1 with probability p and 0 with probability $1 - p$

$$X = \begin{cases} 1 & \text{with probability} \quad p \\ 0 & \text{with probability} \quad 1 - p. \end{cases}$$

Find the mean and variance of X.

Exercise 2.5 Find the mean and variance of X with density $f(x) = \frac{1}{\lambda} \exp\left(-\frac{x}{\lambda}\right)$.

Exercise 2.6 Compute $\mathbb{E}[X]$ and var $[X]$ for the following distributions.

(a) $f(x) = ax^{-a-1}, 0 < x < 1, a > 0$.

(b) $f(x) = \frac{1}{n}, x = 1, 2, \ldots, n$.

(c) $f(x) = \frac{3}{2}(x-1)^2, 0 < x < 2$.

Exercise 2.7 Let X have density

$$f_X(x) = \frac{1}{2^{r/2}\Gamma\left(\frac{r}{2}\right)} x^{r/2-1} \exp\left(-\frac{x}{2}\right)$$

for $x \geq 0$. This is known as the **chi-square** distribution. Let $Y = 1/X$. Show that the density of Y is

$$f_Y(y) = \frac{1}{2^{r/2}\Gamma\left(\frac{r}{2}\right)} y^{-r/2-1} \exp\left(-\frac{1}{2y}\right)$$

for $y \geq 0$. This is known as the **inverse chi-square** distribution.

Exercise 2.8 Show that if the density satisfies $f(x) = f(-x)$ for all $x \in \mathbb{R}$, then the distribution function satisfies $F(-x) = 1 - F(x)$.

Exercise 2.9 Suppose X has density $f(x) = e^{-x}$ on $x > 0$. Set $Y = \lambda X$ for $\lambda > 0$. Find the density of Y.

Exercise 2.10 Suppose X has density $f(x) = \lambda^{-1}e^{-x/\lambda}$ on $x > 0$ for some $\lambda > 0$. Set $Y = X^{1/\alpha}$ for $\alpha > 0$. Find the density of Y.

Exercise 2.11 Suppose X has density $f(x) = e^{-x}$ on $x > 0$. Set $Y = -\log X$. Find the density of Y.

Exercise 2.12 Find the median of the density $f(x) = \frac{1}{2}\exp(-|x|)$, $x \in \mathbb{R}$.

Exercise 2.13 Find a which minimizes $\mathbb{E}\left[(X-a)^2\right]$. Your answer should be a moment of X.

Exercise 2.14 Show that if X is a continuous random variable, then

$$\min_a \mathbb{E}|X-a| = \mathbb{E}|X-m|,$$

where m is the median of X.
Hint: Work out the integral expression of $\mathbb{E}|X-a|$, and notice that it is differentiable.

Exercise 2.15 The **skewness** of a distribution is

$$\text{skew} = \frac{\mu_3}{\sigma^3}$$

where μ_3 is the third central moment.

(a) Show that if the density function is symmetric about some point a, then skew $= 0$.

(b) Calculate skew for $f(x) = \exp(-x)$, $x \geq 0$.

Exercise 2.16 Let X be a random variable with $\mathbb{E}[X]=1$. Show that $\mathbb{E}\left[X^2\right]>1$ if X is not degenerate.
Hint: Use Jensen's inequality.

Exercise 2.17 Let X be a random variable with mean μ and variance σ^2. Show that $\mathbb{E}\left[(X-\mu)^4\right]\geq\sigma^4$.

Exercise 2.18 Suppose the random variable X is a duration (a period of time). Examples include: a spell of unemployment, length of time on a job, length of a labor strike, length of a recession, length of an economic expansion. The **hazard function** $h(x)$ associated with X is

$$h(x)=\lim_{\delta\to 0}\frac{\mathbb{P}\left[x\leq X\leq x+\delta\mid X\geq x\right]}{\delta}.$$

This function can be interpreted as the rate of change of the probability of continued survival. If $h(x)$ increases (decreases) with x, it is described as **increasing (decreasing) hazard**.

(a) Show that if X has distribution F and density f, then $h(x)=f(x)/\left(1-F(x)\right)$.

(b) Suppose $f(x)=\lambda^{-1}\exp\left(-x/\lambda\right)$. Find the hazard function $h(x)$. Is this an increasing or decreasing hazard or neither?

(c) Find the hazard function for the Weibull distribution from Section 3.23. Is this increasing or decreasing hazard or neither?

Exercise 2.19 Let $X\sim U[0,1]$ be uniformly distributed on $[0,1]$. (X has density $f(x)=1$ on $[0,1]$, 0 elsewhere.) Suppose X is truncated to satisfy $X\leq c$ for some $0<c<1$.

(a) Find the density function of the truncated variable X.

(b) Find $\mathbb{E}\left[X\mid X\leq c\right]$.

Exercise 2.20 Let X have density $f(x)=e^{-x}$ for $x\geq 0$. Suppose X is censored to satisfy $X^*\geq c>0$. Find the mean of the censored distribution.

Exercise 2.21 Surveys routinely ask discrete questions when the underlying variable is continuous. For example, *wage* may be continuous, but the survey questions are categorical. Take the following example:

Wage	Frequency
$\$0\leq \text{wage}\leq \$10$	0.1
$\$10< \text{wage}\leq \$20$	0.4
$\$20< \text{wage}\leq \$30$	0.3
$\$30< \text{wage}\leq \$40$	0.2

Assume that \$40 is the maximal wage.

(a) Plot the discrete distribution function, putting the probability mass at the rightmost point of each interval. Repeat putting the probability mass at the leftmost point of each interval. Compare. What can you say about the true distribution function?

(b) Calculate the expected wage using the two discrete distributions from part (a). Compare.

(c) Make the assumption that the distribution is uniform on each interval. Plot this density function, distribution function, and expected wage. Compare with the above results.

Exercise 2.22 First-order stochastic dominance. A distribution $F(x)$ is said to "first-order dominate" distribution $G(x)$ if $F(x) \leq G(x)$ for all x and $F(x) < G(x)$ for at least one x. Show the following proposition: F stochastically dominates G if and only if every utility maximizer with increasing utility in X prefers outcome $X \sim F$ over outcome $X \sim G$.

CHAPTER 3

PARAMETRIC DISTRIBUTIONS

3.1 INTRODUCTION

A **parametric distribution** $F(x \mid \theta)$ is a distribution indexed by a **parameter** $\theta \in \Theta$. For each θ, the function $F(x \mid \theta)$ is a valid distribution function. As θ varies, the distribution function changes. The set Θ is called the **parameter space**. We sometimes call $F(x \mid \theta)$ a **family** of distributions. Parametric distributions are typically simple in shape and functional form, and are selected for their ease of manipulation.

Parametric distributions are often used by economists for economic modeling. A specific distribution may be selected based on appropriateness, convenience, and tractability.

Econometricians use parametric distributions for statistical modeling. A set of observations may be modeled using a specific distribution. The parameters of the distribution are unspecified, and their values chosen (estimated) to match features of the data. It is therefore desirable to understand how variation in the parameters leads to variation in the shape of the distribution.

In this chapter, I list common parametric distributions used by economists and discuss features of these distributions, such as their mean and variance. The list is not exhaustive. It is not necessary to memorize this list or the details. Rather, this information is presented for reference.

3.2 BERNOULLI DISTRIBUTION

A **Bernoulli** random variable is a two-point distribution. It is typically parameterized as

$$\mathbb{P}[X=0]=1-p$$
$$\mathbb{P}[X=1]=p.$$

The probability mass function is

$$\pi(x \mid p)=p^{x}(1-p)^{1-x}$$
$$0<p<1.$$

A Bernoulli distribution is appropriate for any variable which only has two outcomes, such as a coin flip. The parameter p indexes the likelihood of the two events.

$$\mathbb{E}[X]=p$$
$$\operatorname{var}[X]=p(1-p).$$

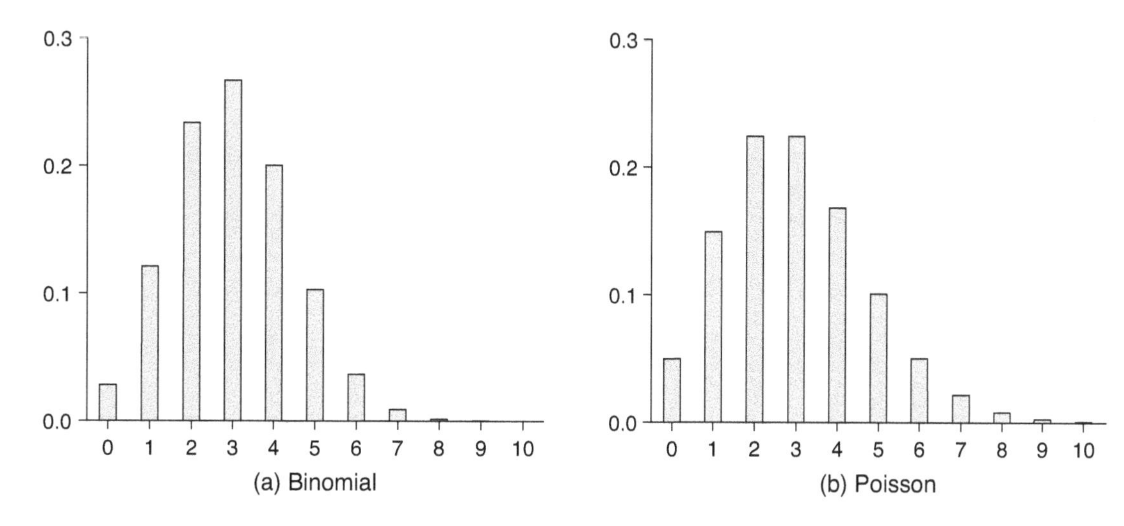

FIGURE 3.1 Discrete distributions

3.3 RADEMACHER DISTRIBUTION

A **Rademacher** random variable is a two-point distribution. It is parameterized as

$$\mathbb{P}[X=-1]=1/2$$
$$\mathbb{P}[X=1]=1/2$$

with

$$\mathbb{E}[X]=0$$
$$\text{var}[X]=1.$$

3.4 BINOMIAL DISTRIBUTION

A **Binomial** random variable has support $\{0, 1, \ldots, n\}$ and probability mass function

$$\pi\left(x \mid n, p\right)=\binom{n}{x}p^{x}\left(1-p\right)^{n-x}, \qquad x=0,1,\ldots,n$$
$$0<p<1.$$

The binomial random variable equals the outcome of n independent Bernoulli trials. If you flip a coin n times, the number of heads has a binomial distribution.

$$\mathbb{E}[X]=np$$
$$\text{var}[X]=np(1-p).$$

The binomial probability mass function is displayed in Figure 3.1(a) for the case $p=0.3$ and $n=10$.

3.5 MULTINOMIAL DISTRIBUTION

The term **multinomial** has two uses in econometrics.

(1) A single **multinomial** random variable or **multinomial trial** is a K-point distribution with support $\{x_1, \ldots, x_K\}$. The probability mass function is

$$\pi\left(x_j \mid p_1, \ldots, p_K\right) = p_j$$
$$\sum_{j=1}^{K} p_j = 1.$$

This can be written as

$$\pi\left(x_j \mid p_1, \ldots, p_K\right) = p_1^{x_1} p_2^{x_2} \cdots p_K^{x_K}.$$

A multinomial can be used to model categorical outcomes (e.g., car, bicycle, bus, or walk), ordered numerical outcomes (a roll of a K-sided die), or numerical outcomes on any set of support points. This is the most common usage of the term "multinomial" in econometrics.

$$\mathbb{E}[X] = \sum_{j=1}^{K} p_j x_j$$
$$\operatorname{var}[X] = \sum_{j=1}^{K} p_j x_j^2 - \left(\sum_{j=1}^{K} p_j x_j\right)^2.$$

(2) A **multinomial** is the set of outcomes from n independent single multinomial trials. It is the sum of outcomes for each category, and is thus a set $(X_1, \ldots, X_K)$ of random variables satisfying $\sum_{j=1}^{K} X_k = n$. The probability mass function is

$$\mathbb{P}\left[X_1 = x_1, \ldots, X_K = x_k \mid n, p_1, \ldots, p_K\right] = \frac{n!}{x_1! \cdots x_K!} p_1^{x_1} p_2^{x_2} \cdots p_K^{x_K}$$
$$\sum_{j=1}^{K} x_k = n$$
$$\sum_{j=1}^{K} p_j = 1.$$

3.6 POISSON DISTRIBUTION

A **Poisson** random variable has support on the nonnegative integers:

$$\pi(x \mid \lambda) = \frac{e^{-\lambda} \lambda^x}{x!}, \qquad x = 0, 1, 2 \ldots,$$
$$\lambda > 0.$$

The parameter λ indexes the mean and spread. In economics, the Poisson distribution is often used for arrival times. Econometricians use it for count (integer-valued) data.

$$\mathbb{E}[X] = \lambda$$

$$\operatorname{var}[X] = \lambda.$$

The Poisson probability mass function is displayed in Figure 3.1(b) for the case $\lambda = 3$.

3.7 NEGATIVE BINOMIAL DISTRIBUTION

A limitation of the Poisson model for count data is that the single parameter λ controls both the mean and variance. An alternative is the **negative binomial**:

$$\pi(x \mid r, p) = \binom{x+r-1}{x} p^x (1-p)^r, \qquad x = 0, 1, 2 \ldots$$

$$0 < p < 1$$

$$r > 0.$$

The distribution has two parameters, so the mean and variance are freely varying.

$$\mathbb{E}[X] = \frac{pr}{1-p}$$

$$\operatorname{var}[X] = \frac{pr}{(1-p)^2}.$$

3.8 UNIFORM DISTRIBUTION

A **uniform** random variable, typically written $U[a, b]$, has density

$$f(x \mid a, b) = \frac{1}{b-a}, \qquad a \leq x \leq b.$$

$$\mathbb{E}[X] = \frac{b+a}{2}$$

$$\operatorname{var}[X] = \frac{(b-a)^2}{12}.$$

3.9 EXPONENTIAL DISTRIBUTION

An **exponential** random variable has density

$$f(x \mid \lambda) = \frac{1}{\lambda} \exp\left(-\frac{x}{\lambda}\right), \qquad x \geq 0$$

$$\lambda > 0.$$

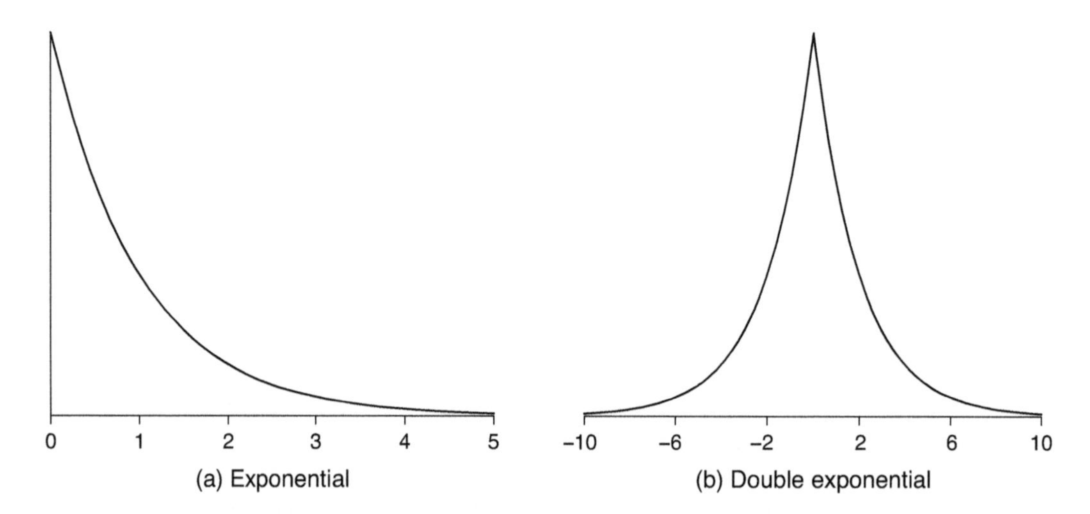

FIGURE 3.2 Exponential and double exponential densities

The exponential is frequently used by economists in theoretical models due to its simplicity. It is not commonly used in econometrics.

$$\mathbb{E}[X] = \lambda$$
$$\operatorname{var}[X] = \lambda^2.$$

If $U \sim U[0,1]$, then $X = -\log U \sim$exponential(1). The exponential density function is displayed in Figure 3.2(a) for the case $\lambda = 1$.

3.10 DOUBLE EXPONENTIAL DISTRIBUTION

A **double exponential** or **Laplace** random variable has density

$$f(x \mid \lambda) = \frac{1}{2\lambda} \exp\left(-\frac{|x|}{\lambda}\right), \qquad x \in \mathbb{R}$$
$$\lambda > 0.$$

The double exponential is used in robust analysis.

$$\mathbb{E}[X] = 0$$
$$\operatorname{var}[X] = 2\lambda^2.$$

The double exponential density function is displayed in Figure 3.2(b) for the case $\lambda = 2$.

3.11 GENERALIZED EXPONENTIAL DISTRIBUTION

A **generalized exponential** random variable has density

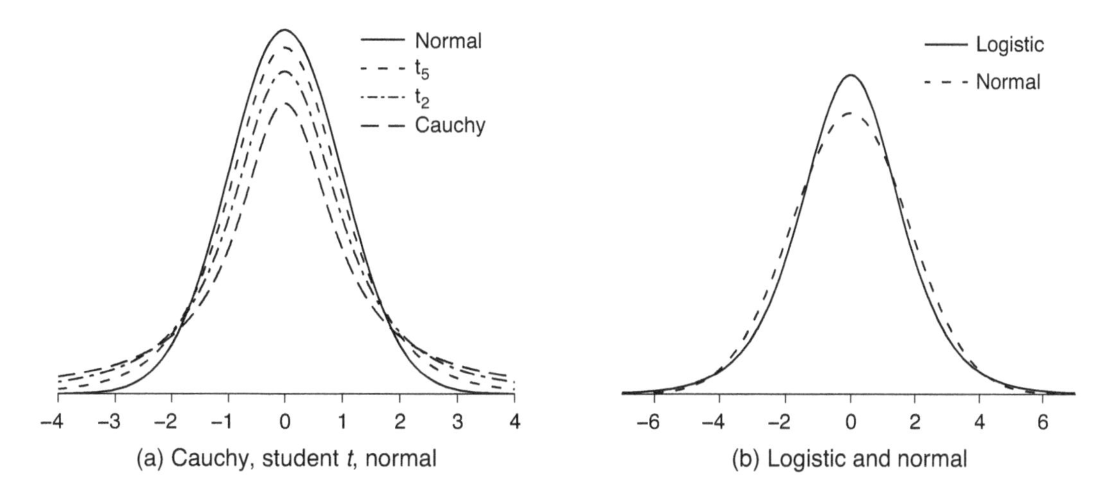

FIGURE 3.3 Normal, cauchy, student t, and logistic densities

$$f(x \mid \lambda, r) = \frac{1}{2\Gamma(1/r)\lambda} \exp\left(-\left|\frac{x}{\lambda}\right|^r\right), \qquad x \in \mathbb{R}$$

$$\lambda > 0$$

$$r > 0$$

where $\Gamma(\alpha)$ is the gamma function (see Definition A.20 in the Appendix). The generalized exponential nests the double exponential and normal distributions.

3.12 NORMAL DISTRIBUTION

A **normal** random variable, typically written $X \sim \mathrm{N}(\mu, \sigma^2)$, has density

$$f\left(x \mid \mu, \sigma^2\right) = \frac{1}{\sqrt{2\pi\sigma^2}} \exp\left(-\frac{(x-\mu)^2}{2\sigma^2}\right), \qquad x \in \mathbb{R}$$

$$\mu \in \mathbb{R}$$

$$\sigma^2 > 0.$$

The normal distribution is the most commonly used distribution in econometrics. When $\mu = 0$ and $\sigma^2 = 1$, it is called the **standard normal**. The standard normal density function is typically written as $\phi(x)$, and the distribution function as $\Phi(x)$. The normal density function can be written as $\phi_\sigma(x-\mu)$, where $\phi_\sigma(u) = \sigma^{-1}\phi(u/\sigma)$. The parameter μ is a location parameter, and σ^2 is a scale parameter.

$$\mathbb{E}[X] = \mu$$

$$\operatorname{var}[X] = \sigma^2.$$

The standard normal density function is displayed in both panels of Figure 3.3.

3.13 CAUCHY DISTRIBUTION

A **Cauchy** random variable has density and distribution function

$$f(x) = \frac{1}{\pi\left(1+x^2\right)}, \qquad x \in \mathbb{R}$$

$$F(x) = \frac{1}{2} + \frac{\arctan(x)}{\pi}.$$

The density is bell-shaped but with thicker tails than the normal distribution. An interesting feature is that it has no finite integer moments.

A Cauchy density function is shown in Figure 3.3(a).

3.14 STUDENT t DISTRIBUTION

A **student *t*** random variable, typically written $X \sim t_r$ or $t(r)$, has density

$$f(x \mid r) = \frac{\Gamma\left(\frac{r+1}{2}\right)}{\sqrt{r\pi}\,\Gamma\left(\frac{r}{2}\right)}\left(1+\frac{x^2}{r}\right)^{-\left(\frac{r+1}{2}\right)}, \qquad -\infty < x < \infty, \tag{3.1}$$

where $\Gamma(\alpha)$ is the gamma function (see Definition A.20 in the Appendix). The parameter r is called the "degrees of freedom". The student t is used for critical values in the normal sampling model.

$$\mathbb{E}[X] = 0, \qquad \text{if } r > 1$$

$$\text{var}[X] = \frac{r}{r-2}, \qquad \text{if } r > 2.$$

The student t distribution has the property that moments below r are finite. Moments greater than or equal to r are undefined.

The student t specializes to the Cauchy distribution when $r = 1$. As a limiting case as $r \to \infty$, it specializes to the normal (as shown in the next result). Thus the student t includes both the Cauchy and normal as limiting special cases.

Theorem 3.1 As $r \to \infty$, $f(x \mid r) \to \phi(x)$.

The proof is presented in Section 3.26.

A **scaled student *t* random variable** has density

$$f(x \mid r, v) = \frac{\Gamma\left(\frac{r+1}{2}\right)}{\sqrt{r\pi v}\,\Gamma\left(\frac{r}{2}\right)}\left(1+\frac{x^2}{rv}\right)^{-\left(\frac{r+1}{2}\right)}, \qquad -\infty < x < \infty$$

where v is a scale parameter. The variance of the scaled student t is $vr/(r-2)$.

Plots of the student t density function are displayed in Figure 3.3(a) for $r = 1$ (Cauchy), 2, 5, and ∞ (Normal). The density function of the student t is bell-shaped like the normal, but the t has thicker tails.

3.15 LOGISTIC DISTRIBUTION

A **logistic** random variable has density and distribution function

$$F(x) = \frac{1}{1+e^{-x}}, \qquad x \in \mathbb{R}$$
$$f(x) = F(x)(1 - F(x)).$$

The density is bell-shaped with a strong resemblance to the normal. It is used frequently in econometrics as a substitute for the normal, because the CDF is available in closed form.

$$\mathbb{E}[X] = 0$$
$$\text{var}[X] = \pi^2/3.$$

If U_1 and U_1 are independent exponential with mean 1, then $X = \log U_1 - \log U_2$ is logistic. If $U \sim U[0, 1]$, then $X = \log(U/(1-U)$ is logistic.

A logistic density function scaled to have unit variance is displayed in Figure 3.3(b). The standard normal density is plotted for contrast.

3.16 CHI-SQUARE DISTRIBUTION

A **chi-square** random variable, written $Q \sim \chi_r^2$ or $\chi^2(r)$, has density

$$f(x \mid r) = \frac{1}{2^{r/2}\Gamma(r/2)} x^{r/2-1} \exp(-x/2), \qquad x \geq 0 \tag{3.2}$$
$$r > 0$$

where $\Gamma(\alpha)$ is the gamma function (see Definition A.20).

$$\mathbb{E}[X] = r$$
$$\text{var}[X] = 2r.$$

The chi-square specializes to the exponential with $\lambda = 2$ when $r = 2$. The chi-square is commonly used for critical values for asymptotic tests.

It will be useful to derive the MGF of the chi-square distribution.

Theorem 3.2 The MGF of $Q \sim \chi_r^2$ is $M(t) = (1-2t)^{-r/2}$.

The proof is presented in Section 3.26.

An interesting calculation (see Exercise 3.8) reveals the inverse moment.

Theorem 3.3 If $Q \sim \chi_r^2$ with $r \geq 2$, then $\mathbb{E}\left[\frac{1}{Q}\right] = \frac{1}{r-2}$.

The chi-square density function is displayed in Figure 3.4(a) for the cases $r = 2$, 3, 4, and 6.

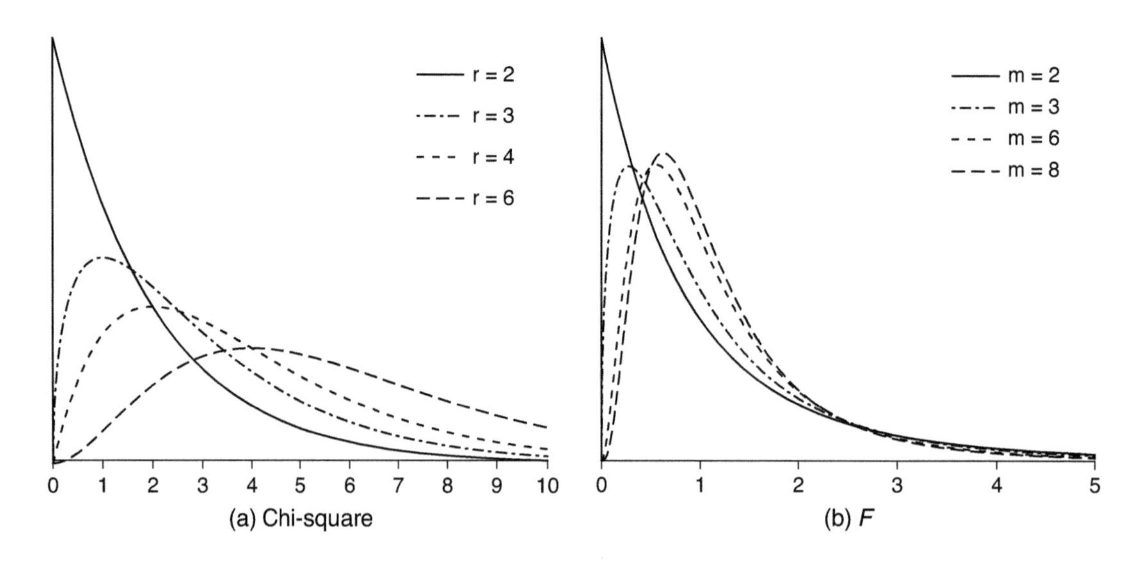

FIGURE 3.4 Chi-Square and F densities

3.17 GAMMA DISTRIBUTION

A **Gamma** random variable has density

$$f(x \mid \alpha, \beta) = \frac{\beta^{\alpha}}{\Gamma(\alpha)} x^{\alpha-1} \exp(-x\beta), \qquad x \geq 0$$

$$\alpha > 0$$

$$\beta > 0$$

where $\Gamma(\alpha)$ is the gamma function (Definition A.20). The gamma distribution includes the chi-square as a special case when $\beta = 1/2$ and $\alpha = r/2$. That is, $\chi_r^2 \sim \text{gamma}(r/2, 1/2)$, and if $Y \sim \text{gamma}(\alpha, \beta)$, then $Y \sim \chi_{2\alpha}^2/2\beta$. The gamma distribution when $\alpha = 1$ is exponential with $\lambda = 1/\beta$.

The gamma distribution is sometimes motivated as a flexible parametric family on the positive real line. For this distribution, α is a shape parameter, while β is a scale parameter. It is also used in Bayesian analysis.

$$\mathbb{E}[X] = \frac{\alpha}{\beta}$$

$$\text{var}[X] = \frac{\alpha}{\beta^2}.$$

3.18 *F* DISTRIBUTION

An F random variable, typically written $X \sim F_{m,r}$ or $F(m, r)$, has density

$$f(x \mid m, r) = \frac{\left(\frac{m}{r}\right)^{m/2} x^{m/2-1} \Gamma\left(\frac{m+r}{2}\right)}{\Gamma\left(\frac{m}{2}\right) \Gamma\left(\frac{r}{2}\right) \left(1 + \frac{m}{r}x\right)^{(m+r)/2}}, \qquad x > 0, \tag{3.3}$$

where $\Gamma(\alpha)$ is the gamma function (Definition A.20). The F is used for critical values in the normal sampling model.

$$\mathbb{E}[X] = \frac{r}{r-2} \qquad \text{if } r > 2.$$

As a limiting case, as $r \to \infty$, the F distribution simplifies to Q_m/m, a normalized χ^2_m. Thus the F distribution is a generalization of the χ^2_m distribution.

Theorem 3.4 Let $X \sim F_{m,r}$. As $r \to \infty$, the density of mX approaches that of χ^2_m.

The proof is presented in Section 3.26.

The F distribution was tabulated by a 1934 paper by George Snedecor. He introduced the notation "F," as the distribution is related to Sir Ronald Fisher's work on the analysis of variance.

Plots of the $F_{m,r}$ density for $m = 2, 3, 6, 8$, and $r = 10$ are displayed in Figure 3.4(b).

3.19 NON-CENTRAL CHI-SQUARE

A **non-central chi-square** random variable, typically written $X \sim \chi^2_r(\lambda)$ or $\chi^2(r, \lambda)$, has density

$$f(x) = \sum_{i=0}^{\infty} \frac{e^{-\lambda/2}}{i!} \left(\frac{\lambda}{2}\right)^i f_{r+2i}(x), \qquad x > 0 \tag{3.4}$$

where $f_r(x)$ is the χ^2_r density function (3.2). This is a weighted average of chi-square densities with Poisson weights. The non-central chi-square is used for theoretical analysis in the multivariate normal model and in asymptotic statistics. The parameter λ is called a **non-centrality parameter**.

The non-central chi-square includes the chi-square as a special case when $\lambda = 0$.

$$\mathbb{E}[X] = r + \lambda$$

$$\text{var}[X] = 2(r + 2\lambda).$$

3.20 BETA DISTRIBUTION

A **beta** random variable has density

$$f(x \mid \alpha, \beta) = \frac{1}{B(\alpha, \beta)} x^{\alpha-1}(1-x)^{\beta-1}, \qquad 0 \le x \le 1$$

$$\alpha > 0$$

$$\beta > 0$$

where

$$B(\alpha, \beta) = \int_0^1 t^{\alpha-1}(1-t)^{\beta-1}\, dt = \frac{\Gamma(\alpha)\Gamma(\beta)}{\Gamma(\alpha+\beta)}$$

is the beta function, and $\Gamma(\alpha)$ is the gamma function (Definition A.20). The beta distribution is used as a flexible parametric family on $[0, 1]$.

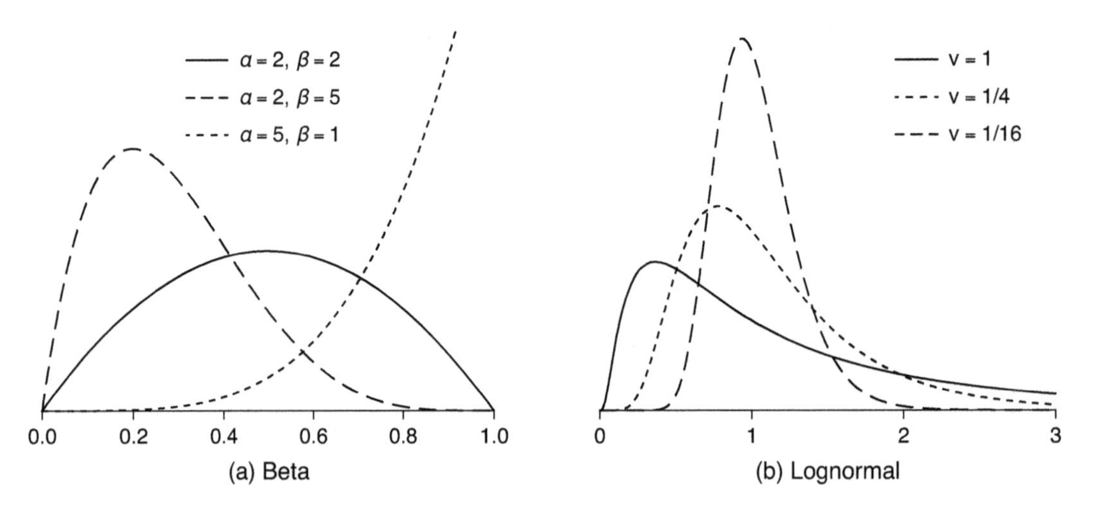

FIGURE 3.5 Beta and lognormal densities

$$\mathbb{E}[X] = \frac{\alpha}{\alpha+\beta}$$

$$\text{var}[X] = \frac{\alpha\beta}{(\alpha+\beta)^2(\alpha+\beta+1)}.$$

The beta density function is displayed in Figure 3.5(a) for the cases $(\alpha, \beta) = (2, 2), (2, 5)$, and $(5, 1)$.

3.21 PARETO DISTRIBUTION

A **Pareto** random variable has density

$$f(x \mid \alpha, \beta) = \frac{\alpha\beta^\alpha}{x^{\alpha+1}}, \qquad x \geq \beta$$

$$\alpha > 0$$

$$\beta > 0.$$

It is used to model thick-tailed distributions. The parameter α controls the rate at which the tail of the density declines to 0.

$$\mathbb{E}[X] = \frac{\alpha\beta}{\alpha-1}, \qquad \text{if } \alpha > 1$$

$$\text{var}[X] = \frac{\alpha\beta^2}{(\alpha-1)^2(\alpha-2)}, \qquad \text{if } \alpha > 2.$$

3.22 LOGNORMAL DISTRIBUTION

A **lognormal** random variable has density

$$f(x \mid \theta, \nu) = \frac{1}{\sqrt{2\pi\nu}} x^{-1} \exp\left(-\frac{(\log x - \theta)^2}{2\nu}\right), \qquad x > 0$$

$$\theta \in \mathbb{R}$$

$$\nu > 0.$$

The name comes from the fact that $\log(X) \sim \mathrm{N}(\theta, \nu)$. It is very common in applied econometrics to apply a normal model to variables after taking logarithms, which implicitly is applying a lognormal model to the levels. The lognormal distribution is highly skewed with a thick right tail.

$$\mathbb{E}[X] = \exp(\theta + \nu/2)$$

$$\mathrm{var}[X] = \exp(2\theta + 2\nu) - \exp(2\theta + \nu).$$

The lognormal density function with $\theta = 0$ and $\nu = 1$, $1/4$, and $1/16$ is displayed in Figure 3.5(b).

3.23 WEIBULL DISTRIBUTION

A **Weibull** random variable has density and distribution function

$$f(x \mid \alpha, \lambda) = \frac{\alpha}{\lambda}\left(\frac{x}{\lambda}\right)^{\alpha-1} \exp\left(-\left(\frac{x}{\lambda}\right)^{\alpha}\right), \qquad x \geq 0$$

$$F(x \mid \alpha, \lambda) = 1 - \exp\left(-\left(\frac{x}{\lambda}\right)^{\alpha}\right)$$

$$\alpha > 0$$

$$\lambda > 0.$$

The Weibull distribution is used in survival analysis. The parameter α controls the shape, and the parameter λ controls the scale.

$$\mathbb{E}[X] = \lambda\Gamma(1 + 1/\alpha)$$

$$\mathrm{var}[X] = \lambda^2\left(\Gamma(1 + 2/\alpha) - (\Gamma(1 + 1/\alpha))^2\right)$$

where $\Gamma(\alpha)$ is the gamma function (Definition A.20).

If $Y \sim$exponential(λ), then $X = Y^{1/\alpha} \sim \mathrm{Weibull}(\alpha, \lambda^{1/\alpha})$.

The Weibull density function with $\lambda = 1$ and $\alpha = 1/2$, 1, 2, and 4 is displayed in Figure 3.6(a).

3.24 EXTREME VALUE DISTRIBUTION

The **type I extreme value** distribution (also known as the **Gumbel**) takes two forms. The density and distribution functions for the most common case are

$$f(x) = \exp(-x)\exp\left(-\exp(-x)\right), \qquad x \in \mathbb{R}$$

$$F(x) = \exp\left(-\exp(-x)\right).$$

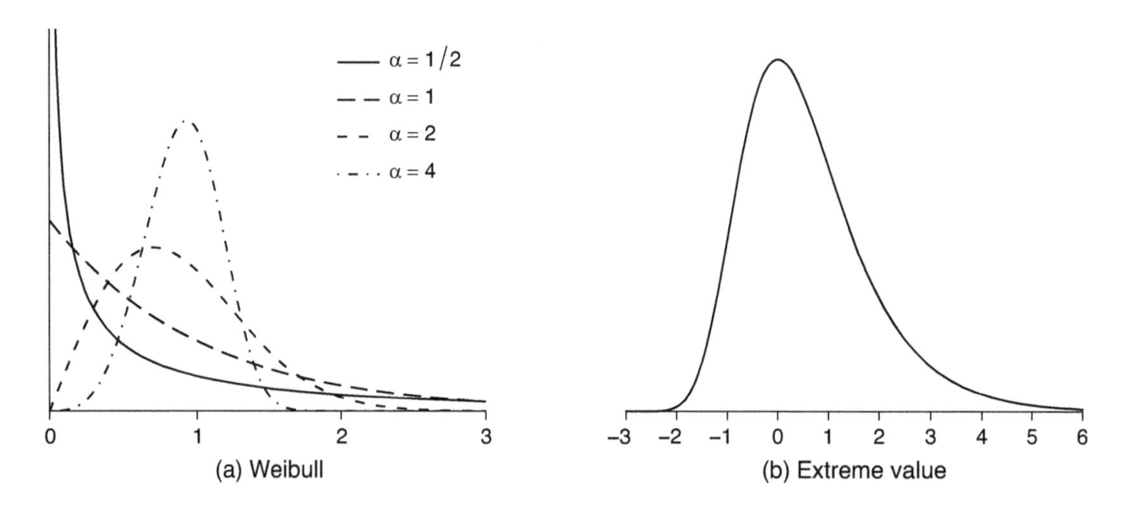

FIGURE 3.6 Weibull and Type I extreme value densities

The density and distriubtion functions for the alternative (minimum) case are

$$f(x) = \exp(x)\exp\left(-\exp(x)\right), \qquad x \in \mathbb{R}$$

$$F(x) = 1 - \exp\left(-\exp(x)\right).$$

The type I extreme value is used in discrete choice modeling.

If $Y \sim$ exponential(1), then $X = -\log Y \sim$ type I extreme value. If X_1 and X_2 are independent type I extreme value, then $Y = X_1 - X_2 \sim$ logistic.

The type I extreme value density function is displayed in Figure 3.6(b).

3.25 MIXTURES OF NORMALS

A **mixture of normals** density function is

$$f\left(x \mid p_1, \mu_1, \sigma_1^2, \ldots, p_M, \mu_M, \sigma_M^2\right) = \sum_{m=1}^{M} p_m \phi_{\sigma_m}(x - \mu_m)$$

$$\sum_{m=1}^{M} p_m = 1.$$

M is the number of mixture components. Mixtures of normals can be motivated by the idea of latent types. The latter means there are M latent types, each with a distinct mean and variance. Mixtures are frequently used in economics to model heterogeneity. Mixtures can also be used to flexibly approximate unknown density shapes. Mixtures of normals can also be convenient for certain theoretical calculations due to their simple structure.

To illustrate the flexibility which can be obtained by mixtures of normals, Figure 3.7 plots six examples of mixture of normals density functions.[1] All are normalized to have mean 0 and variance 1. The labels are descriptive, not formal names.

[1]These are constructed based on examples presented in Marron and Wand (1992), Figure 1 and Table 1.

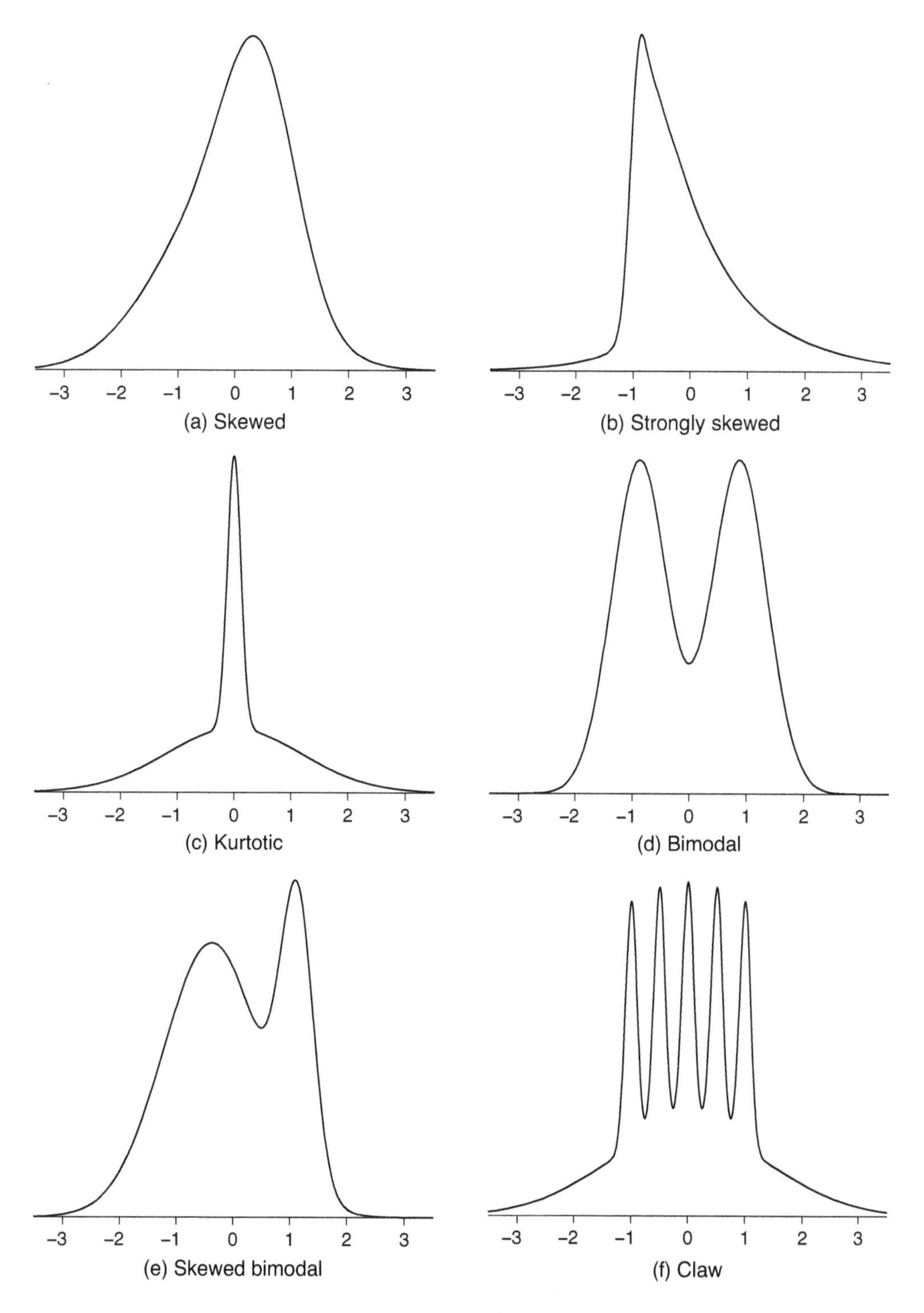

FIGURE 3.7 Mixture of normals densities

3.26 TECHNICAL PROOFS*

Proof of Theorem 3.1 Theorem A.28.6 states

$$\lim_{n\to\infty}\frac{\Gamma(n+x)}{\Gamma(n)\,n^{x}}=1.$$

Setting $n=r/2$ and $x=1/2$, we find

$$\lim_{r\to\infty}\frac{\Gamma\left(\frac{r+1}{2}\right)}{\sqrt{r\pi}\,\Gamma\left(\frac{r}{2}\right)}=\frac{1}{\sqrt{2\pi}}.$$

Using the definition of the exponential function (see Section A.4 in the Appendix), we have

$$\lim_{r\to\infty}\left(1+\frac{x^2}{r}\right)^{r}=\exp\left(x^2\right).$$

Taking the square root, we obtain

$$\lim_{r\to\infty}\left(1+\frac{x^2}{r}\right)^{r/2}=\exp\left(\frac{x^2}{2}\right). \tag{3.5}$$

Furthermore,

$$\lim_{r\to\infty}\left(1+\frac{x^2}{r}\right)^{\frac{1}{2}}=1.$$

Together

$$\lim_{r\to\infty}\frac{\Gamma\left(\frac{r+1}{2}\right)}{\sqrt{r\pi}\,\Gamma\left(\frac{r}{2}\right)}\left(1+\frac{x^2}{r}\right)^{-\left(\frac{r+1}{2}\right)}=\frac{1}{\sqrt{2\pi}}\exp\left(-\frac{x^2}{2}\right)=\phi(x).$$

■

Proof of Theorem 3.2 The MGF of the density (3.2) is

$$\begin{aligned}\int_0^\infty \exp(tq)f(q)dq&=\int_0^\infty \exp(tq)\frac{1}{\Gamma\left(\frac{r}{2}\right)2^{r/2}}q^{r/2-1}\exp(-q/2)\,dq\\&=\int_0^\infty \frac{1}{\Gamma\left(\frac{r}{2}\right)2^{r/2}}q^{r/2-1}\exp\left(-q\,(1/2-t)\right)dq\\&=\frac{1}{\Gamma\left(\frac{r}{2}\right)2^{r/2}}(1/2-t)^{-r/2}\,\Gamma\left(\frac{r}{2}\right)\\&=(1-2t)^{-r/2},\end{aligned} \tag{3.6}$$

the third equality using Theorem A.28.3.

■

Proof of Theorem 3.4 Applying change of variables to the density in (3.3), the density of mF is

$$\frac{x^{m/2-1}\Gamma\left(\frac{m+r}{2}\right)}{r^{m/2}\Gamma\left(\frac{m}{2}\right)\Gamma\left(\frac{r}{2}\right)\left(1+\frac{x}{r}\right)^{(m+r)/2}}. \tag{3.7}$$

Using Theorem A.28.6 with $n=r/2$ and $x=m/2$, we have

$$\lim_{r\to\infty}\frac{\Gamma\left(\frac{m+r}{2}\right)}{r^{m/2}\Gamma\left(\frac{r}{2}\right)}=2^{-m/2}$$

and similarly to (3.5), we have

$$\lim_{r\to\infty}\left(1+\frac{x}{r}\right)^{\left(\frac{m+r}{2}\right)}=\exp\left(\frac{x}{2}\right).$$

Using this result, (3.7) tends to

$$\frac{x^{m/2-1}\exp\left(-\frac{x}{2}\right)}{2^{m/2}\Gamma\left(\frac{m}{2}\right)}$$

which is the χ_m^2 density. ■

3.27 EXERCISES

Exercise 3.1 For the Bernoulli distribution, show

(a) $\sum_{x=0}^{1}\pi\left(x\mid p\right)=1.$

(b) $\mathbb{E}[X]=p.$

(c) $\text{var}[X]=p(1-p).$

Exercise 3.2 For the binomial distribution, show

(a) $\sum_{x=0}^{n}\pi\left(x\mid n,p\right)=1.$
Hint: Use the binomial theorem.

(b) $\mathbb{E}[X]=np.$

(c) $\text{var}[X]=np(1-p).$

Exercise 3.3 For the Poisson distribution, show

(a) $\sum_{x=0}^{\infty}\pi\left(x\mid\lambda\right)=1.$

(b) $\mathbb{E}[X]=\lambda.$

(c) $\text{var}[X]=\lambda.$

Exercise 3.4 For the $U[a,b]$ distribution, show

(a) $\int_a^b f\left(x\mid a,b\right)dx=1.$

(b) $\mathbb{E}[X]=(b-a)/2.$

(c) $\text{var}[X]=(b-a)/12.$

Exercise 3.5 For the exponential distribution, show

(a) $\int_0^{\infty} f\left(x\mid\lambda\right)dx=1.$

(b) $\mathbb{E}[X]=\lambda.$

(c) $\text{var}[X]=\lambda^2.$

Exercise 3.6 For the double exponential distribution, show

(a) $\int_{-\infty}^{\infty} f(x \mid \lambda)\, dx = 1$.
(b) $\mathbb{E}[X] = 0$.
(c) $\operatorname{var}[X] = 2\lambda^2$.

Exercise 3.7 For the chi-square distribution, show

(a) $\int_{0}^{\infty} f(x \mid r)\, dx = 1$.
(b) $\mathbb{E}[X] = r$.
(c) $\operatorname{var}[X] = 2r$.

Exercise 3.8 Show Theorem 3.3. Hint: Show that $x^{-1} f(x \mid r) = \frac{1}{r-2} f(x \mid r-2)$.

Exercise 3.9 For the gamma distribution, show

(a) $\int_{0}^{\infty} f(x \mid \alpha, \beta)\, dx = 1$.
(b) $\mathbb{E}[X] = \frac{\alpha}{\beta}$.
(c) $\operatorname{var}[X] = \frac{\alpha}{\beta^2}$.

Exercise 3.10 Suppose $X \sim \text{gamma}(\alpha, \beta)$. Set $Y = \lambda X$. Find the density of Y. Which distribution is this?

Exercise 3.11 For the Pareto distribution, show

(a) $\int_{\beta}^{\infty} f(x \mid \alpha, \beta)\, dx = 1$.
(b) $F(x \mid \alpha, \beta) = 1 - \frac{\beta^\alpha}{x^\alpha}, x \geq \beta$.
(c) $\mathbb{E}[X] = \frac{\alpha\beta}{\alpha - 1}$.
(d) $\operatorname{var}[X] = \frac{\alpha\beta^2}{(\alpha-1)^2(\alpha-2)}$.

Exercise 3.12 For the logistic distribution, show

(a) $F(x)$ is a valid distribution function.
(b) The density function is $f(x) = \exp(-x) / \left(1 + \exp(-x)\right)^2 = F(x)\,(1 - F(x))$.
(c) The density $f(x)$ is symmetric about 0.

Exercise 3.13 For the lognormal distribution, show

(a) The density is obtained by the transformation $X = \exp(Y)$ with $Y \sim \mathrm{N}(\theta, v)$.
(b) $\mathbb{E}[X] = \exp(\theta + v/2)$.

Exercise 3.14 For the mixture of normals distribution, show

(a) $\int_{-\infty}^{\infty} f(x)\, dx = 1$.

(b) $F(x) = \sum_{m=1}^{M} p_m \Phi\left(\frac{x - \mu_m}{\sigma_m}\right)$.

(c) $\mathbb{E}[X] = \sum_{m=1}^{M} p_m \mu_m$.

(d) $\mathbb{E}\left[X^2\right] = \sum_{m=1}^{M} p_m \left(\sigma_m^2 + \mu_m^2\right)$.

CHAPTER 4

MULTIVARIATE DISTRIBUTIONS

4.1 INTRODUCTION

Chapter 2 introduced the concept of random variables. We now generalize this concept to multiple random variables known as **random vectors**. To make the distinction clear, we will refer to one-dimensional random variables as **univariate**, two-dimensional random pairs as **bivariate**, and vectors of arbitrary dimension as **multivariate**.

We start the chapter with bivariate random variables. Later sections generalize to multivariate random vectors.

4.2 BIVARIATE RANDOM VARIABLES

A pair of bivariate random variables are two random variables with a joint distribution. They are typically represented by a pair of uppercase Latin characters, such as (X, Y) or (X_1, X_2). Specific values are designated by a pair of lowercase characters, such as (x, y) or (x_1, x_2).

Definition 4.1 A pair of **bivariate random variables** is a pair of numerical outcomes; that is, a function from the sample space to $\mathbb{R}^2$.

To illustrate, Figure 4.1 shows a mapping from the two coin flip sample space to $\mathbb{R}^2$, with TT mapped to (0,0), TH mapped to (0,1), HT mapped to (1,0), and HH mapped to (1,1).

For a real-world example, consider the bivariate pair (wage, work experience). We are interested in how wages vary with experience, and therefore in their joint distribution. The mapping is illustrated in Figure 4.2. The ellipse is the sample space with random outcomes a, b, c, d, e. (You can think of outcomes as individual wage earners at a point in time.) The graph is the positive quadrant in $\mathbb{R}^2$ representing the bivariate pairs (wage, work experience). The arrows depict the mapping. Each outcome is a point in the sample space. Each outcome is mapped to a point in $\mathbb{R}^2$. The latter are a pair of random variables (wage and experience). Their values are marked on the plot, with wage measured in dollars per hour and experience in years.

4.3 BIVARIATE DISTRIBUTION FUNCTIONS

The distribution function for bivariate random variables is defined as follows.

Definition 4.2 The **joint distribution function** of (X, Y) is $F(x, y) = \mathbb{P}[X \leq x, Y \leq y] = \mathbb{P}[\{X \leq x\} \cap \{Y \leq y\}]$.

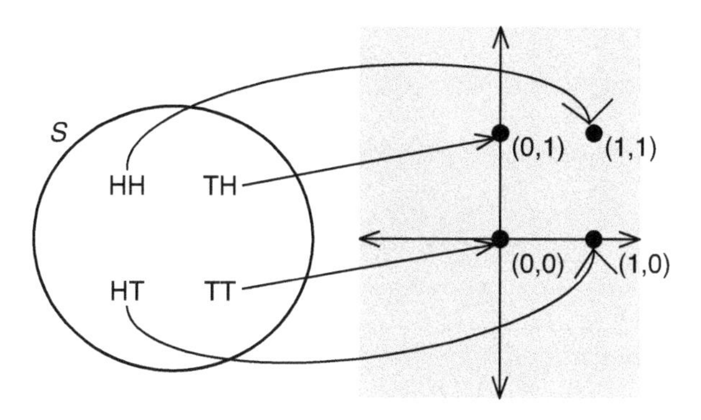

FIGURE 4.1 Two coin flip sample space

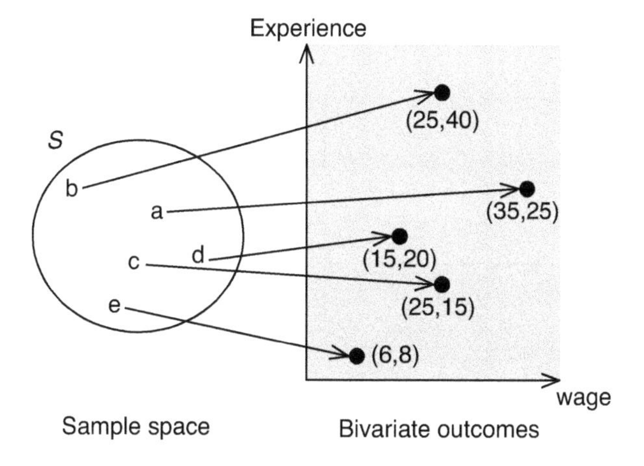

FIGURE 4.2 Bivariate random variables

The term "joint" specifically indicates that this is the distribution of multiple random variables. For simplicity, we omit the term "joint" when the meaning is clear from the context. When we want to be clear that the distribution refers to the pair (X, Y) we add subscripts, for example, $F_{X,Y}(x, y)$. When the variables are clear from the context the subscripts can be omitted.

An example of a joint distribution function is $F(x, y) = \left(1 - e^{-x}\right)\left(1 - e^{-y}\right)$ for $x, y \geq 0$.

The properties of the joint distribution function are similar to the univariate case. The distribution function is weakly increasing in each argument and satisfies $0 \leq F(x, y) \leq 1$.

To illustrate with a real-world example, Figure 4.3 displays the bivariate joint distribution[1] of hourly wages and work experience. Wages are plotted from \$0 to \$60, and experience from 0 to 50 years. The joint distribution function increases from 0 at the origin to near 1 in the upper-right corner. The function is increasing in each argument. To interpret the plot, fix the value of one variable and trace out the curve with respect to the other variable. For example, fix experience at 30 and then trace out the plot with respect to wages. You can see that the function steeply slopes up between \$14 and \$24 and then flattens out. Alternatively, fix hourly wages

[1] Among wage earners in the United States in 2009.

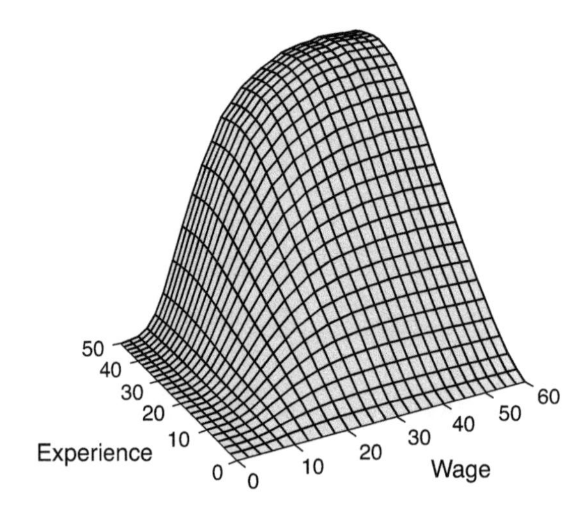

FIGURE 4.3 Bivariate distribution of experience and wages

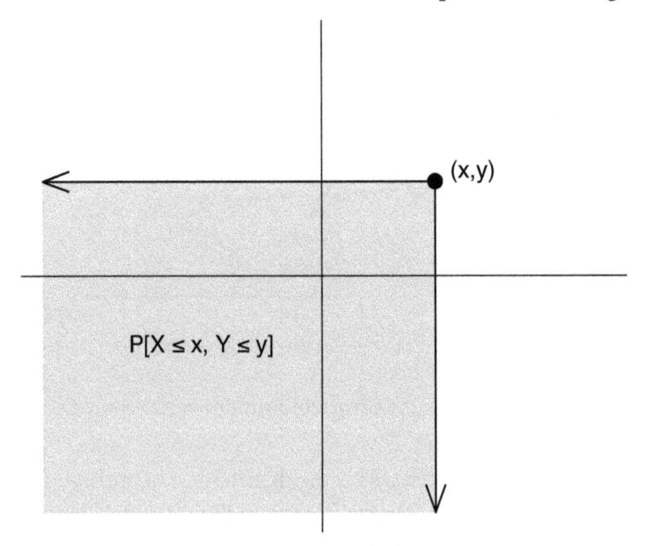

FIGURE 4.4 Bivariate joint distribution calculation

at \$30 and trace the function with respect to experience. In this case, the function has a steady slope up to about 40 years and then flattens.

Figure 4.4 illustrates how the joint distribution function is calculated for a given (x, y). The event $\{X \leq x, Y \leq y\}$ occurs if the pair (X, Y) lies in the shaded region (the region to the lower-left of the point (x, y)). The distribution function is the probability of this event. In our empirical example, if $(x, y) = (30, 30)$, then the calculation is the joint probability that wages are less than or equal to \$30 and experience is less than or equal to 30 years. It is difficult to read this number from the plot in Figure 4.3, but it equals 0.58. This means that 58% of wage earners satisfy these conditions.

The distribution function satisfies the following relationship

$$\mathbb{P}\left[a < X \leq b, c < Y \leq d\right] = F(b, d) - F(b, c) - F(a, d) + F(a, c).$$

See Exercise 4.5. This is illustrated in Figure 4.5.

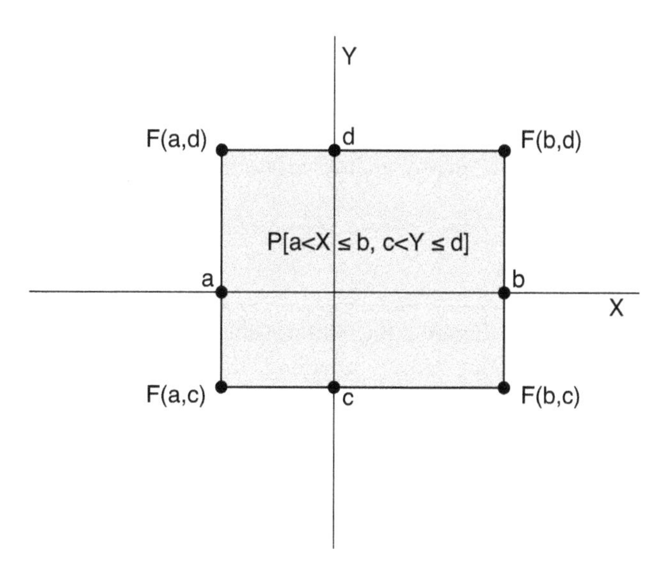

FIGURE 4.5 Probability and distribution functions

The shaded region is the set $\{a < x \leq b, c < y \leq d\}$. The probability that (X, Y) is in the set is the joint probability that X is in $(a, b]$ and Y is in $(c, d]$, and can be calculated from the distribution function evaluated at the four corners. For example,

$$\begin{aligned}\mathbb{P}\left[10 < \text{wage} \leq 20, 10 < \text{experience} \leq 20\right] &= F(20, 20) - F(20, 10) - F(10, 20) + F(10, 10)\\ &= 0.265 - 0.131 - 0.073 + 0.042\\ &= 0.103.\end{aligned}$$

Thus about 10% of wage earners satisfy these conditions.

4.4 PROBABILITY MASS FUNCTION

As for univariate random variables, it is useful to consider separately the case of discrete and continuous bivariate random variables.

A pair of random variables is **discrete** if there is a discrete set $\mathscr{S} \subset \mathbb{R}^2$ such that $\mathbb{P}[(X, Y) \in \mathscr{S}] = 1$. The set $\mathscr{S}$ is the support of (X, Y) and consists of a set of points in $\mathbb{R}^2$. In many cases, the support takes the product form, meaning that the support can be written as $\mathscr{S} = \mathscr{X} \times \mathscr{Y}$, where $\mathscr{X} \subset \mathbb{R}$ and $\mathscr{Y} \subset \mathbb{R}$ are the supports for X and Y. We can write these support points as $\{\tau_1^x, \tau_2^x, \ldots\}$ and $\{\tau_1^y, \tau_2^y, \ldots\}$. The **joint probability mass function** is $\pi(x, y) = \mathbb{P}\left[X = x, Y = y\right]$. At the support points, we set $\pi_{ij} = \pi(\tau_i^x, \tau_j^y)$. There is no loss in generality in assuming the support takes a product form if we allow $\pi_{ij} = 0$ for some pairs.

Example 1: A pizza restaurant caters to students. Each customer purchases either one or two slices of pizza and either one or two drinks during their meal. Let X be the number of pizza slices purchased, and Y be the number of drinks. The joint probability mass function is

$$\pi_{11} = \mathbb{P}[X = 1, Y = 1] = 0.4$$

$$\pi_{12} = \mathbb{P}[X = 1, Y = 2] = 0.1$$

$$\pi_{21} = \mathbb{P}[X=2, Y=1] = 0.2$$
$$\pi_{22} = \mathbb{P}[X=2, Y=2] = 0.3.$$

This is a valid probability function, since all probabilities are nonnegative, and the four probabilities sum to 1.

4.5 PROBABILITY DENSITY FUNCTION

The pair (X, Y) has a **continuous** distribution if the joint distribution function $F(x, y)$ is continuous in (x, y).

In the univariate case, the probability density function is the derivative of the distribution function. In the bivariate case, it is a double partial derivative.

Definition 4.3 When $F(x, y)$ is continuous and differentiable, its **joint density** $f(x, y)$ equals

$$f(x, y) = \frac{\partial^2}{\partial x \partial y} F(x, y).$$

When we want to be clear that the density refers to the pair (X, Y), we add subscripts (e.g., $f_{X,Y}(x, y)$). Joint densities have similar properties to the univariate case. They are nonnegative functions and integrate to 1 over $\mathbb{R}^2$.

Example 2: (X, Y) are continuously distributed on $\mathbb{R}^2_+$ with joint density

$$f(x, y) = \frac{1}{4}(x+y)\,xy \exp(-x-y).$$

The joint density is displayed in Figure 4.6. Let us now verify that this is a valid density by checking that it integrates to 1. The integral is

$$\begin{aligned}
\int_0^\infty \int_0^\infty f(x, y)dxdy &= \int_0^\infty \int_0^\infty \frac{1}{4}(x+y)\,xy \exp(-x-y)\,dxdy \\
&= \frac{1}{4}\left(\int_0^\infty \int_0^\infty x^2 y \exp(-x-y)\,dxdy + \int_0^\infty \int_0^\infty xy^2 \exp(-x-y)\,dxdy\right) \\
&= \frac{1}{4}\left(\int_0^\infty y \exp(-y)\,dy \int_0^\infty x^2 \exp(-x)\,dx + \int_0^\infty y^2 \exp(-y)\,dy \int_0^\infty x \exp(-x)\,dx\right) \\
&= 1.
\end{aligned}$$

Thus $f(x, y)$ is a valid density.

The probability interpretation of a bivariate density is that the probability that the random pair (X, Y) lies in a region in $\mathbb{R}^2$ equals the area under the density over this region. To see this, by the Second Fundamental Theorem of Calculus (Theorem A.20)

$$\int_c^d \int_a^b f(x, y)dxdy = \int_c^d \int_a^b \frac{\partial^2}{\partial x \partial y} F(x, y)dxdy = \mathbb{P}[a \le X \le b, c \le Y \le d].$$

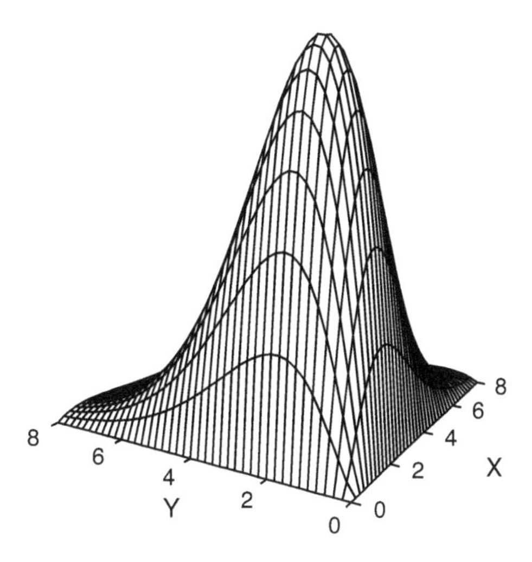

FIGURE 4.6 Joint density

This is similar to the property of univariate density functions, but in the bivariate case, a two-dimensional integration is required. Thus for any $A \subset \mathbb{R}^2$

$$\mathbb{P}\left[(X,Y) \in A\right] = \int_{-\infty}^{\infty} \int_{-\infty}^{\infty} \mathbb{1}\left\{(x,y) \in A\right\} f(x,y) dx dy.$$

In particular, this implies

$$\mathbb{P}\left[a \leq X \leq b, c \leq Y \leq d\right] = \int_c^d \int_a^b f(x,y) dx dy.$$

This is the joint probability that X and Y jointly lie in the intervals $[a, b]$ and $[c, d]$. Take a look at Figure 4.5, which shows this region in the (x, y) plane. The above expression shows that the probability that (X, Y) lies in this region is the integral of the joint density $f(x, y)$ over this region.

As an example, take the density $f(x,y) = 1$ for $0 \leq x, y \leq 1$. We can calculate the probability that $X \leq 1/2$ and $Y \leq 1/2$. It is

$$\begin{aligned}
\mathbb{P}\left[X \leq 1/2, Y \leq 1/2\right] &= \int_0^{1/2} \int_0^{1/2} f(x,y) dx dy \\
&= \int_0^{1/2} dx \int_0^{1/2} dy \\
&= \frac{1}{4}.
\end{aligned}$$

Example 3: Wages and experience. Figure 4.7 displays the bivariate joint probability density of hourly wages and work experience corresponding to the joint distribution in Figure 4.4. Wages are plotted from \$0 to \$60 and experience from 0 to 50 years. Reading bivariate density plots takes some practice. To start, pick some experience level (say, 10 years), and trace out the shape of the density function in wages. You see that the density is bell-shaped with its peak around \$15. The shape is similar to the univariate plot for wages in Figure 2.7.

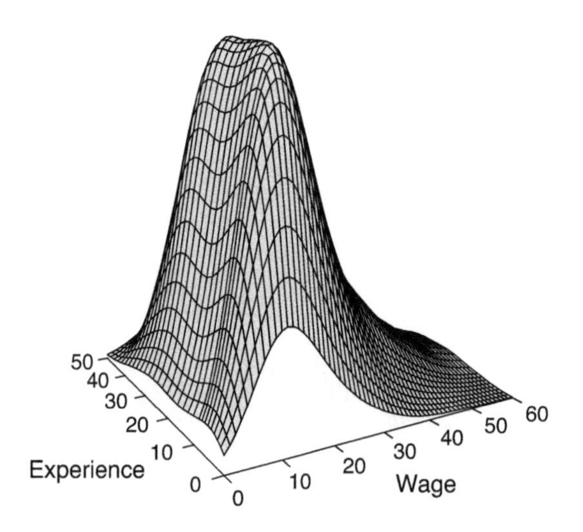

FIGURE 4.7 Bivariate density of experience and log wages

4.6 MARGINAL DISTRIBUTION

The joint distribution of the random vector (X, Y) fully describes the distribution of each component of the random vector.

Definition 4.4 The **marginal distribution** of X is

$$F_X(x) = \mathbb{P}[X \leq x] = \mathbb{P}[X \leq x, Y \leq \infty] = \lim_{y\to\infty} F(x, y).$$

In the continuous case, we can write this as

$$F_X(x) = \lim_{y\to\infty} \int_{-\infty}^{y} \int_{-\infty}^{x} f(u, v) du dv = \int_{-\infty}^{\infty} \int_{-\infty}^{x} f(u, v) du dv.$$

The marginal density of X is the derivative of the marginal distribution, and it equals

$$f_X(x) = \frac{d}{dx} F_X(x) = \frac{d}{dx} \int_{-\infty}^{\infty} \int_{-\infty}^{x} f(u, v) du dv = \int_{-\infty}^{\infty} f(x, y) dy.$$

Similarly, the marginal PDF of Y is

$$f_Y(y) = \frac{d}{dy} F_Y(y) = \int_{-\infty}^{\infty} f(x, y) dx.$$

These marginal PDFs are obtained by "integrating out" the other variable.

Marginal CDFs (PDFs) are simply CDFs (PDFs), but are referred to as "marginal" to distinguish them from the joint CDFs (PDFs) of the random vector. In practice, we treat a marginal PDF the same as a PDF.

Definition 4.5 The **marginal densities** of X and Y, given a joint density $f(x, y)$, are

$$f_X(x) = \int_{-\infty}^{\infty} f(x, y) dy$$

and

$$f_Y(y) = \int_{-\infty}^{\infty} f(x,y)dx.$$

Example 1: (continued). The marginal probabilities are

$$\mathbb{P}[X=1] = \mathbb{P}[X=1, Y=1] + \mathbb{P}[X=1, Y=2] = 0.4 + 0.1 = 0.5$$
$$\mathbb{P}[X=2] = \mathbb{P}[X=2, Y=1] + \mathbb{P}[X=2, Y=2] = 0.2 + 0.3 = 0.5$$

and

$$\mathbb{P}[Y=1] = \mathbb{P}[X=1, Y=1] + \mathbb{P}[X=2, Y=1] = 0.4 + 0.2 = 0.6$$
$$\mathbb{P}[Y=2] = \mathbb{P}[X=1, Y=2] + \mathbb{P}[X=2, Y=2] = 0.1 + 0.3 = 0.4.$$

Thus 50% of the customers order one slice of pizza and 50% order two slices; 60% also order one drink, while 40% order two drinks.

Example 2: (continued). The marginal density of X is

$$\begin{aligned} f_X(x) &= \int_0^{\infty} \frac{1}{4}(x+y)\,xy\exp(-x-y)\,dy \\ &= \left(x^2\int_0^{\infty} y\exp(-y)\,dy + x\int_0^{\infty} y^2\exp(-y)\,dy\right)\frac{1}{4}\exp(-x) \\ &= \frac{x^2+2x}{4}\exp(-x) \end{aligned}$$

for $x \geq 0$.

Example 3: (continued). The marginal density of wages is displayed in Figure 2.8 and can be found from Figure 4.7 by integrating over experience. The marginal density for experience can be similarly found by integrating over wages.

4.7 BIVARIATE EXPECTATION

Definition 4.6 The **expectation** of real-valued $g(X, Y)$ is

$$\mathbb{E}\left[g(X,Y)\right] = \sum_{(x,y)\in\mathbb{R}^2:\pi(x,y)>0} g(x,y)\pi(x,y)$$

for the discrete case, and

$$\mathbb{E}\left[g(X,Y)\right] = \int_{-\infty}^{\infty}\int_{-\infty}^{\infty} g(x,y)f(x,y)dxdy$$

for the continuous case.

If $g(X)$ only depends on one of the variables, the expectation can be written in terms of the marginal density:

$$\mathbb{E}\left[g(X)\right] = \int_{-\infty}^{\infty}\int_{-\infty}^{\infty} g(x)f(x,y)dxdy = \int_{-\infty}^{\infty} g(x)f_X(x)dx.$$

In particular, the expected value of a variable is

$$\mathbb{E}[X] = \int_{-\infty}^{\infty} \int_{-\infty}^{\infty} x f(x, y) dx dy = \int_{-\infty}^{\infty} x f_X(x) dx.$$

Example 1: (continued). We calculate that

$$\mathbb{E}[X] = 1 \times 0.5 + 2 \times 0.5 = 1.5$$
$$\mathbb{E}[Y] = 1 \times 0.6 + 2 \times 0.4 = 1.4.$$

This is the average number of pizza slices and drinks purchased per customer. The second moments are

$$\mathbb{E}\left[X^2\right] = 1^2 \times 0.5 + 2^2 \times 0.5 = 2.5$$
$$\mathbb{E}\left[Y^2\right] = 1^2 \times 0.6 + 2^2 \times 0.4 = 2.2.$$

The variances are

$$\text{var}[X] = \mathbb{E}\left[X^2\right] - (\mathbb{E}[X])^2 = 2.5 - 1.5^2 = 0.25$$
$$\text{var}[Y] = \mathbb{E}\left[Y^2\right] - (\mathbb{E}[Y])^2 = 2.2 - 1.4^2 = 0.24.$$

Example 2: (continued). The expected value of X is

$$\begin{aligned}
\mathbb{E}[X] &= \int_0^{\infty} x f_X(x) dx \\
&= \int_0^{\infty} x \left(\frac{x^2 + 2x}{4}\right) \exp(-x)\, dx \\
&= \int_0^{\infty} \frac{x^3}{4} \exp(-x) dx + \int_0^{\infty} \frac{x^2}{2} \exp(-x) dx \\
&= \frac{5}{2}.
\end{aligned}$$

The second moment is

$$\begin{aligned}
\mathbb{E}\left[X^2\right] &= \int_0^{\infty} x^2 f_X(x) dx \\
&= \int_0^{\infty} x^2 \left(\frac{x^2 + 2x}{4}\right) \exp(-x)\, dx \\
&= \int_0^{\infty} \frac{x^4}{4} \exp(-x) dx + \int_0^{\infty} \frac{x^3}{2} \exp(-x) dx \\
&= 9.
\end{aligned}$$

Its variance is

$$\text{var}[X] = \mathbb{E}\left[X^2\right] - (\mathbb{E}[X])^2 = 9 - \left(\frac{5}{2}\right)^2 = \frac{11}{4}.$$

Example 3: (continued). The means and variances of wages, experience, and experience are presented in the following chart.

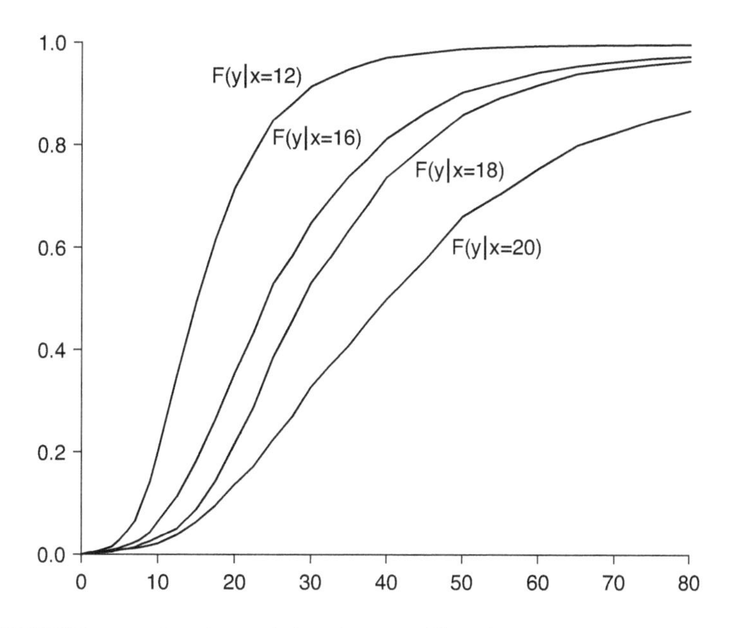

FIGURE 4.8 Conditional distribution of hourly wages given education

Means and Variances

	Mean	Variance
Hourly Wage	23.90	20.7^2
Experience (Years)	22.21	11.7^2
Education (Years)	13.98	2.58^2

4.8 CONDITIONAL DISTRIBUTION FOR DISCRETE X

Here and in Section 4.9, we define the conditional distribution and density of a random variable Y conditional on another random variable X taking a specific value x. In this section, we consider the case where X has a discrete distribution, and Section 4.9, we examine the case where X has a continuous distribution.

Definition 4.7 If X has a discrete distribution, the **conditional distribution function** of Y given $X = x$ is

$$F_{Y|X}(y \mid x) = \mathbb{P}\left[Y \leq y \mid X = x\right]$$

for any x such that $\mathbb{P}[X = x] > 0$.

This is a valid distribution as a function of y. That is, it is weakly increasing in y and asymptotes to 0 and 1. You can think of $F_{Y|X}(y \mid x)$ as the distribution function for the subpopulation where $X = x$. Take the case where Y is hourly wages and X is a worker's gender. Then $F_{Y|X}(y \mid x)$ specifies the distribution of wages separately for the subpopulations of men and women. If X denotes years of education (measured discretely), then $F_{Y|X}(y \mid x)$ specifies the distribution of wages separately for each education level.

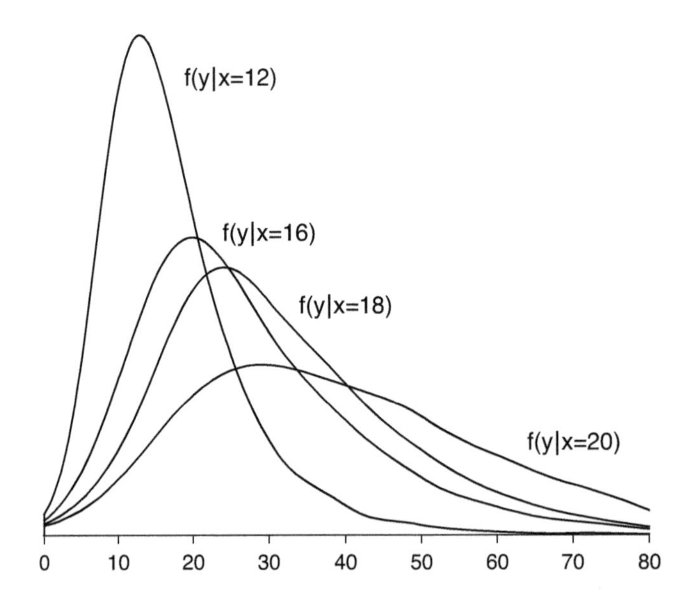

FIGURE 4.9 Conditional density of hourly wages given education

Example 3: (continued). Wages and education. Figure 2.2(b) displays the distribution of education for the population of U.S. wage earners using 10 categories. Each category is a subpopulation of the entirety of wage earners, and for each of these subpopulations, there is a distribution of wages. These conditional distribution functions $F(y \mid x)$ are displayed in Figure 4.8 for four groups defined by education (x) equaling 12, 16, 18, and 20, which correspond to high school degree, college degree, master's degree, and professional/PhD degree. The difference between the distribution functions is large and striking. The distributions shift uniformly to the right with each increase in education level. The largest shifts are between the distributions of those with high school and college degrees, and between those with master's and professional degrees.

If Y is continuously distributed, we can define the conditional density as the derivative of the conditional distribution function.

Definition 4.8 If $F_{Y|X}(y \mid x)$ is differentiable with respect to y, and $\mathbb{P}[X=x]>0$, then the **conditional density function** of Y given $X=x$ is

$$f_{Y|X}(y \mid x)=\frac{\partial}{\partial y}F_{Y|X}(y \mid x).$$

The conditional density $f_{Y|X}(y \mid x)$ is a valid density function, since it is the derivative of a distribution function. You can think of it as the density of Y for the subpopulation with a given value of X.

Example 3: (continued). The conditional density functions $f(y \mid x)$ corresponding to the conditional distributions displayed in Figure 4.8 are displayed in Figure 4.9. By examining the density function, it is easier to see where the probability mass is distributed. Compare the conditional densities for those with a high school $(x=12)$ or college $(x=16)$ degree. The latter density is shifted to the right and is more spread out. Thus college graduates have higher average wages, but they are also more dispersed. While the conditional density for college graduates is substantially shifted to the right, there is a considerable area of overlap between the density functions. Next compare the conditional densities of the college graduates and those with master's degrees ($x = 18$). The latter density is shifted to the right, but rather modestly. Thus these two densities are more similar than dissimilar. Now compare these conditional densities with the final density, that for the highest education level ($x = 20$). This conditional density function is substantially shifted to the right and substantially more dispersed.

4.9 CONDITIONAL DISTRIBUTION FOR CONTINUOUS X

The conditional density for continuous random variables is defined as follows.

Definition 4.9 For continuous X and Y, the **conditional density** of Y given $X = x$ is

$$f_{Y|X}(y \mid x) = \frac{f(x, y)}{f_X(x)}$$

for any x such that $f_X(x) > 0$.

If you are satisfied with this definition, you can skip the remainder of this section. However, if you would like a justification, read on.

Recall that the definition of the conditional distribution function for the case of discrete X is

$$F_{Y|X}(y \mid x) = \mathbb{P}\left[Y \leq y \mid X = x\right].$$

This does not apply for the case of continuous X, because $\mathbb{P}\left[X = x\right] = 0$. Instead we can define the conditional distribution function as a limit. Thus I propose the following definition.

Definition 4.10 For continuous X and Y, the **conditional distribution** of Y given $X = x$ is

$$F_{Y|X}(y \mid x) = \lim_{\epsilon \downarrow 0} \mathbb{P}\left[Y \leq y \mid x - \epsilon \leq X \leq x + \epsilon\right].$$

This expression is the probability that Y is smaller than y, conditional on X being in an arbitrarily small neighborhood of x. It is essentially the same concept as the definition for the discrete case. Fortunately, the expression can be simplified.

Theorem 4.1 If $F(x, y)$ is differentiable with respect to x and $f_X(x) > 0$, then $F_{Y|X}(y \mid x) = \dfrac{\frac{\partial}{\partial x} F(x, y)}{f_X(x)}$.

This result shows that the conditional distribution function is the ratio of a partial derivative of the joint distribution to the marginal density of X.

To prove this theorem, we can use the definition of conditional probability and L'Hôpital's rule (Theorem A.12), which states that the ratio of two limits which each tend to 0 equals the ratio of the two derivatives. We find that

$$\begin{aligned} F_{Y|X}\left(y \mid x\right) &= \lim_{\epsilon \downarrow 0} \mathbb{P}\left[Y \leq y \mid x - \epsilon \leq X \leq x + \epsilon\right] \\ &= \lim_{\epsilon \downarrow 0} \frac{\mathbb{P}\left[Y \leq y, x - \epsilon \leq X \leq x + \epsilon\right]}{\mathbb{P}\left[x - \epsilon \leq X \leq x + \epsilon\right]} \end{aligned}$$

$$
\begin{aligned}
&= \lim_{\epsilon \downarrow 0} \frac{F(x+\epsilon, y) - F(x-\epsilon, y)}{F_X(x+\epsilon) - F_X(x-\epsilon)} \\
&= \lim_{\epsilon \downarrow 0} \frac{\frac{\partial}{\partial x}F(x+\epsilon, y) + \frac{\partial}{\partial x}F(x-\epsilon, y)}{\frac{\partial}{\partial x}F_X(x+\epsilon) + \frac{\partial}{\partial x}F_X(x-\epsilon)} \\
&= \frac{2\frac{\partial}{\partial x}F(x,y)}{2\frac{\partial}{\partial x}F_X(x)} \\
&= \frac{\frac{\partial}{\partial x}F(x,y)}{f_X(x)}.
\end{aligned}
$$

This proves the result.

To find the conditional density, we take the partial derivative of the conditional distribution function:

$$
\begin{aligned}
f_{Y|X}(y \mid x) &= \frac{\partial}{\partial y} F_{Y|X}(y \mid x) \\
&= \frac{\partial}{\partial y} \frac{\frac{\partial}{\partial x}F(x,y)}{f_X(x)} \\
&= \frac{\frac{\partial^2}{\partial y \partial x}F(x,y)}{f_X(x)} \\
&= \frac{f(x,y)}{f_X(x)}.
\end{aligned}
$$

This result is identical to Definition 4.9, given at the beginning of this section.

What we have shown is that this definition (the ratio of the joint density to the marginal density) is the natural generalization of the case of discrete X.

4.10 VISUALIZING CONDITIONAL DENSITIES

To visualize the conditional density $f_{Y|X}(y \mid x)$, start with the joint density $f(x,y)$, which is a two-dimensional surface in three-dimensional space. Fix x. Slicing through the joint density along y creates an unnormalized density in one dimension. It is unnormalized, because it does not integrate to 1. To normalize, divide by $f_X(x)$. To verify that the conditional density integrates to 1 and is thus a valid density, observe that

$$
\int_{-\infty}^{\infty} f_{Y|X}(y \mid x) dy = \int_{-\infty}^{\infty} \frac{f(x,y)}{f_X(x)} dy = \frac{f_X(x)}{f_X(x)} = 1.
$$

Example 2: (continued). The joint density is displayed in Figure 4.6. To visualize the conditional density, select a value of x. Trace the shape of the joint density as a function of y. After renormalization, this is the conditional density function. By varying x, we can obtain different conditional density functions.

To explicitly calculate the conditional density, divide the joint density by the marginal density. We find

$$
f_{Y|X}(y \mid x) = \frac{f(x,y)}{f_X(x)}
$$

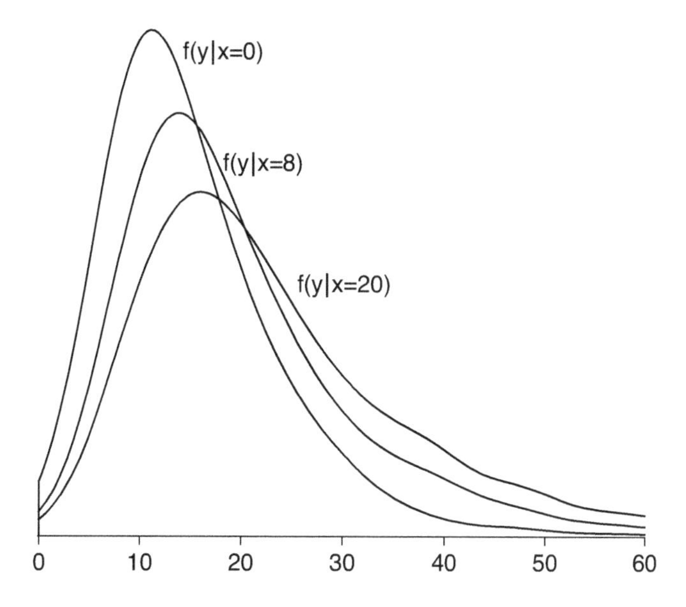

FIGURE 4.10 Conditional density of wages given experience

$$= \frac{\frac{1}{4}(x+y)\,xy\exp(-x-y)}{\frac{1}{4}(x^2+2x)\exp(-x)}$$

$$= \frac{(xy+y^2)\exp(-y)}{x+2}.$$

Example 3: (continued). The conditional density of wages given experience is calculated by taking the joint density in Figure 4.7 and dividing by the marginal density of experience. We do so at three levels of experience (0 years, 8 years, and 30 years). The resulting densities are displayed in Figure 4.10. The shapes of the three conditional densities are similar, but they shift to the right and spread out as experience increases. Thus the distribution of wages shifts upward and widens as work experience increases.

If we compare the conditional densities in Figure 4.10 with those in Figure 4.9, we can see that the effect of education on wages is considerably stronger than the effect of experience.

4.11 INDEPENDENCE

In this section, we define independence between random variables.

Recall that two events A and B are independent if the probability that they both occur equals the product of their probabilities: $\mathbb{P}[A \cap B] = \mathbb{P}[A]\,\mathbb{P}[B]$. Consider the events $A = \{X \leq x\}$ and $B = \{Y \leq y\}$. The probability that they both occur is

$$\mathbb{P}[A \cap B] = \mathbb{P}\left[X \leq x, Y \leq y\right] = F(x, y).$$

The product of their probabilities is

$$\mathbb{P}[A]\,\mathbb{P}[B] = \mathbb{P}[X \leq x]\,\mathbb{P}\left[Y \leq y\right] = F_X(x)F_Y(y).$$

These two expressions are equal if $F(x,y)=F_X(x)F_Y(y)$. Thus, if the events A and B are independent, then the joint distribution function factors as $F(x,y)=F_X(x)F_Y(y)$ for all (x,y). This property can be used as a definition of independence between random variables.

Definition 4.11 The random variables X and Y are **statistically independent** if for all x and y,

$$F(x,y)=F_X(x)F_Y(y).$$

This property is often written as $X \perp\!\!\!\perp Y$.

An implication of statistical independence is that all events of the form $A=\{X\in C_1\}$ and $B=\{Y\in C_2\}$ are independent.

If X and Y fail to satisfy the property of independence, we say that they are **statistically dependent**.

It is more convenient to work with mass functions and densities rather than distributions. In the case of continuous random variables, by differentiating the above expression with respect to x and y, we find $f(x,y)=f_X(x)f_Y(y)$. Thus the definition is equivalent to stating that the joint density function factors into the product of the marginal densities.

In the case of discrete random variables, a similar argument leads to $\pi(x,y)=\pi_X(x)\pi_Y(y)$, which means that the joint probability mass function factors into the product of the marginal mass functions.

Theorem 4.2 The discrete random variables X and Y are statistically independent if for all x and y,

$$\pi(x,y)=\pi_X(x)\pi_Y(y).$$

If X and Y have a differentiable distribution function, they are statistically independent if for all x and y

$$f(x,y)=f_X(x)f_Y(y).$$

An interesting connection arises between the conditional density function and independence.

Theorem 4.3 If X and Y are independent and continuously distributed, then

$$f_{Y|X}(y\mid x)=f_Y(y)$$
$$f_{X|Y}(x\mid y)=f_X(x).$$

Thus the conditional density equals the marginal (unconditional) density. As a result, $X=x$ does not affect the shape of the density of Y. This seems reasonable, given that the two random variables are independent. To see this, note that

$$f_{Y|X}(y\mid x)=\frac{f(x,y)}{f_X(x)}=\frac{f_X(x)f_Y(y)}{f_X(x)}=f_Y(y).$$

Let us now consider a sequence of related results.

First, by rewriting the definition of the conditional density, we can obtain a density version of Bayes Theorem.

Theorem 4.4 Bayes Theorem for Densities.

$$f_{Y|X}(y\mid x)=\frac{f_{X|Y}(x\mid y)f_Y(y)}{f_X(x)}=\frac{f_{X|Y}(x\mid y)f_Y(y)}{\int_{-\infty}^{\infty}f_{X|Y}(x\mid y)f_Y(y)dy}.$$

The next result shows that the expectation of the product of independent random variables is the product of the expectations.

Theorem 4.5 If X and Y are independent, then for any functions $g:\mathbb{R}\to\mathbb{R}$ and $h:\mathbb{R}\to\mathbb{R}$ such that $\mathbb{E}\left|g(X)\right|<\infty$ and $\mathbb{E}\left|h(Y)\right|<\infty$

$$\mathbb{E}\left[g(X)h(Y)\right]=\mathbb{E}\left[g(X)\right]\mathbb{E}\left[h(Y)\right].$$

Proof: For continuous random variables, we have

$$\begin{aligned}\mathbb{E}\left[g(X)h(Y)\right]&=\int_{-\infty}^{\infty}\int_{-\infty}^{\infty}g(x)h(y)f(x,y)dxdy\\&=\int_{-\infty}^{\infty}\int_{-\infty}^{\infty}g(x)h(y)f_X(x)f_Y(y)dxdy\\&=\int_{-\infty}^{\infty}g(x)f_X(x)dx\int_{-\infty}^{\infty}h(y)f_Y(y)dy\\&=\mathbb{E}\left[g(X)\right]\mathbb{E}\left[h(Y)\right].\end{aligned}$$ ∎

Take $g(x)=\mathbb{1}\{x\leq a\}$ and $h(y)=\mathbb{1}\left\{y\leq b\right\}$ for arbitrary constants a and b. Theorem 4.5 shows that independence implies $\mathbb{P}\left[X\leq a,Y\leq b\right]=\mathbb{P}\left[X\leq a\right]\mathbb{P}\left[Y\leq b\right]$, or

$$F(x,y)=F_X(x)F_X(y)$$

for all $(x,y)\in\mathbb{R}^2$. Recall that the latter is the definition of independence. Consequently the definition is "if and only if". That is, X and Y are independent if and only if this equality holds.

A useful application of Theorem 4.5 is to the MGF.

Theorem 4.6 If X and Y are independent with MGFs $M_X(t)$ and $M_Y(t)$, respectively, then the MGF of $Z=X+Y$ is $M_Z(t)=M_X(t)M_Y(t)$.

Proof: By the properties of the exponential function, $\exp\left(t(X+Y)\right)=\exp(tX)\exp(tY)$. By Theorem 4.5, since X and Y are independent,

$$\begin{aligned}M_Z(t)&=\mathbb{E}\left[\exp\left(t(X+Y)\right)\right]\\&=\mathbb{E}\left[\exp(tX))\exp(tY)\right]\\&=\mathbb{E}\left[\exp(tX)\right]\mathbb{E}\left[\exp(tY)\right]\\&=M_X(t)M_Y(t).\end{aligned}$$ ∎

Furthermore, transformations of independent variables are also independent.

Theorem 4.7 If X and Y are independent, then for any functions $g:\mathbb{R}\to\mathbb{R}$ and $h:\mathbb{R}\to\mathbb{R}$, $U=g(X)$ and $V=h(Y)$ are independent.

Proof: For any $u\in\mathbb{R}$ and $v\in\mathbb{R}$, define the sets $A(u)=\left\{x:g(x)\leq u\right\}$ and $B(v)=\left\{y:h(y)\leq v\right\}$. The joint distribution of (U,V) is

$$\begin{aligned}
F_{U,V}(u,v) &= \mathbb{P}[U \le u, V \le v] \\
&= \mathbb{P}\left[g(X) \le u, h(Y) \le v\right] \\
&= \mathbb{P}[X \in A(u), Y \in B(v)] \\
&= \mathbb{P}[X \in A(u)]\,\mathbb{P}[Y \in B(v)] \\
&= \mathbb{P}\left[g(X) \le u\right]\mathbb{P}[h(Y) \le v] \\
&= \mathbb{P}[U \le u]\,\mathbb{P}[V \le v] \\
&= F_U(u)F_V(v).
\end{aligned}$$

Thus the distribution function factors, satisfying the definition of independence. The key is the fourth equality, which uses the fact that the events $\{X \in A(u)\}$ and $\{Y \in B(v)\}$ are independent, which can be shown by an extension of Theorem 4.5. ■

Example 1: (continued). Are the number of pizza slices and drinks independent or dependent? To answer this question, we can check whether the joint probabilities equal the product of the individual probabilities. Recall that

$$\begin{aligned}
\mathbb{P}[X=1] &= 0.5 \\
\mathbb{P}[Y=1] &= 0.6 \\
\mathbb{P}[X=1, Y=1] &= 0.4.
\end{aligned}$$

Since

$$0.5 \times 0.6 = 0.3 \neq 0.4$$

we conclude that X and Y are dependent.

Example 2: (continued). $f_{Y|X}(y\,|\,x) = \frac{(xy+y^2)\exp(-y)}{x+2}$ and $f_Y(y) = \frac{1}{4}\left(y^2+2y\right)\exp\left(-y\right)$. Since these are not equal, we deduce that X and Y are not independent. Another way of seeing this is to notice that $f(x,y) = \frac{1}{4}\left(x+y\right)xy\exp\left(-x-y\right)$ cannot be factored.

Example 3: (continued). Figure 4.10 displays the conditional density of wages given experience. Since the conditional density changes with experience, wages and experience are not independent.

4.12 COVARIANCE AND CORRELATION

A feature of the joint distribution of (X, Y) is the covariance.

Definition 4.12 If X and Y have finite variances, the **covariance** between X and Y is

$$\begin{aligned}
\operatorname{cov}(X,Y) &= \mathbb{E}[(X-\mathbb{E}[X])(Y-\mathbb{E}[Y])] \\
&= \mathbb{E}[XY] - \mathbb{E}[X]\,\mathbb{E}[Y].
\end{aligned}$$

Definition 4.13 If X and Y have finite variances, the **correlation** between X and Y is

$$\operatorname{corr}(X, Y) = \frac{\operatorname{cov}(X, Y)}{\sqrt{\operatorname{var}[X] \operatorname{var}[Y]}}.$$

If $\operatorname{cov}(X, Y) = 0$, then $\operatorname{corr}(X, Y) = 0$, and it is typical to say that X and Y are **uncorrelated**.

Theorem 4.8 If X and Y are independent with finite variances, then X and Y are uncorrelated.

The reverse is not true. For example, suppose that $X \sim U[-1, 1]$. Since it is symmetrically distributed about 0, we see that $\mathbb{E}[X] = 0$ and $\mathbb{E}\left[X^3\right] = 0$. Set $Y = X^2$. Then

$$\operatorname{cov}(X, Y) = \mathbb{E}\left[X^3\right] - \mathbb{E}[X]\,\mathbb{E}\left[X^2\right] = 0.$$

Thus X and Y are uncorrelated yet are fully dependent! This shows that uncorrelated random variables may be dependent.

The following results are quite useful.

Theorem 4.9 If X and Y have finite variances, $\operatorname{var}[X + Y] = \operatorname{var}[X] + \operatorname{var}[Y] + 2\operatorname{cov}(X, Y)$.

To see this, it is sufficient to assume that X and Y have mean 0. Completing the square and using the linear property of expectations gives the desired result:

$$\begin{aligned}
\operatorname{var}[X + Y] &= \mathbb{E}\left[(X + Y)^2\right] \\
&= \mathbb{E}\left[X^2 + Y^2 + 2XY\right] \\
&= \mathbb{E}\left[X^2\right] + \mathbb{E}\left[Y^2\right] + 2\mathbb{E}[XY] \\
&= \operatorname{var}[X] + \operatorname{var}[Y] + 2\operatorname{cov}(X, Y).
\end{aligned}$$

Theorem 4.10 If X and Y are uncorrelated, then $\operatorname{var}[X + Y] = \operatorname{var}[X] + \operatorname{var}[Y]$.

This follows from Theorem 4.10, since uncorrelatedness means that $\operatorname{cov}(X, Y) = 0$.

Example 1: (continued). The cross moment is

$$\mathbb{E}[XY] = 1 \times 1 \times 0.4 + 1 \times 2 \times 0.1 + 2 \times 1 \times 0.2 + 2 \times 2 \times 0.3 = 2.2.$$

The covariance is

$$\operatorname{cov}(X, Y) = \mathbb{E}[XY] - \mathbb{E}[X]\,\mathbb{E}[Y] = 2.2 - 1.5 \times 1.4 = 0.1$$

and the correlation is

$$\operatorname{corr}(X, Y) = \frac{\operatorname{cov}(X, Y)}{\sqrt{\operatorname{var}[X] \operatorname{var}[Y]}} = \frac{0.1}{\sqrt{0.25 \times 0.24}} = 0.41.$$

This is a high correlation. As might be expected, the number of pizza slices and drinks purchased are positively and meaningfully correlated.

Example 2: (continued). The cross moment is

$$\mathbb{E}[XY] = \int_0^\infty \int_0^\infty xy\frac{1}{4}\left(x + y\right) xy \exp\left(-x - y\right) dxdy = 6.$$

The covariance is

$$\text{cov}(X, Y) = 6 - \left(\frac{5}{2}\right)^2 = -\frac{1}{4}$$

and correlation

$$\text{corr}(X, Y) = \frac{-\left(\frac{1}{4}\right)}{\sqrt{\frac{11}{4} \times \frac{11}{4}}} = -\frac{1}{11}.$$

This is a negative correlation, meaning that the two variables co-vary in opposite directions. The magnitude of the correlation, however, is small, indicating that the co-movement is mild.

Example 3 & 4: (continued). Correlations of wages, experience, and education are presented in the following chart, known as a correlation matrix.

Correlation Matrix

	Wage	Experience	Education
Wage	1	0.06	0.40
Experience	0.06	1	−0.17
Education	0.40	−0.17	1

The chart shows that education and wages are highly correlated, wages and experiences mildly correlated, and education and experience negatively correlated. The last feature is likely because the variation in experience at a point in time is mostly due to differences across cohorts (people of different ages). This negative correlation is because education levels are different across cohorts—later generations have higher average education levels.

4.13 CAUCHY-SCHWARZ INEQUALITY

This following inequality is used frequently.

Theorem 4.11 Cauchy-Schwarz Inequality. For any random variables X and Y,

$$\mathbb{E}|XY| \leq \sqrt{\mathbb{E}\left[X^2\right]\mathbb{E}\left[Y^2\right]}.$$

Proof: By the geometric mean inequality (Theorem 2.13),

$$|ab| = \sqrt{a^2}\sqrt{b^2} \leq \frac{a^2+b^2}{2}. \tag{4.1}$$

Set $U = |X| / \sqrt{\mathbb{E}\left[X^2\right]}$ and $V = |Y| / \sqrt{\mathbb{E}\left[Y^2\right]}$. Using equation (4.1), we obtain

$$|UV| \leq \frac{U^2+V^2}{2}.$$

Taking expectations, we find

$$\frac{\mathbb{E}|XY|}{\sqrt{\mathbb{E}\left[X^2\right]\mathbb{E}\left[Y^2\right]}} = \mathbb{E}|UV| \leq \mathbb{E}\left[\frac{U^2+V^2}{2}\right] = 1$$

the final equality holds since $\mathbb{E}\left[U^2\right] = 1$ and $\mathbb{E}\left[V^2\right] = 1$. This is the theorem. ∎

The Cauchy-Schwarz inequality implies bounds on covariances and correlations.

Theorem 4.12 Covariance Inequality. For any random variables X and Y with finite variances

$$|\text{cov}(X,Y)| \leq (\text{var}[X]\,\text{var}[Y]])^{1/2}$$
$$|\text{corr}(X,Y)| \leq 1.$$

Proof. Apply the expectation (Theorem 2.10) and Cauchy-Schwarz (Theorem 4.11) inequalities to find

$$|\text{cov}(X,Y)| \leq \mathbb{E}\,|(X-\mathbb{E}[X])(Y-\mathbb{E}[Y])| \leq (\text{var}[X]\,\text{var}[Y])^{1/2}$$

as stated. ■

The fact that the correlation is bounded below 1 helps us understand how to interpret a correlation. Correlations close to 0 are small. Correlations close to 1 and -1 are large.

4.14 CONDITIONAL EXPECTATION

An important concept in econometrics is conditional expectation. Just as the expectation is the central tendency of a distribution, the conditional expectation is the central tendency of a conditional distribution.

Definition 4.14 The **conditional expectation** of Y given $X=x$ is the expected value of the conditional distribution $F_{Y|X}(y \mid x)$ and is written as $m(x)=\mathbb{E}[Y \mid X=x]$. For discrete random variables, the conditional expectation is

$$\mathbb{E}[Y \mid X=x] = \frac{\sum_{j=1}^{\infty} \tau_j \pi(x,\tau_j)}{\pi_X(x)}.$$

For continuous Y, it is

$$\mathbb{E}[Y \mid X=x] = \int_{-\infty}^{\infty} y f_{Y|X}(y \mid x) dy.$$

We also call $\mathbb{E}[Y \mid X=x]$ the **conditional mean.**

In the continuous case, using the definition of the conditional PDF, we can write $\mathbb{E}[Y \mid X=x]$ as

$$\mathbb{E}[Y \mid X=x] = \frac{\int_{-\infty}^{\infty} y f(y,x) dy}{\int_{-\infty}^{\infty} f(x,y) dy}.$$

The conditional expectation tells us the average value of Y given that X equals the specific value x. When X is discrete, the conditional expectation is the expected value of Y in the subpopulation for which $X=x$. For example, if X is gender, then $\mathbb{E}[Y \mid X=x]$ is the expected value for men and women, separately. If X is education, then $\mathbb{E}[Y \mid X=x]$ is the expected value for each education level. When X is continuous, the conditional expectation is the expected value of Y in the infinitesimally small population for which $X \simeq x$.

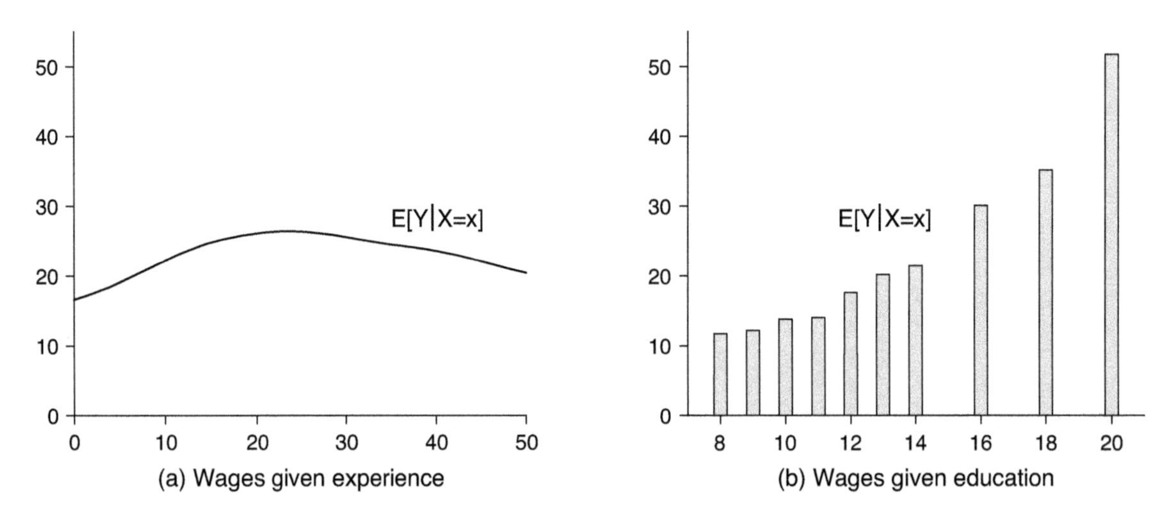

FIGURE 4.11 Conditional expectation functions

Example 1: (continued): The conditional expectation for the number of drinks per customer is

$$\mathbb{E}[Y \mid X=1] = \frac{1 \times 0.4 + 2 \times 0.1}{0.5} = 1.2$$

$$\mathbb{E}[Y \mid X=2] = \frac{1 \times 0.2 + 2 \times 0.3}{0.5} = 1.6.$$

Thus the restaurant can expect to serve (on average) more drinks to customers who purchase two slices of pizza.

Example 2: (continued): The conditional expectation of Y given $X = x$ is

$$\begin{aligned}\mathbb{E}[Y \mid X=x] &= \int_0^\infty y f_{Y|X}(y \mid x) dy \\ &= \int_0^\infty y \frac{(xy + y^2)\exp(-y)}{x+2} dy \\ &= \frac{2x+6}{x+2}.\end{aligned}$$

This conditional expectation function is downward sloping for $x \geq 0$. Thus as x increases, the expected value of Y declines.

Example 3: (continued). The conditional expectation of wages given experience is displayed in Figure 4.11(a). The x-axis is years of experience. The y-axis is wages. You can see that the expected wage is about \$16.50 per hour for 0 years of experience and increases near linearily to about \$26 by 20 years of experience. Above 20 years of experience, the expected wage falls, reaching about \$21 by 50 years of experience. Overall, the shape of the wage-experience profile is an inverted U-shape, increasing for the early years of experience, and decreasing in the more advanced years.

Example 4: (continued). The conditional expectation of wages given education is displayed in Figure 4.11(b). The x-axis is years of education. Since education is discrete, the conditional mean is a discrete function as well. Examining Figure 4.11(b), we see that the conditional expectation is monotonically increasing in years of education. The mean is \$11.75 for an individual with 8 years of education, \$17.61 for an individual with 12 years of education, \$21.49 for an individual with 14 years, \$30.12 for 16 years, \$35.16 for 18 years, and \$51.76 for 20 years.

4.15 LAW OF ITERATED EXPECTATIONS

The function $m(x) = \mathbb{E}[Y \mid X = x]$ is not random. Instead, it is a feature of the joint distribution. Sometimes, however, it is useful to treat the conditional expectation as a random variable. To do so, we evaluate the function $m(x)$ at the random variable X: $m(X) = \mathbb{E}[Y \mid X]$. The expression $m(X)$ is a random variable, a transformation of X.

What does it mean to say that $m(X)$ is a random variable? Take our example 1. We found that $\mathbb{E}[Y \mid X = 1] = 1.2$, and $\mathbb{E}[Y \mid X = 2] = 1.6$. We also know that $\mathbb{P}[X = 1] = \mathbb{P}[X = 2] = 0.5$. Thus $m(X)$ is a random variable with a two-point distribution, equaling 1.2 and 1.6 each with probability 1/2.

This discussion may seem abstract and a bit confusing. Another way of expressing the distinction is that $m(x)$ is the value of the conditional expectation at $X = x$, while $m(X)$ is a function (a transformation) of the random variable X.

By treating $\mathbb{E}[Y \mid X]$ as a random variable, we can do some interesting manipulations. For example, what is the expectation of $\mathbb{E}[Y \mid X]$? Take the continuous case. We find

$$\begin{aligned}
\mathbb{E}[\mathbb{E}[Y \mid X]] &= \int_{-\infty}^{\infty} \mathbb{E}[Y \mid X = x] f_X(x) dx \\
&= \int_{-\infty}^{\infty} \int_{-\infty}^{\infty} y f_{Y|X}(y \mid x) f_X(x) dy dx \\
&= \int_{-\infty}^{\infty} \int_{-\infty}^{\infty} y f(y, x) dy dx \\
&= \mathbb{E}[Y].
\end{aligned}$$

In words, the average across group averages is the grand average.

Now take the discrete case:

$$\begin{aligned}
\mathbb{E}[\mathbb{E}[Y \mid X]] &= \sum_{i=1}^{\infty} \mathbb{E}[Y \mid X = \tau_i] \pi_X(\tau_i) \\
&= \sum_{i=1}^{\infty} \frac{\sum_{j=1}^{\infty} \tau_j \pi(\tau_i, \tau_j)}{\pi_X(\tau_i)} \pi_X(\tau_i) \\
&= \sum_{i=1}^{\infty} \sum_{j=1}^{\infty} \tau_j \pi(\tau_i, \tau_j) \\
&= \mathbb{E}[Y].
\end{aligned}$$

This result is very important.

Theorem 4.13 Law of Iterated Expectations. If $\mathbb{E}|Y| < \infty$, then $\mathbb{E}[\mathbb{E}[Y \mid X]] = \mathbb{E}[Y]$.

Example 1: (continued): We calculated earlier that $\mathbb{E}[Y] = 1.4$. Using the law of iterated expectations gives

$$\begin{aligned}\mathbb{E}[Y] &= \mathbb{E}[Y \mid X=1]\,\mathbb{P}[X=1] + \mathbb{E}[Y \mid X=2]\,\mathbb{P}[X=2] \\ &= 1.2 \times 0.5 + 1.6 \times 0.5 \\ &= 1.4\end{aligned}$$

which is the same result as before.

Example 2: (continued): We calculated that $\mathbb{E}[Y] = 5/2$. Using the law of iterated expectations, we find

$$\begin{aligned}\mathbb{E}[Y] &= \int_0^\infty \mathbb{E}[Y \mid X=x] f_X(x) dx \\ &= \int_0^\infty \left(\frac{2x+6}{x+2}\right)\left(\frac{x^2+2x}{4}\exp(-x)\right) dx \\ &= \int_0^\infty \left(\frac{x^2+3x}{2}\right)\exp(-x)\, dx \\ &= \frac{5}{2}\end{aligned}$$

which is the same result as before.

Example 4: (continued). The conditional expectation of wages given education is displayed in Figure 4.10(b). The marginal probabilities of education are displayed in Figure 2.2(b). By the law of iterated expectations, the unconditional expectation of wages is the sum of the products of these two displays:

$$\begin{aligned}\mathbb{E}\left[\text{wage}\right] = {} & 11.75 \times 0.027 + 12.20 \times 0.011 + 13.78 \times 0.011 + 13.78 \times 0.011 + 14.04 \times 0.026 + 17.61 \times 0.274 \\ & + 20.17 \times 0.182 + 21.49 \times 0.111 + 30.12 \times 0.229 + 35.16 \times 0.092 + 51.76 \times 0.037 \\ = {} & \$23.90.\end{aligned}$$

This equals the average wage.

4.16 CONDITIONAL VARIANCE

Another feature of the conditional distribution is the conditional variance.

Definition 4.15 The **conditional variance** of Y given $X=x$ is the variance of the conditional distribution $F_{Y|X}(y \mid x)$ and is written as $\text{var}[Y \mid X=x]$ or $\sigma^2(x)$. It equals

$$\text{var}[Y \mid X=x] = \mathbb{E}\left[(Y - m(x))^2 \mid X=x\right].$$

The conditional variance $\text{var}[Y \mid X=x]$ is a function of x and can take any nonnegative shape. When X is discrete, $\text{var}[Y \mid X=x]$ is the variance of Y in the subpopulation with $X=x$. When X is continuous, $\text{var}[Y \mid X=x]$ is the variance of Y in the infinitesimally small subpopulation with $X \simeq x$.

By expanding the quadratic, we can re-express the conditional variance as

$$\operatorname{var}[Y \mid X=x] = \mathbb{E}\left[Y^2 \mid X=x\right] - (\mathbb{E}[Y \mid X=x])^2. \tag{4.2}$$

We can also define $\operatorname{var}[Y \mid X] = \sigma^2(X)$, the conditional variance treated as a random variable. We have the following relationship.

Theorem 4.14 $\operatorname{var}[Y] = \mathbb{E}[\operatorname{var}[Y \mid X]] + \operatorname{var}[\mathbb{E}[Y \mid X]]$.

The first term on the right-hand side is often called the **within group variance**, while the second term is called the **across group variance**.

We prove the theorem for the continuous case. Using equation (4.2), we have

$$\begin{aligned}
\mathbb{E}[\operatorname{var}[Y \mid X]] &= \int \operatorname{var}[Y \mid X=x] f_X(x) dx \\
&= \int \mathbb{E}\left[Y^2 \mid X=x\right] f_X(x) dx - \int m(x)^2 f_X(x) dx \\
&= \mathbb{E}\left[Y^2\right] - \mathbb{E}\left[m(X)^2\right] \\
&= \operatorname{var}[Y] - \operatorname{var}[m(X)].
\end{aligned}$$

The third equality uses the law of iterated expectations for Y^2. The fourth uses $\operatorname{var}[Y] = \mathbb{E}\left[Y^2\right] - (\mathbb{E}[Y])^2$ and $\operatorname{var}[m(X)] = \mathbb{E}\left[m(X)^2\right] - (\mathbb{E}[m(X)])^2 = \mathbb{E}\left[m(X)^2\right] - (\mathbb{E}[Y])^2$.

Example 1: (continued). The conditional variance for the number of drinks per customer is

$$\mathbb{E}\left[Y^2 \mid X=1\right] - (\mathbb{E}[Y \mid X=1])^2 = \frac{1^2 \times 0.4 + 2^2 \times 0.1}{0.5} - 1.2^2 = 0.16$$

$$\mathbb{E}\left[Y^2 \mid X=2\right] - (\mathbb{E}[Y \mid X=2])^2 = \frac{1^2 \times 0.2 + 2^2 \times 0.3}{0.5} - 1.6^2 = 0.24.$$

The variability of the number of drinks purchased is greater for customers who purchase two slices of pizza.

Example 2: (continued). The conditional second moment is

$$\begin{aligned}
\mathbb{E}\left[Y^2 \mid X=x\right] &= \int_0^\infty y^2 f_{Y|X}(y \mid x) dy \\
&= \int_0^\infty y^2 \frac{(xy + y^2) \exp(-y)}{x+2} dy \\
&= \frac{6x + 24}{x+2}.
\end{aligned}$$

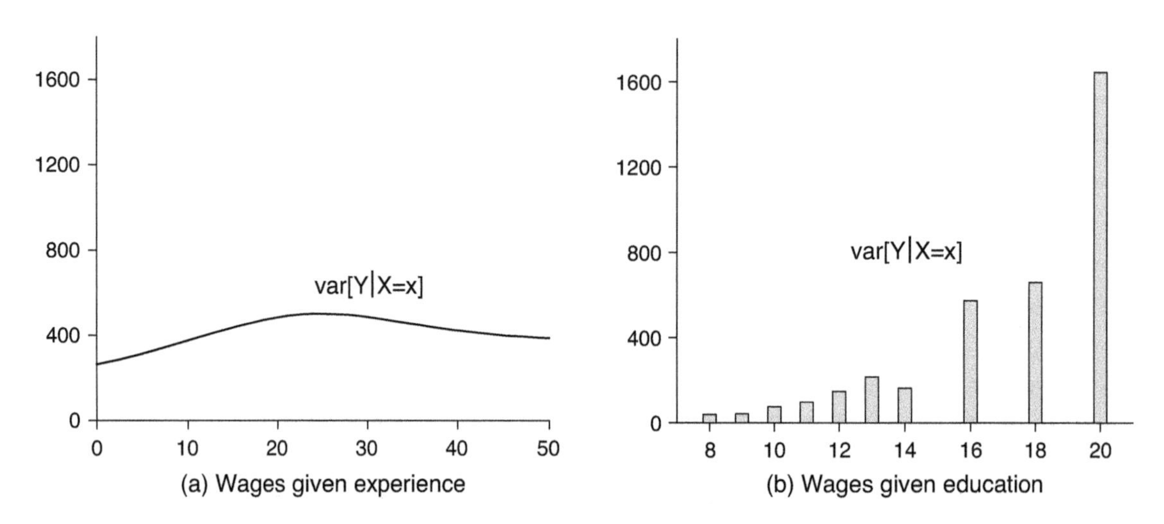

FIGURE 4.12 Conditional variance functions

The conditional variance is

$$\begin{aligned}\operatorname{var}[Y \mid X=x] &= \mathbb{E}\left[Y^2 \mid X=x\right] - (\mathbb{E}[Y \mid X=x])^2 \\ &= \frac{6x+24}{x+2} - \left(\frac{2x+6}{x+2}\right)^2 \\ &= \frac{2x^2+12x+12}{(x+2)^2}\end{aligned}$$

which varies with x.

Example 3: (continued). The conditional variance of wages given experience is displayed in Figure 4.12(a). We see that the variance is a hump-shaped function of experience. The variance substantially increases between 0 and 25 years of experience, and then falls somewhat between 25 and 50 years of experience.

Example 4: (continued). The conditional variance of wages given education is displayed in Figure 4.12(b). The conditional variance is strongly varying as a function of education. The variance for high school graduates is 150; that for college graduates 573; and for those with professional degrees, it is 1646. These are large and meaningful changes, which mean that while the average level of wages increases significantly with education level, so does the spread of the wage distribution. The effect of education on the conditional variance is much stronger than the effect of experience.

4.17 HÖLDER'S AND MINKOWSKI'S INEQUALITIES*

The following inequalities are useful generalizations of the Cauchy-Schwarz inequality.

Theorem 4.15 Hölder's Inequality. For any random variables X and Y and any $p \geq 1$ and $q \geq 1$ satisfying $1/p + 1/q = 1$,

$$\mathbb{E}|XY| \leq \left(\mathbb{E}|X|^p\right)^{1/p}\left(\mathbb{E}|Y|^q\right)^{1/q}.$$

Proof: By the geometric mean inequality (Theorem 2.13), for nonnegative a and b,

$$ab = \left(a^p\right)^{1/p}\left(b^q\right)^{1/q} \leq \frac{a^p}{p} + \frac{b^q}{q}. \tag{4.3}$$

Without loss of generality, assume $\mathbb{E}\,|X|^p = 1$ and $\mathbb{E}\,|Y|^q = 1$. Applying equation (4.3) yields

$$\mathbb{E}\,|XY| \leq \frac{\mathbb{E}\,|X|^p}{p} + \frac{\mathbb{E}\,|Y|^q}{q} = \frac{1}{p} + \frac{1}{q} = 1$$

as needed. ■

Theorem 4.16 Minkowski's Inequality. For any random variables X and Y and any $p \geq 1$,

$$\left(\mathbb{E}\,|X+Y|^p\right)^{1/p} \leq \left(\mathbb{E}\,|X|^p\right)^{1/p} + \left(\mathbb{E}\,|Y|^p\right)^{1/p}.$$

Proof. Using the triangle inequality and then applying Hölder's inequality (Theorem 4.15) to the two expectations gives

$$\begin{aligned}
\mathbb{E}\,|X+Y|^p &= \mathbb{E}\left[|X+Y|\,|X+Y|^{p-1}\right] \\
&\leq \mathbb{E}\left[|X|\,|X+Y|^{p-1}\right] + \mathbb{E}\left(|Y|\,|X+Y|^{p-1}\right) \\
&\leq \left(\mathbb{E}\,|X|^p\right)^{1/p}\left(\mathbb{E}\,|X+Y|^{(p-1)q}\right)^{1/q} \\
&\quad + \left(\mathbb{E}\,|Y|^p\right)^{1/p}\left(\mathbb{E}\,|X+Y|^{(p-1)q}\right)^{1/q} \\
&= \left(\left(\mathbb{E}\,|X|^p\right)^{1/p} + \left(\mathbb{E}\,|Y|^p\right)^{1/p}\right)\left(\mathbb{E}\,|X+Y|^p\right)^{(p-1)/p}
\end{aligned}$$

where the second inequality picks q to satisfy $1/p + 1/q = 1$, and the final equality uses this fact to make the substitution $q = p/(p-1)$ and then collect terms. Dividing both sides by $\left(\mathbb{E}\,|X+Y|^p\right)^{(p-1)/p}$ completes the proof. ■

4.18 VECTOR NOTATION

Write an m-vector as

$$x = \begin{pmatrix} x_1 \\ x_2 \\ \vdots \\ x_m \end{pmatrix}.$$

This is an element of $\mathbb{R}^m$, m-dimensional Euclidean space. In some cases (primarily for matrix algebra), we use boldface $\boldsymbol{x}$ to indicate a vector.

The **transpose** of a column vector x is the row vector

$$x' = \begin{pmatrix} x_1 & x_2 & \cdots & x_m \end{pmatrix}.$$

There is diversity between fields concerning the choice of notation for the transpose. The above notation is the most common in econometrics. In statistics and mathematics, the notation $x^\top$ is typically used.

The **Euclidean norm** is the Euclidean length of the vector x, defined as

$$\|x\| = \left(\sum_{i=1}^{m} x_i^2\right)^{1/2} = \left(x'x\right)^{1/2}.$$

Multivariate random vectors are written as

$$X = \begin{pmatrix} X_1 \\ X_2 \\ \vdots \\ X_m \end{pmatrix}.$$

Some authors use the notation $\vec{X}$ or $\mathbf{X}$ to denote a random vector.

The equality $\{X = x\}$ and inequality $\{X \le x\}$ hold if and only if they hold for all components. Thus $\{X = x\}$ means $\{X_1 = x_1, X_2 = x_2, \ldots, X_m = x_m\}$, and similarly, $\{X \le x\}$ means $\{X_1 \le x_1, X_2 \le x_2, \ldots, X_m \le x_m\}$. The probability notation $\mathbb{P}[X \le x]$ means $\mathbb{P}[X_1 \le x_1, X_2 \le x_2, \ldots, X_m \le x_m]$.

When integrating over $\mathbb{R}^m$, it is convenient to use the following notation. If $f(x) = f(x_1, \ldots, x_m)$, write

$$\int f(x)dx = \int \cdots \int f(x_1, \ldots, x_m)dx_1 \cdots dx_m.$$

Thus an integral with respect to a vector argument dx is shorthand for an m-dimensional integral. The notation on the left of the above equation is more compact and easier to read. We use the notation on the right when we want to be specific about the arguments.

4.19 TRIANGLE INEQUALITIES*

Theorem 4.17 For any real numbers x_j,

$$\left|\sum_{j=1}^{m} x_j\right| \le \sum_{j=1}^{m} |x_j|. \tag{4.4}$$

Proof: Take the case $m = 2$. Observe that

$$-|x_1| \le x_1 \le |x_1|$$
$$-|x_2| \le x_2 \le |x_2|.$$

Adding, we find

$$-|x_1| - |x_2| \le x_1 + x_2 \le |x_1| + |x_2|$$

which is equation (4.4) for $m = 2$. For $m > 2$, we apply equation (4.4) $m - 1$ times. ■

Theorem 4.18 For any vector $x = (x_1, \ldots, x_m)'$,

$$\|x\| \le \sum_{i=1}^{m} |x_i|. \tag{4.5}$$

Proof: Without loss of generality, assume $\sum_{i=1}^m |x_i| = 1$. This implies $|x_i| \leq 1$ and thus $x_i^2 \leq |x_i|$. Hence

$$\|x\|^2 = \sum_{i=1}^m x_i^2 \leq \sum_{i=1}^m |x_i| = 1.$$

Taking the square root of the two sides completes the proof. ■

Theorem 4.19 **Schwarz Inequality**. For any m-vectors x and y,

$$\left|x'y\right| \leq \|x\| \left\|y\right\|. \tag{4.6}$$

Proof: Without loss of generality, assume $\|x\| = 1$ and $\left\|y\right\| = 1$, so our goal is to show that $\left|x'y\right| \leq 1$. Using Theorem 4.17 and then applying equation (4.1) to $\left|x_iy_i\right| = |x_i| \left|y_i\right|$, we find

$$\left|x'y\right| = \left|\sum_{i=1}^m x_iy_i\right| \leq \sum_{i=1}^m \left|x_iy_i\right| \leq \frac{1}{2}\sum_{i=1}^m x_i^2 + \frac{1}{2}\sum_{i=1}^m y_i^2 = 1.$$

the final equality holds since $\|x\| = 1$ and $\left\|y\right\| = 1$. This is expression (4.6). ■

Theorem 4.20 For any m-vectors x and y,

$$\left\|x+y\right\| \leq \|x\| + \left\|y\right\|. \tag{4.7}$$

Proof: We apply expression (4.6):

$$\begin{aligned}\left\|x+y\right\|^2 &= x'x + 2x'y + y'y \\ &\leq \|x\|^2 + 2\|x\| \left\|y\right\| + \left\|y\right\|^2 \\ &= \left(\|x\| + \left\|y\right\|\right)^2.\end{aligned}$$

Taking the square root of the two sides completes the proof. ■

4.20 MULTIVARIATE RANDOM VECTORS

We now consider the case of a random vector $X \in \mathbb{R}^m$.

Definition 4.16 A **multivariate random vector** is a function from the sample space to $\mathbb{R}^m$, written as $X = (X_1, X_2, \ldots, X_m)'$.

The definitions of distribution, mass, and density functions for multivariate random vectors are as follows.

Definition 4.17 The **joint distribution function** is $F(x) = \mathbb{P}[X \leq x] = \mathbb{P}[X_1 \leq x_1, \ldots, X_m \leq x_m]$.

Definition 4.18 For discrete random vectors, the **joint probability mass function** is $\pi(x) = \mathbb{P}[X = x]$.

Definition 4.19 When $F(x)$ is continuous and differentiable, its **joint density** $f(x)$ equals

$$f(x) = \frac{\partial^m}{\partial x_1 \cdots \partial x_m} F(x).$$

Definition 4.20 The **expectation** of $X \in \mathbb{R}^m$ is the vector of expectations of its elements:

$$\mathbb{E}[X] = \begin{pmatrix} \mathbb{E}[X_1] \\ \mathbb{E}[X_2] \\ \vdots \\ \mathbb{E}[X_m] \end{pmatrix}.$$

Definition 4.21 The $m \times m$ **covariance matrix** of $X \in \mathbb{R}^m$ is

$$\operatorname{var}[X] = \mathbb{E}\left[(X - \mathbb{E}[X])(X - \mathbb{E}[X])'\right]$$

and is commonly written as $\operatorname{var}[X] = \Sigma$.

When applied to a random vector, $\operatorname{var}[X]$ is a matrix. It has elements

$$\Sigma = \begin{pmatrix} \sigma_1^2 & \sigma_{12} & \cdots & \sigma_{1m} \\ \sigma_{21} & \sigma_2^2 & \cdots & \sigma_{2m} \\ \vdots & \vdots & \ddots & \vdots \\ \sigma_{m1} & \sigma_{m2} & \cdots & \sigma_m^2 \end{pmatrix}$$

where $\sigma_j^2 = \operatorname{var}\left[X_j\right]$, and $\sigma_{ij} = \operatorname{cov}(X_i, X_j)$ for $i, j = 1, 2, \ldots, m$ and $i \neq j$.

Theorem 4.21 **Properties of the covariance matrix.** Any $m \times m$ covariance matrix Σ is

1. Symmetric: $\Sigma = \Sigma'$.
2. Positive semi-definite: For any $m \times 1$ $a \neq 0$, $a'\Sigma a \geq 0$.

Proof: Symmetry holds, because $\operatorname{cov}(X_i, X_j) = \operatorname{cov}(X_j, X_i)$. For positive semi-definiteness,

$$\begin{aligned} a'\Sigma a &= a'\mathbb{E}\left[(X - \mathbb{E}[X])(X - \mathbb{E}[X])'\right] a \\ &= \mathbb{E}\left[a'(X - \mathbb{E}[X])(X - \mathbb{E}[X])'a\right] \\ &= \mathbb{E}\left[\left(a'(X - \mathbb{E}[X])\right)^2\right] \\ &= \mathbb{E}\left[Z^2\right] \end{aligned}$$

where $Z = a'(X - \mathbb{E}[X])$. Since $Z^2 \geq 0$, we have $\mathbb{E}\left[Z^2\right] \geq 0$ and $a'\Sigma a \geq 0$. ∎

Theorem 4.22 If $X \in \mathbb{R}^m$ with $m \times 1$ expectation μ and $m \times m$ covariance matrix Σ, and $\boldsymbol{A}$ is $q \times m$, then $\boldsymbol{A}X$ is a random vector with mean $\boldsymbol{A}\mu$ and covariance matrix $\boldsymbol{A}\Sigma\boldsymbol{A}'$.

Theorem 4.23 For $X \in \mathbb{R}^m$, $\mathbb{E}\|X\| < \infty$ if and only if $\mathbb{E}\left|X_j\right| < \infty$ for $j = 1, \ldots, m$.

Proof*: Assume $\mathbb{E}\left|X_j\right| \le C < \infty$ for $j=1,\ldots,m$. Applying the triangle inequality (Theorem 4.18) gives

$$\mathbb{E}\|X\| \le \sum_{j=1}^{m} \mathbb{E}\left|X_j\right| \le mC < \infty.$$

For the reverse inequality, the Euclidean norm of a vector is larger than the length of any individual component, so for any j, $\left|X_j\right| \le \|X\|$. Thus, if $\mathbb{E}\|X\| < \infty$, then $\mathbb{E}\left|X_j\right| < \infty$ for $j=1,\ldots,m$. ■

4.21 PAIRS OF MULTIVARIATE VECTORS

Most concepts for pairs of random variables apply to multivariate vectors. Let (X, Y) be a pair of multivariate vectors of dimension m_X and m_Y, respectively. For ease of presentation, we focus on the case of continuous random vectors.

Definition 4.22 The **joint distribution** function of $(X, Y) \in \mathbb{R}^{m_X} \times \mathbb{R}^{m_Y}$ is

$$F(x,y) = \mathbb{P}\left[X \le x, Y \le y\right].$$

The **joint density** function of $(X, Y) \in \mathbb{R}^{m_X} \times \mathbb{R}^{m_Y}$ is

$$f(x,y) = \frac{\partial^{m_x+m_y}}{\partial x_1 \cdots \partial y_{m_y}} F(x,y).$$

The **marginal density** functions of X and Y are

$$f_X(x) = \int f(x,y)dy$$

$$f_Y(y) = \int f(x,y)dx.$$

The **conditional densities** of Y given $X=x$ and X given $Y=y$ are

$$f_{Y|X}(y \mid x) = \frac{f(x,y)}{f_X(x)}$$

$$f_{X|Y}(x \mid y) = \frac{f(x,y)}{f_Y(y)}.$$

The **conditional expectation** of Y given $X=x$ is

$$\mathbb{E}\left[Y \mid X=x\right] = \int y f_{Y|X}(y \mid x)dy.$$

The random vectors Y and X are **independent** if their joint density factors as

$$f(x,y) = f_X(x) f_Y(y).$$

The continuous random variables $(X_1,\ldots,X_m)$ are **mutually independent** if their joint density factors into the products of marginal densities, thus

$$f(x_1,\ldots,x_m) = f_1(x_1)\cdots f_m(x_m).$$

4.22 MULTIVARIATE TRANSFORMATIONS

When $X \in \mathbb{R}^m$ has a density and $Y = g(X) \in \mathbb{R}^q$, where $g(x): \mathbb{R}^m \to \mathbb{R}^q$ is one-to-one, then there is a well-known formula for the joint density of Y.

Theorem 4.24 Suppose X has PDF $f_X(x)$, $g(x)$ is one-to-one, and $h(y) = g^{-1}(y)$ is differentiable. Then $Y = g(X)$ has density

$$f_Y(y) = f_X(h(y))J(y)$$

where

$$J(y) = \left|\det\left(\frac{\partial}{\partial y'}h(y)\right)\right|$$

is the Jacobian of the transformation.

Writing out the derivative matrix in detail, let $h(y) = (h_1(y), h_2(y), \dots, h_m(y))'$ and

$$\frac{\partial}{\partial y'}h(y) = \begin{pmatrix} \partial h_1(y)/\partial y_1 & \partial h_1(y)/\partial y_2 & \cdots & \partial h_1(y)/\partial y_m \\ \partial h_2(y)/\partial y_1 & \partial h_2(y)/\partial y_2 & \cdots & \partial h_2(y)/\partial y_m \\ \vdots & \vdots & \ddots & \vdots \\ \partial h_m(y)/\partial y_1 & \partial h_m(y)/\partial y_2 & \cdots & \partial h_m(y)/\partial y_m \end{pmatrix}.$$

Example: Let X_1 and X_2 be independent with densities e^{-x_1} and e^{-x_2}, respectively. Take the transformation $Y_1 = X_1$ and $Y_2 = X_1 + X_2$. The inverse transformation is $X_1 = Y_1$ and $X_2 = Y_2 - Y_1$ with derivative matrix

$$\frac{\partial}{\partial y'}h(y) = \begin{pmatrix} 1 & 0 \\ -1 & 1 \end{pmatrix}.$$

Thus $J = 1$. The support for Y is $\{0 \le Y_1 \le Y_2 < \infty\}$. The joint density is therefore

$$f_Y(y) = e^{-y_2}\mathbb{1}\left\{y_1 \le y_2\right\}.$$

We can calculate the marginal density of Y_2 by integrating over Y_1. It is

$$f_2(y_2) = \int_0^\infty e^{-y_2}\mathbb{1}\left\{y_1 \le y_2\right\}dy_1 = \int_0^{y_2} e^{-y_2}dy_1 = y_2 e^{-y_2}$$

on $y_2 \in \mathbb{R}$. This is a gamma density with parameters $\alpha = 2$, $\beta = 1$.

4.23 CONVOLUTIONS

A useful method for calculating the distribution of the sum of random variables is by convolution.

Theorem 4.25 Convolution Theorem. If X and Y are independent random variables with densities $f_X(x)$ and $f_Y(y)$, then the density of $Z = X + Y$ is

$$f_Z(z) = \int_{-\infty}^{\infty} f_X(s)f_Y(z-s)ds = \int_{-\infty}^{\infty} f_X(z-s)f_Y(s)ds.$$

Proof: Define the transformation (X, Y) to (W, Z), where $Z = X + Y$ and $W = X$. The Jacobian is 1. The joint density of (W, Z) is $f_X(w)f_Y(z-w)$. The marginal density of Z is obtained by integrating out W, which is the first stated result. The second can be obtained by transformation of variables. ■

The representation $\int_{-\infty}^{\infty} f_X(s)f_Y(z-s)ds$ is known as the **convolution** of f_X and f_Y.

Example: Suppose $X \sim U[0,1]$ and $Y \sim U[0,1]$ are independent, and $Z = X + Y$. Z has support $[0,2]$. By the convolution theorem, Z has density

$$\begin{aligned}
\int_{-\infty}^{\infty} f_X(s)f_Y(z-s)ds &= \int_{-\infty}^{\infty} \mathbb{1}\{0 \le s \le 1\}\, \mathbb{1}\{0 \le z-s \le 1\}\, ds \\
&= \int_0^1 \mathbb{1}\{0 \le z-s \le 1\}\, ds \\
&= \begin{cases} \int_0^z ds & z \le 1 \\ \int_{z-1}^1 ds & 1 \le z \le 2 \end{cases} \\
&= \begin{cases} z & z \le 1 \\ 2-z & 1 \le z \le 2. \end{cases}
\end{aligned}$$

Thus the density of Z has a triangle or "tent" shape on $[0,1]$.

Example: Suppose X and Y are independent, each with density $\lambda^{-1}\exp(-x/\lambda)$ on $x \ge 0$, and $Z = X + Y$. Z has support $[0,\infty)$. By the convolution theorem, we have

$$\begin{aligned}
f_Z(z) &= \int_{-\infty}^{\infty} f_X(t)f_Y(z-t)dt \\
&= \int_{-\infty}^{\infty} \mathbb{1}\{t \ge 0\}\, \mathbb{1}\{z-t \ge 0\} \frac{1}{\lambda}e^{-t/\lambda}\frac{1}{\lambda}e^{-(z-t)/\lambda}dt \\
&= \int_0^z \frac{1}{\lambda^2}e^{-z/\lambda}dt \\
&= \frac{z}{\lambda^2}e^{-z/\lambda}.
\end{aligned}$$

This is a gamma density with parameters $\alpha = 2$ and $\beta = 1/\lambda$.

4.24 HIERARCHICAL DISTRIBUTIONS

Often a useful way to build a probability structure for an economic model is to use a hierarchy. Each stage of the hierarchy is a random variable with a distribution whose parameters are treated as random variables. This can result in a compelling economic structure. The resulting probability distribution can in certain cases equal a known distribution, or it can lead to a new distribution.

For example, suppose we want a model for the number of sales at a retail store. A baseline model is that the store has N customers who each make a binary decision to buy or not with some probability p. This is a binomial model for sales: $X \sim \text{binomial}(N, p)$. If the number of customers N is unobserved, we can also model it as a random variable. A simple model is $N \sim \text{Poisson}(\lambda)$. We examine this model below.

In general, a two-layer hierarchical model takes the form

$$X \mid Y \sim f(x \mid y)$$
$$Y \sim g(y).$$

The joint density of X and Y equals $f(x, y) = f(x \mid y)g(y)$. The marginal density of X is

$$f(x) = \int f(x, y)dy = \int f(x \mid y)g(y)dy.$$

More complicated structures can be built. A three-layer hierarchical model takes the form

$$X \mid Y, Z \sim f(x \mid y, z)$$
$$Y \mid Z \sim g(y \mid z)$$
$$Z \sim h(z).$$

The marginal density of X is

$$f(x) = \iint f(x \mid y, z)g(y \mid z)h(z)dzdy.$$

Binomial-Poisson. This is the retail sales model described above. The distribution of sales X given N customers is binomial, and the distribution of the number of customers is Poisson:

$$X \mid N \sim \text{binomial}(N, p)$$
$$N \sim \text{Poisson}(\lambda).$$

The marginal distribution of X is

$$\begin{aligned}
\mathbb{P}[X = x] &= \sum_{n=x}^{\infty} \binom{n}{x} p^x (1-p)^{n-x} \frac{e^{-\lambda}\lambda^n}{n!} \\
&= \frac{(\lambda p)^x e^{-\lambda}}{x!} \sum_{n=x}^{\infty} \frac{((1-p)\lambda)^{n-x}}{(n-x)!} \\
&= \frac{(\lambda p)^x e^{-\lambda}}{x!} \sum_{t=0}^{\infty} \frac{((1-p)\lambda)^{t}}{t!} \\
&= \frac{(\lambda p)^x e^{-\lambda p}}{x!}.
\end{aligned}$$

The first equality is the sum of the binomial multiplied by the Poisson. The second line writes out the factorials and combines terms. The third line makes the change of variables $t = n - x$. The last line recognizes that the sum is over a Poisson density with parameter $(1 - p)\lambda$. The result in the final line is the probability mass function for the Poisson(λp) distribution. Thus

$$X \sim \text{Poisson}(\lambda p).$$

Hence the binomial-Poisson model implies a Poisson distribution for sales!

This shows the (perhaps surprising) result that if customers arrive with a Poisson distribution and each makes a Bernoulli decision, then total sales are distributed as Poisson.

Beta-Binomial. Returning to the retail sales model, again assume that sales given customers is binomial, but now consider the case that the probability p of a sale is heterogeneous. This can be modeled by treating p as random. A simple model appropriate for a probability is $p \sim \text{beta}(\alpha, \beta)$. The hierarchical model is

$$X \mid N \sim \text{binomial}(N, p)$$
$$p \sim \text{beta}(\alpha, \beta).$$

The marginal distribution of X is

$$\begin{aligned}\mathbb{P}[X=x] &= \int_0^1 \binom{N}{x} p^x (1-p)^{N-x} \frac{p^{\alpha-1}(1-p)^{\beta-1}}{B(\alpha,\beta)} dp \\ &= \frac{B(x+\alpha, N-x+\beta)}{B(\alpha,\beta)} \binom{N}{x}\end{aligned}$$

for $x=0, \ldots, N$. This is different from the binomial, and is known as the **beta-binomial** distribution. It is more dispersed than the binomial distribution. The beta-binomial is used occasionally in economics.

Variance Mixtures. The normal distribution has "thin tails", meaning that the density decays rapidly to 0. We can create a random variable with thicker tails by using a normal variance mixture. Consider the hierarchical model

$$X \mid Q \sim \mathrm{N}(0, Q)$$
$$Q \sim F$$

for some distribution F such that $\mathbb{E}[Q]=1$ and $\mathbb{E}\left[Q^2\right]=\kappa$. The first four moments of X are

$$\begin{aligned}\mathbb{E}[X] &= \mathbb{E}[\mathbb{E}[X \mid Q]] = 0 \\ \mathbb{E}\left[X^2\right] &= \mathbb{E}\left[\mathbb{E}\left[X^2 \mid Q\right]\right] = \mathbb{E}[Q] = 1 \\ \mathbb{E}\left[X^3\right] &= \mathbb{E}\left[\mathbb{E}\left[X^3 \mid Q\right]\right] = 0 \\ \mathbb{E}\left[X^4\right] &= \mathbb{E}\left[\mathbb{E}\left[X^4 \mid Q\right]\right] = \mathbb{E}\left[3Q^2\right] = 3\kappa.\end{aligned}$$

These calculations use the moment properties of the normal, which is introduced in Chapter 5. The first three moments of X match those of the standard normal. The fourth moment is 3κ, while that of the standard normal is 3. Since $\kappa \geq 1$ (see Exercise 2.16), this means that X can have thicker tails than the standard normal.

Normal Mixtures. The model is

$$X \mid T \sim \begin{cases} \mathrm{N}(\mu_1, \sigma_1^2) & \text{if} \quad T=1 \\ \mathrm{N}(\mu_2, \sigma_2^2) & \text{if} \quad T=2 \end{cases}$$

$$\mathbb{P}[T=1] = p$$
$$\mathbb{P}[T=2] = 1-p.$$

Normal mixtures are commonly used in economics to model contexts with multiple "latent types." The random variable T determines the "type" from which the random variable X is drawn. The marginal density of X equals the mixture of normals density:

$$f\left(x \mid p_1, \mu_1, \sigma_1^2, \mu_2, \sigma_2^2\right) = p\phi_{\sigma_1}(x-\mu_1) + (1-p)\phi_{\sigma_2}(x-\mu_2).$$

4.25 EXISTENCE AND UNIQUENESS OF THE CONDITIONAL EXPECTATION*

In Section 4.14 we defined the conditional expectation separately for discrete and continuous random variables. We have explored these cases because the conditional expectation is easiest to describe and understand for them. However, the conditional expectation exists quite generally without appealing to the properties of either discrete or continuous random variables.

To justify this claim, let us consider a deep result from probability theory. It states that the conditional expectation exists for all joint distributions (Y, X) for which Y has a finite expectation.

Theorem 4.26 Existence of the Conditional Expectation. Let (Y, X) have a joint distribution. If $\mathbb{E}|Y| < \infty$, then there exists a function $m(x)$ such that for all sets A for which $\mathbb{P}[X \in A]$ is defined,

$$\mathbb{E}[\mathbb{1}\{X \in A\} Y] = \mathbb{E}[\mathbb{1}\{X \in A\} m(X)]. \tag{4.8}$$

The function $m(x)$ is almost everywhere unique, in the sense that if $h(X)$ satisfies equation (4.8), then there is a set S such that $\mathbb{P}[S] = 1$, and $m(x) = h(x)$ for $x \in S$. The functions $m(x) = \mathbb{E}[Y \mid X = x]$ and $m(X) = \mathbb{E}[Y \mid X]$ are called the **conditional expectation.**

For a proof, see theorem 6.3.3 in Ash (1972).

The conditional expectation $m(x)$ defined by equation (4.8) specializes to our previous definitions when (Y, X) are discrete or have a joint density. The usefulness of definition (4.8) is that Theorem 4.26 shows that the conditional expectation $m(x)$ exists for all finite-mean distributions. This definition allows Y to be discrete or continuous, for X to be scalar or vector valued, and for the components of X to be discrete or continuously distributed.

4.26 IDENTIFICATION

A critical and important issue in structural econometric modeling is identification, meaning that a parameter is uniquely determined by the distribution of the observed variables. It is relatively straightforward in the context of unconditional and conditional expectations, but it is worthwhile to introduce and explore the concept at this point for clarity.

Let F denote a probability distribution, for example, the distribution of the pair (Y, X). Let $\mathscr{F}$ be a collection of distributions. Let θ be a parameter of interest (for example, the mean $\mathbb{E}[Y]$).

Definition 4.23 A parameter $\theta \in \mathbb{R}^k$ is **identified** on $\mathscr{F}$ if for all $F \in \mathscr{F}$ there is a unique value of θ.

Equivalently, θ is identified if we can write it as a mapping $\theta = g(F)$ on the set $\mathscr{F}$. The restriction to the set $\mathscr{F}$ is important. Most parameters are identified only on a strict subset of the space of distributions.

Take, for example, the mean $\mu = \mathbb{E}[Y]$. It is uniquely determined if $\mathbb{E}|Y| < \infty$, so it is clear that μ is identified for the set $\mathscr{F} = \left\{F : \int_{-\infty}^{\infty} |y|\, dF(y) < \infty\right\}$. However, μ is also defined when it is either positive or negative infinity. Hence, defining I_1 and I_2 as in equations (2.8) and (2.9), respectively, we can deduce that μ is identified on the set $\mathscr{F} = \{F : \{I_1 < \infty\} \cup \{I_2 > -\infty\}\}$.

Next, consider the conditional expectation. Theorem 4.26 demonstrated that $\mathbb{E}|Y| < \infty$ is a sufficient condition for identification.

Theorem 4.27 Identification of the Conditional Expectation. Let (Y, X) have a joint distribution. If $\mathbb{E}|Y| < \infty$, then the conditional expectation $m(X) = \mathbb{E}[Y \mid X]$ is identified almost everywhere.

It might seem as if identification is a general property for parameters so long as we exclude degenerate cases. This is true for moments but not necessarily for more complicated parameters. As a case in point, consider censoring. Let Y be a random variable with distribution F. Let Y be censored from above, and let Y^* be defined by the censoring rule

$$Y^* = \begin{cases} Y & \text{if } Y \le \tau \\ \tau & \text{if } Y > \tau. \end{cases}$$

That is, Y^* is capped at the value τ. The variable Y^* has distribution

$$F^*(u) = \begin{cases} F(u) & \text{for } u \le \tau \\ 1 & \text{for } u \ge \tau. \end{cases}$$

Let $\mu = \mathbb{E}[Y]$ be the parameter of interest. The difficulty is that we cannot calculate μ from F^* except in the trivial case of no censoring $\mathbb{P}[Y \ge \tau] = 0$. Thus the mean μ is not identified from the censored distribution.

A typical solution to the identification problem is to assume a parametric distribution. For example, let $\mathscr{F}$ be the set of normal distributions $Y \sim \mathrm{N}(\mu, \sigma^2)$. It is possible to show that the parameters (μ, σ^2) are identified for all $F \in \mathscr{F}$. That is, if we know that the uncensored distribution is normal, we can uniquely determine the parameters from the censored distribution. This is called **parametric identification**, as identification is restricted to a parametric class of distributions. In modern econometrics, parametric identification is viewed as a second-best solution, because identification has been achieved only through the use of an arbitrary and unverifiable parametric assumption.

A pessimistic conclusion might be that it is impossible to identify parameters of interest from censored data without making parametric assumptions. Interestingly, this pessimism is unwarranted. It turns out that we can identify the quantiles $q(\alpha)$ of F for $\alpha \le \mathbb{P}[Y \le \tau]$. For example, if 20% of the distribution is censored, we can identify all quantiles for $\alpha \in (0, 0.8)$. This is called **nonparametric identification**, because the parameters are identified without restriction to a parametric class.

What we have learned from this little exercise is that in the context of censored data, moments can only be parametrically identified, while noncensored quantiles are nonparametrically identified. Part of the message is that a study of identification can help focus attention on what can be learned from the data distributions available.

4.27 EXERCISES

Exercise 4.1 Let $f(x, y) = 1/4$ for $-1 \le x \le 1$ and $-1 \le y \le 1$ (and 0 elsewhere).

(a) Verify that $f(x, y)$ is a valid density function.

(b) Find the marginal density of X.

(c) Find the conditional density of Y given $X = x$.
(d) Find $\mathbb{E}[Y \mid X = x]$.
(e) Determine $\mathbb{P}\left[X^2 + Y^2 < 1\right]$.
(f) Determine $\mathbb{P}[|X + Y| < 2]$.

Exercise 4.2 Let $f(x, y) = x + y$ for $0 \le x \le 1$ and $0 \le y \le 1$ (and 0 elsewhere).

(a) Verify that $f(x, y)$ is a valid density function.
(b) Find the marginal density of X.
(c) Find $\mathbb{E}[Y]$, $\text{var}[X]$, $\mathbb{E}[XY]$, and $\text{corr}(X, Y)$.
(d) Find the conditional density of Y given $X = x$.
(e) Find $\mathbb{E}[Y \mid X = x]$.

Exercise 4.3 Let

$$f(x, y) = \frac{2}{(1 + x + y)^3}$$

for $0 \le x$ and $0 \le y$.

(a) Verify that $f(x, y)$ is a valid density function.
(b) Find the marginal density of X.
(c) Find $\mathbb{E}[Y]$, $\text{var}[Y]$, $\mathbb{E}[XY]$, and $\text{corr}(X, Y)$.
(d) Find the conditional density of Y given $X = x$.
(e) Find $\mathbb{E}[Y \mid X = x]$.

Exercise 4.4 Let the joint PDF of X and Y be given by $f(x, y) = g(x)h(y)$ for some functions $g(x)$ and $h(y)$. Let $a = \int_{-\infty}^{\infty} g(x)dx$, and $b = \int_{-\infty}^{\infty} h(x)dx$.

(a) What conditions should a and b satisfy for $f(x, y)$ to be a bivariate PDF?
(b) Find the marginal PDF of X and Y.
(c) Show that X and Y are independent.

Exercise 4.5 Let $F(x, y)$ be the distribution function of (X, Y). Show that

$$\mathbb{P}[a < X \le b, c < Y \le d] = F(b, d) - F(b, c) - F(a, d) + F(a, c).$$

Hint: Review Figure 4.5.

Exercise 4.6 Let the joint PDf of X and Y be given by

$$f(x, y) = \begin{cases} cxy & \text{if } x, y \in [0, 1], x + y \le 1 \\ 0 & \text{otherwise}. \end{cases}$$

(a) Find the value of c such that $f(x, y)$ is a joint PDF.
(b) Find the marginal distributions of X and Y.
(c) Are X and Y independent?

Exercise 4.7 Let X and Y have density $f(x, y) = \exp(-x - y)$ for $x > 0$ and $y > 0$. Find the marginal density of X and Y. Are X and Y independent or dependent?

Exercise 4.8 Let X and Y have density $f(x, y) = 1$ on $0 < x < 1$ and $0 < y < 1$. Find the density function of $Z = XY$.

Exercise 4.9 Let X and Y have density $f(x, y) = 12xy(1 - y)$ for $0 < x < 1$ and $0 < y < 1$. Are X and Y independent or dependent?

Exercise 4.10 Show that any random variable is uncorrelated with a constant.

Exercise 4.11 Let X and Y be independent random variables with means μ_X, μ_Y, and variances σ_X^2, σ_Y^2, respectively. Find an expression for the correlation of XY and Y in terms of these means and variances.
Hint: "XY" is not a typo.

Exercise 4.12 Prove the following: If $(X_1, X_2, \ldots, X_m)$ are pairwise uncorrelated, then

$$\operatorname{var}\left[\sum_{i=1}^{m} X_i\right] = \sum_{i=1}^{m} \operatorname{var}[X_i].$$

Exercise 4.13 Suppose that X and Y are jointly normal, i.e., they have the joint PDF:

$$f(x, y) = \frac{1}{2\pi\sigma_X\sigma_Y\sqrt{1-\rho^2}} \exp\left(-\frac{1}{2(1-\rho^2)}\left(\frac{x^2}{\sigma_X^2} - 2\frac{\rho xy}{\sigma_X\sigma_Y} + \frac{y^2}{\sigma_Y^2}\right)\right).$$

(a) Derive the marginal distribution of X and Y, and observe that both are normal distributions.
(b) Derive the conditional distribution of Y given $X = x$. Observe that it is also a normal distribution.
(c) Derive the joint distribution of (X, Z), where $Z = (Y/\sigma_Y) - (\rho X/\sigma_X)$, and then show that X and Z are independent.

Exercise 4.14 Let $X_1 \sim \text{gamma}(r, 1)$ and $X_2 \sim \text{gamma}(s, 1)$ be independent. Find the distribution of $Y = X_1 + X_2$.

Exercise 4.15 Suppose that the distribution of Y conditional on $X = x$ is $\mathrm{N}(x, x^2)$, and the marginal distribution of X is $U[0, 1]$.

(a) Find $\mathbb{E}[Y]$.
(b) Find $\operatorname{var}[Y]$.

Exercise 4.16 For any random variables X and Y with finite variances, prove the following:

(a) $\operatorname{cov}(X, Y) = \operatorname{cov}(X, \mathbb{E}[Y \mid X])$.
(b) X and $Y - \mathbb{E}[Y \mid X]$ are uncorrelated.

Exercise 4.17 Suppose that Y conditional on X is $\mathrm{N}(X, X)$, $\mathbb{E}[X] = \mu$, and $\operatorname{var}[X] = \sigma^2$. Find $\mathbb{E}[Y]$ and $\operatorname{var}[Y]$.

Exercise 4.18 Consider the hierarchical distribution

$$X \mid Y \sim \mathrm{N}(Y, \sigma^2)$$
$$Y \sim \mathrm{gamma}(\alpha, \beta).$$

(a) Find $\mathbb{E}[X]$. Hint: Use the law of iterated expectations (Theorem 4.13).

(b) Find $\mathrm{var}[X]$. Hint: Use Theorem 4.14.

Exercise 4.19 Consider the hierarchical distribution

$$X \mid N \sim \chi^2_{2N}$$
$$N \sim \mathrm{Poisson}(\lambda).$$

(a) Find $\mathbb{E}[X]$.

(b) Find $\mathrm{var}[X]$.

Exercise 4.20 Find the covariance and correlation between $a + bX$ and $c + dY$.

Exercise 4.21 If two random variables are independent, are they necessarily uncorrelated? Find an example of random variables which are independent yet not uncorrelated.
Hint: Take a careful look at Theorem 4.8.

Exercise 4.22 Let X be a random variable with finite variance. Find the correlation between

(a) X and X.

(b) X and $-X$.

Exercise 4.23 Use Hölder's inequality (Theorem 4.15) to show the following.

(a) $\mathbb{E}\left|X^3 Y\right| \le \mathbb{E}\left(|X|^4\right)^{3/4} \mathbb{E}\left(|Y|^4\right)^{1/4}$.

(b) $\mathbb{E}\left|X^a Y^b\right| \le \mathbb{E}\left(|X|^{a+b}\right)^{a/(a+b)} \mathbb{E}\left(|Y|^{a+b}\right)^{b/(a+b)}$.

Exercise 4.24 Extend Minkowski's inequality (Theorem 4.16) to to show that if $p \ge 1$,

$$\left(\mathbb{E}\left|\sum_{i=1}^{\infty} X_i\right|^p\right)^{1/p} \le \sum_{i=1}^{\infty} \left(\mathbb{E}\left|X_i\right|^p\right)^{1/p}.$$

CHAPTER 5

NORMAL AND RELATED DISTRIBUTIONS

5.1 INTRODUCTION

The normal distribution is the most important probability distribution in applied econometrics. The regression model is built around the normal model. Many parametric models are derived from normality. Asymptotic approximations use normality as approximation distributions. Many inference procedures use distributions derived from the normal for critical values. These include the normal distribution, the student t distribution, the chi-square distribution, and the F distribution. Consequently it is important to understand the properties of the normal distribution and those derived from it.

5.2 UNIVARIATE NORMAL

Definition 5.1 A random variable Z has the **standard normal distribution**, written $Z \sim \mathrm{N}(0,1)$, if it has the density

$$\phi(x) = \frac{1}{\sqrt{2\pi}} \exp\left(-\frac{x^2}{2}\right), \quad x \in \mathbb{R}.$$

The standard normal density is typically denoted by $\phi(x)$. The distribution function is not available in closed form but is written as

$$\Phi(x) = \int_{-\infty}^{x} \phi(u) du.$$

The standard normal density function is displayed in Figure 3.3.

The fact that $\phi(x)$ integrates to 1 (and is thus a density) is not immediately obvious but is based on a technical calculation.

Theorem 5.1 $\int_{-\infty}^{\infty} \phi(x) dx = 1.$

For the proof, see Exercise 5.1.

The standard normal density $\phi(x)$ is symmetric about 0. Thus $\phi(x) = \phi(-x)$ and $\Phi(x) = 1 - \Phi(-x)$.

Definition 5.2 If $Z \sim \mathrm{N}(0,1)$ and $X = \mu + \sigma Z$ for $\mu \in \mathbb{R}$ and $\sigma \geq 0$ then X has the **normal distribution**, written $X \sim \mathrm{N}(\mu, \sigma^2)$.

Theorem 5.2 If $X \sim \mathrm{N}(\mu, \sigma^2)$ and $\sigma > 0$ then X has the density

$$f\left(x \mid \mu, \sigma^2\right) = \frac{1}{\sqrt{2\pi\sigma^2}} \exp\left(-\frac{(x-\mu)^2}{2\sigma^2}\right) \quad x \in \mathbb{R}.$$

This density is obtained by the change-of-variables formula.

Theorem 5.3 The MGF of $X \sim \mathrm{N}(\mu, \sigma^2)$ is $M(t) = \exp\left(\mu t + \sigma^2 t^2/2\right)$.

For the proof, see Exercise 5.6.

5.3 MOMENTS OF THE NORMAL DISTRIBUTION

All positive integer moments of the standard normal distribution are finite, because the tails of the density decline exponentially.

Theorem 5.4 If $Z \sim \mathrm{N}(0, 1)$, then for any $r > 0$

$$\mathbb{E}|Z|^r = \frac{2^{r/2}}{\sqrt{\pi}}\Gamma\left(\frac{r+1}{2}\right)$$

where $\Gamma(t)$ is the gamma function (Definition A.20).

The proof is presented in Section 5.11.

Since the density is symmetric about 0, all odd moments are 0. The density is normalized so that $\mathrm{var}[Z] = 1$. See Exercise 5.3.

Theorem 5.5 The mean and variance of $Z \sim \mathrm{N}(0, 1)$ are $\mathbb{E}[Z] = 0$ and $\mathrm{var}[Z] = 1$, respectively.

From Theorem 5.4 or by sequential integration by parts, we can calculate the even moments of Z.

Theorem 5.6 For any positive integer m, $\mathbb{E}\left[Z^{2m}\right] = (2m-1)!!$

For a definition of the double factorial $k!!$, see Section A.3. Using Theorem 5.6, we find $\mathbb{E}\left[Z^4\right] = 3$, $\mathbb{E}\left[Z^6\right] = 15$, $\mathbb{E}\left[Z^8\right] = 105$, and $\mathbb{E}\left[Z^{10}\right] = 945$.

5.4 NORMAL CUMULANTS

Recall that the cumulants are polynomial functions of the moments and can be found by a power series expansion of the cumulant generating function, the latter being the natural log of the MGF. Since the MGF of $Z \sim \mathrm{N}(0, 1)$ is $M(t) = \exp\left(t^2/2\right)$, the cumulant generating function is $K(t) = t^2/2$, which has no further power series expansion. We thus find that the cumulants of the standard normal distribution are

$$\kappa_2 = 1$$

$$\kappa_j = 0, \qquad j \neq 2.$$

Thus the normal distribution has the special property that all cumulants except the second are 0.

5.5 NORMAL QUANTILES

The normal distribution is commonly used for statistical inference. Its quantiles are used for hypothesis testing and confidence interval construction. Therefore, having a basic acquaintance with the quantiles of the normal

Table 5.1
Normal probabilities and quantiles

	$\mathbb{P}[Z \leq x]$	$\mathbb{P}[Z > x]$	$\mathbb{P}[\|Z\| > x]$
$x = 0.00$	0.50	0.50	1.00
$x = 1.00$	0.84	0.16	0.32
$x = 1.65$	0.950	0.050	0.100
$x = 1.96$	0.975	0.025	0.050
$x = 2.00$	0.977	0.023	0.046
$x = 2.33$	0.990	0.010	0.020
$x = 2.58$	0.995	0.005	0.010

Table 5.2
Numerical cumulative distribution function

	To calculate $\mathbb{P}[Z \leq x]$ for given x		
	MATLAB	R	Stata
$\mathrm{N}(0,1)$	`normcdf(x)`	`pnorm(x)`	`normal(x)`
χ_r^2	`chi2cdf(x,r)`	`pchisq(x,r)`	`chi2(r,x)`
t_r	`tcdf(x,r)`	`pt(x,r)`	`1-ttail(r,x)`
$F_{r,k}$	`fcdf(x,r,k)`	`pf(x,r,k)`	`F(r,k,x)`
$\chi_r^2(d)$	`ncx2cdf(x,r,d)`	`pchisq(x,r,d)`	`nchi2(r,d,x)`
$F_{r,k}(d)$	`ncfcdf(x,r,k,d)`	`pf(x,r,k,d)`	`1-nFtail(r,k,d,x)`

Table 5.3
Numerical quantile function

	To calculate x which solves $p = \mathbb{P}[Z \leq x]$ for given p		
	MATLAB	R	Stata
$\mathrm{N}(0,1)$	`norminv(p)`	`qnorm(p)`	`invnormal(p)`
χ_r^2	`chi2inv(p,r)`	`qchisq(p,r)`	`invchi2(r,p)`
t_r	`tinv(p,r)`	`qt(p,r)`	`invttail(r,1-p)`
$F_{r,k}$	`finv(p,r,k)`	`qf(p,r,k)`	`invF(r,k,p)`
$\chi_r^2(d)$	`ncx2inv(p,r,d)`	`qchisq(p,r,d)`	`invnchi2(r,d,p)`
$F_{r,k}(d)$	`ncfinv(p,r,k,d)`	`qf(p,r,k,d)`	`invnFtail(r,k,d,1-p)`

distribution can be useful for practical applications. A few important quantiles of the normal distribution are listed in Table 5.1.

Traditionally, statistical and econometrics textbooks would include extensive tables of normal (and other) quantiles. This is unnecessary today, since these calculations are embedded in statistical software. For convenience, Table 5.2 lists the appropriate commands in MATLAB, R, and Stata to compute the cumulative distribution function of commonly used statistical distributions. Table 5.3 lists the appropriate commands to compute the inverse probabilities (quantiles) of the same distributions.

5.6 TRUNCATED AND CENSORED NORMAL DISTRIBUTIONS

For some applications, it is useful to know the moments of the truncated and censored normal distributions. I list the following for reference.

We first consider the moments of the truncated normal distribution.

Definition 5.3 The function $\lambda(x) = \phi(x)/\Phi(x)$ is called[1] the **inverse Mills ratio.**

The name is due to a paper by John P. Mills published in 1926. Theorem 5.7 lists some properties of the inverse Mills ratio. I do not provide a proof, as these results are not essential.

Theorem 5.7 Properties of the Inverse Mills Ratio.

1. $\lambda(-x) = \phi(x)/(1 - \Phi(x))$
2. $\lambda(x) > 0$ for $x \in \mathbb{R}$
3. $\lambda(0) = \sqrt{2/\pi}$
4. For $x > 0$, $\lambda(x) < \sqrt{2/\pi}\exp(-x^2/2)$
5. For $x < 0$, $0 < x + \lambda(x) < 1$
6. For $x \in \mathbb{R}$, $\lambda(x) < 1 + |x|$
7. $\lambda'(x) = -\lambda(x)(x + \lambda(x))$ satisfies $-1 < \lambda'(x) < 0$
8. $\lambda(x)$ is strictly decreasing and convex on $\mathbb{R}$.

For any trunction point c, define the standardized truncation point $c^* = (c - \mu)/\sigma$.

Theorem 5.8 Moments of the Truncated Normal Distribution. If $X \sim \mathrm{N}(\mu, \sigma^2)$, then for $c^* = (c - \mu)/\sigma$, the following properties hold:

1. $\mathbb{E}[\mathbb{1}\{X < c\}] = \Phi(c^*)$
2. $\mathbb{E}[\mathbb{1}\{X > c\}] = 1 - \Phi(c^*)$
3. $\mathbb{E}[X\mathbb{1}\{X < c\}] = \mu\Phi(c^*) - \sigma\phi(c^*)$
4. $\mathbb{E}[X\mathbb{1}\{X > c\}] = \mu(1 - \Phi(c^*)) + \sigma\phi(c^*)$
5. $\mathbb{E}[X \mid X < c] = \mu - \sigma\lambda(c^*)$
6. $\mathbb{E}[X \mid X > c] = \mu + \sigma\lambda(-c^*)$
7. $\mathrm{var}[X \mid X < c] = \sigma^2\left(1 - c^*\lambda(c^*) - \lambda(c^*)^2\right)$
8. $\mathrm{var}[X \mid X > c] = \sigma^2\left(1 + c^*\lambda(-c^*) - \lambda(-c^*)^2\right)$.

The calculations show that when X is truncated from above, the mean of X is reduced. Conversely, when X is truncated from below, the mean is increased.

From properties 5 and 6 of Theorem 5.8, we can see that the effect of truncation on the mean is to shift the mean away from the truncation point. From properties 7 and 8, we can see that the effect of trunction on the variance is to reduce the variance so long as truncation affects less than half of the original distribution. However, the variance increases at sufficiently high truncation levels.

---

[1]The function $\phi(x)/(1 - \Phi(x))$ is also often called the "inverse Mills ratio."

Let us now consider censoring. For any random variable X, define **censoring from below** as

$$X_* = \begin{cases} X & \text{if } X \geq c \\ c & \text{if } X < c. \end{cases}$$

We define **censoring from above** as

$$X^* = \begin{cases} X & \text{if } X \leq c \\ c & \text{if } X > c. \end{cases}$$

Theorem 5.9 lists the moments of the censored normal distribution.

Theorem 5.9 Moments of the Censored Normal Distribution. If $X \sim \mathrm{N}\left(\mu, \sigma^2\right)$, then for $c^* = (c - \mu)/\sigma$, the following properties hold:

1. $\mathbb{E}\left[X^*\right] = \mu + \sigma c^* \left(1 - \Phi\left(c^*\right)\right) - \sigma \phi(c^*)$
2. $\mathbb{E}\left[X_*\right] = \mu + \sigma c^* \Phi\left(c^*\right) + \sigma \phi(c^*)$
3. $\operatorname{var}\left[X^*\right] = \sigma^2 \left(1 + \left(c^{*2} - 1\right)\left(1 - \Phi\left(c^*\right)\right) - c^* \phi(c^*) - \left(c^*\left(1 - \Phi\left(c^*\right)\right) - \phi(c^*)\right)^2\right)$
4. $\operatorname{var}\left[X_*\right] = \sigma^2 \left(1 + \left(c^{*2} - 1\right)\left(1 - \Phi\left(c^*\right)\right) - c^* \phi\left(c^*\right) - \left(c^* \Phi\left(c^*\right) + \phi(c^*)\right)^2\right)$.

5.7 MULTIVARIATE NORMAL

Let $\{Z_1, Z_2, \ldots, Z_m\}$ be independent and identically distributed $\mathrm{N}(0, 1)$. Since the Z_i are mutually independent, the joint density is the product of the marginal densities:

$$\begin{aligned} f(x_1, \ldots, x_m) &= f(x_1) f(x_2) \cdots f(x_m) \\ &= \prod_{i=1}^{m} \frac{1}{\sqrt{2\pi}} \exp\left(-\frac{x_i^2}{2}\right) \\ &= \frac{1}{(2\pi)^{m/2}} \exp\left(-\frac{1}{2} \sum_{i=1}^{m} x_i^2\right) \\ &= \frac{1}{(2\pi)^{m/2}} \exp\left(-\frac{x'x}{2}\right). \end{aligned}$$

This is called the "multivariate standard normal density."

Definition 5.4 An m-vector Z has the **multivariate standard normal distribution**, written $Z \sim \mathrm{N}(0, \boldsymbol{I}_m)$, if it has density:

$$f(x) = \frac{1}{(2\pi)^{m/2}} \exp\left(-\frac{x'x}{2}\right).$$

Theorem 5.10 The mean and the covariance matrix of $Z \sim \mathrm{N}(0, \boldsymbol{I}_m)$ are $\mathbb{E}[Z] = 0$ and $\operatorname{var}[Z] = \boldsymbol{I}_m$, respectively.

Definition 5.5 If $Z \sim \mathrm{N}(0, \boldsymbol{I}_m)$ and $X = \mu + \boldsymbol{B}Z$ for $q \times m$ $\boldsymbol{B}$, then X has the **multivariate normal distribution**, written $X \sim \mathrm{N}(\mu, \Sigma)$, with a $q \times 1$ mean vector μ, and $q \times q$ covariance matrix $\Sigma = \boldsymbol{B}\boldsymbol{B}'$.

Theorem 5.11 If $X \sim \mathrm{N}(\mu, \Sigma)$, where Σ is invertible, then X has PDF

$$f(x) = \frac{1}{(2\pi)^{m/2}(\det \Sigma)^{1/2}} \exp\left(-\frac{(x-\mu)'\Sigma^{-1}(x-\mu)}{2}\right).$$

The density for X is found by the change of variables $X = \mu + \boldsymbol{B}Z$, which has Jacobian $(\det \Sigma)^{-1/2}$.

5.8 PROPERTIES OF THE MULTIVARIATE NORMAL

Theorem 5.12 The mean and covariance martix of $X \sim \mathrm{N}(\mu, \Sigma)$ are $\mathbb{E}[X] = \mu$ and $\operatorname{var}[X] = \Sigma$, respectively.

The proof of this theorem follows from the definition $X = \mu + \boldsymbol{B}Z$ and Theorem 4.22.

In general, uncorrelated random variables are not necessarily independent. An important exception is that uncorrelated multivariate normal random variables are independent.

Theorem 5.13 If (X, Y) are multivariate normal with $\operatorname{cov}(X, Y) = 0$, then X and Y are independent.

The proof is presented in Section 5.11.

To calculate some properties of the normal distribution, it turns out to be useful to calculate its MGF.

Theorem 5.14 The MGF of $m \times 1$ $X \sim \mathrm{N}(\mu, \Sigma)$ is $M(t) = \mathbb{E}\left[\exp(t'X)\right] = \exp\left[t'\mu + \frac{1}{2}t'\Sigma t\right]$, where t is $m \times 1$.

For the proof, see Exercise 5.12.

The following result, showing that affine[2] functions of normal random vectors are normally distributed, is particulary important.

Theorem 5.15 If $X \sim \mathrm{N}(\mu, \Sigma)$, then $Y = \boldsymbol{a} + \boldsymbol{B}X \sim \mathrm{N}\left(\boldsymbol{a} + \boldsymbol{B}\mu, \boldsymbol{B}\Sigma\boldsymbol{B}'\right)$.

The proof is presented in Section 5.11.

Theorem 5.16 If (Y,X) are multivariate normal,

$$\begin{pmatrix} Y \\ X \end{pmatrix} \sim \mathrm{N}\left(\begin{pmatrix} \mu_Y \\ \mu_X \end{pmatrix}, \begin{pmatrix} \Sigma_{YY} & \Sigma_{YX} \\ \Sigma_{XY} & \Sigma_{XX} \end{pmatrix}\right)$$

with $\Sigma_{YY} > 0$ and $\Sigma_{XX} > 0$, then the conditional distributions are

$$Y \mid X \sim \mathrm{N}\left(\mu_Y + \Sigma_{YX}\Sigma_{XX}^{-1}(X - \mu_X), \Sigma_{YY} - \Sigma_{YX}\Sigma_{XX}^{-1}\Sigma_{XY}\right)$$

$$X \mid Y \sim \mathrm{N}\left(\mu_X + \Sigma_{XY}\Sigma_{YY}^{-1}(Y - \mu_Y), \Sigma_{XX} - \Sigma_{XY}\Sigma_{YY}^{-1}\Sigma_{YX}\right).$$

The proof is presented in Section 5.11.

Theorem 5.17 If

$$Y \mid X \sim \mathrm{N}(X, \Sigma_{YY})$$

and

$$X \sim \mathrm{N}(\mu, \Sigma_{XX})$$

[2] A function $f(x)$ of vector-valued x is affine if $f(x) = \boldsymbol{a} + \boldsymbol{B}X$.

then

$$\begin{pmatrix} Y \\ X \end{pmatrix} \sim \mathrm{N}\left(\begin{pmatrix} \mu \\ \mu \end{pmatrix}, \begin{pmatrix} \Sigma_{YY}+\Sigma_{XX} & \Sigma_{XX} \\ \Sigma_{XX} & \Sigma_{XX} \end{pmatrix}\right)$$

and

$$X \mid Y \sim \mathrm{N}\left(\Sigma_{YY}\left(\Sigma_{YY}+\Sigma_{XX}\right)^{-1}\mu+\Sigma_{XX}\left(\Sigma_{YY}+\Sigma_{XX}\right)^{-1}Y, \Sigma_{XX}-\Sigma_{XX}\left(\Sigma_{YY}+\Sigma_{XX}\right)^{-1}\Sigma_{XX}\right).$$

5.9 CHI-SQUARE, *t*, *F*, AND CAUCHY DISTRIBUTIONS

Many important distributions can be derived as transformations of multivariate normal random vectors.

The following seven theorems show that chi-square, student t, F, non-central χ^2, and Cauchy random variables can all be expressed as functions of normal random variables.

Theorem 5.18 Let $Z \sim \mathrm{N}(0, \boldsymbol{I}_r)$ be multivariate standard normal. Then $Z'Z \sim \chi_r^2$.

Theorem 5.19 If $X \sim \mathrm{N}(0, \boldsymbol{A})$ with $\boldsymbol{A} > 0$, $r \times r$, then $X'\boldsymbol{A}^{-1}X \sim \chi_r^2$.

Theorem 5.20 Let $Z \sim \mathrm{N}(0,1)$ and $Q \sim \chi_r^2$ be independent. Then $T = Z/\sqrt{Q/r} \sim t_r$.

Theorem 5.21 Let $Q_m \sim \chi_m^2$ and $Q_r \sim \chi_r^2$ be independent. Then $(Q_m/m)/(Q_r/r) \sim F_{m,r}$.

Theorem 5.22 If $X \sim \mathrm{N}(\mu, \boldsymbol{I}_r)$, then $X'X \sim \chi_r^2(\lambda)$, where $\lambda = \mu'\mu$.

Theorem 5.23 If $X \sim \mathrm{N}(\mu, \boldsymbol{A})$ with $\boldsymbol{A} > 0$, $r \times r$, then $X'\boldsymbol{A}^{-1}X \sim \chi_r^2(\lambda)$, where $\lambda = \mu'\boldsymbol{A}^{-1}\mu$.

Theorem 5.24 If $T = Z_1/Z_2$, where Z_1 and Z_2 are independent normal random variables, then $T \sim$Cauchy.

The proofs of the first six results are presented in Section 5.11. The result for the Cauchy distribution follows from Theorem 5.20, because the Cauchy is a special case of the t with $r = 1$, and the distributions of Z_1/Z_2 and $Z_1/|Z_2|$ are the same by symmetry.

5.10 HERMITE POLYNOMIALS*

The Hermite polynomials are a classical sequence of orthogonal polynomials with respect to the normal density on the real line. They appear occasionally in econometric theory, including the theory of Edgeworth expansions (Section 9.8).

The jth Hermite polynomial is defined as

$$He_j(x) = (-1)^j \frac{\phi^{(j)}(x)}{\phi(x)}.$$

An explicit formula is

$$He_j(x) = j! \sum_{m=0}^{\lfloor j/2 \rfloor} \frac{(-1)^m}{m!\,(j-2m)!} \frac{x^{j-2m}}{2^m}.$$

The first seven Hermite polynomials are

$$\begin{aligned}
He_0(x) &= 1 \\
He_1(x) &= x \\
He_2(x) &= x^2 - 1 \\
He_3(x) &= x^3 - 3x \\
He_4(x) &= x^4 - 6x^2 + 3 \\
He_5(x) &= x^5 - 10x^3 + 15x \\
He_6(x) &= x^6 - 15x^4 + 45x^2 - 15.
\end{aligned}$$

An alternative scaling used in physics is

$$H_j(x) = 2^{j/2} He_j\left(\sqrt{2}x\right).$$

The Hermite polynomials have the following properties.

Theorem 5.25 Properties of the Hermite Polynomials. For nonnegative integers $m < j$,

1. $\int_{-\infty}^{\infty} \left(He_j(x)\right)^2 \phi(x)dx = j!$
2. $\int_{-\infty}^{\infty} He_m(x)He_j(x)\phi(x)dx = 0$
3. $\int_{-\infty}^{\infty} x^m He_j(x)\phi(x)dx = 0$
4. $\dfrac{d}{dx}\left(He_j(x)\phi(x)\right) = -He_{j+1}(x)\phi(x).$

5.11 TECHNICAL PROOFS*

Proof of Theorem 5.4

$$\begin{aligned}
\mathbb{E}|Z|^r &= \int_{-\infty}^{\infty} |x|^r \frac{1}{\sqrt{2\pi}} \exp\left(-x^2/2\right) dx \\
&= \frac{\sqrt{2}}{\sqrt{\pi}} \int_0^{\infty} x^r \exp\left(-x^2/2\right) dx \\
&= \frac{2^{r/2}}{\sqrt{\pi}} \int_0^{\infty} u^{(r-1)/2} \exp(-u)\, dt \\
&= \frac{2^{r/2}}{\sqrt{\pi}} \Gamma\left(\frac{r+1}{2}\right).
\end{aligned}$$

The third equality is the change of variables $u = x^2/2$, and the final one is Definition A.20. ■

Proof of Theorem 5.8 It is convenient to set $Z = (X - \mu)/\sigma$, which is the standard normal with density $\phi(x)$, so that we have the decomposition $X = \mu + \sigma Z$.

For property 1, note that $\mathbb{1}\{X < c\} = \mathbb{1}\{Z < c^*\}$. Then

$$\mathbb{E}[\mathbb{1}\{X < c\}] = \mathbb{E}\left[\mathbb{1}\left\{Z < c^*\right\}\right] = \Phi\left(c^*\right).$$

For property 2, note that $\mathbb{1}\{X > c\} = 1 - \mathbb{1}\{X \le c\}$. Then

$$\mathbb{E}[\mathbb{1}\{X > c\}] = 1 - \mathbb{E}[\mathbb{1}\{X \le c\}] = 1 - \Phi\left(c^*\right).$$

For property 3, use the fact that $\phi'(x) = -x\phi(x)$. (See Exercise 5.2.) Then

$$\mathbb{E}\left[Z\mathbb{1}\left\{Z < c^*\right\}\right] = \int_{-\infty}^{c^*} z\phi(z)dz = -\int_{-\infty}^{c^*} \phi'(z)dz = -\phi(c^*).$$

Using this result, $X = \mu + \sigma Z$, and property 1, we have

$$\begin{aligned}\mathbb{E}[X\mathbb{1}\{X < c\}] &= \mu\,\mathbb{E}\left[\mathbb{1}\left\{Z < c^*\right\}\right] + \sigma\,\mathbb{E}\left[Z\mathbb{1}\left\{Z < c^*\right\}\right]\\ &= \mu\Phi\left(c^*\right) - \sigma\phi(c^*)\end{aligned}$$

as stated.

For property 4, use property 3 and the fact that $\mathbb{E}[X\mathbb{1}\{X < c\}] + \mathbb{E}[X\mathbb{1}\{X \ge c\}] = \mu$ to deduce that

$$\begin{aligned}\mathbb{E}[X\mathbb{1}\{X > c\}] &= \mu - \left(\mu\Phi\left(c^*\right) - \sigma\phi(c^*)\right)\\ &= \mu\left(1 - \Phi\left(c^*\right)\right) + \sigma\phi(c^*).\end{aligned}$$

For property 5, note that

$$\mathbb{P}[X < c] = \mathbb{P}\left[Z < c^*\right] = \Phi\left(c^*\right).$$

Then apply property 3 to find

$$\mathbb{E}[X \mid X < c] = \frac{\mathbb{E}[X\mathbb{1}\{X < c\}]}{\mathbb{P}[X < c]} = \mu - \sigma\frac{\phi\left(c^*\right)}{\Phi\left(c^*\right)} = \mu - \sigma\lambda\left(c^*\right)$$

as stated.

For property 6, we have the similar calculation

$$\mathbb{E}[X \mid X > c] = \frac{\mathbb{E}[X\mathbb{1}\{X > c\}]}{\mathbb{P}[X > c]} = \mu + \sigma\frac{\phi(c^*)}{1 - \Phi(c^*)} = \mu + \sigma\lambda\left(-c^*\right).$$

For property 7, first observe that if we replace c with c^*, the answer will be invariant to μ and proportional to σ^2. Thus without loss of generality, we make the calculation for $Z \sim \mathrm{N}(0,1)$ instead of X. Using $\phi'(x) = -x\phi(x)$ and integration by parts, we find

$$\mathbb{E}\left[Z^2\mathbb{1}\left\{Z < c^*\right\}\right] = \int_{-\infty}^{c^*} x^2\phi(x)dx = -\int_{-\infty}^{c^*} x\phi'(x)dx = \Phi(c^*) - c^*\phi(c^*). \tag{5.1}$$

The truncated variance is

$$\begin{aligned}\operatorname{var}\left[Z \mid Z < c^*\right] &= \frac{\mathbb{E}\left[Z^2\mathbb{1}\{Z < c^*\}\right]}{\mathbb{P}[Z < c^*]} - \left(\mathbb{E}\left[Z \mid Z < c^*\right]\right)^2\\ &= 1 - c^*\frac{\phi\left(c^*\right)}{\Phi\left(c^*\right)} - \left(\frac{\phi(c^*)}{\Phi(c^*)}\right)^2\\ &= 1 - c^*\lambda\left(c^*\right) - \lambda\left(c^*\right)^2.\end{aligned}$$

Scaled by σ^2, this is the stated result. Property 8 follows by a similar calculation. ■

Proof of Theorem 5.9 We can write $X^* = c\mathbb{1}\{X > c\} + X\mathbb{1}\{X \le c\}$. Thus

$$\begin{aligned}\mathbb{E}\left[X^*\right] &= c\mathbb{E}\left[\mathbb{1}\{X > c\}\right] + \mathbb{E}\left[X\mathbb{1}\{X \le c\}\right] \\ &= c\left(1 - \Phi\left(c^*\right)\right) + \mu\Phi\left(c^*\right) - \sigma\phi(c^*) \\ &= \mu + c\left(1 - \Phi\left(c^*\right)\right) - \mu\left(1 - \Phi\left(c^*\right)\right) - \sigma\phi(c_*) \\ &= \mu + \sigma c^*\left(1 - \Phi\left(c^*\right)\right) - \sigma\phi(c^*)\end{aligned}$$

as stated in property 1. Similarly, $X_* = c\mathbb{1}\{X < c\} + X\mathbb{1}\{X \ge c\}$. Thus

$$\begin{aligned}\mathbb{E}\left[X_*\right] &= c\mathbb{E}\left[\mathbb{1}\{X < c\}\right] + \mathbb{E}\left[X\mathbb{1}\{X \ge c\}\right] \\ &= c\Phi\left(c^*\right) + \mu\left(1 - \Phi\left(c^*\right)\right) + \sigma\phi(c^*) \\ &= \mu + \sigma c^*\Phi\left(c^*\right) + \sigma\phi(c^*)\end{aligned}$$

as stated in property 2.

For property 3, first observe that if we replace c with c^*, the answers will be invariant to μ and proportional to σ^2. Thus without loss of generality, we can make the calculation for $Z \sim \mathrm{N}\left(\mu, \sigma^2\right)$ instead of X. We calculate that

$$\begin{aligned}\mathbb{E}\left[Z^{*2}\right] &= c^{*2}\mathbb{E}\left[\mathbb{1}\left\{Z > c^*\right\}\right] + \mathbb{E}\left[Z^2\mathbb{1}\left\{Z \le c^*\right\}\right] \\ &= c^{*2}\left(1 - \Phi\left(c^*\right)\right) + \Phi(c^*) - c^*\phi(c^*).\end{aligned}$$

Thus

$$\begin{aligned}\operatorname{var}\left[Z^*\right] &= \mathbb{E}\left[Z^{*2}\right] - \left(\mathbb{E}\left[Z^*\right]\right)^2 \\ &= c^{*2}\left(1 - \Phi\left(c^*\right)\right) + \Phi(c^*) - c^*\phi(c^*) - \left(c^*\left(1 - \Phi\left(c^*\right)\right) - \phi(c^*)\right)^2 \\ &= 1 + \left(c^{*2} - 1\right)\left(1 - \Phi\left(c^*\right)\right) - c^*\phi(c^*) - \left(c^*\left(1 - \Phi\left(c^*\right)\right) - \phi(c^*)\right)^2\end{aligned}$$

which is property 3. Property 4 follows by a similar calculation. ■

Proof of Theorem 5.13 The fact that $\operatorname{cov}\left(X, Y\right) = 0$ means that Σ is block diagonal:

$$\Sigma = \begin{bmatrix} \Sigma_X & 0 \\ 0 & \Sigma_Y \end{bmatrix}.$$

Thus, $\Sigma^{-1} = \operatorname{diag}\left\{\Sigma_X^{-1}, \Sigma_Y^{-1}\right\}$, and $\det(\Sigma) = \det(\Sigma_X)\det(\Sigma_Y)$. Hence

$$\begin{aligned}f(x,y) &= \frac{1}{(2\pi)^{m/2}(\det(\Sigma_X)\det(\Sigma_Y))^{1/2}}\exp\left(-\frac{(x - \mu_X, y - \mu_Y)'\operatorname{diag}\left\{\Sigma_X^{-1}, \Sigma_Y^{-1}\right\}(x - \mu_X, y - \mu_Y)}{2}\right) \\ &= \frac{1}{(2\pi)^{m_x/2}(\det(\Sigma_X))^{1/2}}\exp\left(-\frac{(x - \mu_X)'\Sigma_X^{-1}(x - \mu_Y)}{2}\right) \\ &\quad \times \frac{1}{(2\pi)^{m_y/2}(\det(\Sigma_Y))^{1/2}}\exp\left(-\frac{(y - \mu_Y)'\Sigma_Y^{-1}(y - \mu_Y)}{2}\right)\end{aligned}$$

which is the product of two multivariate normal densities. Since the joint PDF factors, the random vectors X and Y are independent. ■

Proof of Theorem 5.15 The MGF of Y is

$$\begin{aligned}
\mathbb{E}\left(\exp(t'Y)\right) &= \mathbb{E}\left[\exp(t'(\boldsymbol{a}+\boldsymbol{B}X))\right] \\
&= \exp\left(t'\boldsymbol{a}\right)\mathbb{E}\left[\exp(t'\boldsymbol{B}X)\right] \\
&= \exp\left(t'\boldsymbol{a}\right)\mathbb{E}\left[\exp\left(t'\boldsymbol{B}\mu+\frac{1}{2}t'\boldsymbol{B}\Sigma\boldsymbol{B}'t\right)\right] \\
&= \exp\left(t'\left(\boldsymbol{a}+\boldsymbol{B}\mu\right)+\frac{1}{2}t'\boldsymbol{B}\Sigma\boldsymbol{B}'t\right)
\end{aligned}$$

which is the MGF of $\mathrm{N}\left(\boldsymbol{a}+\boldsymbol{B}\mu, \boldsymbol{B}\Sigma\boldsymbol{B}'\right)$. ■

Proof of Theorem 5.16 By Theorem 5.15 and matrix multiplication, we have

$$\begin{bmatrix} \boldsymbol{I} & -\Sigma_{YX}\Sigma_{XX}^{-1} \\ 0 & \boldsymbol{I} \end{bmatrix}\begin{pmatrix} Y \\ X \end{pmatrix} \sim \mathrm{N}\left(\begin{pmatrix} \mu_Y-\Sigma_{YX}\Sigma_{XX}^{-1}\mu_X \\ \mu_X \end{pmatrix}, \begin{pmatrix} \Sigma_{YY}-\Sigma_{YX}\Sigma_{XX}^{-1}\Sigma_{XY} & 0 \\ 0 & \Sigma_{XX} \end{pmatrix}\right).$$

The zero covariance shows that $Y-\Sigma_{YX}\Sigma_{XX}^{-1}X$ is independent of X and is distributed as

$$\mathrm{N}\left(\mu_Y-\Sigma_{YX}\Sigma_{XX}^{-1}\mu_X, \Sigma_{YY}-\Sigma_{YX}\Sigma_{XX}^{-1}\Sigma_{XY}\right).$$

This implies

$$Y \mid X \sim \mathrm{N}\left(\mu_Y+\Sigma_{YX}\Sigma_{XX}^{-1}\left(X-\mu_X\right), \Sigma_{YY}-\Sigma_{YX}\Sigma_{XX}^{-1}\Sigma_{XY}\right)$$

as stated. The result for $X \mid Y$ is found by symmetry. ■

Proof of Theorem 5.18 The MGF of $Z'Z$ is

$$\begin{aligned}
\mathbb{E}\left[\exp\left(tZ'Z\right)\right] &= \int_{\mathbb{R}^r}\exp\left(tx'x\right)\frac{1}{(2\pi)^{r/2}}\exp\left(-\frac{x'x}{2}\right)dx \\
&= \int_{\mathbb{R}^r}\frac{1}{(2\pi)^{r/2}}\exp\left(-\frac{x'x}{2}(1-2t)\right)dx \\
&= (1-2t)^{-r/2}\int_{\mathbb{R}^r}\frac{1}{(2\pi)^{r/2}}\exp\left(-\frac{u'u}{2}\right)du \\
&= (1-2t)^{-r/2}.
\end{aligned} \tag{5.2}$$

The third equality uses the change of variables $u=(1-2t)^{1/2}x$, and the final equality is the normal probability integral. Equation (5.2) equals the MGF of χ_r^2 from Theorem 3.2. Thus $Z'Z \sim \chi_r^2$. ■

Proof of Theorem 5.19 The fact that $\boldsymbol{A}>0$ means that we can write $\boldsymbol{A}=\boldsymbol{C}\boldsymbol{C}'$, where $\boldsymbol{C}$ is nonsingular (see Section A.11). Then $\boldsymbol{A}^{-1}=\boldsymbol{C}^{-1\prime}\boldsymbol{C}^{-1}$, and by Theorem 5.15,

$$Z=\boldsymbol{C}^{-1}X \sim \mathrm{N}\left(0, \boldsymbol{C}^{-1}\boldsymbol{A}\boldsymbol{C}^{-1\prime}\right)=\mathrm{N}\left(0, \boldsymbol{C}^{-1}\boldsymbol{C}\boldsymbol{C}'\boldsymbol{C}^{-1\prime}\right)=\mathrm{N}\left(0, \boldsymbol{I}_r\right).$$

Thus by Theorem 5.18,

$$X'\boldsymbol{A}^{-1}X=X'\boldsymbol{C}^{-1\prime}\boldsymbol{C}^{-1}X=Z'Z \sim \chi_r^2\left(\mu^{*\prime}\mu^*\right).$$

Since

$$\mu^{*\prime}\mu^*=\mu'\boldsymbol{C}^{-1\prime}\boldsymbol{C}^{-1}\mu=\mu'\boldsymbol{A}^{-1}\mu=\lambda$$

this equals $\chi_r^2(\lambda)$, as claimed. ■

Proof of Theorem 5.20 Using the law of iterated expectations (Theorem 4.13), the distribution of $Z/\sqrt{Q/r}$ is

$$
\begin{aligned}
F(x) &= \mathbb{P}\left[\frac{Z}{\sqrt{Q/r}} \le x\right] \\
&= \mathbb{E}\left[\mathbb{1}\left\{Z \le x\sqrt{\frac{Q}{r}}\right\}\right] \\
&= \mathbb{E}\left[\mathbb{P}\left[Z \le x\sqrt{\frac{Q}{r}} \,\middle|\, Q\right]\right] \\
&= \mathbb{E}\left[\Phi\left(x\sqrt{\frac{Q}{r}}\right)\right].
\end{aligned}
$$

The density is the derivative

$$
\begin{aligned}
f(x) &= \frac{d}{dx}\mathbb{E}\left[\Phi\left(x\sqrt{\frac{Q}{r}}\right)\right] \\
&= \mathbb{E}\left[\frac{d}{dx}\Phi\left(x\sqrt{\frac{Q}{r}}\right)\right] \\
&= \mathbb{E}\left[\phi\left(x\sqrt{\frac{Q}{r}}\right)\sqrt{\frac{Q}{r}}\right] \\
&= \int_0^\infty \left(\frac{1}{\sqrt{2\pi}}\exp\left(-\frac{qx^2}{2r}\right)\right)\sqrt{\frac{q}{r}}\left(\frac{1}{\Gamma\left(\frac{r}{2}\right)2^{r/2}}q^{r/2-1}\exp\left(-q/2\right)\right)dq \\
&= \frac{\Gamma\left(\frac{r+1}{2}\right)}{\sqrt{r\pi}\,\Gamma\left(\frac{r}{2}\right)}\left(1+\frac{x^2}{r}\right)^{-\left(\frac{r+1}{2}\right)}
\end{aligned}
$$

using Theorem A.28.3. This is the student t density of equation (3.1). ■

Proof of Theorem 5.21 Let $f_m(u)$ be the χ^2_m density. By a similar argument as in the proof of Theorem 5.20, Q_m/Q_r has the density function

$$
\begin{aligned}
f_S(s) &= \mathbb{E}\left[f_m\left(sQ_r\right)Q_r\right] \\
&= \int_0^\infty f_m(sv)vf_r(v)dv \\
&= \frac{1}{2^{(m+r)/2}\Gamma\left(\frac{m}{2}\right)\Gamma\left(\frac{r}{2}\right)}\int_0^\infty (sv)^{m/2-1}e^{-sv/2}v^{r/2}e^{-v/2}dv \\
&= \frac{s^{m/2-1}}{2^{(m+r)/2}\Gamma\left(\frac{m}{2}\right)\Gamma\left(\frac{r}{2}\right)}\int_0^\infty v^{(m+r)/2-1}e^{-(s+1)v/2}dv \\
&= \frac{s^{m/2-1}}{\Gamma\left(\frac{m}{2}\right)\Gamma\left(\frac{r}{2}\right)(1+s)^{(m+r)/2}}\int_0^\infty t^{(m+r)/2-1}e^{-t}dt \\
&= \frac{s^{m/2-1}\Gamma\left(\frac{m+r}{2}\right)}{\Gamma\left(\frac{m}{2}\right)\Gamma\left(\frac{r}{2}\right)(1+s)^{(m+r)/2}}.
\end{aligned}
$$

The fifth equality make the change of variables $v = 2t/(1+s)$, and the sixth uses Definition A.20. This is the density of Q_m/Q_r.

To obtain the density of $(Q_m/m)\,/\,(Q_r/r)$, make the change of variables $x = sr/m$. Doing so, we obtain the density function (3.3). ■

Proof of Theorem 5.22 As in the proof of Theorem 5.18, we verify that the MGF of $Q = X'X$ when $X \sim \mathrm{N}(\mu, \boldsymbol{I}_r)$ is equal to the MGF of the density function (3.4).

First, we calculate the MGF of $Q = X'X$ when $X \sim \mathrm{N}(\mu, \boldsymbol{I}_r)$. Construct an orthonormal $r \times r$ matrix $\boldsymbol{H} = [\boldsymbol{h}_1, \boldsymbol{H}_2]$, whose first column equals $\boldsymbol{h}_1 = \mu\,(\mu'\mu)^{-1/2}$. Note that $\boldsymbol{h}_1'\mu = \lambda^{1/2}$ and $\boldsymbol{H}_2'\mu = \boldsymbol{0}$. Define $Z = \boldsymbol{H}'X \sim \mathrm{N}(\mu^*, \boldsymbol{I}_q)$, where

$$\boldsymbol{\mu}^* = \boldsymbol{H}'\boldsymbol{\mu} = \begin{pmatrix} \boldsymbol{h}_1'\mu \\ \boldsymbol{H}_2'\mu \end{pmatrix} = \begin{pmatrix} \lambda^{1/2} \\ 0 \end{pmatrix} \begin{matrix} 1 \\ r-1 \end{matrix}.$$

It follows that $Q = X'X = Z'Z = Z_1^2 + Z_2'Z_2$, where $Z_1 \sim \mathrm{N}\left(\lambda^{1/2}, 1\right)$ and $Z_2 \sim \mathrm{N}\left(0, \boldsymbol{I}_{r-1}\right)$ are independent. Notice that $Z_2'Z_2 \sim \chi^2_{r-1}$ and so has MGF $(1-2t)^{-(r-1)/2}$ by equation (3.6). The MGF of Z_1^2 is

$$\begin{aligned}
\mathbb{E}\left[\exp\left(tZ_1^2\right)\right] &= \int_{-\infty}^{\infty} \exp\left(tx^2\right) \frac{1}{\sqrt{2\pi}} \exp\left(-\frac{1}{2}\left(x-\sqrt{\lambda}\right)^2\right) dx \\
&= \int_{-\infty}^{\infty} \frac{1}{\sqrt{2\pi}} \exp\left(-\frac{1}{2}\left(x^2(1-2t) - 2x\sqrt{\lambda} + \lambda\right)\right) dx \\
&= (1-2t)^{-1/2} \exp\left(-\frac{\lambda}{2}\right) \int_{-\infty}^{\infty} \frac{1}{\sqrt{2\pi}} \exp\left(-\frac{1}{2}\left(u^2 - 2u\sqrt{\frac{\lambda}{1-2t}}\right)\right) du \\
&= (1-2t)^{-1/2} \exp\left(-\frac{\lambda t}{1-2t}\right) \int_{-\infty}^{\infty} \frac{1}{\sqrt{2\pi}} \exp\left(-\frac{1}{2}\left(u - \sqrt{\frac{\lambda}{1-2t}}\right)^2\right) du \\
&= (1-2t)^{-1/2} \exp\left(-\frac{\lambda t}{1-2t}\right)
\end{aligned}$$

where the third equality uses the change of variables $u = (1-2t)^{1/2}x$. Thus the MGF of $Q = Z_1^2 + Z_2'Z_2$ is

$$\begin{aligned}
\mathbb{E}\left[\exp(tQ)\right] &= \mathbb{E}\left[\exp\left(t\left(Z_1^2 + Z_2'Z_2\right)\right)\right] \\
&= \mathbb{E}\left[\exp\left(tZ_1^2\right)\right] \mathbb{E}\left[\exp\left(tZ_2'Z_2\right)\right] \\
&= (1-2t)^{-r/2} \exp\left(-\frac{\lambda t}{1-2t}\right).
\end{aligned} \tag{5.3}$$

Second, we calculate the MGF of equation (3.4). It equals

$$\begin{aligned}
&\int_0^{\infty} \exp(tx) \sum_{i=0}^{\infty} \frac{e^{-\lambda/2}}{i!} \left(\frac{\lambda}{2}\right)^i f_{r+2i}(x)dx \\
&= \sum_{i=0}^{\infty} \frac{e^{-\lambda/2}}{i!} \left(\frac{\lambda}{2}\right)^i \int_0^{\infty} \exp(tx) f_{r+2i}(x)dx \\
&= \sum_{i=0}^{\infty} \frac{e^{-\lambda/2}}{i!} \left(\frac{\lambda}{2}\right)^i (1-2t)^{-(r+2i)/2}
\end{aligned}$$

$$= e^{-\lambda/2}(1-2t)^{-r/2}\sum_{i=0}^{\infty}\frac{1}{i!}\left(\frac{\lambda}{2(1-2t)}\right)^i$$

$$= e^{-\lambda/2}(1-2t)^{-r/2}\exp\left(\frac{\lambda}{2(1-2t)}\right)$$

$$= (1-2t)^{-r/2}\exp\left(\frac{\lambda t}{1-2t}\right) \tag{5.4}$$

where the second equality uses equation (3.6), and the fourth uses the definition of the exponential function. We can see that equation (5.3) equals equation (5.4), verifying that equation (3.4) is the density of Q, as stated. ■

Proof of Theorem 5.23 The fact that $\boldsymbol{A} > 0$ means that we can write $\boldsymbol{A} = \boldsymbol{C}\boldsymbol{C}'$, where $\boldsymbol{C}$ is nonsingular (see Section A.11). Then $\boldsymbol{A}^{-1} = \boldsymbol{C}^{-1\prime}\boldsymbol{C}^{-1}$, and by Theorem 5.15,

$$Y = \boldsymbol{C}^{-1}X \sim \mathrm{N}\left(\boldsymbol{C}^{-1}\mu, \boldsymbol{C}^{-1}\boldsymbol{A}\boldsymbol{C}^{-1\prime}\right) = \mathrm{N}\left(\boldsymbol{C}^{-1}\mu, \boldsymbol{C}^{-1}\boldsymbol{C}\boldsymbol{C}'\boldsymbol{C}^{-1\prime}\right) = \mathrm{N}\left(\mu^*, \boldsymbol{I}_r\right)$$

where $\mu^* = \boldsymbol{C}^{-1}\mu$. Thus by Theorem 5.22,

$$X'\boldsymbol{A}^{-1}X = X'\boldsymbol{C}^{-1\prime}\boldsymbol{C}^{-1}X = Y'Y \sim \chi_r^2\left(\mu^{*\prime}\mu^*\right).$$

Since

$$\mu^{*\prime}\mu^* = \mu'\boldsymbol{C}^{-1\prime}\boldsymbol{C}^{-1}\mu = \mu'\boldsymbol{A}^{-1}\mu = \lambda,$$

this equals $\chi_r^2(\lambda)$, as claimed. ■

5.12 EXERCISES

Exercise 5.1 Verify that $\int_{-\infty}^{\infty}\phi(z)dz = 1$. Use a change of variables and the Gaussian integral (Theorem A.27).

Exercise 5.2 For the standard normal density $\phi(x)$, show that $\phi'(x) = -x\phi(x)$.

Exercise 5.3 Use Exercise 5.2 and integration by parts to show that $\mathbb{E}\left[Z^2\right] = 1$ for $Z \sim \mathrm{N}(0,1)$.

Exercise 5.4 Use Exercises 5.2 and 5.3 plus integration by parts to show that $\mathbb{E}\left[Z^4\right] = 3$ for $Z \sim \mathrm{N}(0,1)$.

Exercise 5.5 Show that the MGF of $Z \sim \mathrm{N}(0,1)$ is $m(t) = \mathbb{E}\left[\exp(tZ)\right] = \exp\left(t^2/2\right)$.

Exercise 5.6 Show that the MGF of $X \sim \mathrm{N}\left(\mu, \sigma^2\right)$ is $m(t) = \mathbb{E}\left[\exp(tX)\right] = \exp\left(t\mu + t^2\sigma^2/2\right)$.
Hint: Write $X = \mu + \sigma Z$.

Exercise 5.7 Use the MGF from Exercise 5.5 to verify that $\mathbb{E}\left[Z^2\right] = m''(0) = 1$, and $\mathbb{E}\left[Z^4\right] = m^{(4)}(0) = 3$.

Exercise 5.8 Find the convolution of the normal density $\phi(x)$ with itself, $\int_{-\infty}^{\infty}\phi(x)\phi(y-x)dx$. Show that it can be written as a normal density.

Exercise 5.9 Show that if T is distributed student t with $r > 2$ degrees of freedom, then $\text{var}[T] = \frac{r}{r-2}$.
Hint: Use Theorems 3.3 and 5.20.

Exercise 5.10 Write the multivariate $\text{N}(0, \boldsymbol{I}_k)$ density as the product of $\text{N}(0,1)$ density functions. That is, show that

$$\frac{1}{(2\pi)^{k/2}} \exp\left(-\frac{x'x}{2}\right) = \phi(x_1)\cdots\phi(x_k).$$

Exercise 5.11 Show that the MGF of $Z \in \mathbb{R}^m$ is $\mathbb{E}\left[\exp\left(t'Z\right)\right] = \exp\left(\frac{1}{2}t't\right)$ for $t \in \mathbb{R}^m$.
Hint: Use Exercise 5.5 and the fact that the elements of Z are independent.

Exercise 5.12 Show that the MGF of $X \sim \text{N}(\mu, \Sigma) \in \mathbb{R}^m$ is

$$M(t) = \mathbb{E}\left[\exp\left(t'X\right)\right] = \exp\left(t'\mu + \frac{1}{2}t'\Sigma t\right).$$

Hint: Write $X = \mu + \Sigma^{1/2}Z$.

Exercise 5.13 Show that the characteristic function of $X \sim \text{N}(\mu, \Sigma) \in \mathbb{R}^m$ is

$$C(t) = \mathbb{E}\left[\exp\left(it'X\right)\right] = \exp\left(i\mu't - \frac{1}{2}t'\Sigma t\right)$$

for $t \in \mathbb{R}^m$.
Hint: Start with $m = 1$. Establish $\mathbb{E}\left[\exp\left(itZ\right)\right] = \exp\left(-\frac{1}{2}t^2\right)$ by integration. Then generalize to $X \sim \text{N}(\mu, \Sigma)$ for $t \in \mathbb{R}^m$ using the same steps as in Exercises 5.11 and 5.12.

Exercise 5.14 A random variable is $Y = \sqrt{Q}e$, where $e \sim \text{N}(0,1)$, $Q \sim \chi_1^2$, and e and Q are independent. It will be helpful to know that $\mathbb{E}[e] = 0$, $\mathbb{E}\left[e^2\right] = 1$, $\mathbb{E}\left[e^3\right] = 0$, $\mathbb{E}\left[e^4\right] = 3$, $\mathbb{E}[Q] = 1$, and $\text{var}[Q] = 2$. Find

(a) $\mathbb{E}[Y]$.
(b) $\mathbb{E}\left[Y^2\right]$.
(c) $\mathbb{E}\left[Y^3\right]$.
(d) $\mathbb{E}\left[Y^4\right]$.
(e) Compare these four moments with those of $\text{N}(0,1)$. What is the difference? (Be explicit.)

Exercise 5.15 Let $X = \sum_{i=1}^n a_i e_i^2$, where the a_i are constants and the e_i are independent $\text{N}(0,1)$. Find the following:

(a) $\mathbb{E}[X]$.
(b) $\text{var}[X]$.

Exercise 5.16 Show that if $Q \sim \chi_k^2(\lambda)$, then $\mathbb{E}[Q] = k + \lambda$.

Exercise 5.17 Suppose X_i are independent $\text{N}\left(\mu_i, \sigma_i^2\right)$. Find the distribution of the weighted sum $\sum_{i=1}^n w_i X_i$.

Exercise 5.18 Show that if $e \sim \text{N}\left(0, \boldsymbol{I}_n\sigma^2\right)$ and $\boldsymbol{H}'\boldsymbol{H} = \boldsymbol{I}_n$, then $u = \boldsymbol{H}'e \sim \text{N}\left(0, \boldsymbol{I}_n\sigma^2\right)$.

Exercise 5.19 Show that if $e \sim \text{N}(0, \Sigma)$ and $\Sigma = \boldsymbol{A}\boldsymbol{A}'$, then $u = \boldsymbol{A}^{-1}e \sim \text{N}(0, \boldsymbol{I}_n)$.

CHAPTER 6
SAMPLING

6.1 INTRODUCTION

We now switch from probability to statistics. Statistics is concerned with inference on parameters from data.

6.2 SAMPLES

In probability theory, we studied the properties of random vectors X. In statistical theory, we extend our study to include collections of such random vectors. The simplest such collection is when random vectors are mutually independent and have the same distribution.

Definition 6.1 The random vectors $\{X_1, \dots, X_n\}$ are **independent and identically distributed (i.i.d.)** if they are mutually independent with identical marginal distributions F.

By "independent", we mean that X_i is independent from X_j for $i \neq j$. By "identical distributions", we mean that the vectors X_i and X_j have the same (joint) distribution $F(x)$.

A dataset is a collection of numbers, typically organized by observation. For example, an individual-level wage dataset might consist of information on individuals, each described as an **observation**, with measurements on each individual's earnings or wages, age, education, and other characteristics. A dataset such as this is typically called a **sample**. The reason for this label is that for statistical analysis, it is typical to view the observations as realizations of random variables obtained by random sampling.

Definition 6.2 A collection of random vectors $\{X_1, \dots, X_n\}$ is a **random sample** from the population F if the X_i are i.i.d. with distribution F.

This definition is somewhat repetitive, but it is merely saying that a random sample is an i.i.d. collection of random variables.

The distribution F is called the **population distribution**, or just the **population**. You can think of it as an infinite population or a mathematical abstraction. The number n is typically used to denote the sample size—the number of observations. Two metaphors are useful for understanding random sampling. The first is that there is an actual or potential population of N individuals, with N much larger than n. Random sampling is then equivalent to drawing a subset of n of these N individuals with equal probability. This metaphor is somewhat consistent with survey sampling. With this metaphor, the randomness and identical distribution property is created by the sampling design. The second useful metaphor is the concept of a **data generating process**. Imagine that there is process by which an observation is created—for example, a controlled

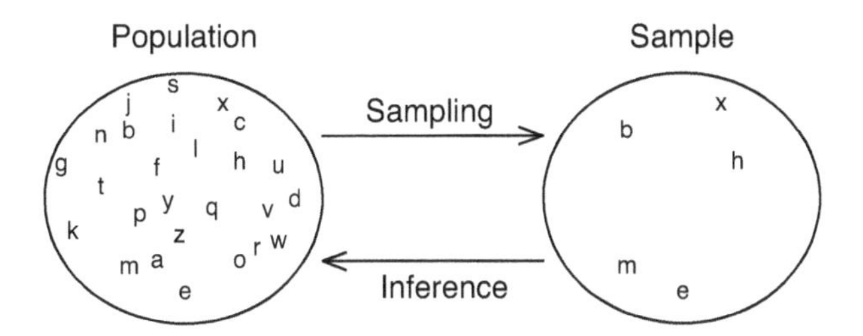

FIGURE 6.1 Sampling and inference

experiment, or alternatively an observational study—and this process is repeated n times. Here the population is the probability model which generates the observations.

An important component of sampling is the number of observations.

Definition 6.3 The **sample size** n is the number of individuals in the sample.

In econometrics, the most common notation for the sample size is n. Other common choices include N and T. Sample sizes can range from 1 up to the millions. When we say that an observation is an "individual", it does not necessarily mean an individual person. In economic datasets, observations can be collected on households, corporations, production plants, machines, patents, goods, stores, countries, states, cities, villages, schools, classrooms, students, years, months, days, or other entities.

Typically, we will use X without the subscript to denote a random variable or vector with distribution F, X_i with a subscript to denote a random observation in the sample, and x_i or x to denote a specific or realized value.

Figure 6.1 illustrates the process of sampling. On the left is the population of letters "a" through "z". Five are selected at random and become the sample indicated on the right.

The problem of inference is to learn about the underlying process—the population distribution or data-generating process—by examining the observations. Figure 6.1 illustrates the problem of inference by the arrow near the bottom of the figure. Inference means that based on the sample (in this case the observations {b, e, h, m, x}), we want to infer properties of the originating population.

Economic datasets can be obtained by a variety of methods, including survey sampling, administrative records, direct observation, web scraping, field experiments, and laboratory experiments. By treating the observations as realizations from a random sample, we are using a specific probability structure to impose order on the data. In some cases, this assumption is reasonably valid; in other cases, less so. In general, how the sample is collected affects the inference method to be used as well as the properties of the estimators of the parameters. We focus on random sampling in this book, as it provides a solid understanding of statistical theory and inference. In econometric applications, it is common to see more complicated forms of sampling. These include: (1) stratified sampling, (2) clustered sampling, (3) panel data, (4) time series data, and (5) spatial data. All of these methods introduce some sort of dependence across observations which complicates inference. A large portion of advanced econometric theory concerns how to explicitly allow these forms of dependence in order to conduct valid inference.

We could also consider i.n.i.d. (independent, not necessarily identically distributed) settings to allow X_i to be drawn from heterogeneous distributions F_i, but for the most part, this approach only complicates the notation without providing meaningful insight.

Table 6.1
Observations from CPS dataset

Observation	Wage	Education	Observation	Wage	Education
1	37.93	18	11	21.63	18
2	40.87	18	12	11.09	16
3	14.18	13	13	10.00	13
4	16.83	16	14	31.73	14
5	33.17	16	15	11.06	12
6	29.81	18	16	18.75	16
7	54.62	16	17	27.35	14
8	43.08	18	18	24.04	16
9	14.42	12	19	36.06	18
10	14.90	16	20	23.08	16

A useful fact implied by probability theory is that transformations of i.i.d. variables are also i.i.d. That is, if $\{X_i : i = 1, \ldots, n\}$ are i.i.d. and if $Y_i = g(X_i)$ for some function $g(x) : \mathbb{R}^m \to \mathbb{R}^q$, then $\{Y_i : i = 1, \ldots, n\}$ are i.i.d.

6.3 EMPIRICAL ILLUSTRATION

I illustrate the concept of sampling with a popular economics dataset. This is the March 2009 Current Population Survey, which has extensive information on the U.S. population. For this illustration, let us use the subsample of married (spouse present) Black female wage earners with 12 years of potential work experience. This subsample has 20 observations. (This subsample was selected mostly to ensure a small sample.)

Table 6.1 displays the observations. Each row is an individual observation, which consists of the data for an individual person. The columns correspond to the variables (measurements) for the individuals. The second column is the reported wage (total annual earnings divided by hours worked). The third column is years of education.

There are two basic ways to view the numbers in Table 6.1. First, we could view them as numbers. These are the wages and education levels of these 20 individuals. Taking this view, we can learn about the wages/education of these specific people—this specific population—but nothing else. Alternatively, we could view the numbers as the result of a random sample from a large population. Taking this view, we can learn about the wages/education of the population. From this viewpoint, we are not interested per se in the wages/education of these specific people; instead we are interested in the wages/education of the population and use the information on these 20 individuals as a window to learn about the broader group. This second view is the statistical view.

As we will learn, it turns out that it is difficult to make inferences about the general population from just 20 observations. Larger samples are needed for reasonable precision. I selected a sample with 20 observations for numerical simplicity and pedagogical value.

We use this example through the next several chapters to illustrate estimation and inference methods.

6.4 STATISTICS, PARAMETERS, AND ESTIMATORS

The goal of statistics is to learn about features of the population. These features are called **parameters**. It is conventional to denote parameters by Greek letters, such as μ, β or θ, though Latin letters can be used as well.

Definition 6.4 A **parameter** θ is any function of the population F.

For example, the population expectation $\mu = \mathbb{E}[X]$ is the first moment of F.
Statistics are constructed from samples.

Definition 6.5 A **statistic** is a function of the sample $\{X_i : 1, \ldots, n\}$.

Recall that there is a distinction between random variables and their realizations. Similarly, there is a distinction between a statistic as a function of a random sample (which is therefore a random variable as well) and a statistic as a function of the realized sample (which is a realized value). When we treat a statistic as random, we are viewing it as a function of a sample of random variables. When we treat it as a realized value, we are viewing it as a function of a set of realized values. One way of grasping the distinction is to think of "before viewing the data" and "after viewing the data". When we think about a statistic "before viewing", we do not know what value it will take. From this vantage point, it is unknown and random. After viewing the data and specifically after computing and viewing the statistic, the latter is a specific number and is therefore a realization. It is what it is, and it is not changing. The randomness is the process by which the data were generated, and the understanding is that if this process were repeated, the sample would be different, and therefore the specific realization would be different.

Some statistics are used to estimate parameters.

Definition 6.6 An **estimator** $\widehat{\theta}$ for a parameter θ is a statistic intended as a guess about θ.

We sometimes call $\widehat{\theta}$ a **point estimator** to distinguish it from an interval estimator. Notice that I define an estimator with the vague phrase "intended as a guess". This vagueness is intentional. At this point, I want to be quite broad and include a wide range of possible estimators.

It is conventional to use the notation $\widehat{\theta}$ ("theta hat") to denote an estimator of the parameter θ. This convention makes it relatively straightforward to understand the intention.

We will sometimes call $\widehat{\theta}$ an "estimator" and sometimes an "estimate". We call $\widehat{\theta}$ an **estimator** when we view it as a function of the random observations and thus as a random variable itself. Since $\widehat{\theta}$ is random, we can use probability theory to calculate its distribution. We call $\widehat{\theta}$ an **estimate** when it is a specific value (or realized value) calculated for a specific sample. Thus when developing a theory of estimation, we will typically refer to $\widehat{\theta}$ as the estimator of θ, but in a specific application will refer to $\widehat{\theta}$ as the estimate of θ.

A standard way to construct an estimator is by using the **analog principle**. The idea is to express the parameter θ as a function of the population F, and then express the estimator $\widehat{\theta}$ as the analog function in the sample. This principle may become more clear by examining the sample mean.

6.5 SAMPLE MEAN

One of the most fundamental parameters in statistics is the population mean $\mu = \mathbb{E}[X]$. Through transformations, many parameters of interest can be written in terms of population means. The population mean is the average value in the (infinitely large) population.

To estimate μ by the analog principle, we apply the same function to the sample. Since μ is the average value of X in the population, the analog estimator is the average value in the sample. This is the sample mean.

Definition 6.7 The **sample mean** is $\overline{X}_n = \frac{1}{n}\sum_{i=1}^{n} X_i$.

It is conventional to denote a sample average by the notation $\overline{X}_n$ (pronounced "X bar"). Sometimes it is written as $\overline{X}$. As discussed, the sample mean $\widehat{\mu} = \overline{X}_n$ is the standard estimator of the population mean μ. We can use either notation—$\overline{X}_n$ or $\widehat{\mu}$—to denote this estimator.

The sample mean is a statistic, since it is a function of the sample. It is random, as discussed in Section 6.4. Thus the sample mean will not have the same value when computed on different random samples.

To illustrate, consider the random sample presented in Table 6.1. Two variables are listed in the table: wage and education. The population means of these variables are μ_{wage} and $\mu_{\text{education}}$. The "population" in this case is the conceptually infinitely large population of married (spouse present) Black female wage earners with 12 years of potential work experience in the United States in 2009. There is no obvious reason that this population expectation should be the same for an arbitrary different population, such as wage earners in Norway, Cambodia, or on the planet Neptune. But they are likely to be similar to population means for similar populations, such as if we look at the similar sub-group with 15 years of work experience (instead of 12 years). It is constructive to understand the population for which the parameter applies, and to understand which generalizations are reasonable and which less reasonable.

For this sample, the sample means are

$$\overline{X}_{\text{wage}} = 25.73$$

$$\overline{X}_{\text{education}} = 15.7.$$

Thus our estimate—based on this information—of the average wage for this population is about $26 dollars per hour, and our estimate of the average education is about 16 years.

6.6 EXPECTED VALUE OF TRANSFORMATIONS

Many parameters can be written as the expectation of a transformation of X. For example, the second moment is $\mu_2' = \mathbb{E}\left[X^2\right]$. In general, we can write this class of parameters as

$$\theta = \mathbb{E}\left[g(X)\right]$$

for some function $g(x)$. The parameter is finite if $\mathbb{E}\left|g(X)\right| < \infty$. The random variable $Y = g(X)$ is a transformation of the random variable X. The parameter θ is the expectation of Y.

The analog estimator of θ is the sample mean of $g(X_i)$. This is

$$\widehat{\theta} = \frac{1}{n}\sum_{i=1}^{n} g(X_i).$$

It is the same as the sample mean of $Y_i = g(X_i)$.

For example, the analog estimator of μ_2' is

$$\widehat{\mu}_2' = \frac{1}{n}\sum_{i=1}^{n} X_i^2.$$

As another example, consider the parameter $\theta = \mu_{\log(X)} = \mathbb{E}\left[\log(X)\right]$. The analog estimator is

$$\widehat{\theta} = \overline{X}_{\log(X)} = \frac{1}{n}\sum_{i=1}^{n} \log(X_i).$$

It is very common in economic applications to work with variables in logarithmic scale. This is done for a number of reasons, including that it makes differences proportionate rather than linear. In our example dataset, the estimate is

$$\overline{X}_{\log(\text{wage})} = 3.13.$$

It is important to recognize that this is not $\log\left(\overline{X}\right)$. Indeed, Jensen's inequality (Theorem 2.12) shows that $\overline{X}_{\log(\text{wage})} < \log\left(\overline{X}\right)$. In this application, $3.13 < 3.25 \simeq \log(25.73)$.

6.7 FUNCTIONS OF PARAMETERS

More generally, many parameters can be written as transformations of moments of the distribution. This class of parameters can be written as

$$\beta = h\left(\mathbb{E}\left[g(X)\right]\right)$$

where g and h are functions.

For example, the population variance of X is

$$\sigma^2 = \mathbb{E}\left[(X - \mathbb{E}\left[X\right])^2\right] = \mathbb{E}\left[X^2\right] - (\mathbb{E}\left[X\right])^2.$$

This is a function of $\mathbb{E}[X]$ and $\mathbb{E}\left[X^2\right]$. In this example, the function g takes the form $g(x) = (x^2, x)$, and h takes the form $h(a, b) = a - b^2$. The population standard deviation is

$$\sigma = \sqrt{\sigma^2} = \sqrt{\mathbb{E}\left[X^2\right] - (\mathbb{E}\left[X\right])^2}.$$

In this case, $h(a, b) = \left(a - b^2\right)^{1/2}$.

As another example, the geometric mean of X is

$$\beta = \exp\left(\mathbb{E}\left[\log(X)\right]\right).$$

In this example, the functions are $g(x) = \log(x)$ and $h(u) = \exp(u)$.

The **plug-in estimator** for $\beta = h\left(\mathbb{E}\left[g(X)\right]\right)$ is

$$\widehat{\beta} = h\left(\widehat{\theta}\right) = h\left(\frac{1}{n}\sum_{i=1}^{n} g(X_i)\right).$$

It is called a "plug-in estimator" because we "plug-in" the estimator $\widehat{\theta}$ into the formula $\beta = h(\theta)$.

For example, the plug-in estimator for σ^2 is

$$\widehat{\sigma}^2 = \frac{1}{n}\sum_{i=1}^{n} X_i^2 - \left(\frac{1}{n}\sum_{i=1}^{n} X_i\right)^2 = \frac{1}{n}\sum_{i=1}^{n}\left(X_i - \overline{X}_n\right)^2.$$

The plug-in estimator of the standard deviation σ is the square root of the sample variance:

$$\widehat{\sigma} = \sqrt{\widehat{\sigma}^2}.$$

The plug-in estimator of the geometric mean is

$$\widehat{\beta} = \exp\left(\frac{1}{n}\sum_{i=1}^{n} \log\left(X_i\right)\right).$$

With the tools of the sample mean and the plug-in estimator, we can construct estimators for any parameter that can be written as an explicit function of moments. This includes quite a broad class of parameters.

To illustrate, the sample standard deviations of the three variables discussed in Section 6.6 are as follows:

$$\widehat{\sigma}_{\text{wage}} = 12.1$$
$$\widehat{\sigma}_{\log(\text{wage})} = 0.490$$
$$\widehat{\sigma}_{\text{education}} = 2.00.$$

It is difficult to interpret these numbers without a further statistical understanding, but a rough rule is to examine the ratio of the mean to the standard deviation. In the case of the wage, this ratio is 2. This means that the spread of the wage distribution is of similar magnitude to the mean of the distribution. Thus, the distribution is spread out: wage rates are diffuse. In contrast, notice that for education, the ratio is about 8. So the years of education is less spread out relative to the mean. This should not be surprising, since most individuals receive at least 12 years of education.

As another example, the sample geometric mean of the wage distribution is

$$\widehat{\beta} = \exp\left(\overline{X}_{\log(\text{wage})}\right) = \exp(3.13) \simeq 22.9.$$

Note that the geometric mean is less than the arithmetic mean (in this example, $\overline{X}_{\text{wage}} = 25.73$). This inequality is a consequence of Jensen's inequality. For strongly skewed distributions (such as wages), the geometric mean is often similar to the median of the distribution, and the latter is a better measure of the "typical" value in the distribution than the arithmetic mean. In this 20-observation sample, the median is 23.55, which is indeed closer to the geometric mean than to the arithmetic mean.

6.8 SAMPLING DISTRIBUTION

Statistics are functions of the random sample and are therefore random variables themselves. Hence statistics have probability distributions of their own. We often call the distribution of a statistic the **sampling distribution**, because it is the distribution induced by samping. When a statistic $\widehat{\theta} = \widehat{\theta}(X_1, \ldots, X_n)$ is a function of an i.i.d. sample (and nothing else), then its distribution is determined by the distribution F of the observations, the sample size n, and the functional form of the statistic. The sampling distribution of $\widehat{\theta}$ is typically unknown, in part because the distribution F is unknown.

The goal of an estimator $\widehat{\theta}$ is to learn about the parameter θ. To make accurate inferences and to measure the accuracy of our measurements, we need to know something about its sampling distribution. Therefore a considerable effort in statistical theory is devoted to understanding the sampling distribution of estimators. We start by studying the distribution of the sample mean $\overline{X}_n$ due to its simplicity and its central importance in statistical methods. It turns out that the theory for many complicated estimators is based on linear approximations to sample means.

There are several approaches to understanding the distribution of $\overline{X}_n$. These include:

1. The exact bias and variance.
2. The exact distribution under normality.
3. The asymptotic distribution as $n \to \infty$.
4. An asymptotic expansion, which is a higher-order asymptotic approximation.
5. Bootstrap approximation.

We will explore the first four approaches in this textbook. In the companion volume, *Econometrics*, bootstrap approximations are presented.

6.9 ESTIMATION BIAS

Let's focus on $\overline{X}_n$ as an estimator of $\mu = \mathbb{E}[X]$ and see if we can find its exact bias. It will be helpful to fix notation. Define the population mean and variance of X as

$$\mu = \mathbb{E}[X]$$

and

$$\sigma^2 = \mathbb{E}\left[(X-\mu)^2\right].$$

We say that an estimator is **biased** if its sampling distribution is incorrectly centered. As we typically measure centering by expectation, bias is typically measured by the expectation of the sampling distribution.

Definition 6.8 The **bias** of an estimator $\widehat{\theta}$ of a parameter θ is

$$\text{bias}\left[\widehat{\theta}\right] = \mathbb{E}\left[\widehat{\theta}\right] - \theta.$$

We say that an estimator is **unbiased** if the bias is 0. However, an estimator could be unbiased for some population distributions F while being biased for others. To be precise with our definitions, it will be useful to introduce a class of distributions. Recall that the population distribution F is one specific distribution. Consider F as one member of a class $\mathscr{F}$ of distributions. For example, $\mathscr{F}$ could be the class of distributions with a finite mean.

Definition 6.9 An estimator $\widehat{\theta}$ of a parameter θ is **unbiased** in $\mathscr{F}$ if bias $\left[\widehat{\theta}\right] = 0$ for every $F \in \mathscr{F}$.

The role of the class $\mathscr{F}$ is that in many cases, an estimator will be unbiased under certain conditions (in a specific class of distributions) but not under other conditions. Hopefully this will become clear by the context.

Returning to the sample mean, let's calculate its expectation:

$$\mathbb{E}\left[\overline{X}_n\right] = \frac{1}{n}\sum_{i=1}^{n}\mathbb{E}[X_i] = \frac{1}{n}\sum_{i=1}^{n}\mu = \mu.$$

The first equality holds by linearity of expectations. The second equality is the assumption that the observations are identically distributed with mean $\mu = \mathbb{E}[X]$, and the third equality is an algebraic simplification. Thus the expectation of the sample mean is the population expectation, which that the former is unbiased. The only assumption needed (beyond i.i.d. sampling) is that the expectation is finite. Thus the class of distributions for which $\overline{X}_n$ is unbiased is the set of distributions with a finite expectation.

Theorem 6.1 $\overline{X}_n$ is unbiased for $\mu = \mathbb{E}[X]$ if $\mathbb{E}|X| < \infty$.

The sample mean is not the only unbiased estimator. For example, X_1 (the first observation) satisfies $\mathbb{E}[X_1] = \mu$, so it is also unbiased. In general, any weighted average

$$\frac{\sum_{i=1}^{n} w_i X_i}{\sum_{i=1}^{n} w_i}$$

with $w_i \geq 0$ is unbiased for μ.

Some estimators are biased. For example, $c\overline{X}_n$ with $c \neq 1$ has expectation $c\mu \neq \mu$ and is therefore biased if $\mu \neq 0$.

It is possible for an estimator to be unbiased for a set of distributions while being biased for other sets. Consider the estimator $\widetilde{\mu} = 0$. It is biased when $\mu \neq 0$ but is unbiased when $\mu = 0$. This may seem like a silly example, but it is useful to be clear about the contexts where an estimator is unbiased or biased.

Affine transformations preserve unbiasedness.

Theorem 6.2 If $\widehat{\theta}$ is an unbiased estimator of θ, then $\widehat{\beta} = a\widehat{\theta} + b$ is an unbiased estimator of $\beta = a\theta + b$.

Nonlinear transformations generally do not preserve unbiasedness. If $h(\theta)$ is nonlinear, then $\widehat{\beta} = h(\widehat{\theta})$ is generally not unbiased for $\beta = h(\theta)$. In certain cases, we can infer the direction of bias. Jensen's inequality (Theorem 2.9) states that if $h(u)$ is concave, then

$$\mathbb{E}\left[\widehat{\beta}\right] = \mathbb{E}\left[h\left(\widehat{\theta}\right)\right] \leq h\left(\mathbb{E}\left[\widehat{\theta}\right]\right) = h(\theta) = \beta$$

so $\widehat{\beta}$ is biased downwards. Similarly, if $h(u)$ is convex, then

$$\mathbb{E}\left[\widehat{\beta}\right] = \mathbb{E}\left[h\left(\widehat{\theta}\right)\right] \geq h\left(\mathbb{E}\left[\widehat{\theta}\right]\right) = h(\theta) = \beta$$

so $\widehat{\beta}$ is biased upwards.

As an example, suppose that we know that $\mu = \mathbb{E}[X] \geq 0$. A reasonable estimator for μ would be

$$\begin{aligned}\widehat{\mu} &= \begin{cases} \overline{X}_n & \text{if } \overline{X}_n \geq 0 \\ 0 & \text{otherwise} \end{cases} \\ &= h\left(\overline{X}_n\right)\end{aligned}$$

where $h(u) = u\mathbb{1}\{u > 0\}$. Since $h(u)$ is convex, we can deduce that $\widehat{\mu}$ is biased upwards: $\mathbb{E}[\widehat{\mu}] \geq \mu$.

This last example shows that "reasonable" and "unbiased" are not necessarily the same thing. While unbiasedness seems like a useful property for an estimator, it is only one of many desirable properties. In many cases, it may be okay for an estimator to possess some bias if it has other good compensating properties.

Let us return to the sample statistics shown in Table 6.1. The sample means of wages, log wages, and education are unbiased estimators for their corresponding population expectations. The other estimators (of variance, standard deviation, and geometric mean) are nonlinear funtions of sample moments, so are not necessarily unbiased estimators of their population counterparts without further study.

6.10 ESTIMATION VARIANCE

I mentioned earlier that the distribution of an estimator $\widehat{\theta}$ is called the "sampling distribution". An important feature of the sampling distribution is its variance. This is the key measure of the dispersion of its distribution. We often call the variance of the estimator the **sampling variance**. Understanding the sampling variance of estimators is one of the core components of the the theory of point estimation. This section shows that the sampling variance of the sample mean is a simple calculation.

Take the sample mean $\overline{X}_n$ and consider its variance. The key is the following equality. The assumption that X_i are mutually independent implies that they are uncorrelated. Uncorrelatedness implies that the variance of the sum equals the sum of the variances:

$$\operatorname{var}\left[\sum_{i=1}^{n} X_i\right] = \sum_{i=1}^{n} \operatorname{var}[X_i].$$

See Exercise 4.12. The assumption of identical distributions implies that $\operatorname{var}[X_i] = \sigma^2$ are identical across i. Hence the above expression equals $n\sigma^2$. Thus

$$\operatorname{var}\left[\overline{X}_n\right] = \operatorname{var}\left[\frac{1}{n}\sum_{i=1}^{n} X_i\right] = \frac{1}{n^2}\operatorname{var}\left[\sum_{i=1}^{n} X_i\right] = \frac{\sigma^2}{n}.$$

Theorem 6.3 If $\mathbb{E}\left[X^2\right] < \infty$, then $\operatorname{var}\left[\overline{X}_n\right] = \sigma^2/n$.

It is interesting that the variance of the sample mean depends on the population variance σ^2 and sample size n. In particular, the variance of the sample mean is declining with sample size. Thus $\overline{X}_n$ as an estimator of μ is more accurate when n is large than when n is small.

Consider affine transformations of statistics, such as $\widehat{\beta} = a\overline{X}_n + b$.

Theorem 6.4 If $\widehat{\beta} = a\widehat{\theta} + b$, then $\operatorname{var}\left[\widehat{\beta}\right] = a^2 \operatorname{var}\left[\widehat{\theta}\right]$.

For nonlinear transformations $h(\overline{X}_n)$, the exact variance is typically not available.

6.11 MEAN SQUARED ERROR

A standard measure of accuracy is the mean squared error (MSE).

Definition 6.10 The **mean squared error** of an estimator $\widehat{\theta}$ for θ is

$$\operatorname{mse}\left[\widehat{\theta}\right] = \mathbb{E}\left[(\widehat{\theta} - \theta)^2\right].$$

By expanding the square, we find that

$$\begin{aligned}
\operatorname{mse}\left[\widehat{\theta}\right] &= \mathbb{E}\left[(\widehat{\theta} - \theta)^2\right] \\
&= \mathbb{E}\left[\left(\widehat{\theta} - \mathbb{E}\left[\widehat{\theta}\right] + \mathbb{E}\left[\widehat{\theta}\right] - \theta\right)^2\right] \\
&= \mathbb{E}\left[\left(\widehat{\theta} - \mathbb{E}\left[\widehat{\theta}\right]\right)^2\right] + 2\mathbb{E}\left[\widehat{\theta} - \mathbb{E}\left[\widehat{\theta}\right]\right]\left(\mathbb{E}\left[\widehat{\theta}\right] - \theta\right) + \left(\mathbb{E}\left[\widehat{\theta}\right] - \theta\right)^2 \\
&= \operatorname{var}\left[\widehat{\theta}\right] + \left(\operatorname{bias}\left[\widehat{\theta}\right]\right)^2.
\end{aligned}$$

Thus the MSE is the variance plus the squared bias. The MSE as a measure of accuracy combines the variance and bias.

Theorem 6.5 For any estimator with a finite variance, $\operatorname{mse}\left[\widehat{\theta}\right] = \operatorname{var}\left[\widehat{\theta}\right] + \left(\operatorname{bias}\left[\widehat{\theta}\right]\right)^2$.

When an estimator is unbiased, its MSE equals its variance. Since $\overline{X}_n$ is unbiased, this means

$$\operatorname{mse}\left[\overline{X}_n\right]=\operatorname{var}\left[\overline{X}_n\right]=\frac{\sigma^2}{n}.$$

6.12 BEST UNBIASED ESTIMATOR

In this section, we derive the best linear unbiased estimator (BLUE) of μ. By "best" we mean "lowest variance". By "linear" we mean "a linear function of X_i". This class of estimators is $\widetilde{\mu}=\sum_{i=1}^n w_iX_i$, where the weights w_i are freely selected. Unbiasedness requires

$$\mathbb{E}\left[\widetilde{\mu}\right]=\mathbb{E}\left[\sum_{i=1}^n w_iX_i\right]=\sum_{i=1}^n w_i\mathbb{E}\left[X_i\right]=\sum_{i=1}^n w_i\mu=\mu$$

which holds if and only if $\sum_{i=1}^n w_i=1$. The variance of $\widetilde{\mu}$ is

$$\operatorname{var}\left[\widetilde{\mu}\right]=\operatorname{var}\left[\sum_{i=1}^n w_iX_i\right]=\sigma^2\sum_{i=1}^n w_i^2.$$

Hence the best (minimum variance) estimator is $\widetilde{\mu}=\sum_{i=1}^n w_iX_i$ with the weights selected to minimize $\sum_{i=1}^n w_i^2$ subject to the restriction $\sum_{i=1}^n w_i=1$.

This can be solved by Lagrangian methods. The problem is to minimize

$$L(w_1,\ldots,w_n)=\sum_{i=1}^n w_i^2-\lambda\left(\sum_{i=1}^n w_i-1\right).$$

The first-order condition with respect to w_i is $2w_i-\lambda=0$, or $w_i=\lambda/2$. This condition implies that the optimal weights are all identical, which implies that they must satisfy $w_i=1/n$ in order to satisfy $\sum_{i=1}^n w_i=1$. We conclude that the BLUE estimator is

$$\sum_{i=1}^n \frac{1}{n}X_i=\overline{X}_n.$$

We have shown that the sample mean is BLUE.

Theorem 6.6 Best Linear Unbiased Estimator (BLUE). If $\sigma^2<\infty$, the sample mean $\overline{X}_n$ has the lowest variance among all linear unbiased estimators of μ.

This BLUE theorem points out that within the class of weighted averages, there is no point in considering an estimator other than the sample mean. However, the result is not deep, because the class of linear estimators is restrictive.

In Section 11.6, Theorem 11.2 will establish a stronger result. We state it here for contrast.

Theorem 6.7 Best Unbiased Estimator. If $\sigma^2<\infty$, the sample mean $\overline{X}_n$ has the lowest variance among all unbiased estimators of μ.

This statement is stronger than Theorem 6.6, because it does not impose a restriction to linear estimators. Because of Theorem 6.7, we can call $\overline{X}_n$ the **best unbiased estimator** of μ.

6.13 ESTIMATION OF VARIANCE

Recall the plug-in estimator of σ^2:

$$\widehat{\sigma}^2 = \frac{1}{n}\sum_{i=1}^{n} X_i^2 - \left(\frac{1}{n}\sum_{i=1}^{n} X_i\right)^2 = \frac{1}{n}\sum_{i=1}^{n} \left(X_i - \overline{X}_n\right)^2.$$

We can calculate whether $\widehat{\sigma}^2$ is unbiased for σ^2. It may be useful to start by considering an idealized estimator. Define

$$\widetilde{\sigma}^2 = \frac{1}{n}\sum_{i=1}^{n} (X_i - \mu)^2$$

which is the estimator we would use if μ were known. This is a sample average of the i.i.d. variables $(X_i - \mu)^2$. Hence it has expectation

$$\mathbb{E}\left[\widetilde{\sigma}^2\right] = \mathbb{E}\left[(X_i - \mu)^2\right] = \sigma^2.$$

Thus $\widetilde{\sigma}^2$ is unbiased for σ^2.

Next, I introduce an algebraic trick. Rewrite the expression for $\widehat{\sigma}^2$ as

$$\begin{aligned}\widehat{\sigma}^2 &= \frac{1}{n}\sum_{i=1}^{n} (X_i - \mu)^2 - \left(\frac{1}{n}\sum_{i=1}^{n} (X_i - \mu)\right)^2 \qquad (6.1)\\ &= \widetilde{\sigma}^2 - (\overline{X}_n - \mu)^2\end{aligned}$$

(see Exercise 6.8). Equation (6.1) shows that the sample variance estimator equals the idealized estimator minus an adjustment for estimation of μ. We can see immediately that $\widehat{\sigma}^2$ is biased towards 0. Furthermore, we can calculate the exact bias. Since $\mathbb{E}\left[\widetilde{\sigma}^2\right] = \sigma^2$ and $\mathbb{E}\left[(\overline{X}_n - \mu)^2\right] = \sigma^2/n$, we find the following:

$$\mathbb{E}\left[\widehat{\sigma}^2\right] = \sigma^2 - \frac{\sigma^2}{n} = \left(1 - \frac{1}{n}\right)\sigma^2 < \sigma^2.$$

Theorem 6.8 If $\mathbb{E}\left[X^2\right] < \infty$, then $\mathbb{E}\left[\widehat{\sigma}^2\right] = \left(1 - \frac{1}{n}\right)\sigma^2$.

One intuition for the downward bias is that $\widehat{\sigma}^2$ centers the observations X_i not at the true mean but at the sample mean $\overline{X}_n$. This implies that the sample-centered variables $X_i - \overline{X}_n$ have less variation than the ideally centered variables $X_i - \mu$.

Since $\widehat{\sigma}^2$ is proportionately biased, it can be corrected by rescaling. Define

$$s^2 = \frac{n}{n-1}\widehat{\sigma}^2 = \frac{1}{n-1}\sum_{i=1}^{n} \left(X_i - \overline{X}_n\right)^2.$$

It is straightforward to see that s^2 is unbiased for σ^2. It is common to call s^2 the **bias-corrected variance estimator**.

Theorem 6.9 If $\mathbb{E}\left[X^2\right] < \infty$, then $\mathbb{E}\left[s^2\right] = \sigma^2$, and so s^2 is unbiased for σ^2.

The estimator s^2 is typically preferred to $\widehat{\sigma}^2$. However, in standard applications, the difference will be minor.

You may notice that the notation s^2 is inconsistent with our standard notation of putting hats on coefficients to indicate an estimator. This is for historical reasons. In earlier eras with hand typesetting, it was difficult to typeset notation of the form $\widehat{\beta}$. It was therefore preferred to use the notation b to denote the estimator of β, s^2 to denote the estimator of σ^2, and so forth. Because of this convention, the notation s^2 for the bias-corrected variance estimator has stuck.

To illustrate, the bias-corrected sample standard deviations of the three variables discussed in Section 6.12 are as follows:

$$s_{\text{wage}} = 12.4$$

$$s_{\log(\text{wage})} = 0.503$$

$$s_{\text{education}} = 2.05.$$

6.14 STANDARD ERROR

Finding the formula for the variance of an estimator is only a partial answer for understanding its accuracy, because the variance depends on unknown parameters. For example, we found that $\operatorname{var}\left[\overline{X}_n\right]$ depends on σ^2. To learn the sampling variance, we need to replace these unknown parameters (e.g., σ^2) by estimators. This allows us to compute an estimator of the sampling variance. To put the latter in the same units as the parameter estimate, we typically take the square root before reporting. We thus arrive at the following concept.

Definition 6.11 A **standard error** $s(\widehat{\theta}) = \widehat{V}^{1/2}$ for an estimator $\widehat{\theta}$ of a parameter θ is the square root of an estimator $\widehat{V}$ of $V = \operatorname{var}\left[\widehat{\theta}\right]$.

This definition is a bit of a mouthful, as it involves two uses of estimation. First, $\widehat{\theta}$ is used to estimate θ. Second, $\widehat{V}$ is used to estimate V. The purpose of $\widehat{\theta}$ is to learn about θ. The purpose of $\widehat{V}$ is to learn about V, which is a measure of the accuracy of $\widehat{\theta}$. We take the square root for convenience of interpretation.

For the estimator $\overline{X}_n$ of the population expectation μ the exact variance is σ^2/n. If we use the unbiased estimator s^2 for σ^2, then an unbiased estimator of $\operatorname{var}\left[\overline{X}_n\right]$ is s^2/n. The standard error for $\overline{X}_n$ is the square root of this estimator, or

$$s(\overline{X}_n) = \frac{s}{\sqrt{n}}.$$

It is useful to observe at this point that there is no reason to expect a standard error to be unbiased for its target $V^{1/2}$. In fact, if $\widehat{V}$ is unbiased for V, then $s(\widehat{\theta}) = \widehat{V}^{1/2}$ is biased downwards by Jensen's inequality. This is ignored in practice, as there is no accepted method to correct for this bias.

Standard errors are reported along with parameter estimates as a way to assess the precision of estimation. Thus $\overline{X}_n$ and $s\left(\overline{X}_n\right)$ will be reported simultaneously.

Example: Mean Wage. Our estimate is 25.73. The standard deviation s equals 12.4. Given the sample size $n = 20$, the standard error is $12.4/\sqrt{20} \simeq 2.78$.

Example: Education. Our estimate is 15.7. The standard deviation is 2.05. The standard error is $2.05/\sqrt{20} \simeq 0.46$.

6.15 MULTIVARIATE MEANS

Let $X \in \mathbb{R}^m$ be a random vector and $\mu = \mathbb{E}\left[X\right] \in \mathbb{R}^m$ be its expectation. Given a random sample $\{X_1, \ldots, X_n\}$, the moment estimator for μ is the $m \times 1$ multivariate sample mean:

$$\overline{X}_n = \frac{1}{n}\sum_{i=1}^{n} X_i = \begin{pmatrix} \overline{X}_{1n} \\ \overline{X}_{2n} \\ \vdots \\ \overline{X}_{mn} \end{pmatrix}.$$

We can write the estimator as $\widehat{\mu} = \overline{X}_n$. In our empirical example, $X_1 =$*wage* and $X_2 =$*education*.

Most properties of the univariate sample mean extend to the multivariate mean.

1. The multivariate mean is unbiased for the population expectation: $\mathbb{E}\left[\overline{X}_n\right] = \mu$, since each element is unbiased.
2. Affine functions of the sample mean are unbiased: The estimator $\widehat{\beta} = \boldsymbol{A}\overline{X}_n + \boldsymbol{c}$ is unbiased for $\beta = \boldsymbol{A}\mu + \boldsymbol{c}$.
3. The exact covariance matrix of $\overline{X}_n$ is $\operatorname{var}\left[\overline{X}_n\right] = \mathbb{E}\left[\left(\overline{X}_n - \mathbb{E}\left[\overline{X}_n\right]\right)\left(\overline{X}_n - \mathbb{E}\left[\overline{X}_n\right]\right)'\right] = n^{-1}\Sigma$, where $\Sigma = \operatorname{var}[X]$.
4. The MSE matrix of $\overline{X}_n$ is $\operatorname{mse}\left[\overline{X}_n\right] = \mathbb{E}\left[\left(\overline{X}_n - \mu\right)\left(\overline{X}_n - \mu\right)'\right] = n^{-1}\Sigma$.
5. $\overline{X}_n$ is the best unbiased estimator for μ.
6. The analog plug-in estimator for Σ is $n^{-1}\sum_{i=1}^{n}\left(X_i - \overline{X}_n\right)\left(X_i - \overline{X}_n\right)'$.
7. This estimator is biased for Σ.
8. An unbiased covariance matrix estimator is $\widehat{\Sigma} = (n-1)^{-1}\sum_{i=1}^{n}\left(X_i - \overline{X}_n\right)\left(X_i - \overline{X}_n\right)'$.

Recall that the covariances between random variables are the off-diagonal elements of the covariance matrix. Therefore an unbiased sample covariance estimator is an off-diagonal element of $\widehat{\Sigma}$. We can write this as

$$s_{XY} = \frac{1}{n-1}\sum_{i=1}^{n}\left(X_i - \overline{X}_n\right)\left(Y_i - \overline{Y}_n\right).$$

The sample correlation between X and Y is

$$\widehat{\rho}_{XY} = \frac{s_{XY}}{s_X s_Y} = \frac{\sum_{i=1}^{n}\left(X_i - \overline{X}_n\right)\left(Y_i - \overline{Y}_n\right)}{\sqrt{\sum_{i=1}^{n}\left(X_i - \overline{X}_n\right)^2}\sqrt{\sum_{i=1}^{n}\left(Y_i - \overline{Y}_n\right)^2}}.$$

For example,

$$\widehat{\operatorname{cov}}\left(\text{wage, education}\right) = 14.8$$
$$\widehat{\operatorname{corr}}\left(\text{wage, education}\right) = 0.58.$$

In this small ($n = 20$) sample, wages and education are positively correlated. Furthermore, the magnitude of the correlation is quite high.

6.16 ORDER STATISTICS*

Definition 6.12 The **order statistics** of $\{X_1, \ldots, X_n\}$ are the ordered values sorted in ascending order. They are denoted by $\{X_{(1)}, \ldots, X_{(n)}\}$.

For example, $X_{(1)} = \min_{1 \le i \le n} X_i$, and $X_{(n)} = \max_{1 \le i \le n} X_i$. Consider the wage observations in Table 6.1. We observe that $wage_{(1)} = 10.00$, $wage_{(2)} = 11.06$, $wage_{(3)} = 11.09, \ldots, wage_{(19)} = 43.08$, and $wage_{(20)} = 54.62$.

When the X_i are i.i.d., it is rather straightforward to characterize the distribution of the order statistics. The easiest to analyze is the maximum $X_{(n)}$. Since $\{X_{(n)} \le x\}$ is the same as $\{X_i \le x\}$ for all i,

$$\mathbb{P}\left[X_{(n)} \le x\right] = \mathbb{P}\left[\bigcap_{i=1}^{n} \{X_i \le x\}\right] = \prod_{i=1}^{n} \mathbb{P}\left[X_i \le x\right] = F(x)^n.$$

Next, observe that the $n-1$ order statistic satisfies $X_{(n-1)} \le x$ if $\{X_i \le x\}$ for $n-1$ observations and $\{X_i > x\}$ for one observation. As there are exactly n such configurations,

$$\mathbb{P}\left[X_{(n-1)} \le x\right] = n\mathbb{P}\left[\bigcap_{i=1}^{n-1} \{X_i \le x\} \cap \{X_n > x\}\right] = n\prod_{i=1}^{n-1} \mathbb{P}\left[X_i \le x\right] \times \mathbb{P}\left[X_n > x\right] = nF(x)^{n-1}\left(1 - F(x)\right).$$

The general result is as follows.

Theorem 6.10 If X_i are i.i.d with a continuous distribution $F(x)$, then the distribution of $X_{(j)}$ is

$$\mathbb{P}\left[X_{(j)} \le x\right] = \sum_{k=j}^{n} \binom{n}{k} F(x)^k \left(1 - F(x)\right)^{n-k}.$$

The density of $X_{(j)}$ is

$$f_{(j)}(x) = \frac{n!}{(j-1)!\,(n-j)!} F(x)^{j-1} \left(1 - F(x)\right)^{n-j} f(x).$$

There is no particular intuition for Theorem 6.10, but it is useful in some problems in applied economics.

Proof of Theorem 6.10 Let Y be the number of X_i less than or equal to x. Since X_i are i.i.d., Y is a binomial random variable with $p = F(x)$. The event $\{X_{(j)} \le x\}$ is the same as the event $\{Y \ge j\}$; that is, at least j of the X_i are less than or equal to x. Applying the binomial model,

$$\mathbb{P}\left[X_{(j)} \le x\right] = \mathbb{P}\left[Y \ge j\right] = \sum_{k=j}^{n} \binom{n}{k} F(x)^k \left(1 - F(x)\right)^{n-k}$$

which is the stated distribution. The density is found by differentiation and a tedious collection of terms. See Casella and Berger (2002, Theorem 5.4.4) for a complete derivation.

6.17 HIGHER MOMENTS OF SAMPLE MEAN*

We have found that the second centered moment of $\overline{X}_n$ is σ^2/n. Higher moments are similarly functions of the moments of X and sample size. It is convenient to rescale the sample mean so that it has constant variance. Define $Z_n = \sqrt{n}\left(\overline{X}_n - \mu\right)$, which has mean 0 and variance σ^2.

To calculate higher moments, for simplicity and without loss of generality, assume $\mu = 0$. The third moment of Z_n is

$$\mathbb{E}\left[Z_n^3\right] = \frac{1}{n^{3/2}} \sum_{i=1}^{n} \sum_{j=1}^{n} \sum_{k=1}^{n} \mathbb{E}\left[X_i X_j X_k\right].$$

Note that

$$\mathbb{E}\left[X_iX_jX_k\right] = \begin{cases} \mathbb{E}\left[X_i^3\right] = \mu_3 & \text{if } i=j=k, \ (n \text{ instances}) \\ 0 & \text{otherwise.} \end{cases}$$

Thus

$$\mathbb{E}\left[Z_n^3\right] = \frac{\mu_3}{n^{1/2}}. \tag{6.2}$$

This shows that the third moment of the normalized sample mean Z_n is a scale of the third central moment of the observations. If X is skewed, then Z_n will skew in the same direction.

The fourth moment of Z_n (again assuming $\mu = 0$) is

$$\mathbb{E}\left[Z_n^4\right] = \frac{1}{n^2} \sum_{i=1}^{n} \sum_{j=1}^{n} \sum_{k=1}^{n} \sum_{\ell=1}^{n} \mathbb{E}\left[X_iX_jX_kX_\ell\right].$$

Note that

$$\mathbb{E}\left[X_iX_jX_kX_\ell\right] = \begin{cases} \mathbb{E}\left[X_i^4\right] = \mu_4 & \text{if } i=j=k=\ell, \ (n \text{ instances}) \\ \mathbb{E}\left[X_i^2\right]\mathbb{E}\left[X_k^2\right] = \sigma^4 & \text{if } i=j\neq k=\ell, \ (n(n-1) \text{ instances}) \\ \mathbb{E}\left[X_i^2\right]\mathbb{E}\left[X_j^2\right] = \sigma^4 & \text{if } i=k\neq j=\ell, \ (n(n-1) \text{ instances}) \\ \mathbb{E}\left[X_i^2\right]\mathbb{E}\left[X_j^2\right] = \sigma^4 & \text{if } i=\ell\neq j=k, \ (n(n-1) \text{ instances}) \\ 0 & \text{otherwise.} \end{cases}$$

Thus

$$\mathbb{E}\left[Z_n^4\right] = \frac{\mu_4}{n} + 3\sigma^4\left(\frac{n-1}{n}\right) = 3\sigma^4 + \frac{\kappa_4}{n} \tag{6.3}$$

where $\kappa_4 = \mu_4 - 3\sigma^4$ is the fourth cumulant of the distribution of X (see Section 2.24 for the definition of the cumulants). Recall that the fourth central moment of $Z \sim \mathrm{N}\left(0,\sigma^2\right)$ is $3\sigma^4$. Thus the fourth moment of Z_n is close to that of the normal distribution, with a deviation depending on the fourth cumulant of X.

For higher moments, we can make similar direct yet tedious calculations. A simpler though less intuitive method calculates the moments of Z_n using the cumulant generating function $K(t) = \log(M(t))$, where $M(t)$ is the MGF of X (see Section 2.24). Since the observations are independent, the cumulant generating function of $S_n = \sum_{i=1}^n X_i$ is $\log(M(t)^n) = nK(t)$. It follows that the rth cumulant of S_n is $nK^{(r)}(0) = n\kappa_r$ where $\kappa_r = K^{(r)}(0)$ is the rth cumulant of X. Rescaling, we find that the rth cumulant of $Z_n = \sqrt{n}\left(\overline{X}_n - \mu\right)$ is $\kappa_r/n^{r/2-1}$. Using the relations between central moments and cumulants described in Section 2.24, we deduce that the third through sixth moments of Z_n are

$$\begin{aligned} \mathbb{E}\left[Z_n^3\right] &= \kappa_3/n^{1/2} && (6.4) \\ \mathbb{E}\left[Z_n^4\right] &= \kappa_4/n + 3\kappa_2^2 \\ \mathbb{E}\left[Z_n^5\right] &= \kappa_5/n^{3/2} - 10\kappa_3\kappa_2/n^{1/2} \\ \mathbb{E}\left[Z_n^6\right] &= \kappa_6/n^2 + \left(15\kappa_4\kappa_2 + 10\kappa_3^2\right)/n + 15\kappa_2^3. && (6.5) \end{aligned}$$

Since $\kappa_2 = \sigma^2$ and $\mu_3 = \kappa_3$, the first two expressions are identical to equation (6.2) and equation (6.3). The last two give the exact fifth and sixth moments of Z_n, expressed in terms of the cumulants of X and the sample size n.

This technique can be used to calculate any nonnegative integer moment of Z_n. For odd r, the moments take the form

$$\mathbb{E}\left[Z_n^r\right] = \sum_{j=0}^{(r-3)/2} a_{rj} n^{-1/2-j}$$

and for even r, they are

$$\mathbb{E}\left[Z_n^r\right] = (r-1)!!\sigma^{2r} + \sum_{j=1}^{(r-2)/2} b_{rj} n^{-j}$$

where a_{rj} and b_{rj} are the sum of constants multiplied by the products of cumulants whose indices sum to r. These are the exact (finite sample) moments of Z_n. The centered moments of $\overline{X}_n$ are found by rescaling.

6.18 NORMAL SAMPLING MODEL

While we have found the expectation and variance of $\overline{X}_n$ under quite broad conditions, we are also interested in the distribution of $\overline{X}_n$. There are various approaches to assess the distribution of $\overline{X}_n$. One traditional approach is to assume that the observations are normally distributed. This is convenient, because an elegant exact theory can be derived.

Let $\{X_1, X_2, \ldots, X_n\}$ be a random sample from $\mathrm{N}(\mu, \sigma^2)$. This is called the **normal sampling model**. Consider the sample mean $\overline{X}_n$. It is a linear function of the observations. Theorem 5.15 showed that linear functions of multivariate normal random vectors are normally distributed. Thus, $\overline{X}_n$ is normally distributed. We already know that the sample mean $\overline{X}_n$ has expectation μ and variance σ^2/n. We conclude that $\overline{X}_n$ is distributed as $\mathrm{N}\left(\mu, \sigma^2/n\right)$.

Theorem 6.11 If X_i are i.i.d. $\mathrm{N}\left(\mu, \sigma^2\right)$, then $\overline{X}_n \sim \mathrm{N}\left(\mu, \sigma^2/n\right)$.

This is an important finding. It shows that when our observations are normally distributed, we have an exact expression for the sampling distribution of the sample mean. This is much more powerful than just knowing the distribution's mean and variance.

6.19 NORMAL RESIDUALS

Define the **residuals** $\widehat{e}_i = X_i - \overline{X}_n$. These are the observations centered at their mean. Since the residuals are a linear function of the normal vector $X = (X_1, X_2, \ldots, X_n)'$, they are also normally distributed. They have mean 0, since

$$\mathbb{E}\left[\widehat{e}_i\right] = \mathbb{E}\left[X_i\right] - \mathbb{E}\left[\overline{X}_n\right] = \mu - \mu = 0.$$

Their covariance with $\overline{X}_n$ is 0, since

$$\begin{aligned}
\operatorname{cov}\left(\widehat{e}_i, \overline{X}_n\right) &= \mathbb{E}\left[\widehat{e}_i\left(\overline{X}_n - \mu\right)\right] \\
&= \mathbb{E}\left[(X_i - \mu)(\overline{X}_n - \mu)\right] - \mathbb{E}\left[(\overline{X}_n - \mu)^2\right] \\
&= \frac{\sigma^2}{n} - \frac{\sigma^2}{n} = 0.
\end{aligned}$$

Since $\widehat{e}_i$ and $\overline{X}_n$ are jointly normal, this implies they are independent. Thus any function of the residuals (including the variance estimator) is independent of $\overline{X}_n$.

6.20 NORMAL VARIANCE ESTIMATION

Recall the variance estimator $s^2 = (n-1)^{-1} \sum_{i=1}^{n} (X_i - \overline{X}_n)^2$ defined in equation (6.1). It equals

$$s^2 = (n-1)^{-1} \sum_{i-1}^{n} (X_i - \mu)^2 - (\overline{X}_n - \mu)^2.$$

Let $Z_i = (X_i - \mu)/\sigma$. Then

$$Q = \frac{(n-1)\, s^2}{\sigma^2} = \sum_{i-1}^{n} Z_i^2 - n\overline{Z}_n^2 = Q_n - Q_1,$$

say, or $Q_n = Q + Q_1$. Since Z_i are i.i.d. $\mathrm{N}(0,1)$, the sum of their squares Q_n is distributed as χ_n^2, the chi-square distribution with n degrees of freedom. Since $\sqrt{n}\,\overline{Z}_n \sim \mathrm{N}(0,1)$, $n\overline{Z}_n^2 \sim \chi_1^2$. The random variables Q and Q_1 are independent, because Q is a function only of the normal residuals, while Q_1 is a function only of $\overline{X}_n$, and the two are independent.

Compute the MGF of $Q_n = Q + Q_1$ using Theorem 4.6. Given the independence of Q and Q_1, we find

$$\mathbb{E}\left[\exp(tQ_n)\right] = \mathbb{E}\left[\exp(tQ)\right] \mathbb{E}\left[\exp(tQ_1)\right].$$

This implies

$$\mathbb{E}\left[\exp(tQ)\right] = \frac{\mathbb{E}\left[\exp(tQ_n)\right]}{\mathbb{E}\left[\exp(tQ_1)\right]} = \frac{(1-2t)^{-n/2}}{(1-2t)^{-1/2}} = (1-2t)^{-(n-1)/2}.$$

The second equality uses the expression for the chi-square MGF from Theorem 3.2. The final term is the MGF of χ_{n-1}^2. This establishes that Q is distributed as χ_{n-1}^2. Recall that Q is independent of $\overline{X}_n$, since Q is a function only of the normal residuals.

We have established the following deep result.

Theorem 6.12 If X_i are i.i.d. $\mathrm{N}(\mu, \sigma^2)$, then

1. $\overline{X}_n \sim \mathrm{N}(\mu, \sigma^2/n)$.
2. $\dfrac{n\widehat{\sigma}^2}{\sigma^2} = \dfrac{(n-1)s^2}{\sigma^2} \sim \chi_{n-1}^2$.
3. The above two statistics are independent.

This is an important theorem. The derivation may have been a bit rushed, so it is time to catch a breath. While the derivation is not particularly important, the result is important. One again, it states that in the normal sampling model, the sample mean is normally distributed, the sample variance has a chi-square distribution, and these two statistics are independent.

6.21 STUDENTIZED RATIO

An important statistic is the **studentized ratio** or t ratio

$$T = \frac{\sqrt{n}(\overline{X}_n - \mu)}{s}.$$

Sometime authors describe this as a **t-statistic**. Others use the term **z-statistic**.

We can write it as

$$\frac{\sqrt{n}(\overline{X}_n - \mu)/\sigma}{\sqrt{s^2/\sigma^2}}.$$

From Theorem 6.12, $\sqrt{n}(\overline{X}_n - \mu)/\sigma \sim \mathrm{N}(0,1)$, and $s^2/\sigma^2 \sim \chi^2_{n-1}/(n-1)$, and the two are independent. Therefore T has the distribution

$$T \sim \frac{\mathrm{N}(0,1)}{\sqrt{\chi^2_{n-1}/(n-1)}} \sim t_{n-1}$$

which is a student t distribution with $n-1$ degrees of freedom.

Theorem 6.13 If X_i are i.i.d. $\mathrm{N}\left(\mu, \sigma^2\right)$, then

$$T = \frac{\sqrt{n}(\overline{X}_n - \mu)}{s} \sim t_{n-1}$$

which is a student t distribution with $n-1$ degrees of freedom.

This is one of the most famous results in statistical theory, discovered by Gosset (1908). An interesting historical note is that at the time, Gosset worked at Guinness Brewery, which prohibited its employees from publishing in order to prevent the possible loss of trade secrets. To circumvent this barrier, Gosset published under the pseudonym "Student". Consequently, this famous distribution is known as the student t rather than Gosset's t! You should remember this when you next enjoy a pint of Guinness stout.

6.22 MULTIVARIATE NORMAL SAMPLING

Let $\{X_1, X_2, \ldots, X_n\}$ be a random sample from $\mathrm{N}(\mu, \Sigma) \in \mathbb{R}^m$. This is called the **multivariate normal sampling model**. The sample mean $\overline{X}_n$ is a linear function of independent normal random vectors. Therefore it is normally distributed as well.

Theorem 6.14 If X_i are i.i.d. $\mathrm{N}(\mu, \Sigma) \in \mathbb{R}^m$, then $\overline{X}_n \sim \mathrm{N}\left(\mu, \frac{1}{n}\Sigma\right)$.

The distribution of the sample covariance matrix estimator $\widehat{\Sigma}$ has a multivariate generalization of the chi-square distribution called the **Wishart** distribution. See Section 11.15 of *Econometrics*.

6.23 EXERCISES

Most of the problems assume a random sample $\{X_1, X_2, \ldots, X_n\}$ from a common distribution F with density f such that $\mathbb{E}[X] = \mu$ and $\operatorname{var}[X] = \sigma^2$ for a generic random variable $X \sim F$. The sample mean and variances are denoted $\overline{X}_n$ and $\widehat{\sigma}^2 = n^{-1}\sum_{i=1}^n (X_i - \overline{X}_n)^2$, with the bias-corrected variance $s^2 = (n-1)^{-1}\sum_{i=1}^n (X_i - \overline{X}_n)^2$.

Exercise 6.1 Suppose that another observation X_{n+1} becomes available. Show that

(a) $\overline{X}_{n+1} = (n\overline{X}_n + X_{n+1})/(n+1)$.
(b) $s_{n+1}^2 = \left((n-1)s_n^2 + (n/(n+1))(X_{n+1} - \overline{X}_n)^2\right)/n$.

Exercise 6.2 For some integer k, set $\mu'_k = \mathbb{E}\left[X^k\right]$. Construct an unbiased estimator $\widehat{\mu}'_k$ for μ'_k. (Show that it is unbiased.)

Exercise 6.3 Consider the central moment $\mu_k = \mathbb{E}\left[(X-\mu)^k\right]$. Construct an estimator $\widehat{\mu}_k$ for μ_k without assuming a known mean μ. In general, do you expect $\widehat{\mu}_k$ to be biased or unbiased?

Exercise 6.4 Calculate the variance $\text{var}\left[\widehat{\mu}'_k\right]$ of the estimator $\widehat{\mu}'_k$ that you proposed in Exercise 6.2.

Exercise 6.5 Propose an estimator of $\text{var}\left[\widehat{\mu}'_k\right]$. Does an unbiased version exist?

Exercise 6.6 Show that $\mathbb{E}[s] \leq \sigma$, where $s = \sqrt{s^2}$.
Hint: Use Jensen's inequality (Theorem 2.9).

Exercise 6.7 Calculate $\mathbb{E}\left[(\overline{X}_n - \mu)^3\right]$, the skewness of $\overline{X}_n$. Under what condition is it 0?

Exercise 6.8 Show algebraically that $\widehat{\sigma}^2 = n^{-1}\sum_{i=1}^n (X_i - \mu)^2 - (\overline{X}_n - \mu)^2$.

Exercise 6.9 Propose estimators for

(a) $\theta = \exp(\mathbb{E}[X])$.
(b) $\theta = \log\left(\mathbb{E}\left[\exp(X)\right]\right)$.
(c) $\theta = \sqrt{\mathbb{E}\left[X^4\right]}$.
(d) $\theta = \text{var}\left[X^2\right]$.

Exercise 6.10 Let $\theta = \mu^2$.

(a) Propose a plug-in estimator $\widehat{\theta}$ for θ.
(b) Calculate $\mathbb{E}\left[\widehat{\theta}\right]$.
(c) Propose an unbiased estimator for θ.

Exercise 6.11 Let p be the unknown probability that a given basketball player makes a free throw attempt. The player takes n random free throws, of which she makes X of the attempts.

(a) Find an unbiased estimator $\widehat{p}$ of p.
(b) Find $\text{var}\left[\widehat{p}\right]$.

Exercise 6.12 We know that $\text{var}\left[\overline{X}_n\right] = \sigma^2/n$.

(a) Find the standard deviation of $\overline{X}_n$.
(b) Suppose we know σ^2 and want our estimator to have a standard deviation smaller than a tolerance τ. How large does n need to be to make this happen?

Exercise 6.13 Find the covariance of $\widehat{\sigma}^2$ and $\overline{X}_n$. Under what condition is this 0? Hint: Use the form obtained in Exercise 6.8. This exercise shows that the zero correlation between the numerator and the denominator of the $t-$ ratio does not always hold when the random sample is not from a normal distribution.

Exercise 6.14 Suppose that X_i are i.n.i.d. (independent, not necessarily identically distributed) with $\mathbb{E}[X_i] = \mu_i$ and $\text{var}[X_i] = \sigma_i^2$.

(a) Find $\mathbb{E}\left[\overline{X}_n\right]$.
(b) Find $\text{var}\left[\overline{X}_n\right]$.

Exercise 6.15 Suppose that $X_i \sim \text{N}(\mu_X, \sigma_X^2) : i = 1, \ldots, n_1$ and $Y_i \sim \text{N}(\mu_Y, \sigma_Y^2)$, $i = 1, \ldots, n_2$ are mutually independent. Set $\overline{X}_{n_1} = n_1^{-1} \sum_{i=1}^{n_1} X_i$ and $\overline{Y}_{n_2} = n_2^{-1} \sum_{i=1}^{n_2} Y_i$.

(a) Find $\mathbb{E}\left[\overline{X}_{n_1} - \overline{Y}_{n_2}\right]$.
(b) Find $\text{var}\left[\overline{X}_{n_1} - \overline{Y}_{n_2}\right]$.
(c) Find the distribution of $\overline{X}_{n_1} - \overline{Y}_{n_2}$.
(d) Propose an estimator of $\text{var}\left[\overline{X}_{n_1} - \overline{Y}_{n_2}\right]$.

Exercise 6.16 Let $X \sim U[0, 1]$ and set $Y = \max_{1 \leq i \leq n} X_i$.

(a) Find the distribution and density function of Y.
(b) Find $\mathbb{E}[Y]$.
(c) Suppose that there is a sealed-bid highest-price auction[1] for a contract. There are n individual bidders with independent random valuations, each drawn from $U[0, 1]$. Assume for simplicity that each bidder bids their private valuation. What is the expected winning bid?

Exercise 6.17 Let $X \sim U[0, 1]$, and set $Y = X_{(n-1)}$, the $n-1$ order statistic.

(a) Find the distribution and density function of Y.
(b) Find $\mathbb{E}[Y]$.
(c) Suppose that there is a sealed-bid second-price auction[2] for a book on econometrics. There are n individual bidders with independent random valuations, each drawn from $U[0, 1]$. Auction theory predicts that each bidder bids their private valuation. What is the expected winning bid?

[1] The winner is the bidder who makes the highest bid. The winner pays the amount they bid.
[2] The winner is the bidder who makes the highest bid. The winner pays the amount of the second-highest bid.

CHAPTER 7

LAW OF LARGE NUMBERS

7.1 INTRODUCTION

In Chapter 6, we derived the expectation and variance of the sample mean $\overline{X}_n$ and derived the exact distribution under the additional assumption of normal sampling. These are useful results but are not complete. What is the distribution of $\overline{X}_n$ when the observations are not normal? What is the distribution of estimators that are nonlinear functions of sample means?

One approach to answer these questions is to use large-sample asymptotic approximations. These are a set of sampling distributions obtained under the approximation that the sample size n diverges to positive infinity. The primary tools of asymptotic theory are the weak law of large numbers (WLLN), the central limit theorem (CLT), and the continuous mapping theorem (CMT). With these tools, we can approximate the sampling distributions of most econometric estimators.

In this chapter, I introduce laws of large numbers and the associated CMT.

7.2 ASYMPTOTIC LIMITS

"Asymptotic analysis" is a method of approximation obtained by taking a suitable limit. There is more than one method to take limits, but the most common is to take the limit of the sequence of sampling distributions as the sample size tends to positive infinity, written "as $n \to \infty$." It is not meant to be interpreted literally, but rather as an approximating device.

The first building block for asymptotic analysis is the concept of a limit of a sequence.

Definition 7.1 A sequence a_n has the **limit** a, written $a_n \to a$ as $n \to \infty$, or alternatively, as $\lim_{n\to\infty} a_n = a$, if for all $\delta > 0$ there is some $n_\delta < \infty$ such that for all $n \geq n_\delta$, $|a_n - a| \leq \delta$.

In other words, a_n has the limit a if the sequence gets closer and closer to a as n gets larger. If a sequence has a limit, that limit is unique (a sequence cannot have two distinct limits). If a_n has the limit a, we also say that a_n **converges** to a as $n \to \infty$.

To illustrate, Figure 7.1 displays a sequence a_n against n. Also plotted are the limit a and two bands of width δ about a. The point n_δ at which the sequence satisfies $|a_n - a| \leq \delta$ is marked.

Example 1: $a_n = n^{-1}$. Pick $\delta > 0$. If $n \geq 1/\delta$, then $|a_n| = n^{-1} \leq \delta$. This means that the definition of convergence is satisfied with $a = 0$ and $n_\delta = 1/\delta$. Hence a_n converges to $a = 0$.

Example 2: $a_n = n^{-1}(-1)^n$. Again, if $n \geq 1/\delta$, then $|a_n| = n^{-1} \leq \delta$. Hence $a_n \to 0$.

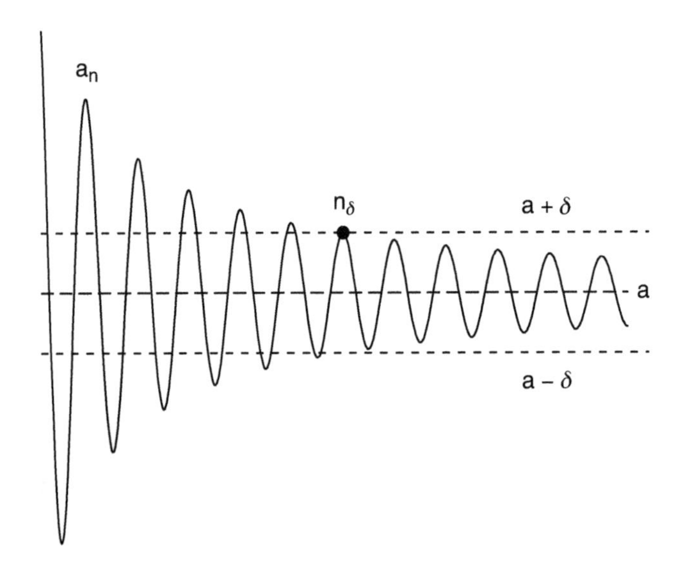

FIGURE 7.1 Limit of a sequence

Example 3: $a_n = \sin(\pi n/8)/n$. Since the sine function is bounded below 1, $|a_n| \le n^{-1}$. Thus if $n \ge 1/\delta$, then $|a_n| = n^{-1} \le \delta$. We conclude that a_n converges to 0.

In general, to show that a sequence a_n converges to 0, the proof technique is as follows: Fix $\delta > 0$. Find n_δ such that $|a_n| \le \delta$ for $n \ge n_\delta$. If this can be done for any δ, then a_n converges to 0. Often the solution for n_δ is found by setting $|a_n| = \delta$ and solving for n.

7.3 CONVERGENCE IN PROBABILITY

A sequence of numbers may converge, but what about a sequence of random variables? For example, consider a sample mean $\overline{X}_n = n^{-1}\sum_{i=1}^{n} X_i$. As n increases, the distribution of $\overline{X}_n$ changes. In what sense can we describe the "limit" of $\overline{X}_n$? In what sense does it converge?

Since $\overline{X}_n$ is a random variable, we cannot directly apply the deterministic concept of a sequence of numbers. Instead, we require a definition of convergence which is appropriate for random variables.

It may help to start by considering three simple examples of a sequence of random variables Z_n.

Example 1: Z_n has a two-point distribution with $\mathbb{P}[Z_n = 0] = 1 - p_n$ and $\mathbb{P}[Z_n = a_n] = p_n$. It seems sensible to describe Z_n as converging to 0 if either $p_n \to 0$ or $a_n \to 0$.

Example 2: $Z_n = b_n Z$, where Z is a random variable. It seems sensible to describe Z_n as converging to 0 if $b_n \to 0$.

Example 3: Z_n has variance σ_n^2. It seems sensible to describe Z_n as converging to 0 if $\sigma_n^2 \to 0$.

Multiple convergence concepts can be used for random variables. The one most commonly used in econometrics is the following.

Definition 7.2 A sequence of random variables $Z_n \in \mathbb{R}$ **converges in probability** to c as $n \to \infty$, denoted $Z_n \underset{p}{\longrightarrow} c$ or $\text{plim}_{n\to\infty} Z_n = c$, if for all $\delta > 0$,

$$\lim_{n\to\infty} \mathbb{P}\left[|Z_n - c| \leq \delta\right] = 1. \tag{7.1}$$

We call c the **probability limit** (or **plim**) of Z_n.

The condition (7.1) can alternatively be written as

$$\lim_{n\to\infty} \mathbb{P}\left[|Z_n - c| > \delta\right] = 0.$$

The definition is rather technical. Nevertheless, it concisely formalizes the concept of a sequence of random variables concentrating about a point. For any $\delta > 0$, the event $\{|Z_n - c| \leq \delta\}$ occurs when Z_n is within δ of the point c. The expression $\mathbb{P}\left[|Z_n - c| \leq \delta\right]$ is the probability of this event. Equation (7.1) states that this probability approaches 1 as the sample size n increases. The definition requires that this holds for any δ. So for any small interval about c, the distribution of Z_n concentrates within this interval for large n.

You may notice that the definition concerns the distribution of the random variables Z_n, not their realizations. Furthermore, notice that the definition uses the concept of a conventional (deterministic) limit, but the latter is applied to a sequence of probabilities, not directly to the random variables Z_n or their realizations.

Three comments about the notation are worth mentioning. First, it is conventional to write the convergence symbol as $\underset{p}{\longrightarrow}$, where the "$p$" below the arrow indicates that the convergence is "in probability". You can also write it as $\overset{p}{\longrightarrow}$ or as $\to_p$. You should try to adhere to one of these choices and not simply write $Z_n \to c$. It is important to distinguish convergence in probability from conventional (nonrandom) convergence. Second, it is important to include the phrase "as $n \to \infty$" to be specific about how the limit is obtained. This is because while $n \to \infty$ is the most common asymptotic limit, it is not the only asymptotic limit. Third, the expression to the right of the arrow "$\underset{p}{\longrightarrow}$" must be free of dependence on the sample size n. Thus expressions of the form "$Z_n \underset{p}{\longrightarrow} c_n$" are notationally meaningless and should not be used.

Let us consider the first two examples introduced in Section 7.2.

Example 1: Take any $\delta > 0$. If $a_n \to 0$, then for n large enough, $a_n < \delta$. Then $\mathbb{P}\left[|Z_n| \leq \delta\right] = 1$, so (7.1) holds. If $p_n \to 0$, then regardless of the sequence a_n, $\mathbb{P}\left[|Z_n| \leq \delta\right] \geq 1 - p_n \to 1$. Hence we see that either $a_n \to 0$ or $p_n \to 0$ is sufficient for (7.1), as intuitively expected.

Example 2: Take any $\delta > 0$ and any $\epsilon > 0$. Since Z is a random variable, it has a distribution function $F(x)$. A distribution function has the property that there is a number B sufficiently large so that $F(B) \geq 1 - \epsilon$ and $F(-B) \leq \epsilon$. If $b_n \to 0$, then there is an n large enough so that $\delta / b_n \geq B$. Then

$$\begin{aligned}
\mathbb{P}\left[|Z_n| \leq \delta\right] &= \mathbb{P}\left[|Z| \leq \delta / b_n\right] \\
&= F(\delta/b_n) - F(-\delta/b_n) \\
&\geq F(B) - F(-B) \\
&\geq 1 - 2\epsilon.
\end{aligned}$$

Since ϵ is arbitrary, this means $\mathbb{P}\left[|Z_n| \leq \delta\right] \to 1$, so (7.1) holds.

We defer the discussion of Example 3 to the following section.

7.4 CHEBYSHEV'S INEQUALITY

Take any random variable X with finite mean μ and finite variance. We want to bound the probability $\mathbb{P}[|X-\mu|>\delta]$ for all distributions and all δ. We can calculate the probability for known distributions. If X is standard normal, then $\mathbb{P}[|X|>\delta]=2(1-\Phi(\delta))$. If logistic, then $\mathbb{P}[|X|>\delta]=2(1+\exp(\delta))^{-1}$. If Pareto, then $\mathbb{P}[|X|>\delta]=\delta^{-\alpha}$. The last example has the slowest rate of convergence as δ increases, because it is a power law rather than an exponential. The rate is slowest for small α. The restriction to distributions with a finite variance requires the Pareto parameter to satisfy $\alpha>2$. We deduce that the slowest rate of convergence is attained at the bound. Thus among the examples considered, the worst case for the probability bound is $\mathbb{P}[|X|>\delta]\simeq\delta^{-2}$. It turns out that this is indeed the worst case. This bound is known as Chebyshev's Inequality, and it is the key to the WLLN.

To derive the inequality, set $Z=X-\mu$ and write the two-sided probability as

$$\mathbb{P}[|Z|\geq\delta]=\int_{\{|x|\geq\delta\}}f(x)dx$$

where $f(x)$ is the density of Z. Over the region $\{|x|\geq\delta\}$, the inequality $1\leq x^2/\delta^2$ holds. Thus the above integral is smaller than

$$\int_{\{|x|\geq\delta\}}\frac{x^2}{\delta^2}f(x)dx\leq\int_{-\infty}^{\infty}\frac{x^2}{\delta^2}f(x)dx=\frac{\text{var}[Z]}{\delta^2}=\frac{\text{var}[X]}{\delta^2}.$$

The inequality expands the region of integration to the real line, and the following equality is the definition of the variance of Z (since Z has 0 mean). We have established the following result.

Theorem 7.1 **Chebyshev's Inequality**. For any random variable X and any $\delta>0$

$$\mathbb{P}[|X-\mathbb{E}[X]|\geq\delta]\leq\frac{\text{var}[X]}{\delta^2}.$$

Now return to Example 3 from Section 7.3. By Chebyshev's inequality and the assumption $\sigma_n^2\to 0$

$$\mathbb{P}[|Z_n-\mathbb{E}[Z_n]|>\delta]\leq\frac{\sigma_n^2}{\delta^2}\to 0.$$

This is the same as (7.1). Hence $\sigma_n^2\to 0$ is sufficient for that equation to hold.

Theorem 7.2 For any sequence of random variables Z_n such that $\text{var}[Z_n]\to 0$, $Z_n-\mathbb{E}[Z_n]\underset{p}{\longrightarrow}0$ as $n\to\infty$.

We close this section with a generalization of Chebyshev's inequality.

Theorem 7.3 **Markov's Inequality**. For any random variable X and any $\delta>0$

$$\mathbb{P}[|X|\geq\delta]\leq\frac{\mathbb{E}[|X|\,\mathbb{1}\{|X|\geq\delta\}]}{\delta}\leq\frac{\mathbb{E}|X|}{\delta}.$$

Markov's inequality can be established by an argument similar to that for Chebyshev's.

7.5 WEAK LAW OF LARGE NUMBERS

Consider the sample mean $\overline{X}_n$. We have learned that $\overline{X}_n$ is unbiased for $\mu = \mathbb{E}[X]$ and has variance σ^2/n. Since the latter limits to 0 as $n \to \infty$, Theorem 7.2 shows that $\overline{X}_n \underset{p}{\longrightarrow} \mu$. This is called the **weak law of large numbers** (WLLN). It turns out that with a more detailed technical proof, we can show that the WLLN holds without requiring that X has a finite variance.

Theorem 7.4 **Weak Law of Large Numbers (WLLN).** If X_i are i.i.d. and $\mathbb{E}|X| < \infty$, then as $n \to \infty$

$$\overline{X}_n = \frac{1}{n}\sum_{i=1}^{n} X_i \underset{p}{\longrightarrow} \mathbb{E}[X].$$

The proof is in Section 7.14.

The WLLN states that the sample mean converges in probability to the population expectation. In general, an estimator that converges in probability to the population value is called **consistent**.

Definition 7.3 An estimator $\widehat{\theta}$ of a parameter θ is **consistent** if $\widehat{\theta} \underset{p}{\longrightarrow} \theta$ as $n \to \infty$.

Theorem 7.5 If X_i are i.i.d. and $\mathbb{E}|X| < \infty$, then $\widehat{\mu} = \overline{X}_n$ is consistent for the population expectation $\mu = \mathbb{E}[X]$.

Consistency is a good property for an estimator to possess. It means that for any given data distribution, there is a sample size n sufficiently large such that the estimator $\widehat{\theta}$ will be arbitrarily close to the true value θ with high probability. The property does not tell us, however, how large this n has to be. Thus the property does not give practical guidance for empirical practice. Still, it is a minimal property for an estimator to be considered a "good" estimator, and it provides a foundation for more useful approximations.

7.6 COUNTEREXAMPLES

To understand the WLLN, it is helpful to understand situations where it does not hold.

Example 1: Assume the observations have the structure $X_i = Z + U_i$, where Z is a common component, and U_i is an individual-specific component. Suppose that Z and U_i are independent, and that the U_i are i.i.d. with $\mathbb{E}[U] = 0$. Then

$$\overline{X}_n = Z + \overline{U}_n \underset{p}{\longrightarrow} Z.$$

The sample mean converges in probability but not to the population mean. Instead it converges to the common component Z. This holds regardless of the relative variances of Z and U.

This example shows the importance of the assumption of independence across the observations. Violations of independence can lead to a violation of the WLLN.

Example 2: Assume the observations are independent but have different variances. Suppose that $\operatorname{var}[X_i] = 1$ for $i \leq n/2$, and $\operatorname{var}[X_i] = n$ for $i > n/2$. (Thus half of the observations have a much larger variance than the other half.) Then $\operatorname{var}\left[\overline{X}_n\right] = (n/2 + n^2/2)/n^2 \to 1/2$ does not converge to 0. $\overline{X}_n$ does not converge in probability.

This example shows the importance of the "identical distribution" assumption. With sufficient heterogeneity, the WLLN can fail.

Example 3: Assume the observations are i.i.d. Cauchy with density $f(x) = 1/\left(\pi(1+x^2)\right)$. The assumption $\mathbb{E}|X| < \infty$ fails. The characteristic function of the Cauchy distribution[1] is $C(t) = \exp(-|t|)$. This result means that the characteristic function of $\overline{X}_n$ is

$$\mathbb{E}\left[\exp\left(it\overline{X}_n\right)\right] = \prod_{i=1}^{n} \mathbb{E}\left[\exp\left(\frac{itX_i}{n}\right)\right] = C\left(\frac{t}{n}\right)^n = \exp\left(-\left|\frac{t}{n}\right|\right)^n = \exp(-|t|) = C(t).$$

This is the characteristic function of X. Hence $\overline{X}_n \sim$ Cauchy has the same distribution as X. Thus $\overline{X}_n$ does not converge in probability.

This example shows the importance of the finite mean assumption. If $\mathbb{E}|X| < \infty$ fails, the sample mean will not converge in probability to a finite value.

7.7 EXAMPLES

The WLLN can be used to establish the convergence in probability of many sample moments.

1. If $\mathbb{E}|X| < \infty$, $\frac{1}{n}\sum_{i=1}^{n} X_i \underset{p}{\longrightarrow} \mathbb{E}[X]$
2. If $\mathbb{E}\left[X^2\right] < \infty$, $\frac{1}{n}\sum_{i=1}^{n} X_i^2 \underset{p}{\longrightarrow} \mathbb{E}\left[X^2\right]$
3. If $\mathbb{E}|X|^m < \infty$, $\frac{1}{n}\sum_{i=1}^{n} |X_i|^m \underset{p}{\longrightarrow} \mathbb{E}|X|^m$
4. If $\mathbb{E}\left[\exp(X)\right] < \infty$, $\frac{1}{n}\sum_{i=1}^{n} \exp(X_i) \underset{p}{\longrightarrow} \mathbb{E}\left[\exp(X)\right]$
5. If $\mathbb{E}\left|\log(X)\right| < \infty$, $\frac{1}{n}\sum_{i=1}^{n} \log(X_i) \underset{p}{\longrightarrow} \mathbb{E}\left[\log(X)\right]$
6. For $X_i \geq 0$, $\frac{1}{n}\sum_{i=1}^{n} \frac{1}{1+X_i} \underset{p}{\longrightarrow} \mathbb{E}\left[\frac{1}{1+X}\right]$
7. $\frac{1}{n}\sum_{i=1}^{n} \sin(X_i) \underset{p}{\longrightarrow} \mathbb{E}[\sin(X)]$.

7.8 ILLUSTRATING CHEBYSHEV'S INEQUALITY

Consider the WLLN based on Chebyshev's inequality. The beauty is that it applies to any distribution with a finite variance. A limitation is that the implied probability bounds can be quite imprecise.

Consider the U.S. hourly wage distribution, for which the variance is $\sigma^2 = 430$. Suppose we take a random sample of n individuals, take the sample mean of the wages $\overline{X}_n$, and use this as an estimator of the expectation μ. How accurate is the estimator? For example, how large would the sample size need to be for the probability to be less than 1% that $\overline{X}_n$ differs from the true value by less than \$1?

We know that $\mathbb{E}\left[\overline{X}_n\right] = \mu$ and $\operatorname{var}\left[\overline{X}_n\right] = \sigma^2/n = 430/n$. Chebyshev's inequality states that

$$\mathbb{P}\left[\left|\overline{X}_n - \mu\right| \geq 1\right] = \mathbb{P}\left[\left|\overline{X}_n - \mathbb{E}\left[\overline{X}_n\right]\right| \geq 1\right] \leq \operatorname{var}\left[\overline{X}_n\right] = \frac{430}{n}.$$

[1] Finding the characteristic function of the Cauchy distribution is an exercise in advanced calculus and so is not presented here.

This implies the following probability bounds

$$\mathbb{P}\left[\left|\overline{X}_n - \mu\right| \geq 1\right] \leq \begin{cases} 1 & \text{if} \quad n = 430 \\ 0.5 & \text{if} \quad n = 860 \\ 0.25 & \text{if} \quad n = 1720 \\ 0.01 & \text{if} \quad n = 43,000. \end{cases}$$

For the sample mean to be within \$1 of the true value with 99% probability by Chebyshev's inequality, the sample size would have to be 43,000!

The reason the sample size needs to be so large is that Chebyshev's inequality needs to hold for all possible distributions and threshold values. One way to intrepret this is that Chebyshev's inequality is conservative. The inequality is true, but it can considerably overstate the probability.

Still, our calculations do illustrate the WLLN. The probability bound that $\overline{X}_n$ differs from μ by \$1 is steadily decreasing as n increases.

7.9 VECTOR-VALUED MOMENTS

The preceding discussion focused on the case where X is real-valued (a scalar), but nothing important changes if we generalize to the case where $X \in \mathbb{R}^m$ is a vector.

The $m \times m$ covariance matrix of X is

$$\Sigma = \operatorname{var}[X] = \mathbb{E}\left[(X - \mu)(X - \mu)'\right].$$

It can be shown that the elements of Σ are finite if $\mathbb{E}\|X\|^2 < \infty$.

A random sample $\{X_1, \ldots, X_n\}$ consists of n observations of i.i.d. draws from the distribution of X. (Each draw is an m-vector.) The vector sample mean

$$\overline{X}_n = \frac{1}{n}\sum_{i=1}^{n} X_i = \begin{pmatrix} \overline{X}_{n,1} \\ \overline{X}_{n,2} \\ \vdots \\ \overline{X}_{n,m} \end{pmatrix}$$

is the vector of sample means of the individual variables.

Convergence in probability of a vector is defined as convergence in probability of all elements in the vector. Thus $\overline{X}_n \underset{p}{\longrightarrow} \mu$ if and only if $\overline{X}_{n,j} \underset{p}{\longrightarrow} \mu_j$ for $j = 1, \ldots, m$. Since the latter holds if $\mathbb{E}\left|X_j\right| < \infty$ for $j = 1, \ldots, m$, or equivalently, $\mathbb{E}\|X\| < \infty$, we can state this formally.

Theorem 7.6 WLLN for Random Vectors. If $X_i \in \mathbb{R}^m$ are i.i.d. and $\mathbb{E}\|X\| < \infty$, then as $n \to \infty$,

$$\overline{X}_n = \frac{1}{n}\sum_{i=1}^{n} X_i \underset{p}{\longrightarrow} \mathbb{E}[X].$$

7.10 CONTINUOUS MAPPING THEOREM

Recall the definition of a continuous function.

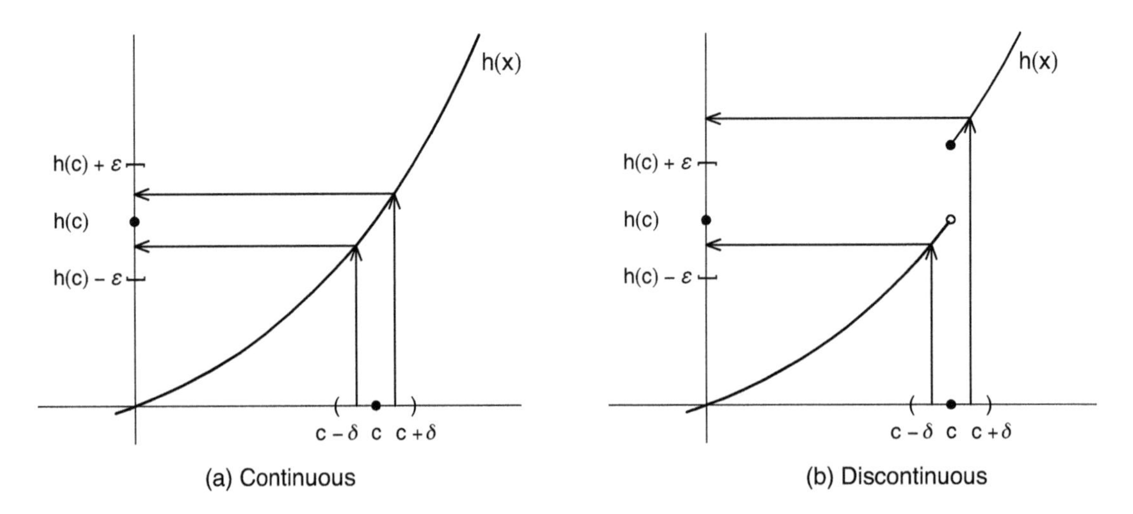

FIGURE 7.2 Continuity

Definition 7.4 A function $h(x)$ is **continuous** at $x = c$ if for all $\epsilon > 0$, there is some $\delta > 0$ such that $\|x - c\| \leq \delta$ implies $\|h(x) - h(c)\| \leq \epsilon$.

In words, a function is continuous if small changes in an input result in small changes in the output. A function is discontinuous if small changes in an input result in large changes in the output. This is illustrated in Figure 7.2. Each panel displays a function; the one in panel (a) is continuous, the one in panel (b) is discontinuous.

Examine panel (a), which displays a continuous function $h(x)$. On the x-axis, the point c is marked, and the set $\{|x - c| \leq \delta\}$ is marked by parentheses. On the y-axis, the point $h(c)$ is marked, and the set $|h(x) - h(c)| \leq \epsilon$ is marked by brackets. The arrows show how a point in $[c - \delta, c + \delta]$ on the x-axis is mapped to a point on the y-axis inside the set $[h(c) - \epsilon, h(c) + \epsilon]$.

Examine panel (b), which displays a discontinuous function $h(x)$. The same sets are marked on the axes as in panel (a). The arrows show that points in $[c, c + \delta)$ on the x-axis are mapped to points on the y-axis outside the set $[h(c) - \epsilon, h(c) + \epsilon]$. This holds for any δ due to the discontinuity at c. Thus, small changes in x result in large changes in $h(x)$.

When a continuous function is applied to a random variable which converges in probability, the result is a new random variable which converges in probability. Thus convergence in probability is preserved by continuous mappings.

Theorem 7.7 Continuous Mapping Theorem (CMT). If $Z_n \underset{p}{\longrightarrow} c$ as $n \to \infty$ and $h(\cdot)$ is continuous at c, then $h(Z_n) \underset{p}{\longrightarrow} h(c)$ as $n \to \infty$.

The proof is in Section 7.14.

For example, if $Z_n \underset{p}{\longrightarrow} c$ as $n \to \infty$, then

$$Z_n + a \underset{p}{\longrightarrow} c + a$$

$$aZ_n \underset{p}{\longrightarrow} ac$$

$$Z_n^2 \underset{p}{\longrightarrow} c^2$$

because the functions $h(u) = u + a$, $h(u) = au$, and $h(u) = u^2$ are continuous. Also

$$\frac{a}{Z_n} \xrightarrow[p]{} \frac{a}{c}$$

if $c \neq 0$. The condition $c \neq 0$ is important, because the function $h(u) = a/u$ is not continuous at $u = 0$.

We can again use Figure 7.2 to illustrate the point. Supose that $Z_n \xrightarrow[p]{} c$ is displayed the x-axis and $Y_n = h(Z_n)$ on the y-axis. In panel (a), we see that for large n, Z_n is within δ of the plim c. Thus the map $Y_n = h(Z_n)$ is within ϵ of $h(c)$. Hence $Y_n \xrightarrow[p]{} h(c)$. In panel (b), however, this is not the case. While Z_n is within δ of the plim c, the discontinuity at c means that Y_n is not within ϵ of $h(c)$. Hence Y_n does not converge in probability to $h(c)$.

Consider the plug-in estimator $\widehat{\beta} = h(\widehat{\theta})$ of the parameter $\beta = h(\theta)$ when $h(\cdot)$ is continuous. Under the conditions of Theorem 7.6, $\widehat{\theta} \xrightarrow[p]{} \theta$. Applying the CMT, $\widehat{\beta} = h(\widehat{\theta}) \xrightarrow[p]{} h(\theta) = \beta$.

Theorem 7.8 If the X_i are i.i.d., $\mathbb{E}\|g(X)\| < \infty$, and $h(u)$ is continuous at $u = \theta = \mathbb{E}[g(X)]$, then for $\widehat{\theta} = \frac{1}{n}\sum_{i=1}^n g(X_i)$, $\widehat{\beta} = h(\widehat{\theta}) \xrightarrow[p]{} \beta = h(\theta)$ as $n \to \infty$.

A convenient property of the CMT is that it does not require that the transformation possess any moments. Suppose that X has a finite mean and variance but does not possess finite higher order moments. Consider the parameter $\beta = \mu^3$ and its estimator $\widehat{\beta} = \overline{X}_n^3$. Since $\mathbb{E}|X|^3 = \infty$, $\mathbb{E}|\widehat{\beta}| = \infty$. But since the sample mean $\overline{X}_n$ is consistent, $\overline{X}_n \xrightarrow[p]{} \mu = \mathbb{E}[X]$. By an application of the CMT, we obtain $\widehat{\beta} \xrightarrow[p]{} \mu^3 = \beta$. Thus $\widehat{\beta}$ is consistent, even though it does not have a finite mean.

7.11 EXAMPLES

1. If $\mathbb{E}|X| < \infty$, $\overline{X}_n^2 \xrightarrow[p]{} \mu^2$.
2. If $\mathbb{E}|X| < \infty$, $\exp(\overline{X}_n) \xrightarrow[p]{} \exp(\mu)$.
3. If $\mathbb{E}|X| < \infty$, $\log(\overline{X}_n) \xrightarrow[p]{} \log(\mu)$.
4. If $\mathbb{E}[X^2] < \infty$, $\widehat{\sigma}^2 = \frac{1}{n}\sum_{i=1}^n X_i^2 - \overline{X}_n^2 \xrightarrow[p]{} \mathbb{E}[X^2] - \mu^2 = \sigma^2$.
5. If $\mathbb{E}[X^2] < \infty$, $\widehat{\sigma} = \sqrt{\widehat{\sigma}^2} \xrightarrow[p]{} \sqrt{\sigma^2} = \sigma$.
6. If $\mathbb{E}[X^2] < \infty$ and $\mathbb{E}[X] \neq 0$, $\widehat{\sigma}/\overline{X}_n \xrightarrow[p]{} \sigma/\mathbb{E}[X]$.
7. If $\mathbb{E}|X| < \infty$, $\mathbb{E}|Y| < \infty$, and $\mathbb{E}[Y] \neq 0$, then $\overline{X}_n/\overline{Y}_n \xrightarrow[p]{} \mathbb{E}[X]/\mathbb{E}[Y]$.

7.12 UNIFORMITY OVER DISTRIBUTIONS*

The weak law of law numbers (Theorem 7.4) states that for any distribution with a finite expectation, the sample mean approaches the population expectation in probability as the sample size gets large. One difficulty with this result, however, is that it is not uniform across distributions. Specifically, for any sample size n no matter how large, there is a valid data distribution such that the sample mean is far from the population expectation with high probability:

The WLLN states that if $\mathbb{E}|X| < \infty$, then for any $\epsilon > 0$

$$\mathbb{P}\left[\left|\overline{X}_n - \mathbb{E}[X]\right| > \epsilon\right] \to 0$$

as $n \to \infty$. However, the following is also true.

Theorem 7.9 Let $\mathscr{F}$ denote the set of distributions for which $\mathbb{E}|X| < \infty$. Then for any sample size n and any $\epsilon > 0$

$$\sup_{F \in \Im} \mathbb{P}\left[\left|\overline{X}_n - \mathbb{E}[X]\right| > \epsilon\right] = 1. \tag{7.2}$$

Equation (7.2) shows that for any n, there is a probability distribution such that the probability that the sample mean differs from the population expectation by ϵ is 1. This is a failure of **uniform convergence.** While $\overline{X}_n$ converges pointwise to $\mathbb{E}[X]$ it does not converge uniformly across all distributions for which $\mathbb{E}|X| < \infty$.

Proof of Theorem 7.9 For any $p \in (0, 1)$, define the two-point distribution

$$X = \begin{cases} 1 & \text{with probability} \quad 1-p \\ \dfrac{p-1}{p} & \text{with probability} \quad p. \end{cases}$$

This random variable satisfies $\mathbb{E}|X| = 2(1-p) < \infty$, so $F \in \Im$. It also satisfies $\mathbb{E}[X] = 0$.

The probability that the entire sample consists of all 1's is $(1-p)^n$. When this occurs, $\overline{X}_n = 1$. Then for $\epsilon < 1$

$$\mathbb{P}\left[\left|\overline{X}_n\right| > \epsilon\right] \ge \mathbb{P}\left[\overline{X}_n = 1\right] = (1-p)^n.$$

Thus

$$\sup_{F \in \Im} \mathbb{P}\left[\left|\overline{X}_n - \mathbb{E}[X]\right| > \epsilon\right] \ge \sup_{0<p<1} \mathbb{P}\left[\left|\overline{X}_n\right| > \epsilon\right] \ge \sup_{0<p<1} (1-p)^n = 1.$$

The first inequality is the statement that $\Im$ is larger than the class of two-point distriubtions, which demonstrates (7.2). ■

This result may seem to be a contradiction. The WLLN says that for any distribution with a finite expectation, the sample mean gets close to the population expectation as the sample size gets large. Equation (7.2) says that for any sample size, there is a distribution such that the sample mean is not close to the population expectation. The difference is that the WLLN holds the distribution fixed as $n \to \infty$, while equation (7.2) holds the sample size fixed as we search for the worst-case distribution.

The solution is to restrict the set of distributions. It turns out that a sufficient condition is to bound a moment greater than 1. Consider a bounded second moment. That is, assume $\text{var}[X] \le B$ for some $B < \infty$. This excludes, for example, two-point distributions with arbitrarily small probabilities p. Under this restriction, the WLLN holds uniformly across the set of distributions.

Theorem 7.10 Let $\mathscr{F}_2$ denote the set of distributions which satisfy $\text{var}[X] \le B$ for some $B < \infty$. Then for all $\epsilon > 0$, as $n \to \infty$,

$$\sup_{F \in \Im_2} \mathbb{P}\left[\left|\overline{X}_n - \mathbb{E}[X]\right| > \epsilon\right] \to 0.$$

Proof of Theorem 7.10 The restriction $\operatorname{var}[X] \leq B$ means that for all $F \in \Im_2$,

$$\operatorname{var}\left[\overline{X}_n\right] = \frac{1}{n} \operatorname{var}[X] \leq \frac{1}{n} B. \tag{7.3}$$

By an application of Chebyshev's inequality (7.1) and then (7.3)

$$\mathbb{P}\left[\left|\overline{X}_n - \mathbb{E}[X]\right| > \epsilon\right] \leq \frac{\operatorname{var}\left[\overline{X}_n\right]}{\epsilon^2} \leq \frac{B}{\epsilon^2 n}.$$

The right-hand side does not depend on the distribution, only on the bound B. Thus

$$\sup_{F \in \Im_2} \mathbb{P}\left[\left|\overline{X}_n - \mathbb{E}[X]\right| > \epsilon\right] \leq \frac{B}{\epsilon^2 n} \to 0$$

as $n \to \infty$. ■

Theorem 7.10 shows that uniform convergence holds under a stronger moment condition. Notice that it is not sufficient merely to assume that the variance is finite; the proof uses the fact that the variance is bounded.

Another solution is to apply the WLLN to rescaled sample means.

Theorem 7.11 Let $\mathscr{F}_*$ denote the set of distributions for which $\sigma^2 < \infty$. Define $X_i^* = X_i/\sigma$ and $\overline{X}_n^* = n^{-1} \sum_{i=1}^n X_i^*$. Then for all $\epsilon > 0$, as $n \to \infty$

$$\sup_{F \in \Im_*} \mathbb{P}\left[\left|\overline{X}_n^* - \mathbb{E}\left[X^*\right]\right| > \epsilon\right] \to 0.$$

7.13 ALMOST SURE CONVERGENCE AND THE STRONG LAW*

Convergence in probability is sometimes called **weak convergence**. A related concept is **almost sure convergence**, also known as **strong convergence**. (In probability theory, the term "almost sure" means "with probability equal to 1". An event which is random but occurs with probability equal to 1 is said to be **almost sure**.)

Definition 7.5 A sequence of random variables $Z_n \in \mathbb{R}$ **converges almost surely** to c as $n \to \infty$, denoted $Z_n \underset{a.s.}{\longrightarrow} c$, if

$$\mathbb{P}\left[\lim_{n \to \infty} Z_n = c\right] = 1. \tag{7.4}$$

The convergence (7.4) is stronger than (7.1) because it computes the probability of a limit rather than the limit of a probability. Almost sure convergence is stronger than convergence in probability in the sense that $Z_n \underset{a.s.}{\longrightarrow} c$ implies $Z_n \underset{p}{\longrightarrow} c$.

Recall from Section 7.3 the example of a two-point distribution $\mathbb{P}[Z_n = 0] = 1 - p_n$ and $\mathbb{P}[Z_n = a_n] = p_n$. We showed that Z_n converges in probability to 0 if either $p_n \to 0$ or $a_n \to 0$. This allows, for example, for $a_n \to \infty$ so long as $p_n \to 0$. In contrast, for Z_n to converge to zero almost surely, it is necessary that $a_n \to 0$.

In the random sampling context, the sample mean can be shown to converge almost surely to the population expectation. This is called the "strong law of large numbers".

Theorem 7.12 Strong Law of Large Numbers (SLLN). If X_i are i.i.d. and $\mathbb{E}|X| < \infty$, then as $n \to \infty$,

$$\overline{X}_n = \frac{1}{n}\sum_{i=1}^{n} X_i \underset{a.s.}{\longrightarrow} \mathbb{E}[X].$$

The SLLN is more elegant than the WLLN and so is preferred by probabilists and theoretical statisticians. However, for most practical purposes, the WLLN is sufficient. Thus in econometrics, we primarily use the WLLN.

Proving the SLLN uses more advanced tools. For fun we provide the details for interested readers. Others can safely skip the proof.

To simplify the proof, we strengthen the moment bound to $\mathbb{E}\left[X^2\right] < \infty$. This restriction can be avoided by even more advanced steps. See a probability text, such as Ash (1972) or Billingsley (1995), for a complete argument.

The proof of the SLLN is based on the following strengthening of Chebyshev's inequality, with its proof provided in Section 7.14.

Theorem 7.13 Kolmogorov's Inequality. If X_i are independent, $\mathbb{E}[X] = 0$, and $\mathbb{E}\left[X^2\right] < \infty$, then for all $\epsilon > 0$

$$\mathbb{P}\left[\max_{1\le j\le n}\left|\sum_{i=1}^{j} X_i\right| > \epsilon\right] \le \epsilon^{-2}\sum_{i=1}^{n}\mathbb{E}\left[X_i^2\right].$$

7.14 TECHNICAL PROOFS*

Proof of Theorem 7.4 Without loss of generality we can assume $\mathbb{E}[X] = 0$. We need to show that for all $\delta > 0$ and $\eta > 0$, there is some $N < \infty$ so that for all $n \ge N$, $\mathbb{P}\left[\left|\overline{X}_n\right| > \delta\right] \le \eta$. Fix δ and η. Set $\epsilon = \delta\eta/3$. Pick $C < \infty$ large enough so that

$$\mathbb{E}\left[|X|\,\mathbb{1}\{|X| > C\}\right] \le \epsilon \tag{7.5}$$

which is possible since $\mathbb{E}|X| < \infty$. Define the random variables

$$W_i = X_i\mathbb{1}\{|X_i| \le C\} - \mathbb{E}\left[X\mathbb{1}\{|X| \le C\}\right]$$

$$Z_i = X_i\mathbb{1}\{|X_i| > C\} - \mathbb{E}\left[X\mathbb{1}\{|X| > C\}\right]$$

so that

$$\overline{X}_n = \overline{W}_n + \overline{Z}_n$$

and

$$\mathbb{E}\left|\overline{X}_n\right| \le \mathbb{E}\left|\overline{W}_n\right| + \mathbb{E}\left|\overline{Z}_n\right|. \tag{7.6}$$

We now show that sum of the expectations on the right-hand side can be bounded below 3ϵ.

First, by the triangle inequality and the expectation inequality (Theorem 2.10),

$$\begin{aligned}
\mathbb{E}|Z| &= \mathbb{E}\left|X\mathbb{1}\{|X| > C\} - \mathbb{E}\left[X\mathbb{1}\{|X| > C\}\right]\right| \\
&\le \mathbb{E}\left|X\mathbb{1}\{|X| > C\}\right| + \left|\mathbb{E}\left[X\mathbb{1}\{|X| > C\}\right]\right| \\
&\le 2\mathbb{E}\left|X\mathbb{1}\{|X| > C\}\right| \\
&\le 2\epsilon
\end{aligned} \tag{7.7}$$

and thus by the triangle inequality and (7.7),

$$\mathbb{E}\left|\overline{Z}_n\right| = \mathbb{E}\left|\frac{1}{n}\sum_{i=1}^{n} Z_i\right| \leq \frac{1}{n}\sum_{i=1}^{n}\mathbb{E}\left|Z_i\right| \leq 2\epsilon. \tag{7.8}$$

Second, by a similar argument

$$\begin{aligned}|W_i| &= |X_i \mathbb{1}\{|X_i| \leq C\} - \mathbb{E}[X\mathbb{1}\{|X| \leq C\}]| \\ &\leq |X_i \mathbb{1}\{|X_i| \leq C\}| + |\mathbb{E}[X\mathbb{1}\{|X| \leq C\}]| \\ &\leq 2|X_i \mathbb{1}\{|X_i| \leq C\}| \\ &\leq 2C \end{aligned} \tag{7.9}$$

where the final inequality is (7.5). Then by Jensen's inequality (Theorem 2.9), the fact that the W_i are i.i.d. and mean 0, and (7.9),

$$\left(\mathbb{E}\left|\overline{W}_n\right|\right)^2 \leq \mathbb{E}\left[\left|\overline{W}_n\right|^2\right] = \frac{\mathbb{E}\left[W^2\right]}{n} \leq \frac{4C^2}{n} \leq \epsilon^2 \tag{7.10}$$

the final inequality holding for $n \geq 4C^2/\epsilon^2 = 36C^2/\delta^2\eta^2$. Equations (7.6), (7.8), and (7.10) together show that

$$\mathbb{E}\left|\overline{X}_n\right| \leq 3\epsilon \tag{7.11}$$

as desired.

Finally, by Markov's inequality (Theorem 7.3) and (7.11),

$$\mathbb{P}\left[\left|\overline{X}_n\right| > \delta\right] \leq \frac{\mathbb{E}\left|\overline{X}_n\right|}{\delta} \leq \frac{3\epsilon}{\delta} = \eta,$$

the final equality by the definition of ϵ. We have shown that for any $\delta > 0$ and $\eta > 0$, then for all $n \geq 36C^2/\delta^2\eta^2$, $\mathbb{P}\left[\left|\overline{X}_n\right| > \delta\right] \leq \eta$, as needed. ■

Proof of Theorem 7.7 Fix $\epsilon > 0$. Continuity of $h(u)$ at c means that there exists $\delta > 0$ such that $\|u - c\| \leq \delta$ implies $\|h(u) - h(c)\| \leq \epsilon$. Evaluated at $u = Z_n$ we find

$$\mathbb{P}\left[\|h(Z_n) - h(c)\| \leq \epsilon\right] \geq \mathbb{P}\left[\|Z_n - c\| \leq \delta\right] \longrightarrow 1$$

where the final convergence holds as $n \to \infty$ by the assumption that $Z_n \underset{p}{\longrightarrow} c$. This implies $h(Z_n) \underset{p}{\longrightarrow} h(c)$. ■

Proof of Theorem 7.12 Without loss of generality, assume $\mathbb{E}[X] = 0$. We will use the algebraic fact

$$\sum_{i=m+1}^{\infty} i^{-2} \leq \frac{1}{m} \tag{7.12}$$

which can be deduced by noting that $\sum_{i=m+1}^{\infty} i^{-2}$ is the sum of the unit width rectangles under the curve x^{-2} from m to infinity, whose summed area is less than the integral.

Define $S_n = \sum_{i=1}^{n} i^{-1}X_i$. By the Kronecker lemma (Theorem A.6), if S_n converges, then $\lim_{n\to\infty} \overline{X}_n = 0$. The Cauchy criterion (Theorem A.2) states that S_n converges if for all $\epsilon > 0$

$$\inf_{m} \sup_{j>m} \left|S_j - S_m\right| \leq \epsilon.$$

Let A_ϵ be this event. Its complement is

$$A_\epsilon^c = \bigcap_{m=1}^{\infty} \left\{ \sup_{j>m} \left| \sum_{i=m+1}^{j} \frac{1}{i} X_i \right| > \epsilon \right\}.$$

This has probability

$$\mathbb{P}\left[A_\epsilon^c\right] \leq \lim_{m\to\infty} \mathbb{P}\left[\sup_{j>m} \left| \sum_{i=m+1}^{j} \frac{1}{i} X_i \right| > \epsilon \right] \leq \lim_{m\to\infty} \frac{1}{\epsilon^2} \sum_{i=m+1}^{\infty} \frac{1}{i^2} \mathbb{E}\left[X^2\right] \leq \lim_{m\to\infty} \frac{\mathbb{E}\left[X^2\right]}{\epsilon^2} \frac{1}{m} = 0.$$

The second inequality is Kolmogorov's (Theorem 7.13) and the third one is (7.12). Hence for all $\epsilon > 0$, $\mathbb{P}\left[A_\epsilon^c\right] = 0$, and $\mathbb{P}\left[A_\epsilon\right] = 1$. Hence with probability 1, S_n converges. ■

Proof of Theorem 7.13 Set $S_i = \sum_{j=1}^{i} X_j$. Let A_i be the indicator of the event that $|S_i|$ is the first $|S_j|$ which exceeds ϵ. Formally

$$A_i = \mathbb{1}\left\{ |S_i| > \epsilon, \max_{j<i} |S_j| \leq \epsilon \right\}. \tag{7.13}$$

Then

$$A = \sum_{i=1}^{n} A_i = \mathbb{1}\left\{ \max_{i\leq n} |S_i| > \epsilon \right\}.$$

By writing $S_n = S_i + (S_n - S_i)$, we can deduce that

$$\begin{aligned}
\mathbb{E}\left[S_i^2 A_i\right] &= \mathbb{E}\left[S_n^2 A_i\right] - \mathbb{E}\left[(S_n - S_i)^2 A_i\right] - 2\mathbb{E}\left[(S_n - S_i) S_i A_i\right] \\
&= \mathbb{E}\left[S_n^2 A_i\right] - \mathbb{E}\left[(S_n - S_i)^2 A_i\right] \\
&\leq \mathbb{E}\left[S_n^2 A_i\right].
\end{aligned}$$

The first equality is algebraic. The second equality uses the fact that $(S_n - S_i)$ and $S_i A_i$ are independent and mean 0. The final inequality is the fact that $(S_n - S_i)^2 A_i \geq 0$. It follows that

$$\sum_{i=1}^{n} \mathbb{E}\left[S_i^2 A_i\right] \leq \sum_{i=1}^{n} \mathbb{E}\left[S_n^2 A_i\right] = \mathbb{E}\left[S_n^2 A\right] \leq \mathbb{E}\left[S_n^2\right].$$

Similarly to the proof of Theorem 7.3 (Markov's inequality) and using the above deduction, we have

$$\epsilon^2 \mathbb{P}\left[\max_{1\leq j\leq n} \left| \sum_{i=1}^{j} X_i \right| > \epsilon \right] = \epsilon^2 \mathbb{E}\left[A\right] = \sum_{i=1}^{n} \mathbb{E}\left[\epsilon^2 A_i\right] \leq \sum_{i=1}^{n} \mathbb{E}\left[S_i^2 A_i\right] \leq \mathbb{E}\left[S_n^2\right] = \sum_{i=1}^{n} \mathbb{E}\left[X_i^2\right].$$

The final equality uses the fact that S_n is the sum of independent mean 0 random variables. ■

7.15 EXERCISES

Exercise 7.1 For the following sequences, show that $a_n \to 0$ as $n \to \infty$.

(a) $a_n = 1/n^2$.

(b) $a_n = \dfrac{1}{n^2} \sin\left(\dfrac{\pi}{8} n\right)$.

(c) $a_n = \sigma^2/n$.

Exercise 7.2 Does the sequence $a_n = \sin\left(\frac{\pi}{2}n\right)$ converge?

Exercise 7.3 Assume that $a_n \to 0$. Show that

(a) $a_n^{1/2} \longrightarrow 0$ (assume that $a_n \geq 0$).
(b) $a_n^2 \longrightarrow 0$.

Exercise 7.4 Consider a random variable Z_n with the probability distribution

$$Z_n = \begin{cases} -n & \text{with probability } 1/n \\ 0 & \text{with probability } 1-2/n \\ n & \text{with probability } 1/n. \end{cases}$$

(a) Does $Z_n \to_p 0$ as $n \to \infty$?
(b) Calculate $\mathbb{E}[Z_n]$.
(c) Calculate $\text{var}[Z_n]$.
(d) Now suppose the distribution is

$$Z_n = \begin{cases} 0 & \text{with probability } 1-1/n \\ n & \text{with probability } 1/n. \end{cases}$$

Calculate $\mathbb{E}[Z_n]$.
(e) Conclude that $Z_n \to_p 0$ as $n \to \infty$ and $\mathbb{E}[Z_n] \to 0$ are unrelated.

Exercise 7.5 Find the probability limits (if they exist) of the following sequences of random variables.

(a) $Z_n \sim U[0, 1/n]$.
(b) $Z_n \sim$Bernoulli(p_n) with $p_n = 1 - 1/n$.
(c) $Z_n \sim$Poisson(λ_n) with $\lambda_n = 1/n$.
(d) $Z_n \sim$exponential(λ_n) with $\lambda_n = 1/n$.
(e) $Z_n \sim$Pareto(α_n, β) with $\beta = 1$ and $\alpha_n = n$.
(f) $Z_n \sim$gamma(α_n, β_n) with $\alpha_n = n$ and $\beta_n = n$.
(g) $Z_n = X_n/n$ with $X_n \sim$binomial(n, p).
(h) Z_n has mean μ and variance a/n^r for $a > 0$ and $r > 0$.
(i) $Z_n = X_{(n)}$, the nth order statistic when $X \sim U[0, 1]$.
(j) $Z_n = X_n/n$, where $X_n \sim \chi_n^2$.
(k) $Z_n = 1/X_n$, where $X_n \sim \chi_n^2$.

Exercise 7.6 Take a random sample $\{X_1, \ldots, X_n\}$. Which of the following statistics converge in probability by the WLLN and the CMT? For each, which moments are needed to exist?

(a) $\frac{1}{n}\sum_{i=1}^n X_i^2$.
(b) $\frac{1}{n}\sum_{i=1}^n X_i^3$.
(c) $\max_{i \leq n} X_i$.
(d) $\frac{1}{n}\sum_{i=1}^n X_i^2 - \left(\frac{1}{n}\sum_{i=1}^n X_i\right)^2$.

(e) $\dfrac{\sum_{i=1}^{n} X_i^2}{\sum_{i=1}^{n} X_i}$ assuming $\mathbb{E}[X] > 0$.

(f) $\mathbb{1}\left\{\frac{1}{n}\sum_{i=1}^{n} X_i > 0\right\}$.

(g) $\frac{1}{n}\sum_{i=1}^{n} X_i Y_i$.

Exercise 7.7 A weighted sample mean takes the form $\overline{X}_n^* = \frac{1}{n}\sum_{i=1}^{n} w_i X_i$ for some nonnegative constants w_i satisfying $\frac{1}{n}\sum_{i=1}^{n} w_i = 1$. Assume that X_i is i.i.d.

(a) Show that $\overline{X}_n^*$ is unbiased for $\mu = \mathbb{E}[X]$.

(b) Calculate $\operatorname{var}\left[\overline{X}_n^*\right]$.

(c) Show that a sufficient condition for $\overline{X}_n^* \underset{p}{\longrightarrow} \mu$ is that $n^{-2}\sum_{i=1}^{n} w_i^2 \to 0$.

(d) Show that a sufficient condition for the condition in part c is $\max_{i \le n} w_i \to 0$ as $n \to \infty$.

Exercise 7.8 Take a random sample $\{X_1, \ldots, X_n\}$ and randomly split the sample into two equal subsamples 1 and 2. (For simplicity, assume that n is even.) Let $\overline{X}_{1n}$ and $\overline{X}_{2n}$ the subsample averages. Are $\overline{X}_{1n}$ and $\overline{X}_{2n}$ consistent for the mean $\mu = \mathbb{E}[X]$? (Show this.)

Exercise 7.9 Show that the bias-corrected variance estimator s^2 is consistent for the population variance σ^2.

Exercise 7.10 Find the probability limit for the standard error $s\left(\overline{X}_n\right) = s/\sqrt{n}$ for the sample mean.

Exercise 7.11 Take a random sample $\{X_1, \ldots, X_n\}$, where $X > 0$ and $\mathbb{E}\left|\log X\right| < \infty$. Consider the sample geometric mean

$$\widehat{\mu} = \left(\prod_{i=1}^{n} X_i\right)^{1/n}$$

and population geometric mean

$$\mu = \exp\left(\mathbb{E}\left[\log X\right]\right).$$

Show that $\widehat{\mu} \underset{p}{\longrightarrow} \mu$ as $n \to \infty$.

Exercise 7.12 Take a random variable Z such that $\mathbb{E}[Z] = 0$ and $\operatorname{var}[Z] = 1$. Use Chebyshev's inequality (Theorem 7.1) to find a δ such that $\mathbb{P}[|Z| > \delta] \le 0.05$. Contrast this with the exact δ, which solves $\mathbb{P}[|Z| > \delta] = 0.05$ when $Z \sim \mathrm{N}(0, 1)$. Comment on the difference.

Exercise 7.13 Suppose $Z_n \underset{p}{\longrightarrow} c$ as $n \to \infty$. Show that $Z_n^2 \underset{p}{\longrightarrow} c^2$ as $n \to \infty$ by using the definition of convergence in probability but not appealing to the CMT.

Exercise 7.14 What does the WLLN imply about sample statistics (sample means and variances) calculated on very large samples, such as an administrative database or a census?

CHAPTER 8

CENTRAL LIMIT THEORY

8.1 INTRODUCTION

In Chapter 7, we investigated laws of large numbers, which show that sample averages converge to their population averages. These results are first steps toward but do not provide distributional approximations. In this chapter, we extend asymptotic theory to the next level and obtain asymptotic approximations to the distributions of sample averages. These results are the foundation for inference (hypothesis testing and confidence intervals) for most econometric estimators.

8.2 CONVERGENCE IN DISTRIBUTION

Our primary goal is to obtain the sampling distribution of the sample mean $\overline{X}_n$. The sampling distribution is a function of the distribution F of the observations and the sample size n. An asymptotic approximation is obtained as the limit as $n \to \infty$ of this sampling distribution after suitable standardization.

To develop this theory, we first need to define and understand what is meant by the asymptotic limit of the sampling distribution. The general concept we use is **convergence in distribution**. A sequence of random variables Z_n converges in distribution if the sequence of distribution functions $G_n(u) = \mathbb{P}[Z_n \le u]$ converges to a limit distribution function $G(u)$. We will use the notation G_n and G for the distribution function of Z_n and Z, respectively, to distinguish the sampling distributions from the distribution F of the observations.

Definition 8.1 Let Z_n be a sequence of random variables or vectors with distribution $G_n(u) = \mathbb{P}[Z_n \le u]$. We say that Z_n **converges in distribution** to Z as $n \to \infty$, denoted by $Z_n \underset{d}{\longrightarrow} Z$, if for all u at which $G(u) = \mathbb{P}[Z \le u]$ is continuous, $G_n(u) \to G(u)$ as $n \to \infty$.

Under these conditions, it is also said that G_n **converges weakly** to G. It is common to refer to Z and its distribution $G(u)$ as the **asymptotic distribution, large sample distribution**, or **limit distribution** of Z_n. The caveat in the definition concerning "for all u at which $G(u)$ is continuous" is a technicality and can be safely ignored. Most asymptotic distribution results of interest concern the case where the limit distribution $G(u)$ is everywhere continuous, so this caveat does not apply in such contexts.

When the limit distribution G is degenerate (that is, $\mathbb{P}[Z = c] = 1$ for some c), we can write the convergence as $Z_n \underset{d}{\longrightarrow} c$, which is equivalent to convergence in probability, $Z_n \underset{p}{\longrightarrow} c$.

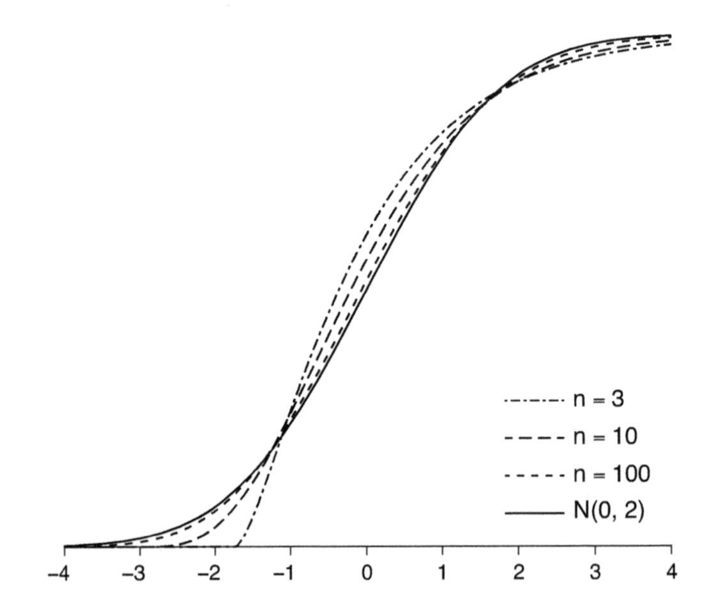

FIGURE 8.1 Convergence of sample mean with chi-square variables

8.3 SAMPLE MEAN

As discussed in Section 8.2, the sampling distribution of the sample mean $\overline{X}_n$ is a function of the distribution F of the observations and the sample size n. We would like to find the asymptotic distribution of $\overline{X}_n$. This means we want to show $\overline{X}_n \underset{d}{\longrightarrow} Z$ for some random variable Z. From the WLLN, we know that $\overline{X}_n \underset{p}{\longrightarrow} \mu$. Since convergence in probability to a constant is the same as convergence in distribution, we have that $\overline{X}_n \underset{d}{\longrightarrow} \mu$ as well. This is not a useful distributional result, as the limit distribution is degenerate. To obtain a nondegenerate distribution, we need to rescale $\overline{X}_n$. Recall that $\operatorname{var}\left[\overline{X}_n - \mu\right] = \sigma^2/n$. Thus we have that $\operatorname{var}\left[\sqrt{n}\left(\overline{X}_n - \mu\right)\right] = \sigma^2$. This result suggests renormalizing the statistic as

$$Z_n = \sqrt{n}\left(\overline{X}_n - \mu\right).$$

Doing so, we find that $\mathbb{E}\left[Z_n\right] = 0$ and $\operatorname{var}\left[Z_n\right] = \sigma^2$. Hence the mean and variance have been stabilized. We now seek to determine the asymptotic distribution of Z_n.

In most cases, it is effectively impossible to calculate the exact distribution of Z_n, even when F is known, but there are some exceptions. The first (and notable) exception is when $X \sim \mathrm{N}(\mu, \sigma^2)$. In this case, we know that $Z_n \sim \mathrm{N}(0, \sigma^2)$. Since this distribution is not changing with n, it means that $\mathrm{N}(0, \sigma^2)$ is the asymptotic distribution of Z_n.

We can calculate the exact distribution of Z_n when $X \sim \chi_1^2$. In this case, the sum $\sum_{i=1}^n X_i$ has an exact chi-square distribution. The distribution of Z_n takes the known form $\sqrt{n}\left(\left(\chi_n^2/n\right) - 1\right)$. This distribution function is displayed in Figure 8.1 for $n = 3$, 10, and 100. We also display the $\mathrm{N}(0, 2)$, which has the same mean and variance. We see that the distribution of Z_n is very different from the normal for small n, but as n increases the distribution approaches the $\mathrm{N}(0, 2)$ distribution. For $n = 100$, it appears to be quite close. This plot provides the intriguing suggestion that the asymptotic distribution of Z_n is normal. The following sections provide further evidence.

8.4 A MOMENT INVESTIGATION

Consider the moments of $Z_n = \sqrt{n}\left(\overline{X}_n - \mu\right)$. We know that the mean is 0 and variance is σ^2. In Section 6.17, we calculated the exact finite sample moments of $Z_n = \sqrt{n}\left(\overline{X}_n - \mu\right)$. Let's look at the third and fourth moments and calculate their limits as $n \to \infty$. They are

$$\mathbb{E}\left[Z_n^3\right] = \kappa_3/n^{1/2} \to 0$$
$$\mathbb{E}\left[Z_n^4\right] = \kappa_4/n + 3\sigma^4 \to 3\sigma^4.$$

The limits are the third and fourth moments of the $\mathrm{N}(0, \sigma^2)$ distribution! (See Section 5.3.) Thus as the sample size increases, these moments converge to those of the normal distribution.

Now let's look at the fifth and sixth moments. They are

$$\mathbb{E}\left[Z_n^5\right] = \kappa_5/n^{3/2} - 10\kappa_3\kappa_2/n^{1/2} \to 0$$
$$\mathbb{E}\left[Z_n^6\right] = \kappa_6/n^2 + \left(15\kappa_4\kappa_2 + 10\kappa_3^2\right)/n + 15\sigma_2^6 \to 15\sigma_2^6.$$

These limits are also the fifth and sixth moments of $\mathrm{N}(0, \sigma^2)$.

Indeed, if X has a finite r^{th} moment, then $\mathbb{E}\left[Z_n^r\right]$ converges to $\mathbb{E}\left[Z^r\right]$ where $Z \sim \mathrm{N}(0, \sigma^2)$.

Convergence of moments is suggestive that we have convergence in distribution, but it is not sufficient. Still, it provides considerable evidence that the asymptotic distribution of Z_n is normal.

8.5 CONVERGENCE OF THE MOMENT GENERATING FUNCTION

It turns out that for most cases of interest (including the sample mean), it is nearly impossible to demonstrate convergence in distribution by working directly with the distribution function of Z_n. Instead it is easier to use the MGF $M_n(t) = \mathbb{E}\left[\exp(tZ_n)\right]$. (See Section 2.23 for the definition.) The MGF is a transformation of the distribution of Z_n and completely characterizes the distribution.

It seems reasonable to expect that if $M_n(t)$ converges to a limit function $M(t)$, then the the distribution of Z_n converges as well. This turns out to be true and is known as Lévy's continuity theorem.

Theorem 8.1 Lévy's Continuity Theorem $Z_n \underset{d}{\longrightarrow} Z$ if $\mathbb{E}\left[\exp(tZ_n)\right] \to \mathbb{E}\left[\exp(tZ)\right]$ for every $t \in \mathbb{R}$.

We now calculate the MGF of $Z_n = \sqrt{n}\left(\overline{X}_n - \mu\right)$. Without loss of generality assume $\mu = 0$. Recall that $M(t) = \mathbb{E}\left[\exp(tX)\right]$ is the MGF of X, and $K(t) = \log M(t)$ is its cumulant generating function. We can write

$$Z_n = \sum_{i=1}^{n} \frac{X_i}{\sqrt{n}}$$

as the sum of independent random variables. From Theorem 4.6, the MGF of a sum of independent random variables is the product of the MGFs. The MGF of $X_i/\sqrt{n}$ is $\mathbb{E}\left[\exp\left(tX/\sqrt{n}\right)\right] = M\left(t/\sqrt{n}\right)$, so that of Z_n is

$$M_n(t) = \prod_{i=1}^{n} M\left(\frac{t}{\sqrt{n}}\right) = M\left(\frac{t}{\sqrt{n}}\right)^n.$$

Thus the cumulant generating function of Z_n takes the simple form

$$K_n(t) = \log M_n(t) = nK\left(\frac{t}{\sqrt{n}}\right). \tag{8.1}$$

By the definition of the cumulant generating function, its power series expansion is

$$K(s) = \kappa_0 + s\kappa_1 + s^2\frac{\kappa_2}{2} + s^3\frac{\kappa_3}{6} + \cdots \tag{8.2}$$

where κ_j are the cumulants of X. Recall that $\kappa_0 = 0$, $\kappa_1 = 0$, and $\kappa_2 = \sigma^2$. Thus

$$K(s) = \frac{s^2}{2}\sigma^2 + \frac{s^3}{6}\kappa_3 + \cdots.$$

Placing this into the expression for $K_n(t)$, we find

$$\begin{aligned} K_n(t) &= n\left(\left(\frac{t}{\sqrt{n}}\right)^2\frac{\sigma^2}{2} + \left(\frac{t}{\sqrt{n}}\right)^3\frac{\kappa_3}{6} + \cdots\right) \\ &= \frac{t^2\sigma^2}{2} + \frac{t^3\kappa_3}{\sqrt{n}6} + \cdots \\ &\to \frac{t^2\sigma^2}{2} \end{aligned}$$

as $n \to \infty$. Hence the cumulant generating function of Z_n converges to $t^2\sigma^2/2$. Thus we have

$$M_n(t) = \exp\left(K_n(t)\right) \to \exp\left(\frac{t^2\sigma^2}{2}\right).$$

Theorem 8.2 For random variables with a finite MGF,

$$M_n(t) = \mathbb{E}\left[\exp\left(tZ_n\right)\right] \to \exp\left(\frac{t^2\sigma^2}{2}\right).$$

For technically interested students, the above proof is heuristic, because it is unclear if truncation of the power series expansion is valid. A rigorous proof is provided in Section 8.15.

8.6 CENTRAL LIMIT THEOREM

Theorem 8.2 shows that the MGF of Z_n converges to $\exp\left(t^2\sigma^2/2\right)$, which is the MGF of the $\mathrm{N}(0, \sigma^2)$ distribution. Combined with Lévy's continuity theorem, this proves that Z_n converges in distribution to $\mathrm{N}(0, \sigma^2)$. Theorem 8.2 applies to any random variable that has a finite MGF, which includes many distributions but excludes others, such as the Pareto and student t. To avoid this restriction, the proof can be updated using the characteristic function $C_n(t) = \mathbb{E}\left[\exp\left(itZ_n\right)\right]$, which exists for all random variables and yields the same asymptotic approximation. It turns out that a sufficient condition is a finite variance. No other condition is needed.

Theorem 8.3 Lindeberg-Lévy Central Limit Theorem (CLT) If X_i are i.i.d. and $\mathbb{E}\left[X^2\right] < \infty$, then as $n \to \infty$,

$$\sqrt{n}\left(\overline{X}_n - \mu\right) \underset{d}{\longrightarrow} \mathrm{N}\left(0, \sigma^2\right)$$

where $\mu = \mathbb{E}\left[X\right]$ and $\sigma^2 = \mathbb{E}\left[(X-\mu)^2\right]$.

As we discussed above, in finite samples, the standardized sum $Z_n = \sqrt{n}\left(\overline{X}_n - \mu\right)$ has mean 0 and variance σ^2. What the CLT adds is that Z_n is also approximately normally distributed and that the normal approximation improves as n increases.

The CLT is one of the most powerful and mysterious results in statistical theory. It shows that the simple process of averaging induces normality. The first version of the CLT (for the number of heads resulting from many tosses of a fair coin) was established by the French mathematician Abraham de Moivre in a private manuscript circulated in 1733. The most general statements for independent variables are credited to work by the Russian mathematician Aleksandr Lyapunov and the Finnish mathematician Jarl Waldemar Lindeberg. The above statement of the theorem is known as the classic (or Lindeberg-Lévy) CLT due to contributions by Lindeberg and the French mathematician Paul Pierre Lévy.

The core of the argument from Section 8.6 is that the MGF of the sample mean is a local transformation of the MGF of X. Locally the cumulant generating function of X is approximately quadratic, so the cumulant generating function of the sample mean is approximately quadratic for large n. The quadratic is the cumulant generating function of the normal distribution. Thus asymptotic normality occurs because the normal distribution has a quadratic cumulant generating function.

8.7 APPLYING THE CENTRAL LIMIT THEOREM

The CLT states that

$$\sqrt{n}\left(\overline{X}_n - \mu\right) \underset{d}{\longrightarrow} \mathrm{N}\left(0, \sigma^2\right) \tag{8.3}$$

which means that the distribution of the random variable $\sqrt{n}\left(\overline{X}_n - \mu\right)$ is approximately the same as $\mathrm{N}\left(0, \sigma^2\right)$ when n is large. We sometimes use the CLT, however, to approximate the distribution of the unnormalized statistic $\overline{X}_n$. In this case, we may write

$$\overline{X}_n \underset{a}{\sim} \mathrm{N}\left(\mu, \frac{\sigma^2}{n}\right). \tag{8.4}$$

Technically, this means nothing different from (8.3). Equation (8.4) is an informal interpretive statement. Its usage, however, is to suggest that $\overline{X}_n$ is approximately distributed as $\mathrm{N}\left(\mu, \sigma^2/n\right)$. The "$a$" means that the left-hand side is approximately (asymptotically upon rescaling) distributed as the right-hand side.

Example: If $X \sim$exponential(λ), then $\overline{X}_n \underset{a}{\sim} \mathrm{N}\left(\lambda, \lambda^2/n\right)$. Furthermore, if $\lambda = 5$ and $n = 100$, $\overline{X}_n \underset{a}{\sim} \mathrm{N}\left(5, 1/4\right)$.

Example: If $X \sim$gamma(2, 1) and $n = 500$, then $\overline{X}_n \underset{a}{\sim} \mathrm{N}\left(2, 1/250\right)$.

Example: U.S. wages. Since $\mu = 24$ and $\sigma^2 = 430$, $\overline{X}_n \underset{a}{\sim} \mathrm{N}\left(24, 0.43\right)$ if $n = 1,000$. We can also use the calculation to revisit the question of Section 7.8: How large does n need to be so that $\overline{X}_n$ is within \$1 of the true mean? The asymptotic approximation states that if $n = 2,862$, then $1/\operatorname{sd}\left(\overline{X}_n\right) = \sqrt{2862/430} = 2.58$. Then

$$\mathbb{P}\left[\left|\overline{X}_n - \mu\right| \geq 1\right] \underset{a}{\sim} \mathbb{P}\left[\left|\mathrm{N}\left(0,1\right)\right| \geq 2.58\right] = 0.01$$

from the normal table. Thus the asymptotic approximation indicates that a sample size of $n = 2,862$ is sufficient for this degree of accuracy, which is considerably different from the $n = 43,000$ obtained by using Chebyshev's inequality.

8.8 MULTIVARIATE CENTRAL LIMIT THEOREM

When $Z_n \in \mathbb{R}^k$ is a random vector, we define convergence in distribution just as for random variables: the joint distribution function converges to a limit distribution.

Multivariate convergence can be reduced to univariate convergence by the following result.

Theorem 8.4 Cramér-Wold Device $Z_n \underset{d}{\longrightarrow} Z$ if and only if $\lambda' Z_n \underset{d}{\longrightarrow} \lambda' Z$ for every $\lambda \in \mathbb{R}^k$ with $\lambda'\lambda = 1$.

Consider vector-valued observations X_i and sample averages $\overline{X}_n$. The mean of $\overline{X}_n$ is $\mu = \mathbb{E}[X]$, and its variance is $n^{-1}\Sigma$, where $\Sigma = \mathbb{E}\left[(X-\mu)(X-\mu)'\right]$. To transform $\overline{X}_n$ so that its mean and variance do not depend on n, we set $Z_n = \sqrt{n}\left(\overline{X}_n - \mu\right)$. This has mean 0 and variance Σ, which are independent of n, as desired.

Using Theorem 8.4, we can derive a multivariate version of Theorem 8.3.

Theorem 8.5 Multivariate Lindeberg-Lévy Central Limit Theorem If $X_i \in \mathbb{R}^k$ are i.i.d. and $\mathbb{E}\|X\|^2 < \infty$ then as $n \to \infty$,

$$\sqrt{n}\left(\overline{X}_n - \mu\right) \underset{d}{\longrightarrow} \mathrm{N}(0, \Sigma)$$

where $\mu = \mathbb{E}[X]$ and $\Sigma = \mathbb{E}\left[(X-\mu)(X-\mu)'\right]$.

The proof of Theorems 8.4 and 8.5 are presented in Section 8.15.

8.9 DELTA METHOD

This section introduces two tools—an extended version of the CMT and the delta method—which allow us to calculate the asymptotic distribution of the plug-in estimator $\widehat{\beta} = h\left(\widehat{\theta}\right)$. We allow for both θ and β to be vector-valued, meaning that β consists of multiple transformations of multiple inputs.

I first present an extended version of the CMT, which allows convergence in distribution. The following result is very important. The wording is technical, so be patient.

Theorem 8.6 Continuous Mapping Theorem (CMT) If $Z_n \underset{d}{\longrightarrow} Z$ as $n \to \infty$, and $h : \mathbb{R}^m \to \mathbb{R}^k$ has the set of discontinuity points D_h such that $\mathbb{P}[Z \in D_h] = 0$, then $h(Z_n) \underset{d}{\longrightarrow} h(Z)$ as $n \to \infty$.

A proof is not provided, as the details are rather advanced. See Theorem 2.3 of van der Vaart (1998). In econometrics, the theorem is typically referred to as the **Continuous Mapping Theorem**. It was first proved by Mann and Wald (1943) and is therefore also known as the **Mann-Wald Theorem**.

The conditions of Theorem 8.6 seem technical. More simply stated, if h is continuous, then $h(Z_n) \underset{d}{\longrightarrow} h(Z)$. That is, convergence in distribution is preserved by continuous transformations. This property is really useful in practice. Most statistics can can be written as functions of sample averages or approximately so. Since sample averages are asymptotically normal by the CLT, the continuous mapping theorem allows us to deduce the asymptotic distribution of the functions of the sample averages.

The technical nature of the conditions allow the function h to be discontinuous if the probability of a discontinuity point is 0. For example, consider the function $h(u) = u^{-1}$. It is discontinuous at $u = 0$. But if the limit distribution equals 0 with probability 0 (as is the case with the CLT), then $h(u)$ satisfies the conditions of Theorem 8.6. Thus if $Z_n \underset{d}{\longrightarrow} Z \sim \mathrm{N}(0,1)$, then $Z_n^{-1} \underset{d}{\longrightarrow} Z^{-1}$.

A special case of the continuous mapping theorem is known as Slutsky's theorem.

Theorem 8.7 Slutsky's theorem If $Z_n \underset{d}{\longrightarrow} Z$ and $c_n \underset{p}{\longrightarrow} c$ as $n \to \infty$, then

1. $Z_n + c_n \underset{d}{\longrightarrow} Z + c$
2. $Z_n c_n \underset{d}{\longrightarrow} Zc$
3. $\frac{Z_n}{c_n} \underset{d}{\longrightarrow} \frac{Z}{c}$ if $c \neq 0$.

Even though Slutsky's theorem is a special case of the CMT, it is a useful statement, as it focuses on the most common applications—addition, multiplication, and division.

Even though the plug-in estimator $\widehat{\beta} = h\left(\widehat{\theta}\right)$ is a function of $\widehat{\theta}$ for which we have an asymptotic distribution, Theorem 8.6 does not directly give us an asymptotic distribution for $\widehat{\beta}$. This is because $\widehat{\beta} = h\left(\widehat{\theta}\right)$ is written as a function of $\widehat{\theta}$, not of the standardized sequence $\sqrt{n}\left(\widehat{\theta} - \theta\right)$. We need an intermediate step: a first-order Taylor series expansion. This step is so critical to statistical theory that it has its own name: **the delta method**.

Theorem 8.8 Delta Method If $\sqrt{n}\left(\widehat{\theta} - \theta\right) \underset{d}{\longrightarrow} \xi$ and $h(u)$ is continuously differentiable in a neighborhood of θ, then as $n \to \infty$

$$\sqrt{n}\left(h\left(\widehat{\theta}\right) - h(\theta)\right) \underset{d}{\longrightarrow} \boldsymbol{H}'\xi \tag{8.5}$$

where $\boldsymbol{H}(u) = \frac{\partial}{\partial u} h(u)'$ and $\boldsymbol{H} = \boldsymbol{H}(\theta)$. In particular, if $\xi \sim \mathrm{N}\left(0, \boldsymbol{V}\right)$, then as $n \to \infty$

$$\sqrt{n}\left(h\left(\widehat{\theta}\right) - h(\theta)\right) \underset{d}{\longrightarrow} \mathrm{N}\left(0, \boldsymbol{H}'\boldsymbol{V}\boldsymbol{H}\right). \tag{8.6}$$

When θ and h are scalar, the expression (8.6) can be written as

$$\sqrt{n}\left(h\left(\widehat{\theta}\right) - h(\theta)\right) \underset{d}{\longrightarrow} \mathrm{N}\left(0, \left(h'(\theta)\right)^2 V\right).$$

The proof is given Section 8.15.

8.10 EXAMPLES

For univariate X with finite variance, $\sqrt{n}\left(\overline{X}_n - \mu\right) \underset{d}{\longrightarrow} \mathrm{N}\left(0, \sigma^2\right)$. The delta method implies the following results:

1. $\sqrt{n}\left(\overline{X}_n^2 - \mu^2\right) \underset{d}{\longrightarrow} \mathrm{N}\left(0, (2\mu)^2 \sigma^2\right)$.
2. $\sqrt{n}\left(\exp\left(\overline{X}_n\right) - \exp(\mu)\right) \underset{d}{\longrightarrow} \mathrm{N}\left(0, \exp(2\mu)\sigma^2\right)$.
3. For $X > 0$, $\sqrt{n}\left(\log\left(\overline{X}_n\right) - \log(\mu)\right) \underset{d}{\longrightarrow} \mathrm{N}\left(0, \frac{\sigma^2}{\mu^2}\right)$.

For bivariate X and Y with finite variances,

$$\sqrt{n}\begin{pmatrix} \overline{X}_n - \mu_X \\ \overline{Y}_n - \mu_Y \end{pmatrix} \underset{d}{\longrightarrow} \mathrm{N}\left(0, \Sigma\right).$$

The delta method implies the following results:

1. $\sqrt{n}\left(\overline{X}_n\overline{Y}_n - \mu_X\mu_Y\right) \underset{d}{\longrightarrow} \mathrm{N}\left(0, h'\Sigma h\right)$, where $h = \begin{pmatrix}\mu_Y\\ \mu_X\end{pmatrix}$.
2. If $\mu_Y \neq 0$, $\sqrt{n}\left(\dfrac{\overline{X}_n}{\overline{Y}_n} - \dfrac{\mu_X}{\mu_Y}\right) \underset{d}{\longrightarrow} \mathrm{N}\left(0, h'\Sigma h\right)$, where $h = \begin{pmatrix}1/\mu_Y\\ -\mu_X/\mu_Y^2\end{pmatrix}$.
3. $\sqrt{n}\left(\overline{X}_n^2 + \overline{X}_n\overline{Y}_n - \left(\mu_X^2 + \mu_X\mu_Y\right)\right) \underset{d}{\longrightarrow} \mathrm{N}\left(0, h'\Sigma h\right)$, where $h = \begin{pmatrix}2\mu_X + \mu_Y\\ \mu_X\end{pmatrix}$.

8.11 ASYMPTOTIC DISTRIBUTION FOR PLUG-IN ESTIMATOR

The delta method allows us to complete our derivation of the asymptotic distribution of the plug-in estimator $\widehat{\beta}$ of β:

Theorem 8.9 If $X_i \in \mathbb{R}^m$ are i.i.d., $\theta = \mathbb{E}\left[g(X)\right] \in \mathbb{R}^\ell$, $\beta = h(\theta) \in \mathbb{R}^k$, $\mathbb{E}\left[\left\|g(X)\right\|^2\right] < \infty$, and $\boldsymbol{H}(u) = \dfrac{\partial}{\partial u} h(u)'$ is continuous in a neighborhood of θ, for $\widehat{\beta} = h\left(\widehat{\theta}\right)$ with $\widehat{\theta} = \frac{1}{n}\sum_{i=1}^n g(X_i)$, then as $n \to \infty$,

$$\sqrt{n}\left(\widehat{\beta} - \beta\right) \underset{d}{\longrightarrow} \mathrm{N}\left(0, \boldsymbol{V}_\beta\right)$$

where $\boldsymbol{V}_\beta = \boldsymbol{H}'\boldsymbol{V}\boldsymbol{H}$, $\boldsymbol{V} = \mathbb{E}\left[\left(g(X) - \theta\right)\left(g(X) - \theta\right)'\right]$, and $\boldsymbol{H} = \boldsymbol{H}(\theta)$.

Theorem 7.8 establishes the consistency of $\widehat{\beta}$ for β, and Theorem 8.9 establishes its asymptotic normality. It is instructive to compare the conditions required for these results. Consistency requires that $g(X)$ have a finite mean, while asymptotic normality requires that this variable has a finite variance. Consistency requires that $h(u)$ be continuous, while our proof of asymptotic normality uses the assumption that $h(u)$ is continuously differentiable.

8.12 COVARIANCE MATRIX ESTIMATION

To use the asymptotic distribution in Theorem 8.9, we need an estimator of the asymptotic covariance matrix $\boldsymbol{V}_\beta = \boldsymbol{H}'\boldsymbol{V}\boldsymbol{H}$. The natural plug-in estimator is

$$\widehat{\boldsymbol{V}}_\beta = \widehat{\boldsymbol{H}}'\widehat{\boldsymbol{V}}\widehat{\boldsymbol{H}}$$
$$\widehat{\boldsymbol{H}} = \boldsymbol{H}\left(\widehat{\theta}\right)$$
$$\widehat{\boldsymbol{V}} = \frac{1}{n-1}\sum_{i=1}^n \left(g(X_i) - \widehat{\theta}\right)\left(g(X_i) - \widehat{\theta}\right)'.$$

Under the assumptions of Theorem 8.9, the WLLN implies $\widehat{\theta} \underset{p}{\longrightarrow} \theta$ and $\widehat{\boldsymbol{V}} \underset{p}{\longrightarrow} \boldsymbol{V}$. The CMT implies $\widehat{\boldsymbol{H}} \underset{p}{\longrightarrow} \boldsymbol{H}$ and with a second application, $\widehat{\boldsymbol{V}}_\beta = \widehat{\boldsymbol{H}}'\widehat{\boldsymbol{V}}\widehat{\boldsymbol{H}} \underset{p}{\longrightarrow} \boldsymbol{H}'\boldsymbol{V}\boldsymbol{H} = \boldsymbol{V}_\beta$. We have established that $\widehat{\boldsymbol{V}}_\beta$ is consistent for $\boldsymbol{V}_\beta$.

Theorem 8.10 Under the assumptions of Theorem 8.9, $\widehat{\boldsymbol{V}}_\beta \underset{p}{\longrightarrow} \boldsymbol{V}_\beta$ as $n \to \infty$.

8.13 *t*-RATIOS

When $k = 1$, we can combine Theorems 8.9 and 8.10 to obtain the asymptotic distribution of the studentized statistic

$$T = \frac{\sqrt{n}\left(\widehat{\beta} - \beta\right)}{\sqrt{\widehat{V}_\beta}} \underset{d}{\longrightarrow} \frac{\mathrm{N}\left(0, V_\beta\right)}{\sqrt{V_\beta}} \sim \mathrm{N}\left(0, 1\right). \tag{8.7}$$

The final equality is by the property that affine functions of normal random variables are normally distributed (Theorem 5.15).

Theorem 8.11 Under the assumptions of Theorem 8.9, $T \underset{d}{\longrightarrow} \mathrm{N}(0, 1)$ as $n \to \infty$.

8.14 STOCHASTIC ORDER SYMBOLS

It is convenient to have simple symbols for random variables and vectors that converge in probability to 0 or are stochastically bounded. In this section, we introduce some of the most common notation.

It might be useful to review the common notation for nonrandom convergence and boundedness. Let x_n and a_n, $n = 1, 2, \ldots$, be nonrandom sequences. The notation

$$x_n = o(1)$$

(pronounced "small oh-one") is equivalent to $x_n \to 0$ as $n \to \infty$. The notation

$$x_n = o(a_n)$$

is equivalent to $a_n^{-1} x_n \to 0$ as $n \to \infty$. The notation

$$x_n = O(1)$$

(pronounced "big oh-one") means that x_n is bounded uniformly in n: There exists an $M < \infty$ such that $|x_n| \leq M$ for all n. The notation

$$x_n = O(a_n)$$

is equivalent to $a_n^{-1} x_n = O(1)$.

I now introduce similar concepts for sequences of random variables. Let Z_n and a_n, $n = 1, 2, \ldots$ be sequences of random variables and constants, respectively. The notation

$$Z_n = o_p(1)$$

("small oh-P-one") means that $Z_n \underset{p}{\longrightarrow} 0$ as $n \to \infty$. For example, for any consistent estimator $\widehat{\theta}$ for θ, we can write

$$\widehat{\theta} = \theta + o_p(1).$$

We also write

$$Z_n = o_p(a_n)$$

if $a_n^{-1} Z_n = o_p(1)$.

Similarly, the notation $Z_n = O_p(1)$ ("big oh-P-one") means that Z_n is bounded in probability. More precisely, for any $\epsilon > 0$, there is a constant $M_\epsilon < \infty$ such that

$$\limsup_{n\to\infty} \mathbb{P}\left[|Z_n| > M_\epsilon\right] \leq \epsilon.$$

Furthermore, we write

$$Z_n = O_p(a_n)$$

if $a_n^{-1} Z_n = O_p(1)$.

$O_p(1)$ is weaker than $o_p(1)$ in the sense that $Z_n = o_p(1)$ implies $Z_n = O_p(1)$ but not the reverse. However, if $Z_n = O_p(a_n)$, then $Z_n = o_p(b_n)$ for any b_n such that $a_n/b_n \to 0$.

If a random variable Z_n converges in distribution, then $Z_n = O_p(1)$. It follows that for estimators $\widehat{\beta}$ which satisfy the convergence of Theorem 8.9, we can write

$$\widehat{\beta} = \beta + O_p(n^{-1/2}). \tag{8.8}$$

In words, this statement says that the estimator $\widehat{\beta}$ equals the true coefficient β plus a random component that is bounded when scaled by $n^{1/2}$. An equivalent way to write equation (8.8) is

$$n^{1/2}\left(\widehat{\beta} - \beta\right) = O_p(1).$$

Another useful observation is that a random sequence with a bounded moment is stochastically bounded.

Theorem 8.12 If Z_n is a random variable which satisfies

$$\mathbb{E}\,|Z_n|^\delta = O\left(a_n\right)$$

for some sequence a_n and $\delta > 0$, then

$$Z_n = O_p(a_n^{1/\delta}).$$

Similarly, $\mathbb{E}\,|Z_n|^\delta = o\left(a_n\right)$ implies $Z_n = o_p(a_n^{1/\delta})$.

Proof: The assumptions imply that there is some $M < \infty$ such that $\mathbb{E}\,|Z_n|^\delta \leq M a_n$ for all n. For any $\epsilon > 0$, set $B = \left(\frac{M}{\epsilon}\right)^{1/\delta}$. Then using Markov's inequality (Theorem 7.3),

$$\mathbb{P}\left[a_n^{-1/\delta}\,|Z_n| > B\right] = \mathbb{P}\left[|Z_n|^\delta > \frac{M a_n}{\epsilon}\right] \leq \frac{\epsilon}{M a_n}\mathbb{E}\,|Z_n|^\delta \leq \epsilon$$

as required. ∎

Many simple rules for manipulating $o_p(1)$ and $O_p(1)$ sequences can be deduced from the CMT or Slutsky's theorem. For example,

$$o_p(1) + o_p(1) = o_p(1)$$

$$o_p(1) + O_p(1) = O_p(1)$$

$$O_p(1) + O_p(1) = O_p(1)$$

$$o_p(1)o_p(1) = o_p(1)$$

$$o_p(1)O_p(1) = o_p(1)$$

$$O_p(1)O_p(1) = O_p(1).$$

8.15 TECHNICAL PROOFS*

Proof of Theorem 8.2 Instead of the power series expansion (8.2), use the mean-value expansion

$$K(s) = K(0) + sK'(0) + \frac{s^2}{2}K''(s^*)$$

where s^* lies between 0 and s. Since $\mathbb{E}\left[X^2\right] < \infty$, the second derivative $K''(s)$ is continuous at 0, so this expansion is valid for small s. Since $K(0) = K'(0) = 0$, this implies $K(s) = \frac{s^2}{2}K''(s^*)$. Inserted into (8.1), this implies

$$K_n(t) = nK\left(\frac{t}{\sqrt{n}}\right) = \frac{t^2}{2}K''\left(\frac{t^*}{\sqrt{n}}\right)$$

where t^* lies between 0 and $t/\sqrt{n}$. For any t, $t/\sqrt{n} \to 0$ as $n \to \infty$ and thus $K''\left(t^*/\sqrt{n}\right) \to K''(0) = \sigma^2$. It follows that

$$K_n(t) \to \frac{t^2\sigma^2}{2}$$

and hence $M_n(t) \to \exp\left(t^2\sigma^2/2\right)$, as stated. ∎

Proof of Theorem 8.4 By Lévy's continuity theorem (Theorem 8.1), $Z_n \underset{d}{\longrightarrow} Z$ if and only if $\mathbb{E}\left[\exp\left(is'Z_n\right)\right] \to \mathbb{E}\left[\exp\left(is'Z\right)\right]$ for every $s \in \mathbb{R}^k$. We can write $s = t\lambda$, where $t \in \mathbb{R}$ and $\lambda \in \mathbb{R}^k$ with $\lambda'\lambda = 1$. Thus the convergence holds if and only if $\mathbb{E}\left[\exp\left(it\lambda'Z_n\right)\right] \to \mathbb{E}\left[\exp\left(it\lambda'Z\right)\right]$ for every $t \in \mathbb{R}$ and $\lambda \in \mathbb{R}^k$ with $\lambda'\lambda = 1$. Again by Lévy's continuity theorem, this holds if and only if $\lambda'Z_n \underset{d}{\longrightarrow} \lambda'Z$ for every $\lambda \in \mathbb{R}^k$ and with $\lambda'\lambda = 1$. ∎

Proof of Theorem 8.5 Set $\lambda \in \mathbb{R}^k$ with $\lambda'\lambda = 1$, and define $U_i = \lambda'\left(X_i - \mu\right)$. The U_i are i.i.d. with $\mathbb{E}\left[U^2\right] = \lambda'\Sigma\lambda < \infty$. By Theorem 8.3,

$$\lambda'\sqrt{n}\left(\overline{X}_n - \mu\right) = \frac{1}{\sqrt{n}}\sum_{i=1}^{n} U_i \underset{d}{\longrightarrow} \mathrm{N}\left(0, \lambda'\Sigma\lambda\right).$$

Notice that if $Z \sim \mathrm{N}(0, \Sigma)$, then $\lambda'Z \sim \mathrm{N}\left(0, \lambda'\Sigma\lambda\right)$. Thus

$$\lambda'\sqrt{n}\left(\overline{X}_n - \mu\right) \underset{d}{\longrightarrow} \lambda'Z.$$

Since this holds for all λ, the conditions of Theorem 8.4 are satisfied, and we deduce that

$$\sqrt{n}\left(\overline{X}_n - \mu\right) \underset{d}{\longrightarrow} Z \sim \mathrm{N}(0, \Sigma)$$

as stated. ∎

Proof of Theorem 8.8 By a vector Taylor series expansion, for each element of h,

$$h_j(\theta_n) = h_j(\theta) + h_{j\theta}(\theta^*_{jn})(\theta_n - \theta)$$

where θ^*_{nj} lies on the line segment between θ_n and θ and therefore converges in probability to θ. It follows that $a_{jn} = h_{j\theta}(\theta^*_{jn}) - h_{j\theta} \underset{p}{\longrightarrow} 0$. Stacking across elements of h, we find

$$\sqrt{n}\left(h\left(\theta_n\right) - h(\theta)\right) = \left(\boldsymbol{H} + a_n\right)'\sqrt{n}\left(\theta_n - \theta\right) \underset{d}{\longrightarrow} \boldsymbol{H}'\xi. \tag{8.9}$$

The convergence is by Theorem 8.6, as $\boldsymbol{H}+a_n \underset{d}{\longrightarrow} \boldsymbol{H}$, $\sqrt{n}\left(\theta_n-\theta\right) \underset{d}{\longrightarrow} \xi$, and their product is continuous. This establishes (8.5)

When $\xi \sim \mathrm{N}(0, \boldsymbol{V})$, the right-hand side of (8.9) equals

$$\boldsymbol{H}'\xi = \boldsymbol{H}'\,\mathrm{N}(0, \boldsymbol{V}) = \mathrm{N}\left(0, \boldsymbol{H}'\boldsymbol{V}\boldsymbol{H}\right)$$

establishing (8.6). ■

8.16 EXERCISES

All exercises assume a sample of independent random variables with n observations.

Exercise 8.1 Let X be distributed Bernoulli $\mathbb{P}[X=1]=p$ and $\mathbb{P}[X=0]=1-p$.

(a) Show that $p=\mathbb{E}[X]$.
(b) Write down the moment estimator $\widehat{p}$ of p.
(c) Find $\operatorname{var}\left[\widehat{p}\right]$.
(d) Find the asymptotic distribution of $\sqrt{n}\left(\widehat{p}-p\right)$ as $n \to \infty$.

Exercise 8.2 Find the moment estimator $\widehat{\mu}_2'$ of $\mu_2'=\mathbb{E}\left[X^2\right]$. Show that $\sqrt{n}\left(\widehat{\mu}_2'-\mu_2'\right) \underset{d}{\longrightarrow} \mathrm{N}\left(0, v^2\right)$ for some v^2. Write v^2 as a function of the moments of X.

Exercise 8.3 Find the moment estimator $\widehat{\mu}_3'$ of $\mu_3'=\mathbb{E}\left[X^3\right]$, and show that $\sqrt{n}\left(\widehat{\mu}_3'-\mu_3'\right) \underset{d}{\longrightarrow} \mathrm{N}\left(0, v^2\right)$ for some v^2. Write v^2 as a function of the moments of X.

Exercise 8.4 Let $\mu_k'=\mathbb{E}\left[X^k\right]$ for some integer $k \geq 1$. Assume $\mathbb{E}\left[X^{2k}\right]<\infty$.

(a) Write down the moment estimator $\widehat{\mu}_k'$ of μ_k'.
(b) Find the asymptotic distribution of $\sqrt{n}\left(\widehat{\mu}_k'-\mu_k'\right)$ as $n \to \infty$.

Exercise 8.5 Let X have an exponential distribution with $\lambda=1$.

(a) Find the first four cumulants of the distribution of X.
(b) Use the equations in Section 8.4 to calculate the third and fourth moments of $Z_n=\sqrt{n}\left(\overline{X}_n-\mu\right)$ for $n=10$, $n=100$, $n=1,000$.
(c) How large does n have to be for the third and fourth moments to be "close" to those of the normal approximation? (Use your judgment to assess closeness.)

Exercise 8.6 Let $m_k=\left(\mathbb{E}\,|X|^k\right)^{1/k}$ for some integer $k \geq 1$.

(a) Write down an estimator $\widehat{m}_k$ of m_k.
(b) Find the asymptotic distribution of $\sqrt{n}\left(\widehat{m}_k-m_k\right)$ as $n \to \infty$.

Exercise 8.7 Assume that $\sqrt{n}\left(\widehat{\theta}-\theta\right) \underset{d}{\longrightarrow} \mathrm{N}\left(0, v^2\right)$. Use the delta method to find the asymptotic distribution of the following statistics:

(a) $\widehat{\theta^2}$.
(b) $\widehat{\theta^4}$.
(c) $\widehat{\theta^k}$.
(d) $\widehat{\theta^2} + \widehat{\theta^3}$.
(e) $\dfrac{1}{1+\widehat{\theta^2}}$.
(f) $\dfrac{1}{1+\exp\left(-\widehat{\theta}\right)}$.

Exercise 8.8 Assume that

$$\sqrt{n}\begin{pmatrix}\widehat{\theta}_1 - \theta_1 \\ \widehat{\theta}_2 - \theta_2\end{pmatrix} \underset{d}{\longrightarrow} \mathrm{N}(0, \Sigma).$$

Use the delta method to find the asymptotic distribution of the following statistics:

(a) $\widehat{\theta}_1\widehat{\theta}_2$.
(b) $\exp\left(\widehat{\theta}_1 + \widehat{\theta}_2\right)$.
(c) If $\theta_2 \neq 0$, $\widehat{\theta}_1/\widehat{\theta}_2^2$.
(d) $\widehat{\theta}_1^3 + \widehat{\theta}_1\widehat{\theta}_2^2$.

Exercise 8.9 Suppose $\sqrt{n}\left(\widehat{\theta} - \theta\right) \underset{d}{\longrightarrow} \mathrm{N}\left(0, v^2\right)$, and set $\beta = \theta^2$ and $\widehat{\beta} = \widehat{\theta}^2$.

(a) Use the delta method to obtain an asymptotic distribution for $\sqrt{n}\left(\widehat{\beta} - \beta\right)$.
(b) Now suppose $\theta = 0$. Describe what happens to the asymptotic distribution from part (a).
(c) Improve on the your answer in part (b). Under the assumption $\theta = 0$, find the asymptotic distribution for $n\widehat{\beta} = n\widehat{\theta}^2$.
(d) Comment on the differences between the answers in parts (a) and (c).

Exercise 8.10 Let $X \sim U[0, b]$ and $M_n = \max_{i\le n} X_i$. Derive the asymptotic distribution using the following the steps.

(a) Calculate the distribution $F(x)$ of $U[0, b]$.
(b) Show that

$$Z_n = n(M_n - b) = n\left(\max_{1\le i\le n} X_i - b\right) = \max_{1\le i\le n} n(X_i - b).$$

(c) Show that the CDF of Z_n is

$$G_n(x) = \mathbb{P}[Z_n \le x] = \left(F\left(b + \frac{x}{n}\right)\right)^n.$$

(d) Derive the limit of $G_n(x)$ as $n \to \infty$ for $x < 0$.
Hint: Use $\lim_{n\to\infty}\left(1 + \frac{x}{n}\right)^n = \exp(x)$.
(e) Derive the limit of $G_n(x)$ as $n \to \infty$ for $x \ge 0$.
(f) Find the asymptotic distribution of Z_n as $n \to \infty$.

Exercise 8.11 Let $X \sim$exponential(1) and $M_n = \max_{i\le n} X_i$. Derive the asymptotic distribution of $Z_n = M_n - \log n$ using similar steps as in exercise 8.10.

CHAPTER 9

ADVANCED ASYMPTOTIC THEORY*

9.1 INTRODUCTION

This chapter presents some advanced results in asymptotic theory. These results are useful for those interested in econometric theory but are less useful for practitioners. Some readers may simply skim this chapter. Students interested in econometric theory should read the chapter in more detail. The chapter presents central limit theory for heterogeneous random variables, a uniform CLT, uniform stochastic bounds, convergence of moments, and higher-order expansions. The latter results will not be used in this volume but are central for the asymptotic theory of bootstrap inference, which is covered in *Econometrics*.

This chapter has no exercises. All proofs are presented in Section 9.11 unless otherwise indicated.

9.2 HETEROGENEOUS CENTRAL LIMIT THEORY

Some versions of the CLT allow heterogeneous distributions.

Theorem 9.1 Lindeberg Central Limit Theorem Suppose that for each n, X_{ni}, $i=1,\ldots,r_n$ are independent but not necessarily identically distributed with expectations $\mathbb{E}[X_{ni}]=0$ and variances $\sigma_{ni}^2=\mathbb{E}\left[X_{ni}^2\right]$. Suppose $\overline{\sigma}_n^2=\sum_{i=1}^{r_n}\sigma_{ni}^2>0$, and for all $\epsilon>0$,

$$\lim_{n\to\infty}\frac{1}{\overline{\sigma}_n^2}\sum_{i=1}^{r_n}\mathbb{E}\left[X_{ni}^2\mathbb{1}\left\{X_{ni}^2\geq\epsilon\overline{\sigma}_n^2\right\}\right]=0. \tag{9.1}$$

Then as $n\to\infty$,

$$\frac{\sum_{i=1}^{r_n}X_{ni}}{\overline{\sigma}_n}\underset{d}{\longrightarrow}\mathrm{N}(0,1).$$

The proof of the Lindeberg CLT is advanced, so I do not present it here. See Billingsley (1995, Theorem 27.2).

The Lindeberg CLT is used by econometricians in some theoretical arguments where we need to treat random variables as arrays (indexed by i and n) or as heterogeneous. For applications, it provides no special insight.

The Lindeberg CLT is quite general, as it puts minimal conditions on the sequence of expectations and variances. The key assumption is equation (9.1), which is known as **Lindeberg's condition.** In its raw form this equation is difficult to interpret. The intuition for (9.1) is that it excludes any single observation from dominating the asymptotic distribution. Since (9.1) is quite abstract, in many contexts we use more elementary conditions that are simpler to interpret. All of the following conditions assume $r_n=n$.

One such alternative is called **Lyapunov's condition**: For some $\delta > 0$,

$$\lim_{n\to\infty} \frac{1}{\overline{\sigma}_n^{2+\delta}} \sum_{i=1}^{n} \mathbb{E}\,|X_{ni}|^{2+\delta} = 0. \tag{9.2}$$

Lyapunov's condition implies Lindeberg's condition and hence the CLT. Indeed, the left side of (9.1) is bounded by

$$\begin{aligned}
&\lim_{n\to\infty} \frac{1}{\overline{\sigma}_n^2} \sum_{i=1}^{n} \mathbb{E}\left[\frac{|X_{ni}|^{2+\delta}}{|X_{ni}|^{\delta}} \mathbb{1}\left\{X_{ni}^2 \geq \epsilon \overline{\sigma}_n^2\right\}\right] \\
&\leq \lim_{n\to\infty} \frac{1}{\epsilon^{\delta/2} \overline{\sigma}_n^{2+\delta}} \sum_{i=1}^{n} \mathbb{E}\,|X_{ni}|^{2+\delta} \\
&= 0
\end{aligned}$$

by (9.2).

Lyapunov's condition is still awkward to interpret. A still simpler condition is a uniform moment bound: For some $\delta > 0$,

$$\sup_{n,i} \mathbb{E}\,|X_{ni}|^{2+\delta} < \infty. \tag{9.3}$$

This is typically combined with the lower variance bound:

$$\liminf_{n\to\infty} \frac{\overline{\sigma}_n^2}{n} > 0. \tag{9.4}$$

These bounds together imply Lyapunov's condition. To see this, note that expressions (9.3) and (9.4) imply that there is some $C < \infty$ such that $\sup_{n,i} \mathbb{E}\left[|X_{ni}|^{2+\delta}\right] \leq C$ and $\liminf_{n\to\infty} n^{-1}\overline{\sigma}_n^2 \geq C^{-1}$. Without loss of generality, assume $\mu_{ni} = 0$. Then the left side of (9.2) is bounded by

$$\lim_{n\to\infty} \frac{C^{2+\delta/2}}{n^{\delta/2}} = 0.$$

Lyapunov's condition holds, and hence the CLT does, too.

An alternative to (9.4) is to assume that the average variance converges to a constant:

$$\frac{\overline{\sigma}_n^2}{n} = n^{-1} \sum_{i=1}^{n} \sigma_{ni}^2 \to \sigma^2 < \infty. \tag{9.5}$$

This assumption is reasonable in many applications.

We now state the simplest and most commonly used version of a heterogeneous CLT based on the Lindeberg CLT.

Theorem 9.2 Suppose that X_{ni} are independent but not necessarily identically distributed, and conditions (9.3) and (9.5) hold. Then as $n \to \infty$,

$$\sqrt{n}\left(\overline{X}_n - \mathbb{E}\left[\overline{X}_n\right]\right) \underset{d}{\longrightarrow} \mathrm{N}\left(0, \sigma^2\right).$$

One advantage of Theorem 9.2 is that it allows $\sigma^2 = 0$ (unlike Theorem 9.1).

9.3 MULTIVARIATE HETEROGENEOUS CENTRAL LIMIT THEORY

The following is a multivariate version of Theorem 9.1.

Theorem 9.3 Multivariate Lindeberg CLT Suppose that for all n, $X_{ni} \in \mathbb{R}^k$, $i = 1, \ldots, r_n$, are independent but not necessarily identically distributed with expectation $\mathbb{E}[X_{ni}] = 0$ and covariance matrices $\Sigma_{ni} = \mathbb{E}\left[X_{ni}X'_{ni}\right]$. Set $\overline{\Sigma}_n = \sum_{i=1}^n \Sigma_{ni}$ and $\nu_n^2 = \lambda_{\min}(\overline{\Sigma}_n)$. Suppose that $\nu_n^2 > 0$, and for all $\epsilon > 0$,

$$\lim_{n\to\infty} \frac{1}{\nu_n^2} \sum_{i=1}^{r_n} \mathbb{E}\left[\|X_{ni}\|^2 \mathbb{1}\left\{\|X_{ni}\|^2 \geq \epsilon \nu_n^2\right\}\right] = 0. \tag{9.6}$$

Then as $n \to \infty$,

$$\overline{\Sigma}_n^{-1/2} \sum_{i=1}^{r_n} X_{ni} \underset{d}{\longrightarrow} \mathrm{N}(0, \boldsymbol{I}_k).$$

The following is a multivariate version of Theorem 9.2.

Theorem 9.4 Suppose that $X_{ni} \in \mathbb{R}^k$ are independent but not necessarily identically distributed with expectations $\mathbb{E}[X_{ni}] = 0$ and covariance matrices $\Sigma_{ni} = \mathbb{E}\left[X_{ni}X'_{ni}\right]$. Set $\overline{\Sigma}_n = n^{-1}\sum_{i=1}^n \Sigma_{ni}$. Suppose that

$$\overline{\Sigma}_n \to \Sigma > 0$$

and for some $\delta > 0$,

$$\sup_{n,i} \mathbb{E}\,\|X_{ni}\|^{2+\delta} < \infty. \tag{9.7}$$

Then as $n \to \infty$,

$$\sqrt{n}\overline{X} \underset{d}{\longrightarrow} \mathrm{N}(0, \Sigma).$$

Similarly to Theorem 9.2, an advantage of Theorem 9.4 is that it allows the covariance matrix Σ to be singular.

9.4 UNIFORM CENTRAL LIMIT THEORY

The Lindeberg-Lévy CLT (Theorem 8.3) states that for any distribution with a finite variance, the sample mean is approximately normally distributed for sufficiently large n. This does not mean, however, that a large sample size implies that the normal distribution is necessarily a good approximation. There is always a finite-variance distribution for which the normal approximation can be made to be arbitrarily poor.

Consider the examples in Section 7.12. Recall that $\overline{X}_n \leq 1$. Thus the standardized sample mean satisfies

$$\frac{\overline{X}_n}{\sqrt{\operatorname{var}\left[\overline{X}_n\right]}} \leq \sqrt{\frac{np}{1-p}}.$$

Suppose that $p = 1/(n+1)$. The statistic is truncated at 1. It follows that the $\mathrm{N}(0,1)$ approximation is not accurate, in particular in the right tail.

The problem is a failure of uniform convergence. While the standardized sample mean converges to the normal distribution for every sampling distribution, it does not do so uniformly. For every sample size, there is a distribution that can cause arbitrary failure in the asymptotic normal approximation.

Similar to a uniform law of large numbers, the solution is to restrict the set of distributions. Unlike the uniform WLLN, however, it is not sufficient to impose an upper moment bound. We also need to prevent asymptotically degenerate variances. A sufficient set of conditions is given in Theorem 9.5.

Theorem 9.5 Let $\mathscr{F}$ denote the set of distributions such that for some $r > 2$, $B < \infty$, and $\delta > 0$, $\mathbb{E}|X|^r \leq B$ and $\operatorname{var}[X] \geq \delta$. Then for all x, as $n \to \infty$

$$\sup_{F \in \mathfrak{F}} \left| \mathbb{P}\left[\frac{\sqrt{n}\left(\bar{X}_n - \mathbb{E}[X]\right)}{\sqrt{\operatorname{var}[X]}} \leq x \right] - \Phi(x) \right| \to 0$$

where $\Phi(x)$ is the standard normal distribution function.

Theorem 9.5 states that the standardized sample mean converges uniformly over $\mathscr{F}$ to the normal distribution. This is a much stronger result than the classic CLT (Theorem 8.3) which is pointwise in $\mathscr{F}$.

The proof of Theorem 9.5 is refreshingly straightforward. If the theorem were false, there would be a sequence of distributions $F_n \in \mathscr{F}$ and some x for which

$$\mathbb{P}_n\left[\frac{\sqrt{n}\left(\bar{X}_n - \mathbb{E}[X]\right)}{\sqrt{\operatorname{var}[X]}} \leq x \right] \to \Phi(x) \tag{9.8}$$

fails, where $\mathbb{P}_n$ denotes the probability calculated under the distribution F_n. However, as discussed after Theorem 9.1, the assumptions of Theorem 9.5 imply that under the sequence F_n, the Lindeberg condition (9.1) holds and hence the Lindeberg CLT holds. Thus (9.8) holds for all x.

9.5 UNIFORM INTEGRABILITY

To allow for nonidentically distributed random variables, let us introduce a new concept called **uniform integrability**. A random variable X is said to be **integrable** if $\mathbb{E}|X| = \int_{-\infty}^{\infty} |x|\, dF < \infty$, or equivalently, if

$$\lim_{M\to\infty} \mathbb{E}\left[|X|\, \mathbb{1}\{|X| > M\}\right] = \lim_{M\to\infty} \int_M^{\infty} |x|\, dF = 0.$$

A sequence of random variables is said to be uniformly integrable if the limit is 0 uniformly over the sequence.

Definition 9.1 The sequence of random variables Z_n is **uniformly integrable** as $n \to \infty$ if

$$\lim_{M\to\infty} \limsup_{n\to\infty} \mathbb{E}\left[|Z_n|\, \mathbb{1}\{|Z_n| > M\}\right] = 0.$$

If X_i is i.i.d. and $\mathbb{E}|X| < \infty$, then the sequence X_i is uniformly integrable. Uniform integrability is more general than integrability, allowing for nonidentically distributed random variables, yet imposing enough homogeneity so that many results for i.i.d. random variables hold as well for uniformly integrable random variables. For example, if X_i is an independent and uniformly integrable sequence, then the WLLN holds for the sample mean.

Uniform integrability is a rather abstract concept. It is implied by a uniformly bounded moment larger than 1.

Theorem 9.6 If for some $r > 1$, $\mathbb{E}|Z_n|^r \leq C < \infty$, then Z_n is uniformly integrable.

We can apply uniform integrability to powers of random variables. In particular, we say Z_n is **uniformly square integrable** if $|Z_n|^2$ is uniformly integrable, thus if

$$\lim_{M\to\infty} \limsup_{n\to\infty} \mathbb{E}\left[|Z_n|^2 \mathbb{1}\left\{|Z_n|^2 > M\right\}\right] = 0. \tag{9.9}$$

Uniform square integrability is similar (but slightly stronger) to the Lindeberg condition (9.1) when $\overline{\sigma}_n^2 \geq \delta > 0$. To see this, assume that (9.9) holds for $Z_n = X_{ni}$. Then for any $\epsilon > 0$, there is an M large enough so that $\limsup_{n\to\infty} \mathbb{E}\left[Z_n^2 \mathbb{1}\left\{Z_n^2 > M\right\}\right] \leq \epsilon\delta$. Since $\epsilon n \overline{\sigma}_n^2 \to \infty$, we have

$$\frac{1}{n\overline{\sigma}_n^2} \sum_{i=1}^n \mathbb{E}\left[X_{ni}^2 \mathbb{1}\left\{X_{ni}^2 \geq \epsilon n \overline{\sigma}_n^2\right\}\right] \leq \frac{\epsilon\delta}{\overline{\sigma}_n^2} \leq \epsilon$$

which implies (9.1).

9.6 UNIFORM STOCHASTIC BOUNDS

For some applications, it can be useful to obtain the stochastic order of the random variable

$$\max_{1\leq i\leq n} |X_i| .$$

This is the magnitude of the largest observation in the sample $\{X_1, \ldots, X_n\}$. If the support of the distribution of X_i is unbounded, then as the sample size n increases, the largest observation will also tend to increase. It turns out that there is a simple characterization of this behavior under uniform integrability.

Theorem 9.7 If $|X_i|^r$ is uniformly integrable, then as $n \to \infty$

$$n^{-1/r} \max_{1\leq i\leq n} |X_i| \underset{p}{\longrightarrow} 0. \tag{9.10}$$

Equation (9.10) says that under the assumed condition, the largest observation will diverge at a rate slower than $n^{1/r}$. As r increases, this rate slows. Thus the higher the moments, the slower the rate of divergence.

9.7 CONVERGENCE OF MOMENTS

Sometimes we are interested in moments (often the mean and variance) of a statistic Z_n. When Z_n is a normalized sample mean, we have explicit expressions for the integer moments of Z_n (as presented in Section 6.17). But for other statistics, such as nonlinear functions of the sample mean, such expressions are not available.

The statement $Z_n \underset{d}{\longrightarrow} Z$ means that we can approximate the distribution of Z_n with that of Z. In this case, we may think of approximating the moments of Z_n with those of Z. This approach can be rigorously justified if the moments of Z_n converge to the corresponding moments of Z. In this section, we explore conditions under which this convergence holds.

We first give a sufficient condition for the existence of the mean of the asymptotic distribution.

Theorem 9.8 If $Z_n \underset{d}{\longrightarrow} Z$ and $\mathbb{E}|Z_n| \leq C$, then $\mathbb{E}|Z| \leq C$.

We next consider conditions under which $\mathbb{E}[Z_n]$ converges to $\mathbb{E}[Z]$. One might guess that the conditions of Theorem 9.8 would be sufficient, but a counterexample demonstrates that this is incorrect. Let Z_n be a random variable that equals n with probability $1/n$ and equals 0 with probability $1-1/n$. Then $Z_n \underset{d}{\longrightarrow} Z$, where $\mathbb{P}[Z=0]=1$. We can also calculate that $\mathbb{E}[Z_n]=1$. Thus the assumptions of Theorem 9.8 are satisfied. However, $\mathbb{E}[Z_n]=1$ does not converge to $\mathbb{E}[Z]=0$. Thus the boundedness of moments $\mathbb{E}|Z_n| \leq C < \infty$ is not sufficient to ensure the convergence of moments. The problem is due to a lack of what is called "tightness" of the sequence of distributions. The culprit is the small probability mass[1] which "escapes to infinity".

The solution is uniform integrability, which is the key condition that allows us to establish the convergence of moments.

Theorem 9.9 If $Z_n \underset{d}{\longrightarrow} Z$ and Z_n is uniformly integrable, then $\mathbb{E}[Z_n] \to \mathbb{E}[Z]$.

We complete this section by giving conditions under which moments of $Z_n = \sqrt{n}\left(\overline{X}_n - \mathbb{E}\left[\overline{X}_n\right]\right)$ converge to those of the normal distribution. Section 6.17 presents exact expressions for the integer moments of Z_n. We now consider noninteger moments as well.

Theorem 9.10 If X_{ni} satisfies the conditions of Theorem 9.2, and $\sup_{n,i} \mathbb{E}|X_{ni}|^r < \infty$ for some $r>2$, then for any $0<s<r$, $\mathbb{E}|Z_n|^s \longrightarrow \mathbb{E}|Z|^s$, where $Z \sim \mathrm{N}\left(0,\sigma^2\right)$.

9.8 EDGEWORTH EXPANSION FOR THE SAMPLE MEAN

The CLT shows that normalized estimators are approximately normally distributed if the sample size n is sufficiently large. In practice, how good is this approximation? One way to measure the discrepancy between the actual distribution and the asymptotic distribution is by **higher-order expansions**. Higher-order expansions of the distribution function are known as **Edgeworth expansions**.

Let $G_n(x)$ be the distribution function of the normalized mean $Z_n = \sqrt{n}\left(\overline{X}_n - \mu\right)/\sigma$ for a random sample. An Edgeworth expansion is a series representation for $G_n(x)$ expressed as powers of $n^{-1/2}$. It equals

$$G_n(x) = \Phi(x) - n^{-1/2}\frac{\kappa_3}{6}He_2(x)\phi(x) - n^{-1}\left(\frac{\kappa_4}{24}He_3(x) + \frac{\kappa_3^2}{72}He_5(x)\right)\phi(x) + o\left(n^{-1}\right) \tag{9.11}$$

where $\Phi(x)$ and $\phi(x)$ are the standard normal distribution and density functions, κ_3 and κ_4 are the third and fourth cumulants of X, and $He_j(x)$ is the jth Hermite polynomial (see Section 5.10).

Below we give a justification for (9.11).

Sufficient regularity conditions for the validity of the Edgeworth expansion (9.11) are $\mathbb{E}\left[X^4\right] < \infty$ and that the characteristic function of X is bounded below 1. The latter—known as Cramer's condition—requires X to have an absolutely continuous distribution.

The expression (9.11) shows that the exact distribution $G_n(x)$ can be written as the sum of the normal distribution, an $n^{-1/2}$ correction for the main effect of skewness, and an n^{-1} correction for the main effect of kurtosis and the secondary effect of skewness. The $n^{-1/2}$ skewness correction is an even function[2] of x which means that it changes the distribution function symmetrically about zero. This means that this term captures

---

[1] The probability mass at n.

[2] A function $f(x)$ is **even** if $f(-x) = f(x)$.

skewness in the distribution of Z_n. The n^{-1} correction is an odd function[3] of x, which means that this term moves probability mass symmetrically either away from, or toward, the center. This term captures kurtosis in the distribution of Z_n.

We now derive (9.11) using the MGF $M_n(t) = \mathbb{E}\left[\exp(tZ_n)\right]$ of the normalized mean Z_n. For a more rigorous argument, the characteristic funtion could be used with minimal change in details. For simplicity, assume $\mu = 0$ and $\sigma^2 = 1$. In the proof of the CLT (Theorem 8.3), we showed that

$$M_n(t) = \exp\left(nK\left(\frac{t}{\sqrt{n}}\right)\right)$$

where $K(t) = \log\left(\mathbb{E}\left[\exp(tX)\right]\right)$ is the cumulant generating function of X (see Section 2.24). By a series expansion about $t = 0$, and using the facts $K(0) = K^{(1)}(0) = 0$, $K^{(2)}(0) = 1$, $K^{(3)}(0) = \kappa_3$ and $K^{(4)}(0) = \kappa_4$, the above expression equals

$$\begin{aligned} M_n(t) &= \exp\left(\frac{t^2}{2} + n^{-1/2}\frac{\kappa_3}{6}t^3 + n^{-1}\frac{\kappa_4}{24}t^4 + o\left(n^{-1}\right)\right) \\ &= \exp\left(t^2/2\right) + n^{-1/2}\exp\left(t^2/2\right)\frac{\kappa_3}{6}t^3 + n^{-1}\exp\left(t^2/2\right)\left(\frac{\kappa_4}{24}t^4 + \frac{\kappa_3^2}{72}t^6\right) + o\left(n^{-1}\right). \end{aligned} \tag{9.12}$$

The second line holds by a second-order expansion of the exponential function.

By the formula for the normal MGF, the fact $He_0(x) = 1$, and repeated integration by parts applying property 4 from Theorem 5.25, we find

$$\begin{aligned} \exp\left(t^2/2\right) &= \int_{-\infty}^{\infty} e^{tx}\phi(x)dx \\ &= \int_{-\infty}^{\infty} e^{tx}He_0(x)\phi(x)dx \\ &= t^{-1}\int_{-\infty}^{\infty} e^{tx}He_1(x)\phi(x)dx \\ &= t^{-2}\int_{-\infty}^{\infty} e^{tx}He_2(x)\phi(x)dx \\ &\ \ \vdots \\ &= t^{-j}\int_{-\infty}^{\infty} e^{tx}He_j(x)\phi(x)dx. \end{aligned}$$

This implies that for any $j \geq 0$,

$$\exp\left(t^2/2\right)t^j = \int_{-\infty}^{\infty} e^{tx}He_j(x)\phi(x)dx.$$

Substituting into (9.12), we find

$$\begin{aligned} M_n(t) = &\int_{-\infty}^{\infty} e^{tx}\phi(x)dx + n^{-1/2}\frac{\kappa_3}{6}\int_{-\infty}^{\infty} e^{tx}He_3(x)\phi(x)dx \\ &+ n^{-1}\left(\frac{\kappa_4}{24}\int_{-\infty}^{\infty} e^{tx}He_4(x)\phi(x)dx + \frac{\kappa_3^2}{72}\int_{-\infty}^{\infty} e^{tx}He_6(x)\phi(x)dx\right) + o\left(n^{-1}\right) \end{aligned}$$

[3] A function $f(x)$ is **odd** if $f(-x) = -f(x)$.

$$= \int_{-\infty}^{\infty} e^{tx} \left(\phi(x) + n^{-1/2} \frac{\kappa_3}{6} He_3(x)\phi(x) + n^{-1} \left(\frac{\kappa_4}{24} He_4(x)\phi(x) + \frac{\kappa_3^2}{72} He_6(x)\phi(x) \right) \right) dx$$

$$= \int_{-\infty}^{\infty} e^{tx} d \left(\Phi(x) - n^{-1/2} \frac{\kappa_3}{6} He_2(x)\phi(x) - n^{-1} \left(\frac{\kappa_4}{24} He_3(x) + \frac{\kappa_3^2}{72} He_5(x) \right) \phi(x) \right)$$

where the third equality uses property 4 from Theorem 5.25. The final line shows that this is the MGF of the distribution in brackets, which is (9.11). We have shown that the MGF expansion of Z_n equals that of (9.11), so they are identical, as claimed.

9.9 EDGEWORTH EXPANSION FOR SMOOTH FUNCTION MODEL

Most applications of Edgeworth expansions concern statistics that are more complicated than sample means. The following result applies to general smooth functions of means that includes most estimators. This result includes that of Section 9.8 as a special case.

Theorem 9.11 Assume $X_i \in \mathbb{R}^m$ are i.i.d., $\beta = h(\theta) \in \mathbb{R}$, $\theta = \mathbb{E}[g(X)] \in \mathbb{R}^\ell$, $\mathbb{E}\|g(X)\|^4 < \infty$, $h(u)$ has four continuous derivatives in a neighborhood of θ, and $\mathbb{E}[\exp(t\|g(X)\|)] \le B < 1$. Then for $\widehat{\beta} = h(\widehat{\theta})$, $\widehat{\theta} = \frac{1}{n}\sum_{i=1}^n g(X_i)$, $\boldsymbol{V} = \mathbb{E}\left[(g(X) - \theta)(g(X) - \theta)'\right]$, and $\boldsymbol{H} = \boldsymbol{H}(\theta)$, as $n \to \infty$

$$\mathbb{P}\left[\frac{\sqrt{n}(\widehat{\beta} - \beta)}{\sqrt{\boldsymbol{H}'\boldsymbol{V}\boldsymbol{H}}} \le x\right] = \Phi(x) + n^{-1/2} p_1(x)\phi(x) + n^{-1} p_2(x)\phi(x) + o\left(n^{-1}\right)$$

uniformly in x, where $p_1(x)$ is an even polynomial of order 2, and $p_2(x)$ is an odd polynomial of degree 5, with coefficients depending on the moments of $g(X)$ up to order 4.

For a proof, see Theorem 2.2 of Hall (1992).

This Edgeworth expansion is identical in form to equation (9.11) derived in Section 9.8 for the sample mean. The only difference is in the coefficients of the polynomials.

We are also interested in expansions for studentized statistics, such as the t-ratio. Theorem 9.11 applies to such cases as well, so long as the variance estimator can be written as a function of sample means.

Theorem 9.12 Under the asssumptions of Theorem 9.11, if in addition, $\mathbb{E}\|g(X)\|^8 < \infty$, $h(u)$ has five continuous derivatives in a neighborhood of θ, $\boldsymbol{H}'\boldsymbol{V}\boldsymbol{H} > 0$, and $\mathbb{E}\left[\exp\left(t\|g(X)\|^2\right)\right] \le B < 1$, for

$$T = \frac{\sqrt{n}(\widehat{\beta} - \beta)}{\sqrt{\widehat{V}_\beta}}$$

and $\widehat{V}_\beta = \widehat{\boldsymbol{H}}'\widehat{\boldsymbol{V}}\widehat{\boldsymbol{H}}$ as defined in Section 8.12, then as $n \to \infty$

$$\mathbb{P}[T \le x] = \Phi(x) + n^{-1/2} p_1(x)\phi(x) + n^{-1} p_2(x)\phi(x) + o\left(n^{-1}\right)$$

uniformly in x, where $p_1(x)$ is an even polynomial of order 2, and $p_2(x)$ is an odd polynomial of degree 5, with coefficients depending on the moments of $g(X)$ up to order 8.

This Edgeworth expansion is nearly identical in form to the others presented, with the only difference appearing in the coefficients of the polynomials.

To see that Theorem 9.11 implies Theorem 9.12, define $\overline{g}(X_i)$ as the vector stacking $g(X_i)$ and all squares and cross products of the elements of $g(X_i)$. Set

$$\overline{\mu}_n = \frac{1}{n}\sum_{i=1}^{n} \overline{g}(X_i).$$

Notice that $h(\widehat{\theta})$, $\widehat{\boldsymbol{H}} = \boldsymbol{H}(\widehat{\theta})$, and $\widehat{\boldsymbol{V}} = \frac{1}{n}\sum_{i=1}^{n} g(X_i)\, g(X_i)' - \widehat{\theta}\widehat{\theta}'$ are all functions of $\overline{\mu}_n$. Apply Theorem 9.11 to $\sqrt{n}\overline{h}(\overline{\mu}_n)$, where

$$\overline{h}(\overline{\mu}_n) = \frac{h(\widehat{\mu}) - h(\mu)}{\sqrt{\widehat{\boldsymbol{H}}'\widehat{\boldsymbol{V}}\widehat{\boldsymbol{H}}}}.$$

The assumption $\mathbb{E}\left\|g(X)\right\|^8 < \infty$ implies $\mathbb{E}\left\|\overline{g}(X)\right\|^4 < \infty$, and the assumptions that $h(u)$ has five continuous derivatives and $\boldsymbol{H}'\boldsymbol{V}\boldsymbol{H} > 0$ imply that $\overline{h}(u)$ has four continuous derivatives. Thus the conditions of Theorem 9.11 are satisfied. Hence Theorem 9.12 follows.

Theorem 9.12 is an Edgeworth expansion for a standard t-ratio. One implication is that when the normal distribution $\Phi(x)$ is used as an approximation to the actual distribution $\mathbb{P}[T \le x]$, the error is

$$\mathbb{P}[T \le x] - \Phi(x) = n^{-1/2} p_1(x)\phi(x) + O\left(n^{-1}\right) = O\left(n^{-1/2}\right).$$

Sometimes we are interested in the distribution of the absolute value of the t-ratio $|T|$. It has the distribution

$$\mathbb{P}[|T| \le x] = \mathbb{P}[-x \le T \le x] = \mathbb{P}[T \le x] - \mathbb{P}[T < x].$$

From Theorem 9.12, we find that this equals

$$\begin{aligned}
&\Phi(x) + n^{-1/2} p_1(x)\phi(x) + n^{-1} p_2(x)\phi(x) \\
&- \left(\Phi(-x) + n^{-1/2} p_1(-x)\phi(-x) + n^{-1} p_2(-x)\phi(-x)\right) + o\left(n^{-1}\right) \\
&= 2\Phi(x) - 1 + n^{-1} 2 p_2(x)\phi(x) + o\left(n^{-1}\right)
\end{aligned}$$

where the equality holds since $\Phi(-x) = 1 - \Phi(x)$, $\phi(-x) = \phi(x)$, $p_1(-x) = p_1(x)$ (since ϕ and p_1 are even functions), and $p_2(-x) = -p_2(x)$ (since p_2 is an odd function). Thus when the normal distribution $2\Phi(x) - 1$ is used as an approximation to the actual distribution $\mathbb{P}[|T| \le x]$, the error is

$$\mathbb{P}[|T| \le x] - (2\Phi(x) - 1) = n^{-1} 2 p_2(x)\phi(x) + o\left(n^{-1}\right) = O\left(n^{-1}\right).$$

What is occurring is that the $O\left(n^{-1/2}\right)$ skewness term affects the two distributional tails equally and is offsetting. One tail has extra probability, and the other has too little (relative to the normal approximation), so they offset. In contrast, the $O\left(n^{-1}\right)$ kurtosis term affects the two tails equally with the same sign, so the effect doubles (either both tails have too much probability or both have too little probability relative to the normal).

There is also a version of the delta method for Edgeworth expansions. Essentially, if two random variables differ by $O_p(a_n)$, then they have the same Edgeworth expansions up to $O(a_n)$.

Theorem 9.13 Suppose the distribution of a random variable T has the Edgeworth expansion

$$\mathbb{P}[T \le x] = \Phi(x) + a_n^{-1} p_1(x)\phi(x) + o\left(a_n^{-1}\right)$$

and a random variable X satisfies $X = T + o_p\left(a_n^{-1}\right)$. Then X has the Edgeworth expansion

$$\mathbb{P}[X \le x] = \Phi(x) + a_n^{-1} p_1(x)\phi(x) + o\left(a_n^{-1}\right).$$

To prove this result, note that the assumption $X = T + o_p\left(a_n^{-1}\right)$ means that for any $\epsilon > 0$, there is n sufficiently large such that $\mathbb{P}\left[|X - T| > a_n^{-1}\epsilon\right] \le \epsilon$. Then

$$\begin{aligned}
\mathbb{P}[X \le x] &\le \mathbb{P}\left[X \le x, |X - T| \le a_n^{-1}\epsilon\right] + \epsilon \\
&\le \mathbb{P}\left[T \le x + a_n^{-1}\epsilon\right] + \epsilon \\
&= \Phi(x + a_n^{-1}\epsilon) + a_n^{-1}p_1(x + a_n^{-1}\epsilon)\phi(x + a_n^{-1}\epsilon) + \epsilon + o\left(a_n^{-1}\right) \\
&\le \Phi(x) + a_n^{-1}p_1(x)\phi(x) + \frac{a_n^{-1}\epsilon}{\sqrt{2\pi}} + \epsilon + o\left(a_n^{-1}\right) \\
&\le \Phi(x) + a_n^{-1}p_1(x)\phi(x) + o\left(a_n^{-1}\right).
\end{aligned}$$

The last inequality holds since ϵ is arbitrary. Similarly, $\mathbb{P}[X \le x] \ge \Phi(x) + n^{-1/2}p_1(x)\phi(x) + o\left(a_n^{-1}\right)$.

9.10 CORNISH-FISHER EXPANSIONS

The Sections 9.8 and 9.9 describe expansions for distribution functions. For some purposes, it is useful to have similar expansions for the quantiles of the distribution function. Such expansions are known as Cornish-Fisher expansions. Recall that the αth quantile of a continuous distribution $F(x)$ is the solution to $F(q) = \alpha$. Suppose that a statistic T has distribution $G_n(x) = \mathbb{P}[T \le x]$. For any $\alpha \in (0, 1)$, its αth quantile q_n is the solution to $G_n(q_n) = \alpha$. A Cornish-Fisher expansion expresses q_n in terms of the αth normal quantile q plus approximation error terms.

Theorem 9.14 Suppose the distribution of a random variable T has the Edgeworth expansion

$$G_n(x) = \mathbb{P}[T \le x] = \Phi(x) + n^{-1/2}p_1(x)\phi(x) + n^{-1}p_2(x)\phi(x) + o\left(n^{-1}\right)$$

uniformly in x. For some $\alpha \in (0, 1)$, let q_n and q be the αth quantiles of $G_n(x)$ and $\Phi(x)$, respectively, that is, the solutions to $G_n(q_n) = \alpha$ and $\Phi(q) = \alpha$. Then

$$q_n = q + n^{-1/2}p_{11}(q) + n^{-1}p_{21}(q) + o(n^{-1}) \tag{9.13}$$

where

$$p_{11}(x) = -p_1(x) \tag{9.14}$$

$$p_{21}(x) = -p_2(x) + p_1(x)p_1'(x) - \frac{1}{2}xp_1(x)^2. \tag{9.15}$$

Under the conditions of Theorem 9.12, the functions $p_{11}(x)$ and $p_{21}(x)$ are even and odd functions of x, with coefficients depending on the moments of $h(X)$ up to order 4.

Theorem 9.14 can be derived from the Edgeworth expansion using Taylor expansions. Evaluating the Edgeworth expansion at q_n and substituting in (9.13), we have

$$\begin{aligned}
\alpha &= G_n(q_n) \\
&= \Phi(q_n) + n^{-1/2}p_1(q_n)\phi(q_n) + n^{-1}p_2(q_n)\phi(q_n) + o\left(n^{-1}\right) \\
&= \Phi\left(q + n^{-1/2}p_{11}(q) + n^{-1}p_{21}(q)\right) \\
&\quad + n^{-1/2}p_1(q + n^{-1/2}p_{11}(q))\phi(z_\alpha + n^{-1/2}p_{11}(q)) \\
&\quad + n^{-1}p_{21}(q) + o(n^{-1}).
\end{aligned}$$

Next, expand $\Phi(x)$ in a second-order Taylor expansion about q, and $p_1(x)$ and $\phi(x)$ in first-order expansions. We find that the above expression equals

$$\Phi(q)+n^{-1/2}\phi(q)\left(p_{11}(q)+p_1(q)\right)$$
$$+n^{-1}\phi(q)\left(p_{21}(q)-\frac{qp_1(q)^2}{2}+p_1'(q)p_{11}(q)-qp_1(q)p_{11}(q)+p_2(q)\right)+o(n^{-1}).$$

For this expression to equal α, we deduce that $p_{11}(x)$ and $p_{21}(x)$ must take the values given in (9.14) and (9.15).

9.11 TECHNICAL PROOFS*

Proof of Theorem 9.2 Without loss of generality, suppose $\mathbb{E}[X_{ni}]=0$. First, suppose that $\sigma^2=0$. Then $\mathrm{var}\left[\sqrt{n}\,\overline{X}_n\right]=\overline{\sigma}_n^2\to\sigma^2=0$, so $\sqrt{n}\,\overline{X}_n\xrightarrow[p]{}0$ and hence $\sqrt{n}\,\overline{X}_n\xrightarrow[d]{}0$. The random variable $\mathrm{N}\left(0,\sigma^2\right)=\mathrm{N}(0,0)$ is 0 with probability 1, so this is $\sqrt{n}\,\overline{X}_n\xrightarrow[d]{}\mathrm{N}\left(0,\sigma^2\right)$, as stated.

Now suppose that $\sigma^2>0$. This implies (9.4). Together with (9.3) this implies Lyapunov's condition, and hence Lindeberg's condition, and hence Theorem 9.1, which states $\overline{\sigma}_n^{-1/2}\sqrt{n}\,\overline{X}_n\xrightarrow[d]{}\mathrm{N}(0,1)$. Combined with (9.5), we deduce $\sqrt{n}\,\overline{X}_n\xrightarrow[d]{}\mathrm{N}\left(0,\sigma^2\right)$, as stated. ■

Proof of Theorem 9.3 Set $\lambda\in\mathbb{R}^k$ with $\lambda'\lambda=1$, and define $U_{ni}=\lambda_n'\overline{\Sigma}^{-1/2}X_{ni}$. The matrix $\boldsymbol{A}^{1/2}$ is called a "matrix square root of $\boldsymbol{A}$" and is defined as a solution to $\boldsymbol{A}^{1/2}\boldsymbol{A}^{1/2\prime}=\boldsymbol{A}$. Notice that the U_{ni} are independent and have variance $\sigma_{ni}^2=\lambda'\overline{\Sigma}_n^{-1/2}\Sigma_{ni}\overline{\Sigma}_n^{-1/2}\lambda$ and $\overline{\sigma}_n^2=\sum_{i=1}^{r_n}\sigma_{ni}^2=1$. It is sufficient to verify (9.1). By the Schwarz inequality (Theorem 4.19),

$$\begin{aligned}U_{ni}^2&=\left(\lambda'\overline{\Sigma}_n^{-1/2}X_{ni}\right)^2\\&\le\lambda'\overline{\Sigma}_n^{-1}\lambda\,\|X_{ni}\|^2\\&\le\frac{\|X_{ni}\|^2}{\lambda_{\min}\left(\overline{\Sigma}_n\right)}\\&=\frac{\|X_{ni}\|^2}{\nu_n^2}.\end{aligned}$$

Then

$$\begin{aligned}\frac{1}{\overline{\sigma}_n^2}\sum_{i=1}^{r_n}\mathbb{E}\left[U_{ni}^2\mathbb{1}\left\{U_{ni}^2\ge\epsilon\overline{\sigma}_n^2\right\}\right]&=\sum_{i=1}^{r_n}\mathbb{E}\left[U_{ni}^2\mathbb{1}\left\{U_{ni}^2\ge\epsilon\right\}\right]\\&\le\frac{1}{\nu_n^2}\sum_{i=1}^{r_n}\mathbb{E}\left[\|X_{ni}\|^2\,\mathbb{1}\left\{\|X_{ni}\|^2\ge\epsilon\nu_n^2\right\}\right]\\&\to 0\end{aligned}$$

by (9.6). This establishes (9.1). We deduce from Theorem 9.1 that

$$\sum_{i=1}^{r_n} u_{ni} = \lambda' \overline{\Sigma}_n^{-1/2} \sum_{i=1}^{r_n} X_{ni} \underset{d}{\longrightarrow} \mathrm{N}(0,1) = \lambda' Z$$

where $Z \sim \mathrm{N}(0, \boldsymbol{I}_k)$. Since this holds for all λ, the conditions of Theorem 8.4 are satisfied, and we deduce that

$$\overline{\Sigma}_n^{-1/2} \sum_{i=1}^{r_n} X_{ni} \underset{d}{\longrightarrow} \mathrm{N}(0, \boldsymbol{I}_k)$$

as stated. ■

Proof of Theorem 9.4 Set $\lambda \in \mathbb{R}^k$ with $\lambda'\lambda = 1$, and define $U_{ni} = \lambda' X_{ni}$. Using the Schwarz inequality (Theorem 4.19) and Assumption (9.7), we obtain

$$\sup_{n,i} \mathbb{E} |U_{ni}|^{2+\delta} = \sup_{n,i} \mathbb{E} \left|\lambda' X_{ni}\right|^{2+\delta} \leq \|\lambda\|^{2+\delta} \sup_{n,i} \mathbb{E} \|X_{ni}\|^{2+\delta} = \sup_{n,i} \mathbb{E} \|X_{ni}\|^{2+\delta} < \infty$$

which is (9.3). Notice that

$$\frac{1}{n}\sum_{i=1}^{n} \mathbb{E}\left[U_{ni}^2\right] = \lambda' \frac{1}{n}\sum_{i=1}^{n} \Sigma_{ni}\lambda = \lambda' \overline{\Sigma}_n \lambda \to \lambda' \Sigma \lambda$$

which is (9.5). Since the U_{ni} are independent, by Theorem 8.5,

$$\lambda' \sqrt{n}\, \overline{X}_n = \frac{1}{\sqrt{n}} \sum_{i=1}^{n} U_{ni} \underset{d}{\longrightarrow} \mathrm{N}\left(0, \lambda'\Sigma\lambda\right) = \lambda' Z$$

where $Z \sim \mathrm{N}(0, \Sigma)$. Since this holds for all λ, the conditions of Theorem 8.4 are satisfied, and we deduce that

$$\sqrt{n}\, \overline{X}_n \underset{d}{\longrightarrow} \mathrm{N}(0, \Sigma)$$

as stated. ■

Proof of Theorem 9.6 Fix $\epsilon > 0$ and set $M \geq (C/\epsilon)^{1/(r-1)}$. Then

$$\begin{aligned}
\mathbb{E}\left[|Z_n| \mathbb{1}\{|Z_n| > M\}\right] &= \mathbb{E}\left[\frac{|Z_n|^r}{|Z_n|^{r-1}} \mathbb{1}\{|Z_n| > M\}\right] \\
&\leq \frac{\mathbb{E}\left[|Z_n|^r \mathbb{1}\{|Z_n| > M\}\right]}{M^{r-1}} \\
&\leq \frac{\mathbb{E}|Z_n|^r}{M^{r-1}} \\
&\leq \frac{C}{M^{r-1}} \\
&\leq \epsilon.
\end{aligned}$$

■

Proof of Theorem 9.7 Take any $\delta > 0$. The event $\left\{\max_{1\leq i\leq n} |X_i| > \delta n^{1/r}\right\}$ means that at least one of the $|X_i|$ exceeds $\delta n^{1/r}$, which is the same as the event $\bigcup_{i=1}^{n} \left\{|X_i| > \delta n^{1/r}\right\}$ or equivalently, $\bigcup_{i=1}^{n} \left\{|X_i|^r > \delta^r n\right\}$. By

Booles' inequality (Theorem 1.2, property 6) and Markov's inequality (Theorem 7.3),

$$\begin{aligned}\mathbb{P}\left[n^{-1/r}\max_{1\le i\le n}|X_i|>\delta\right]&=\mathbb{P}\left[\bigcup_{i=1}^{n}\left\{|X_i|^r>\delta^r n\right\}\right]\\&\le\sum_{i=1}^{n}\mathbb{P}\left[|X_i|^r>n\delta^r\right]\\&\le\frac{1}{n\delta^r}\sum_{i=1}^{n}\mathbb{E}\left[|X_i|^r\mathbb{1}\left\{|X_i|^r>n\delta^r\right\}\right]\\&=\frac{1}{\delta^r}\max_{i\le n}\mathbb{E}\left[|X_i|^r\mathbb{1}\left\{|X_i|^r>n\delta^r\right\}\right].\end{aligned}$$

Since $|X_i|^r$ is uniformly integrable, the final expression converges to 0 as $n\delta^r\to\infty$. This establishes (9.10). ■

Proof of Theorem 9.8 We first establish the following expectation equality. For any nonnegative random variable X with $\mathbb{E}[X]<\infty$,

$$\mathbb{E}[X]=\int_0^\infty \mathbb{P}[X>u]\,du. \tag{9.16}$$

To show this, let $F(u)$ be the distribution function of X, and set $F^*(u)=1-F(u)$. By integration by parts,

$$\mathbb{E}[X]=\int_0^\infty udF(u)=-\int_0^\infty udF^*(u)=-\left[uF^*(u)\right]_0^\infty+\int_0^\infty F^*(u)du=\int_0^\infty \mathbb{P}[X>u]\,du.$$

Let $F_n(u)$ and $F(u)$ be the distribution functions of $|Z_n|$ and $|Z|$. Using (9.16), Definition 8.1, Fatou's lemma (Theorem A.24), again (9.16), and the bound $\mathbb{E}|Z_n|\le C$, we have

$$\begin{aligned}\mathbb{E}|Z|&=\int_0^\infty(1-F(x))\,dx\\&=\int_0^\infty\lim_{n\to\infty}(1-F_n(x))\,dx\\&\le\liminf_{n\to\infty}\int_0^\infty(1-F_n(x))\,dx\\&=\liminf_{n\to\infty}\mathbb{E}|Z_n|\le C\end{aligned}$$

as required. ■

Proof of Theorem 9.9 Without loss of generality, assume Z_n is scalar and $Z_n\ge 0$. Let $a\wedge b=\min(a,b)$. Fix $\epsilon>0$. By Theorem 9.8, Z is integrable, and by assumption, Z_n is uniformly integrable. Thus we can find an $M<\infty$ such that for all large n,

$$\mathbb{E}[Z-(Z\wedge M)]=\mathbb{E}[(Z-M)\mathbb{1}\{Z>M\}]\le\mathbb{E}[Z\mathbb{1}\{Z>M\}]\le\epsilon$$

and

$$\mathbb{E}[Z_n-(Z_n\wedge M)]=\mathbb{E}[(Z_n-M)\mathbb{1}\{Z_n>M\}]\le\mathbb{E}[Z_n\mathbb{1}\{Z_n>M\}]\le\epsilon.$$

The function $(Z_n \wedge M)$ is continuous and bounded. Since $Z_n \xrightarrow[d]{} Z$, boundedness implies $\mathbb{E}[Z_n \wedge M] \to \mathbb{E}[Z \wedge M]$. Thus for n sufficiently large,

$$|\mathbb{E}[(Z_n \wedge M) - (Z \wedge M)]| \leq \epsilon.$$

Applying the triangle inequality and the above three inequalities, we find

$$|\mathbb{E}[Z_n - Z]| \leq |\mathbb{E}[Z_n - (Z_n \wedge M)]| + |\mathbb{E}[(Z_n \wedge M) - (Z \wedge M)]| + |\mathbb{E}[Z - (Z \wedge M)]| \leq 3\epsilon.$$

Since ϵ is arbitrary, we conclude $|\mathbb{E}[Z_n - Z]| \to 0$, as required. ■

Proof of Theorem 9.10 Theorem 9.2 establishes $Z_n \xrightarrow[d]{} Z$, and the CMT establishes $Z_n^s \xrightarrow[d]{} Z^s$. We now establish that Z_n^s is uniformly integrable. By Lyapunov's inequality (Theorem 2.11), Minkowski's inequality (Theorem 4.16), and $\sup_{n,i} \mathbb{E}|X_{ni}|^r = B < \infty$, we have

$$\left(\mathbb{E}|X_{ni} - \mathbb{E}[X_{ni}]|^2\right)^{1/2} \leq \left(\mathbb{E}|X_{ni} - \mathbb{E}[X_{ni}]|^r\right)^{1/r} \leq 2\left(\mathbb{E}|X_{ni}|^r\right)^{1/r} \leq 2B^{1/r}. \tag{9.17}$$

The Rosenthal inequality (see *Econometrics*, Appendix B) establishes that there is a constant $R_r < \infty$ such that

$$\mathbb{E}\left|\sum_{i=1}^n (X_{ni} - \mathbb{E}[X_{ni}])\right|^r \leq R_r \left\{\left(\sum_{i=1}^n \mathbb{E}|X_{ni} - \mathbb{E}[X_{ni}]|^2\right)^{r/2} + \sum_{i=1}^n \mathbb{E}|X_{ni} - \mathbb{E}[X_{ni}]|^r\right\}.$$

Therefore,

$$\begin{aligned}
\mathbb{E}|Z_n|^r &= \frac{1}{n^{r/2}}\mathbb{E}\left(\left|\sum_{i=1}^n (X_{ni} - \mathbb{E}[X_{ni}])\right|^r\right) \\
&\leq \frac{1}{n^{r/2}} R_r \left\{\left(\sum_{i=1}^n \mathbb{E}|X_{ni} - \mathbb{E}[X_{ni}]|^2\right)^{r/2} + \sum_{i=1}^n \mathbb{E}|X_{ni} - \mathbb{E}[X_{ni}]|^r\right\} \\
&\leq \frac{1}{n^{r/2}} R_r \left\{\left(n4B^{2/r}\right)^{r/2} + n2^r B\right\} \\
&\leq 2^{r+1} R_r B.
\end{aligned}$$

The second inequality is (9.17). This shows that $\mathbb{E}|Z_n|^r$ is uniformly bounded, so $|Z_n|^s$ is uniformly integrable for any $s < r$ by Theorem 9.6. Since $|Z_n|^s \xrightarrow[d]{} |Z_n|^s$ and $|Z_n|^s$ is uniformly integrable, by Theorem 9.9, we conclude that $\mathbb{E}|Z_n|^s \to \mathbb{E}|Z|^s$, as stated. ■

CHAPTER 10

MAXIMUM LIKELIHOOD ESTIMATION

10.1 INTRODUCTION

A major class of statistical inference concerns maximum likelihood estimation of parametric models. These are statistical models that are complete probability functions. Parametric models are especially popular in structural economic modeling.

10.2 PARAMETRIC MODEL

A **parametric model** for X is a complete probability function depending on an unknown parameter vector θ. (In many of the examples in this chapter, the parameter θ will be scalar; but in most actual applications, the parameter will be a vector.) In the discrete case, we can write a parametric model as a probability mass function $\pi(X \mid \theta)$. In the continuous case, we can write it as a density function $f(x \mid \theta)$. The parameter θ belongs to a set Θ called the **parameter space**.

A parametric model specifies that the population distribution is a member of a specific collection of distributions. We often call the collection a **parametric family**.

For example, one parametric model is $X \sim \mathrm{N}\left(\mu, \sigma^2\right)$, which has density $f(x \mid \mu, \sigma^2) = \sigma^{-1}\phi((x-\mu)/\sigma)$. The parameters are $\mu \in \mathbb{R}$ and $\sigma^2 > 0$.

As another example, a parametric model is that X is exponentially distributed with density $f(x \mid \lambda) = \lambda^{-1}\exp(-x/\lambda)$ and parameter $\lambda > 0$.

A parametric model does not need to be one of the textbook functional forms listed in Chapter 3. A parametric model can be developed by a user for a specific application. What is common is that the model is a complete probability specification.

We focus on unconditional distributions, meaning that the probability function does not depend on conditioning variables. In *Econometrics*, we will study conditional models, where the probability function depends on conditioning variables.

The main advantage of the parametric modeling approach is that $f(x \mid \theta)$ is a full description of the population distribution of X. This makes it useful for causal inference and policy analysis. A disadvantage is that it may be sensitive to misspecification.

A parametric model specifies the distribution of all observations. In this book, we are focused on random samples; thus the observations are i.i.d. A parametric model is specified as follows.

Definition 10.1 A **model** for a random sample is the assumption that X_i, $i = 1, \ldots, n$, are i.i.d. with known density function $f(x \mid \theta)$ or mass function $\pi(x \mid \theta)$ with unknown parameter $\theta \in \Theta$.

The model is correctly specified when there exists a parameter such that the model corresponds to the true data distribution.

Definition 10.2 A model is **correctly specified** when there exists a unique parameter value $\theta_0 \in \Theta$ such that $f(x \mid \theta_0) = f(x)$, the true data distribution. This parameter value θ_0 is called the **true parameter value**. The parameter θ_0 is **unique** if there is no other θ such that $f(x \mid \theta_0) = f(x \mid \theta)$. A model is **misspecified** if there is no parameter value $\theta \in \Theta$ such that $f(x \mid \theta) = f(x)$.

For example, suppose the true density is $f(x) = 2e^{-2x}$. The exponential model $f(x \mid \lambda) = \lambda^{-1} \exp(-x/\lambda)$ is a correctly specified model with $\lambda_0 = 1/2$. The gamma model is also correctly specified with $\beta_0 = 1/2$ and $\alpha_0 = 1$. The lognormal model, in contrast, is misspecified, as there are no parameters such that the lognormal density equals $f(x) = 2e^{-2x}$.

As another example, suppose that the true density is $f(x) = \phi(x)$. In this case, correctly specified models include the normal and the student t, but not, for example, the logistic. Consider the mixture of normals model $f\left(x \mid p, \mu_1, \sigma_1^2, \mu_2, \sigma_2^2\right) = p\phi_{\sigma_1}(x - \mu_1) + (1 - p)\phi_{\sigma_2}(x - \mu_2)$. This includes $\phi(x)$ as a special case, so it is a correct model, but the "true" parameter is not unique. True parameter values include $\left(p, \mu_1, \sigma_1^2, \mu_2, \sigma_2^2\right) = \left(p, 0, 1, 0, 1\right)$ for any p, $\left(1, 0, 1, \mu_2, \sigma_2^2\right)$ for any μ_2 and σ_2^2, and $\left(0, \mu_1, \sigma_1^2, 0, 1\right)$ for any μ_1 and σ_1^2. Thus although this model is correct, it does not meet the above definition of correctly specified.

Likelihood theory is typically developed under the assumption that the model is correctly specified. It is possible, however, that a given parametric model is misspecified. In this case, it is useful to discuss the properties of estimation under the assumption of misspecification. We do so in Sections 10.16–10.19.

The primary tools of likelihood analysis are the likelihood, the maximum likelihood estimator, and the Fisher information. We derive these sequentially and discuss their use.

10.3 LIKELIHOOD

The likelihood is the joint density of the observations calculated using the model. Independence of the observations means that the joint density is the product of the individual densities. "Identical distributions" means that all the densities are identical, so that the joint density equals the following expression:

$$f(x_1, \ldots, x_n \mid \theta) = f(x_1 \mid \theta) f(x_2 \mid \theta) \cdots f(x_n \mid \theta) = \prod_{i=1}^{n} f(x_i \mid \theta).$$

The joint density evaluated at the observed data and viewed as a function of θ is called the **likelihood function**.

Definition 10.3 The **likelihood function** is

$$L_n(\theta) \equiv f(X_1, \ldots, X_n \mid \theta) = \prod_{i=1}^{n} f(X_i \mid \theta)$$

for continuous random variables and

$$L_n(\theta) \equiv \prod_{i=1}^{n} \pi(X_i \mid \theta)$$

for discrete random variables.

In probability theory, we typically use a density (or distribution) to describe the probability that X takes specified values. In likelihood analysis, we flip the usage. As the data are given to us, we use the likelihood function to describe which values of θ are most compatible with the data.

The goal of estimation is to find the value θ which best describes the data—and ideally, is closest to the true value θ_0. As the density function $f(x \mid \theta)$ shows us which values of X are most likely to occur given a specific value of θ, the likelihood function $L_n(\theta)$ shows us the values of θ most likely to have generated the observations. The value of θ most compatible with the observations is the value which maximizes the likelihood. This is a reasonable estimator of θ.

Definition 10.4 The **maximum likelihood estimator** $\widehat{\theta}$ of θ is the value which maximizes $L_n(\theta)$:

$$\widehat{\theta} = \underset{\theta \in \Theta}{\operatorname{argmax}}\, L_n(\theta).$$

Example: $f(x \mid \lambda) = \lambda^{-1} \exp(-x/\lambda)$. The likelihood function is

$$L_n(\lambda) = \prod_{i=1}^{n} \left(\frac{1}{\lambda} \exp\left(-\frac{X_i}{\lambda} \right) \right) = \frac{1}{\lambda^n} \exp\left(-\frac{n\overline{X}_n}{\lambda} \right).$$

The first-order condition for maximization is

$$0 = \frac{d}{d\lambda} L_n(\lambda) = -n \frac{1}{\lambda^{n+1}} \exp\left(-\frac{n\overline{X}_n}{\lambda} \right) + \frac{1}{\lambda^n} \exp\left(-\frac{n\overline{X}_n}{\lambda} \right) \frac{n\overline{X}_n}{\lambda^2}.$$

Canceling the common terms and solving, we find the unique solution which is the maximum likelihood estimator (MLE) for λ:

$$\widehat{\lambda} = \overline{X}_n.$$

This likelihood function is displayed in Figure 10.1(a) for $\overline{X}_n = 2$. The maximizer as marked is the MLE $\widehat{\lambda}$.

In most cases, it is more convenient to calculate and maximize the logarithm of the likelihood function than the level of the likelihood.

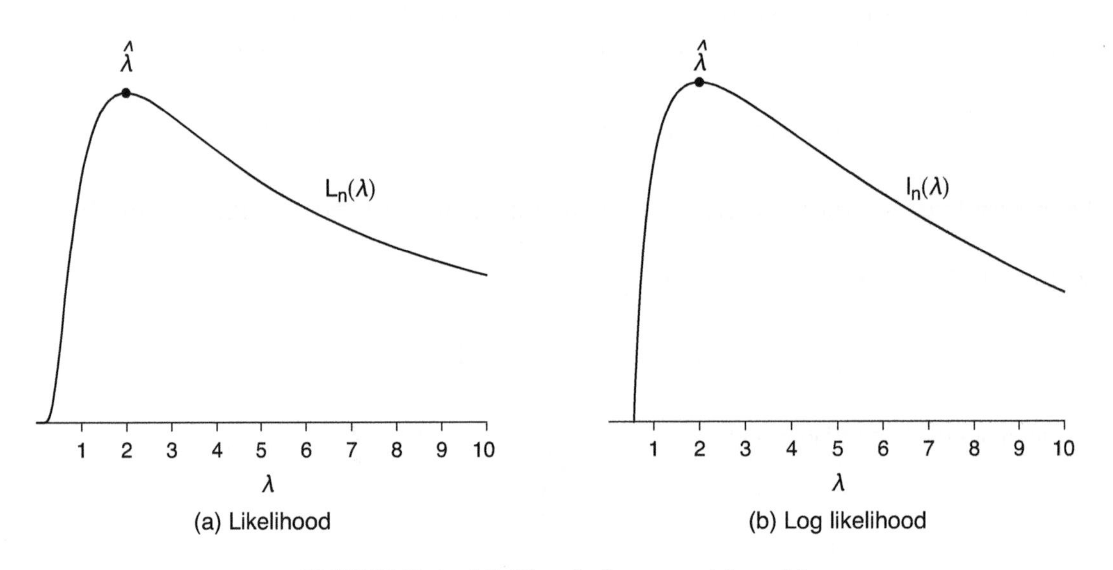

FIGURE 10.1 Likelihood of exponential model

Definition 10.5 The **log-likelihood function** is

$$\ell_n(\theta) \equiv \log L_n(\theta) = \sum_{i=1}^{n} \log f(X_i \mid \theta).$$

One reason this construct is more convenient is because the log likelihood is a sum of the individual log densities rather than the product. A second reason is because for many parametric models, the log density is computationally more robust than the level density.

Example: $f(x \mid \lambda) = \lambda^{-1} \exp(-x/\lambda)$. The log density is $\log f(x \mid \lambda) = -\log \lambda - x/\lambda$. The log-likelihood function is

$$\ell_n(\lambda) \equiv \log L_n(\lambda) = \sum_{i=1}^{n} \left(-\log \lambda - \frac{X_i}{\lambda} \right) = -n \log \lambda - \frac{n\overline{X}_n}{\lambda}.$$

Example: $\pi\left(x \mid p\right) = p^x(1-p)^{1-x}$. The log mass function is $\log \pi(x) = x \log p + (1-x)\log(1-p)$. The log-likelihood function is

$$\ell_n(p) = \sum_{i=1}^{n} X_i \log p + (1 - X_i)\log(1-p) = n\overline{X}_n \log p + n\left(1 - \overline{X}_n\right)\log(1-p).$$

The maximizer of the likelihood and log-likelihood functions are the same, since the logarithm is a monotonically increasing function.

Theorem 10.1 $\widehat{\theta} = \underset{\theta \in \Theta}{\operatorname{argmax}}\, \ell_n(\theta)$.

Example: $f(x \mid \lambda) = \lambda^{-1} \exp(-x/\lambda)$. The first-order condition is

$$0 = \frac{d}{d\lambda}\ell_n(\lambda) = -\frac{n}{\lambda} + \frac{n\overline{X}_n}{\lambda^2}.$$

The unique solution is $\widehat{\lambda} = \overline{X}_n$. The second-order condition is

$$\frac{d^2}{d\lambda^2}\ell_n(\widehat{\lambda}) = \frac{n}{\widehat{\lambda}^2} - 2\frac{n\overline{X}_n}{\widehat{\lambda}^3} = -\frac{n}{\overline{X}_n^2} < 0.$$

This condition verifies that $\widehat{\lambda}$ is a maximizer rather than a minimizer. This log-likelihood function is displayed in Figure 10.1(b) for $\overline{X}_n = 2$. The maximizer as marked is the MLE $\widehat{\lambda}$, and it equals the maximizer in panel (a). You can see that the likelihood $L_n(\lambda)$ in panel (a) and log-likelihood $\ell_n(\lambda)$ in panel (b) have similar shapes.

Example: $\pi\left(x \mid p\right) = p^x(1-p)^{1-x}$. The first-order condition is

$$0 = \frac{d}{dp}\ell_n(p) = \frac{n\overline{X}_n}{p} - \frac{n\left(1 - \overline{X}_n\right)}{1-p}.$$

The unique solution is $\widehat{p} = \overline{X}_n$. The second-order condition is

$$\frac{d^2}{dp^2}\ell_n(\widehat{p}) = -\frac{n\overline{X}_n}{\widehat{p}^2} - \frac{n\left(1-\overline{X}_n\right)}{\left(1-\widehat{p}\right)^2} = -\frac{n}{\widehat{p}\left(1-\widehat{p}\right)} < 0.$$

Thus $\widehat{p}$ is the maximizer.

10.4 LIKELIHOOD ANALOG PRINCIPLE

I have suggested that a general way to construct estimators is as sample analogs of population parameters. Let us now show that the MLE is a sample analog of the true parameter.

Define the **expected log density** function

$$\ell(\theta) = \mathbb{E}\left[\log f(X \mid \theta)\right]$$

which is a function of the parameter θ.

Theorem 10.2 When the model is correctly specified, the true parameter θ_0 maximizes the expected log density $\ell(\theta)$:

$$\theta_0 = \underset{\theta\in\Theta}{\operatorname{argmax}}\, \ell(\theta).$$

For a proof, see Section 10.20.

This theorem shows that the true parameter θ_0 maximizes the expectation of the log density. The sample analog of $\ell(\theta)$ is the average log likelihood

$$\overline{\ell}_n(\theta) = \frac{1}{n}\ell_n(\theta) = \frac{1}{n}\sum_{i=1}^{n}\log f(X_i \mid \theta)$$

which has maximizer $\widehat{\theta}$, because normalizing by n^{-1} does not alter the maximizer. In this sense, the MLE $\widehat{\theta}$ is an analog estimator of θ_0, because θ_0 maximizes $\ell(\theta)$ and $\widehat{\theta}$ maximizes the sample analog $\overline{\ell}_n(\theta)$.

Example: $f(x \mid \lambda) = \lambda^{-1}\exp(-x/\lambda)$. The log density is $\log f(x \mid \lambda) = -\log\lambda - x/\lambda$, which has expected value

$$\ell(\lambda) = \mathbb{E}\left[\log f(X \mid \lambda)\right] = \mathbb{E}\left[-\log\lambda - X/\lambda\right] = -\log\lambda - \frac{\mathbb{E}\left[X\right]}{\lambda} = -\log\lambda - \frac{\lambda_0}{\lambda}.$$

The first-order condition for maximization is $-\lambda^{-1} + \lambda_0\lambda^{-2} = 0$, which has the unique solution $\lambda = \lambda_0$. The second-order condition is negative, so this is a maximum. This shows that the true parameter λ_0 is the maximizer of $\mathbb{E}\left[\log f(X \mid \lambda)\right]$.

Example: $\pi\left(x \mid p\right) = p^x(1-p)^{1-x}$. The log density is $\log\pi(x \mid p) = x\log p + (1-x)\log\left(1-p\right)$, which has expected value

$$\ell(p) = \mathbb{E}\left[\log\pi(X \mid p)\right] = \mathbb{E}\left[X\log p + (1-X)\log\left(1-p\right)\right] = p_0\log p + (1-p_0)\log\left(1-p\right).$$

The first-order condition for maximization is

$$\frac{p_0}{p} - \frac{1-p_0}{1-p} = 0$$

which has the unique solution $p = p_0$. The second-order condition is negative. Hence the true parameter p_0 is the maximizer of $\mathbb{E}\left[\log \pi(X \mid p)\right]$.

Example: Normal distribution with known variance. The density of X is

$$f(x \mid \mu) = \frac{1}{\left(2\pi\sigma_0^2\right)^{1/2}} \exp\left(-\frac{(x-\mu)^2}{2\sigma_0^2}\right).$$

The log density is

$$\log f(x \mid \mu) = -\frac{1}{2}\log(2\pi\sigma_0^2) - \frac{(x-\mu)^2}{2\sigma_0^2}.$$

Thus

$$\ell(\mu) = -\frac{1}{2}\log(2\pi\sigma_0^2) - \frac{\mathbb{E}\left[(X-\mu)^2\right]}{2\sigma_0^2} = -\frac{1}{2}\log(2\pi\sigma_0^2) - \frac{(\mu_0-\mu)^2}{2\sigma_0^2} - \frac{1}{2\sigma_0^2}.$$

As this expression is a quadratic in μ, we can see by inspection that it is maximized at $\mu = \mu_0$.

10.5 INVARIANCE PROPERTY

A special property of the MLE (not shared by all estimators) is that it is invariant to transformations.

Theorem 10.3 If $\widehat{\theta}$ is the MLE of $\theta \in \mathbb{R}^m$, then for any transformation $\beta = h(\theta) \in \mathbb{R}^\ell$, the MLE of β is $\widehat{\beta} = h\left(\widehat{\theta}\right)$.

For a proof, see Section 10.20.

Example: $f(x \mid \lambda) = \lambda^{-1}\exp(-x/\lambda)$. We know the MLE is $\widehat{\lambda} = \overline{X}_n$. Set $\beta = 1/\lambda$, so $h(\lambda) = 1/\lambda$. The log density of the reparameterized model is $\log f(x \mid \beta) = \log \beta - x\beta$. The log-likelihood function is

$$\ell_n(\beta) = n \log \beta - \beta n \overline{X}_n$$

which has maximizer $\widehat{\beta} = 1/\overline{X}_n = h(\overline{X}_n)$, as claimed.

Invariance is an important property, as applications typically involve calculations which are effectively transformations of the estimators. The invariance property shows that these calculations are MLE of their population counterparts.

10.6 EXAMPLES

To find the MLE, take the following steps:

1. Construct $f(x \mid \theta)$ as a function of x and θ.
2. Take the logarithm $\log f(x \mid \theta)$.
3. Evaluate at $x = X_i$ and sum over i: $\ell_n(\theta) = \sum_{i=1}^n \log f(X_i \mid \theta)$.
4. If possible, solve the first-order condition (FOC) to find the maximum.
5. Check the second-order condition to verify that it is a maximum.
6. If solving the FOC is not possible, use numerical methods to maximize $\ell_n(\theta)$.

Example: Normal distribution with known variance. The density of X is

$$f(x \mid \mu) = \frac{1}{\left(2\pi\sigma_0^2\right)^{1/2}} \exp\left(-\frac{(x-\mu)^2}{2\sigma_0^2}\right).$$

The log density is

$$\log f(x \mid \mu) = -\frac{1}{2}\log(2\pi\sigma_0^2) - \frac{(x-\mu)^2}{2\sigma_0^2}.$$

The log likelihood is

$$\ell_n(\mu) = -\frac{n}{2}\log(2\pi\sigma_0^2) - \frac{1}{2\sigma_0^2}\sum_{i=1}^{n}(X_i-\mu)^2.$$

The FOC for $\widehat{\mu}$ is

$$\frac{\partial}{\partial\mu}\ell_n(\widehat{\mu}) = \frac{1}{\sigma_0^2}\sum_{i=1}^{n}(X_i-\widehat{\mu}) = 0.$$

The solution is

$$\widehat{\mu} = \overline{X}_n.$$

The second-order condition is

$$\frac{\partial^2}{\partial\mu^2}\ell_n(\widehat{\mu}) = -\frac{n}{\sigma_0^2} < 0$$

as required. The log likelihood is displayed in Figure 10.2 for $\overline{X}_n = 1$. The maximizer is marked as the MLE.

Example: Normal distribution with known mean. The density of X is

$$f\left(x \mid \sigma^2\right) = \frac{1}{\left(2\pi\sigma^2\right)^{1/2}} \exp\left(-\frac{(x-\mu_0)^2}{2\sigma^2}\right).$$

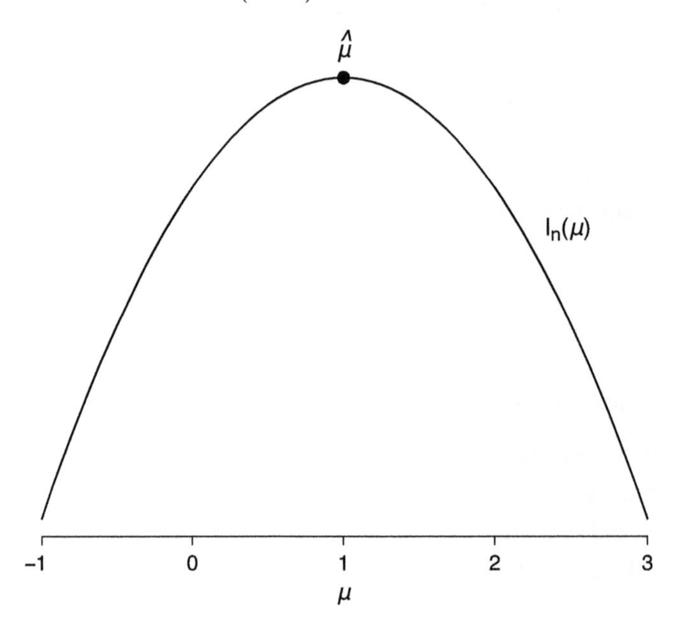

FIGURE 10.2 N(μ, 1) log-likelihood function

The log density is

$$\log f\left(x \mid \sigma^2\right) = -\frac{1}{2}\log(2\pi) - \frac{1}{2}\log(\sigma^2) - \frac{(x-\mu_0)^2}{2\sigma^2}.$$

The log likelihood is

$$\ell_n\left(\sigma^2\right) = -\frac{n}{2}\log(2\pi) - \frac{n}{2}\log(\sigma^2) - \frac{\sum_{i=1}^n (X_i-\mu_0)^2}{2\sigma^2}.$$

The FOC for $\widehat{\sigma}^2$ is

$$0 = -\frac{n}{2\widehat{\sigma}^2} + \frac{\sum_{i=1}^n (X_i-\mu_0)^2}{2\widehat{\sigma}^4}.$$

The solution is

$$\widehat{\sigma}^2 = \frac{1}{n}\sum_{i=1}^n (X_i-\mu_0)^2.$$

The second-order condition is

$$\frac{\partial^2}{\partial\left(\sigma^2\right)^2}\ell_n(\widehat{\sigma}^2) = \frac{n}{2\widehat{\sigma}^4} - \frac{\sum_{i=1}^n (X_i-\mu_0)^2}{\widehat{\sigma}^6} = -\frac{n}{2\widehat{\sigma}^4} < 0$$

as required. The log likelihood is displayed in Figure 10.3 for $\widehat{\sigma}^2 = 1$. The maximizer is marked as the MLE.

Example: $X \sim U[0,\theta]$. This is a tricky case. The density of X is

$$f(x \mid \theta) = \frac{1}{\theta}, \qquad 0 \le x \le \theta.$$

The log density is

$$\log f(x \mid \theta) = \begin{cases} -\log(\theta) & 0 \le x \le \theta \\ -\infty & \text{otherwise.} \end{cases}$$

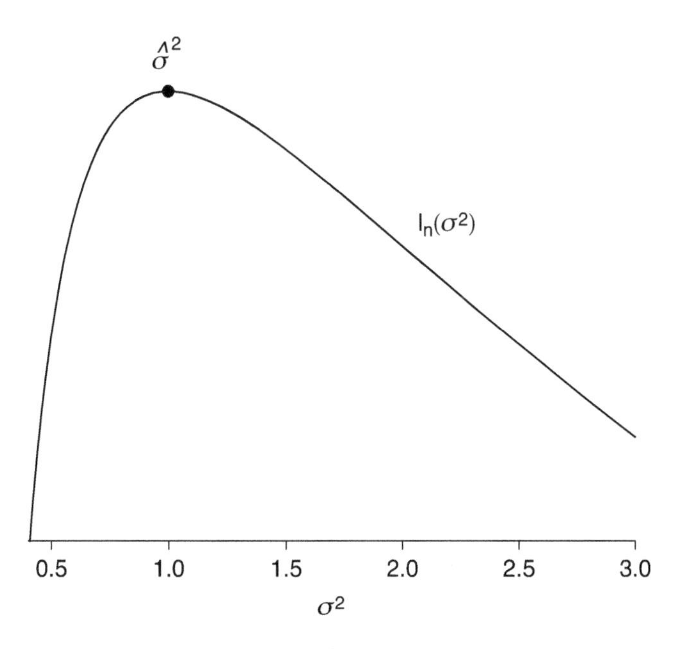

FIGURE 10.3 N$(0,\sigma^2)$ log-likelihood function

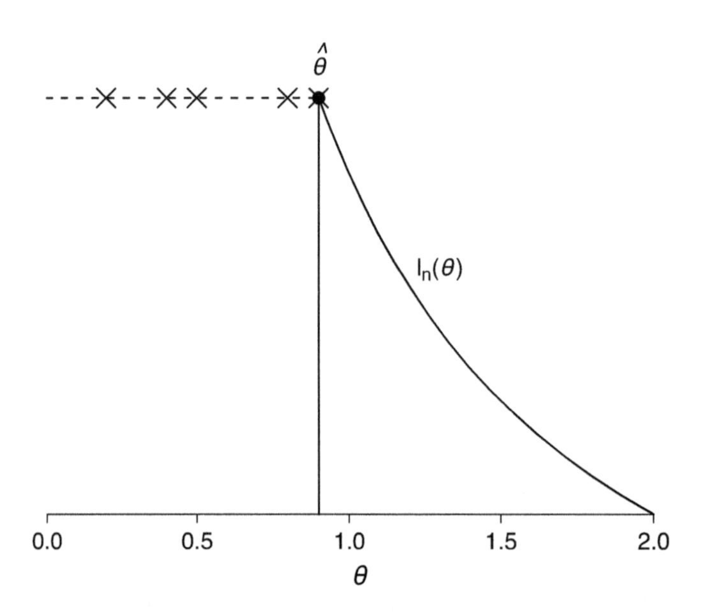

FIGURE 10.4 Uniform log-likelihood function

Let $M_n = \max_{i \le n} X_i$. The log likelihood is

$$\ell_n(\theta) = \begin{cases} -n\log(\theta) & M_n \le \theta \\ -\infty & \text{otherwise.} \end{cases}$$

This is an unusually shaped log likelihood. It is negative infinity for $\theta < M_n$ and finite for $\theta \ge M_n$. It takes its maximum at $\theta = M_n$ and is negatively sloped for $\theta > M_n$. Thus the log likelihood is maximized at M_n, which means

$$\widehat{\theta} = \max_{i \le n} X_i.$$

Perhaps this is not surprising. By setting $\widehat{\theta} = \max_{i \le n} X_i$, the density $U[0, \widehat{\theta}]$ is consistent with the observed data. Among all densities consistent with the observed data, this density has the highest value. It achieves the highest likelihood and so is the MLE.

An interesting and different feature of this likelihood is that the likelihood is not differentiable at the maximum. Thus the MLE does not satisfy a first-order condition. Hence the MLE cannot be found by solving first-order conditions.

The log likelihood is displayed in Figure 10.4 for $M_n = 0.9$. The maximum is marked as the MLE. The sample points are indicated by the "X" marks.

Example: Mixture. Suppose that $X \sim f_1(x)$ with probability p and $X \sim f_2(X)$ with probability $1 - p$, where the densities f_1 and f_2 are known. The density of X is

$$f(x \mid p) = f_1(x)p + f_2(x)(1 - p).$$

The log-likelihood function is

$$\ell_n(p) = \sum_{i=1}^{n} \log(f_1(X_i)p + f_2(X_i)(1 - p)).$$

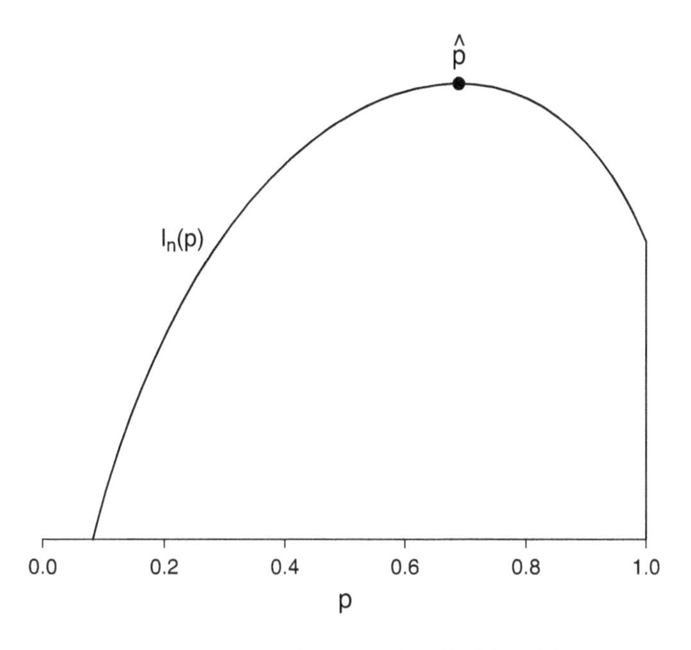

FIGURE 10.5 Normal mixture log-likelihood function

The FOC for $\widehat{p}$ is

$$0 = \sum_{i=1}^{n} \frac{f_1(X_i) - f_2(X_i)}{f_1(X_i)\widehat{p} + f_2(X_i)(1-\widehat{p})}.$$

This equation does not have an algebraic solution. Instead, $\widehat{p}$ must be found numerically.

The log likelihood is displayed in Figure 10.5 for $f_1(x) = \phi(x-1)$ and $f_2(x) = \phi(x+1)$ and given three data points $\{-1, 0.5, 1.5\}$. The maximizer $\widehat{p} = 0.69$ is marked as the MLE.

Example: Double Exponential. The density is $f(x \mid \theta) \sim 2^{-1}\exp(-|x-\theta|)$. The log density is $\log f(x \mid \theta) = -\log(2) - |x-\theta|$. The log-likelihood is

$$\ell_n(\theta) = -n\log(2) - \sum_{i=1}^{n} |X_i - \theta|.$$

This function is continuous and piecewise linear in θ, with kinks at the n sample points (when the X_i have no ties). The derivative of $\ell_n(\theta)$ for θ not at a kink point is

$$\frac{d}{d\theta}\ell_n(\theta) = \sum_{i=1}^{n} \operatorname{sgn}(X_i - \theta)$$

where $\operatorname{sgn}(x) = \mathbb{1}\{x > 0\} - \mathbb{1}\{x < 0\}$ is "the sign of x". The function $\frac{d}{d\theta}\ell_n(\theta)$ is a decreasing step function with steps of size -2 at each of the the n sample points. The function $\frac{d}{d\theta}\ell_n(\theta)$ equals n for $\theta > \max_i X_i$ and equals $-n$ for $\theta < \min X_i$. The FOC for $\widehat{\theta}$ is

$$\frac{d}{d\theta}\ell_n(\widehat{\theta}) = \sum_{i=1}^{n} \operatorname{sgn}(X_i - \widehat{\theta}) = 0.$$

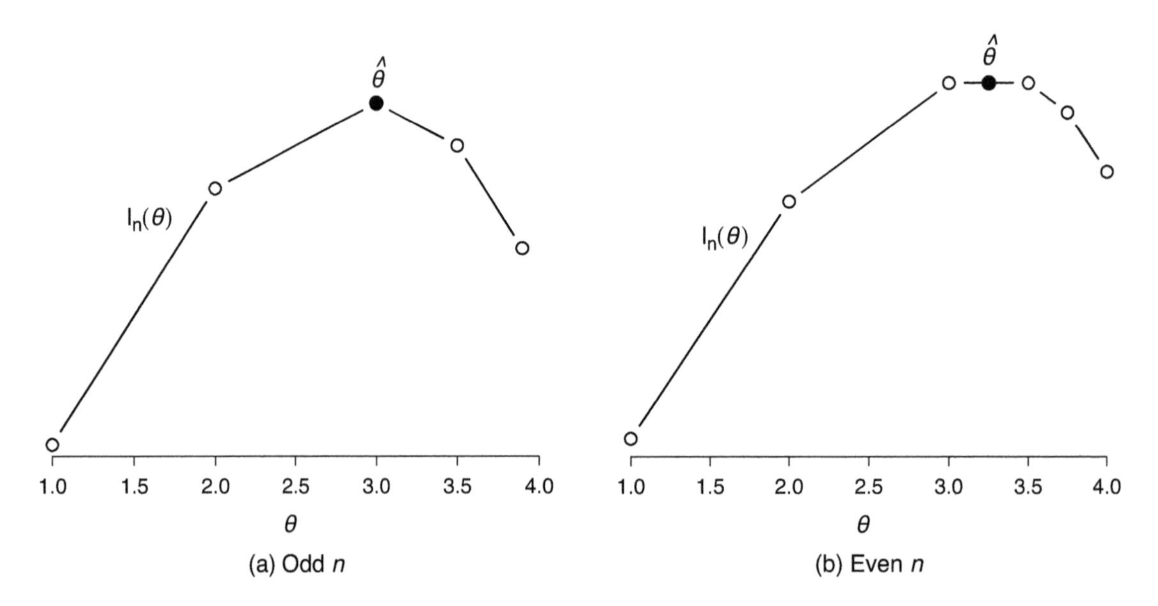

FIGURE 10.6 Double exponential log likelihood

To find the solution, we separately consider the cases of odd and even n. If n is odd, $\frac{d}{d\theta}\ell_n(\theta)$ crosses 0 at the $(n+1)/2$ ordered observation (the $(n+1)/2$ order statistic) which corresponds to the sample mean. Thus the MLE is the sample median. If n is even, $\frac{d}{d\theta}\ell_n(\theta)$ equals 0 on the interval between the $n/2$ and $n/2+1$ ordered observations. The FOC holds for all θ on this interval. Thus any value in this interval is a valid MLE. It is typical to define the MLE as the midpoint, which also corresponds to the sample median. So for both odd and even n, the MLE $\widehat{\theta}$ equals the sample median.

Two examples of the log likelihood are displayed in Figure 10.6. Panel (a) shows the log likelihood for a sample with observations $\{1, 2, 3, 3.5, 3.75\}$, and panel (b) shows the case of observations $\{1, 2, 3, 3.5, 3.75, 4\}$. Panel (a) has an odd number of observations, so it has a unique MLE $\widehat{\theta} = 3$, as marked. Panel (b) has an even number of observations, so the maximum is achieved by an interval. The midpoint $\widehat{\theta} = 3.25$ is marked. In both cases, the log likelihood is continuous piecewise linear and concave with kinks at the observations.

10.7 SCORE, HESSIAN, AND INFORMATION

Recall the log-likelihood function

$$\ell_n(\theta) = \sum_{i=1}^{n} \log f(X_i \mid \theta).$$

Assume that $f(x \mid \theta)$ is differentiable with respect to θ. The **likelihood score** is the derivative of the likelihood function. When θ is a vector, the score is a vector of partial derivatives:

$$S_n(\theta) = \frac{\partial}{\partial \theta}\ell_n(\theta) = \sum_{i=1}^{n} \frac{\partial}{\partial \theta} \log f(X_i \mid \theta).$$

The score tells us how sensitive the log likelihood is to the parameter vector. A property of the score is that it is algebraically 0 at the MLE, $S_n(\widehat{\theta}) = 0$, when $\widehat{\theta}$ is an interior solution.

The **likelihood Hessian** is the negative second derivative of the likelihood function (when it exists). When θ is a vector, the Hessian is a matrix of second partial derivatives

$$\mathscr{H}_n(\theta) = -\frac{\partial^2}{\partial\theta\partial\theta'}\ell_n(\theta) = -\sum_{i=1}^{n}\frac{\partial^2}{\partial\theta\partial\theta'}\log f(X_i \mid \theta).$$

The Hessian indicates the degree of curvature in the log likelihood. Larger values indicate that the likelihood is more curved, smaller values indicate a flatter likelihood.

The **efficient score** is the derivative of the log likelihood for a single observation, evaluated at the random vector X and the true parameter vector:

$$S = \frac{\partial}{\partial\theta}\log f(X \mid \theta_0).$$

The efficient score plays important roles in asymptotic distribution and testing theory. An important property of the efficient score is that it is a mean 0 random vector.

Theorem 10.4 Assume that the model is correctly specified, the support of X does not depend on θ, and θ_0 lies in the interior of Θ. Then the efficient score S satisfies $\mathbb{E}[S] = 0$.

Proof: By the Leibniz Rule (Theorem A.22) for interchange of integration and differentiation,

$$\begin{aligned}
\mathbb{E}[S] &= \mathbb{E}\left[\frac{\partial}{\partial\theta}\log f(X \mid \theta_0)\right] \\
&= \frac{\partial}{\partial\theta}\mathbb{E}\left[\log f(X \mid \theta_0)\right] \\
&= \frac{\partial}{\partial\theta}\ell(\theta_0) \\
&= 0.
\end{aligned}$$

The final equality holds because θ_0 maximizes $\ell(\theta)$ at an interior point. The function $\ell(\theta)$ is differentiable because $f(x \mid \theta)$ is differentiable. ■

The assumptions that the support of X does not depend on θ and that θ_0 lies in the interior of Θ are examples of **regularity conditions.**[1] The assumption that the support does not depend on the parameter is needed for interchanging integration and differentiation. The assumption that θ_0 lies in the interior of Θ is needed to ensure that the expected log likelihood $\ell(\theta)$ satisfies a first-order condition. In contrast, if the maximum occurs on the boundary, then the first-order condition may not hold.

Most econometric models satisfy the conditions of Theorem 10.4, but some do not. The latter models are called **nonregular**. For example, a model which fails the support assumption is the uniform distribution $U[0,\theta]$, because the support $[0,\theta]$ depends on θ. A model which fails the boundary condition is the mixture of normals $f\left(x \mid p, \mu_1, \sigma_1^2, \mu_2, \sigma_2^2\right) = p\phi_{\sigma_1}(x-\mu_1) + (1-p)\phi_{\sigma_2}(x-\mu_2)$ when $p = 0$ or $p = 1$.

Definition 10.6 The **Fisher information** is the variance of the efficient score:

$$\mathscr{I}_\theta = \mathbb{E}\left[SS'\right].$$

[1] These conditions hold in typical (regular) models and are used as sufficient assumptions to ensure typical sampling behavior.

Definition 10.7 The **expected Hessian** is

$$\mathscr{H}_\theta = -\frac{\partial^2}{\partial\theta\partial\theta'}\ell(\theta_0).$$

When $f(x \mid \theta)$ is twice differentiable in θ and the support of X does not depend on θ, the expected Hessian equals the expectation of the likelihood Hessian for a single observation:

$$\mathscr{H}_\theta = -\mathbb{E}\left[\frac{\partial^2}{\partial\theta\partial\theta'}\log f(X \mid \theta_0)\right].$$

Theorem 10.5 Information Matrix Equality. Assume that the model is correctly specified and the support of X does not depend on θ. Then the Fisher information equals the expected Hessian: $\mathscr{I}_\theta = \mathscr{H}_\theta$.

For a proof, see Section 10.20.

This is a fascinating result. It says that the curvature in the likelihood and the variance of the score are identical. There is no particular intuitive story for this information matrix equality. It is important mostly because it is used to simplify the formula for the asymptotic variance of the MLE and similarly for estimation of the asymptotic variance.

10.8 EXAMPLES

Example: $f(x \mid \lambda) = \lambda^{-1}\exp(-x/\lambda)$. We know that $\mathbb{E}[X] = \lambda_0$ and $\operatorname{var}[X] = \lambda_0^2$. The log density is $\log(x \mid \theta) = -\log(\lambda) - x/\lambda$ with expectation $\ell(\lambda) = -\log(\lambda) - \lambda_0/\lambda$. The derivative and second derivative of the log density are

$$\frac{d}{d\lambda}\log f(x \mid \lambda) = -\frac{1}{\lambda} + \frac{x}{\lambda^2}$$

$$\frac{d^2}{d\lambda^2}\log f(x \mid \lambda) = \frac{1}{\lambda^2} - 2\frac{x}{\lambda^3}.$$

The efficient score is the first derivative evaluated at X and λ_0:

$$S = \frac{d}{d\lambda}\log f(X \mid \lambda_0) = -\frac{1}{\lambda_0} + \frac{X}{\lambda_0^2}.$$

It has expectation

$$\mathbb{E}[S] = -\frac{1}{\lambda_0} + \frac{\mathbb{E}[S]}{\lambda_0^2} = -\frac{1}{\lambda_0} + \frac{\lambda_0}{\lambda_0^2} = 0$$

and variance

$$\operatorname{var}[S] = \operatorname{var}\left[-\frac{1}{\lambda_0} + \frac{X}{\lambda_0^2}\right] = \frac{1}{\lambda_0^4}\operatorname{var}[X] = \frac{\lambda_0^2}{\lambda_0^4} = \frac{1}{\lambda_0^2}.$$

The expected Hessian is

$$\mathscr{H}_\lambda = -\frac{d^2}{d\lambda^2}\ell(\lambda_0) = -\frac{1}{\lambda_0^2} + 2\frac{\lambda_0}{\lambda_0^3} = \frac{1}{\lambda_0^2}.$$

An alternative calculation is

$$\mathscr{H}_\lambda = \mathbb{E}\left[-\frac{d^2}{d\lambda^2}\log f(X \mid \lambda_0)\right] = -\frac{1}{\lambda_0^2} + 2\frac{\mathbb{E}[X]}{\lambda_0^3} = -\frac{1}{\lambda_0^2} + 2\frac{\lambda_0}{\lambda_0^3} = \frac{1}{\lambda_0^2}.$$

Thus

$$\mathscr{I}_\lambda = \text{var}\,[S] = \frac{1}{\lambda_0^2} = \mathscr{H}_\lambda$$

and the information equality holds.

Example: $X \sim \mathrm{N}(0,\theta)$. We parameterize the variance as θ rather than σ^2, so that it is notationally easier to take derivatives. We know $\mathbb{E}\,[X]=0$ and $\text{var}\,[X]=\mathbb{E}\left[X^2\right]=\theta_0$. The log density is

$$\log f(x \mid \theta) = -\frac{1}{2}\log(2\pi) - \frac{1}{2}\log(\theta) - \frac{x^2}{2\theta}$$

with expectation

$$\ell\,(\theta) = -\frac{1}{2}\log(2\pi) - \frac{1}{2}\log(\theta) - \frac{\theta_0}{2\theta}$$

and with first and second derivatives

$$\frac{d}{d\theta}\log f(x \mid \theta) = -\frac{1}{2\theta} + \frac{x^2}{2\theta^2} = \frac{x^2-\theta}{2\theta^2}$$

$$\frac{d^2}{d\theta^2}\log f(x \mid \theta) = \frac{1}{2\theta^2} - \frac{x^2}{\theta^3} = \frac{\theta - 2x^2}{2\theta^3}.$$

The efficient score is

$$S = \frac{d}{d\theta}\log f(X \mid \theta_0) = \frac{X^2-\theta_0}{2\theta_0^2}.$$

It has mean

$$\mathbb{E}\,[S] = \mathbb{E}\left[\frac{X^2-\theta_0}{2\theta_0^2}\right] = 0$$

and variance

$$\text{var}\,[S] = \frac{\mathbb{E}\left[\left(X^2-\theta_0\right)^2\right]}{4\theta_0^4} = \frac{\mathbb{E}\left[X^4 - 2X^2\theta_0 + \theta_0^2\right]}{4\theta_0^4} = \frac{3\theta_0^2 - 2\theta_0^2 + \theta_0^2}{4\theta_0^4} = \frac{1}{2\theta_0^2}.$$

The expected Hessian is

$$\mathscr{H}_\theta = -\frac{d^2}{d\theta^2}\ell\,(\theta_0) = -\frac{1}{2\theta_0^2} + \frac{\theta_0}{\theta_0^3} = \frac{1}{2\theta_0^2}.$$

Equivalently, it can be found by calculating

$$\mathscr{H}_\theta = \mathbb{E}\left[-\frac{d^2}{d\theta^2}\log f(X \mid \theta_0)\right] = \mathbb{E}\left[\frac{2X^2-\theta_0}{2\theta_0^3}\right] = \frac{1}{2\theta_0^2}.$$

Thus,

$$\mathscr{I}_\theta = \text{var}\,[S] = \frac{1}{2\theta_0^2} = \mathscr{H}_\theta$$

and the information equality holds.

Example: $X \sim 2^{-1}\exp\left(-\left|x-\theta\right|\right)$ for $x \in \mathbb{R}$. The log density is $\log f(x \mid \theta) = -\log(2) - |x-\theta|$. The derivative of the log density is

$$\frac{d}{d\theta}\log f(x \mid \theta) = \text{sgn}\,(x-\theta)\,.$$

The second derivative does not exist at $x = \theta$. The efficient score is the first derivative evaluated at X and θ_0:

$$S = \frac{d}{d\theta} \log f(X \mid \theta_0) = \operatorname{sgn}(X - \theta_0).$$

Since X is symmetrically distributed about θ_0 and $S^2 = 1$, we deduce that $\mathbb{E}[S] = 0$ and $\mathbb{E}\left[S^2\right] = 1$. The expected log density is

$$\begin{aligned} \ell(\theta) &= -\log(2) - \mathbb{E}|X - \theta| \\ &= -\log(2) - \int_{-\infty}^{\infty} |x - \theta| f(x \mid \theta_0)\, dx \\ &= -\log(2) + \int_{-\infty}^{\theta} (x - \theta) f(x \mid \theta_0)\, dx - \int_{\theta}^{\infty} (x - \theta) f(x \mid \theta_0) dx. \end{aligned}$$

By the Leibniz Rule (Theorem A.22),

$$\begin{aligned} \frac{d}{d\theta}\ell(\theta) &= (\theta - \theta) f(x \mid \theta_0) - \int_{-\infty}^{\theta} f(x \mid \theta_0)\, dx - (\theta - \theta) f(x \mid \theta_0) + \int_{\theta}^{\infty} f(x \mid \theta_0) dx \\ &= 1 - 2F(\theta \mid \theta_0). \end{aligned}$$

The expected Hessian is

$$\mathscr{H}_\theta = -\frac{d^2}{d\theta^2}\ell(\theta_0) = -\frac{d}{d\theta}(1 - 2F(\theta \mid \theta_0)) = 2f(x_0 \mid \theta_0) = 1.$$

Hence

$$\mathscr{I}_\theta = \operatorname{var}[S] = 1 = \mathscr{H}_\theta$$

and the information equality holds.

10.9 CRAMÉR-RAO LOWER BOUND

The information matrix provides a lower bound for the variance among unbiased estimators.

Theorem 10.6 Cramér-Rao Lower Bound. Assume that the model is correctly specified, the support of X does not depend on θ, and θ_0 lies in the interior of Θ. If $\widetilde{\theta}$ is an unbiased estimator of θ, then $\operatorname{var}\left[\widetilde{\theta}\right] \geq (n\mathscr{I}_\theta)^{-1}$.

For a proof, see Section 10.20.

Definition 10.8 The **Cramér-Rao Lower Bound (CRLB)** is $(n\mathscr{I}_\theta)^{-1}$.

Definition 10.9 An estimator $\widetilde{\theta}$ is **Cramér-Rao efficient** if it is unbiased for θ and $\operatorname{var}\left[\widetilde{\theta}\right] = (n\mathscr{I}_\theta)^{-1}$.

The CRLB is a famous result. It says that in the class of unbiased estimators, the lowest possible variance is the inverse Fisher information scaled by sample size. Thus the Fisher information provides a bound on the precision of estimation.

When θ is a vector, the CRLB states that the covariance matrix is bounded from below by the matrix inverse of the matrix Fisher information, meaning that the difference is positive semi-definite.

10.10 EXAMPLES

Example: $f(x \mid \lambda) = \lambda^{-1}\exp(-x/\lambda)$. We have calculated that $\mathscr{I}_\lambda = \lambda^{-2}$. Thus the Cramér-Rao lower bound is λ^2/n. We have also found that the MLE for θ is $\widehat{\theta} = \overline{X}_n$. This is an unbiased estimator, and $\operatorname{var}\left[\overline{X}_n\right] = n^{-1}\operatorname{var}[X] = \lambda^2/n$, which equals the CRLB. Thus this MLE is Cramér-Rao efficient.

Example: $X \sim \mathrm{N}(\mu, \sigma^2)$ with known σ^2. We first calculate the information. The second derivative of the log density is

$$\frac{d^2}{d\mu^2}\log f(x \mid \mu) = \frac{d^2}{d\mu^2}\left(-\frac{1}{2}\log(2\pi\sigma^2) - \frac{(x-\mu)^2}{2\sigma^2}\right) = -\frac{1}{\sigma^2}.$$

Hence $\mathscr{I}_\mu = \sigma^{-2}$. Thus the CRLB is σ^2/n. The MLE is $\widehat{\mu} = \overline{X}_n$, which is unbiased and has variance $\operatorname{var}[\widehat{\mu}] = \sigma^2/n$, which equals the CRLB. Hence this MLE is Cramér-Rao efficient.

Example: $X \sim \mathrm{N}(\mu, \sigma^2)$ with unknown μ and σ^2. We need to calculate the information matrix for the parameter vector $\theta = (\mu, \sigma^2)$. The log density is

$$\log f\left(x \mid \mu, \sigma^2\right) = -\frac{1}{2}\log(2\pi) - \frac{1}{2}\log\sigma^2 - \frac{(x-\mu)^2}{2\sigma^2}.$$

The first derivatives are

$$\frac{\partial}{\partial\mu}\log f\left(x \mid \mu, \sigma^2\right) = \frac{x-\mu}{\sigma^2}$$

$$\frac{\partial}{\partial\sigma^2}\log f\left(x \mid \mu, \sigma^2\right) = -\frac{1}{2\sigma^2} + \frac{(x-\mu)^2}{2\sigma^4}.$$

The second derivatives are

$$\frac{\partial^2}{\partial\mu^2}\log f\left(x \mid \mu, \sigma^2\right) = -\frac{1}{\sigma^2}$$

$$\frac{\partial}{\partial\left(\sigma^2\right)^2}\log f\left(x \mid \mu, \sigma^2\right) = \frac{1}{2\sigma^4} - \frac{(x-\mu)^2}{\sigma^6}$$

$$\frac{\partial^2}{\partial\mu\partial\sigma^2}\log f\left(x \mid \mu, \sigma^2\right) = -\frac{x-\mu}{2\sigma^4}.$$

The expected Fisher information is

$$\mathscr{I}_\theta = \mathbb{E}\begin{bmatrix} \dfrac{1}{\sigma^2} & \dfrac{X-\mu}{2\sigma^4} \\ \dfrac{X-\mu}{2\sigma^4} & \dfrac{(X-\mu)^2}{\sigma^6} - \dfrac{1}{2\sigma^4} \end{bmatrix} = \begin{pmatrix} \dfrac{1}{\sigma^2} & 0 \\ 0 & \dfrac{1}{2\sigma^4} \end{pmatrix}.$$

The lower bound is

$$\mathrm{CRLB} = \left(n\mathscr{I}_\theta\right)^{-1} = \begin{pmatrix} \sigma^2/n & 0 \\ 0 & 2\sigma^4/n \end{pmatrix}.$$

Two things are interesting about this result. First, the information matrix is diagonal. This means that the information for μ and σ^2 are unrelated. Second, the diagonal terms are identical to the CRLB from the simpler cases where σ^2 or μ is known.

Consider the estimation of μ. The CRLB is σ^2/n, which equals the variance of the sample mean. Thus the sample mean is Cramér-Rao efficient.

Now consider the estimation of σ^2. The CRLB is $2\sigma^4/n$. The moment estimator $\widehat{\sigma}^2$ (which equals the MLE) is biased. The unbiased estimator is $s^2 = (n-1)^{-1}\sum_{i=1}^{n}\left(X_i - \overline{X}\right)^2$. In the normal sampling model, this has the exact distribution $\sigma^2\chi^2_{n-1}/(n-1)$, which has variance

$$\operatorname{var}\left[s^2\right] = \operatorname{var}\left[\frac{\sigma^2\chi^2_{n-1}}{n-1}\right] = \frac{2\sigma^4}{n-1} > \frac{2\sigma^4}{n}.$$

Thus the unbiased estimator is not Cramér-Rao efficient.

As illustrated in the last example, many MLEs are neither unbiased nor Cramér-Rao efficient.

10.11 CRAMÉR-RAO BOUND FOR FUNCTIONS OF PARAMETERS

In this section, we derive the Cramér-Rao bound for the transformation $\beta = h(\theta)$. Set $\boldsymbol{H} = \frac{\partial}{\partial\theta}h(\theta_0)'$.

Theorem 10.7 Cramér-Rao Lower Bound for Transformations. Assume the conditions of Theorem 10.6, $\mathscr{I}_\theta > 0$, and that $h(u)$ is continuously differentiable at θ_0. If $\widetilde{\beta}$ is an unbiased estimator of β, then $\operatorname{var}\left[\widetilde{\beta}\right] \geq n^{-1}\boldsymbol{H}'\mathscr{I}_\theta^{-1}\boldsymbol{H}$.

For a proof, see Section 10.20. This provides a lower bound for estimation of any smooth function of the parameters.

10.12 CONSISTENT ESTIMATION

In this section, we discuss conditions for (asymptotic) consistency of the MLE $\widehat{\theta}$. Recall that $\widehat{\theta}$ is defined as the maximizer of the log-likelihood function. When there is no explicit algebraic solution for $\widehat{\theta}$, we demonstrate the consistency of $\widehat{\theta}$ through the properties of the likelihood function.

Write the average log likelihood as

$$\overline{\ell}_n(\theta) = \frac{1}{n}\ell_n(\theta) = \frac{1}{n}\sum_{i=1}^{n}\log f(X_i \mid \theta).$$

Since $\log f(X_i \mid \theta)$ is a transformation of X_i, it is also i.i.d. The WLLN (Theorem 7.4) implies $\overline{\ell}_n(\theta) \underset{p}{\longrightarrow} \ell(\theta) = \mathbb{E}\left[\log f(X \mid \theta)\right]$. It seems reasonable to expect that the maximizer of $\overline{\ell}_n(\theta)$ (which is $\widehat{\theta}$) will converge in probability to the maximizer of $\ell(\theta)$ (which is θ_0). This is generally true, though a few extra conditions need to be satisfied to ensure this convergence.

Theorem 10.8 Assume the following:

1. X_i are i.i.d.
2. $\mathbb{E}\left|\log f(X \mid \theta)\right| \leq G(X)$, and $\mathbb{E}\left[G(X)\right] < \infty$.
3. $\log f(X \mid \theta)$ is continuous in θ with probability 1.

4. Θ is compact.
5. For all $\theta \neq \theta_0$, $\ell(\theta) < \ell(\theta_0)$.

Then $\widehat{\theta} \underset{p}{\longrightarrow} \theta_0$ as $n \to \infty$.

Theorem 10.8 shows that the MLE is consistent for the true parameter value under the stated assumptions. The latter are weak and apply to most econometric models.

Assumption 1 states that the observations are i.i.d., which is the sampling framework used in this textbook. Assumption 2 states that the log density has an envelope with a finite expectation. This implies that the log density has a finite expectation.This is necessary in order to apply the WLLN.

Assumption 3 states that the log density $\log f(X \mid \theta)$ is almost surely continuous in the parameter. It allows for discontinuities but only at points of 0 probability. This condition is used to establish the uniform law of large numbers for the sample average of the log density. Other technical conditions can be used in place of this assumption; see Section 18.5.

Assumption 4 is technical. It states that the parameter space is compact. This assumption can be omitted by a more detailed argument and a strengthening of assumption 3.

Assumption 5 is the most critical condition. It states that θ_0 is the unique maximizer of $\ell(\theta)$. This is an identification assumption.

10.13 ASYMPTOTIC NORMALITY

The sample mean is asymptotically normally distributed. In this section, we show that the MLE is asymptotically normally distributed as well. This holds even though we do not have an explicit expression for the MLE in most models.

The reason the MLE is asymptotically normal is that in large samples, the MLE can be approximated by (a matrix scale of) the sample average of the efficient scores. The latter are i.i.d., mean 0, and have a covariance matrix corresponding to the Fisher information. Therefore, the CLT reveals that the MLE is asymptotically normal.

One technical challenge is to construct the linear approximation to the MLE. A classical approach is to apply a mean value expansion to the first-order condition of the MLE. This produces an approximation for the MLE in terms of the first and second derivatives of the log likelihood (the likelihood score and Hessian). A modern approach is to instead apply a mean value expansion to the first-order condition for maximization of the expected log density. It turns out that this second (modern) approach is valid under broader conditions but is technically more demanding. I sketch the classical argument but state the result under the modern conditions.

The MLE $\widehat{\theta}$ maximizes the average log likelihood $\overline{\ell}_n(\theta)$, so it satisfies the first-order condition

$$0 = \frac{\partial}{\partial \theta} \overline{\ell}_n\left(\widehat{\theta}\right).$$

Since $\widehat{\theta}$ is consistent for θ_0 (Theorem 10.8), $\widehat{\theta}$ is close to θ_0 for n sufficiently large. This permits us to make a first-order Taylor approximation of the above first-order condition about $\theta = \theta_0$. For simplicity, we ignore the remainder terms. This Taylor expansion leads to the expression

$$0 = \frac{\partial}{\partial \theta} \overline{\ell}_n\left(\widehat{\theta}\right) \simeq \frac{\partial}{\partial \theta} \overline{\ell}_n\left(\theta_0\right) + \frac{\partial^2}{\partial \theta \partial \theta'} \overline{\ell}_n\left(\theta_0\right)\left(\widehat{\theta} - \theta_0\right)$$

which can be rewritten as

$$\sqrt{n}\left(\widehat{\theta}-\theta_0\right) \simeq \left(-\frac{\partial^2}{\partial\theta\partial\theta'}\overline{\ell}_n\left(\theta_0\right)\right)^{-1}\left(\sqrt{n}\frac{\partial}{\partial\theta}\overline{\ell}_n\left(\theta_0\right)\right). \tag{10.1}$$

The term in the inverse equals

$$\frac{1}{n}\sum_{i=1}^{n}-\frac{\partial^2}{\partial\theta\partial\theta'}\log f\left(X_i \mid \theta_0\right) \underset{p}{\longrightarrow} \mathbb{E}\left[-\frac{\partial^2}{\partial\theta\partial\theta'}\log f\left(X_i \mid \theta_0\right)\right] = \mathscr{H}_\theta.$$

The second term in expression (10.1) equals

$$\frac{1}{\sqrt{n}}\sum_{i=1}^{n}\frac{\partial}{\partial\theta}\log f\left(X_i \mid \theta_0\right) \underset{d}{\longrightarrow} \mathrm{N}\left(0, \mathscr{I}_\theta\right)$$

by the CLT, since the vectors $\frac{\partial}{\partial\theta}\log f\left(X_i \mid \theta_0\right)$ are mean 0 (Theorem 10.4), i.i.d., and have variance $\mathscr{I}_\theta$. We deduce that

$$\sqrt{n}\left(\widehat{\theta}-\theta_0\right) \underset{d}{\longrightarrow} \mathscr{H}_\theta\,\mathrm{N}\left(0, \mathscr{I}_\theta\right) = \mathrm{N}\left(0, \mathscr{H}_\theta^{-1}\mathscr{I}_\theta\mathscr{H}_\theta^{-1}\right) = \mathrm{N}\left(0, \mathscr{I}_\theta^{-1}\right)$$

where the final equality is the information equality theorem (Theorem 10.5).

This shows that the MLE converges at rate $n^{-1/2}$, is asymptotically normal with no bias term, and has an asymptotic variance equal to the inverse Fisher information.

We now consider the asymptotic distribution under full regularity conditions. Define

$$\mathscr{H}_\theta\left(\theta\right) = -\frac{\partial^2}{\partial\theta\partial\theta'}\mathbb{E}\left[\log f\left(X \mid \theta\right)\right]. \tag{10.2}$$

Let $\mathscr{N}$ be a neighborhood of θ_0.

Theorem 10.9 Assume that the conditions of Theorem 10.8 hold, plus

1. $\mathbb{E}\left\|\frac{\partial}{\partial\theta}\log f\left(X \mid \theta_0\right)\right\|^2 < \infty$.
2. $\mathscr{H}_\theta(\theta)$ is continuous in $\theta \in \mathscr{N}$.
3. $\frac{\partial}{\partial\theta}\log f\left(X \mid \theta\right)$ is Lipschitz continuous in $\mathscr{N}$, meaning that for all $\theta_1, \theta_2 \in \mathscr{N}$,

$$\left\|\frac{\partial}{\partial\theta}\log f\left(x \mid \theta_1\right) - \frac{\partial}{\partial\theta}\log f\left(x \mid \theta_1\right)\right\| \le B(x)\left\|\theta_1-\theta_2\right\|$$

   where $\mathbb{E}\left[B(X)^2\right] < \infty$.
4. $\mathscr{H}_\theta > 0$.
5. θ_0 is in the interior of Θ.
6. $\mathscr{I}_\theta = \mathscr{H}_\theta$.

Then as $n \to \infty$,

$$\sqrt{n}\left(\widehat{\theta}-\theta_0\right) \underset{d}{\longrightarrow} \mathrm{N}\left(0, \mathscr{I}_\theta^{-1}\right).$$

This shows that the MLE is asymptotically normally distributed with an asymptotic variance that is the matrix inverse of the Fisher information. The proof is deferred to the chapter on M-estimation in *Econometrics*.

Assumption 1 states that the efficient score has a finite second moment. This is required to apply the central limit theory. Assumption 2 states that the second derivative of the expected log density is continuous near θ_0, which is required to ensure that the remainder terms from the expansion are negligible. Assumption 3 states that the score is Lipschitz-continuous in the parameter. This is satisfied[2] in typical applications.

Assumption 4 states that the expected Hessian is invertible, which is required; otherwise, the inversion step of the proof is invalid. This condition excludes the possibility of redundant parameters. This assumption is connected with identification. Singularity of the matrix $\mathscr{H}_\theta$ occurs under a lack of identification, or under weak identification. Assumption 5 is required to justify the mean value expansion. If the true parameter is on the boundary of the parameter space, the MLE typically has a non-normal asymptotic distribution. Assumption 6 is the information matrix equality, which holds under Theorem 10.5 when the model is correctly specified and the support of X does not depend on θ.

10.14 ASYMPTOTIC CRAMÉR-RAO EFFICIENCY

The Cramér-Rao theorem shows that no unbiased estimator has a lower covariance matrix than $(n\mathscr{I}_\theta)^{-1}$. Thus the variance of a centered and scaled unbiased estimator $\sqrt{n}\left(\widetilde{\theta}-\theta_0\right)$ cannot be less than $\mathscr{I}_\theta^{-1}$. We describe an estimator as being asymptotically Cramér-Rao efficient if it attains this same distribution.

Definition 10.10 An estimator $\widetilde{\theta}$ is **asymptotically Cramér-Rao efficient** if $\sqrt{n}\left(\widetilde{\theta}-\theta_0\right) \underset{d}{\longrightarrow} Z$, where $\mathbb{E}[Z]=0$ and $\operatorname{var}[Z]=\mathscr{I}_\theta^{-1}$.

Theorem 10.10 Under the conditions of Theorem 10.9, the MLE is asymptotically Cramér-Rao efficient.

Together with the Cramér-Rao theorem, this theorem is an important efficiency result. What we have shown is that the MLE has an asymptotic variance that is the best possible among unbiased estimators. The MLE is not (in general) unbiased, but it is asymptotically close to unbiased in the sense that when centered and rescaled, the asymptotic distribution is free of bias. Thus the distribution is properly centered and (approximately) of low bias.

The theorem as stated does have limitations. It does not show that the MSE of the rescaled MLE converges to the best-possible variance. Indeed in some cases, the MLE does not have a finite variance! The asymptotic unbiasedness and low variance are properties of the asymptotic distribution, not limiting properties of the finite sample distribution. The Cramér-Rao bound is also developed under the restriction of unbiased estimators, so it leaves open the possibility that biased estimators could have better overall performance. The theory also relies on parametric likelihood models and correct specification, which excludes many important econometric models. Regardless, the theorem is a major step forward.

10.15 VARIANCE ESTIMATION

The knowledge that the asymptotic distribution is normal is incomplete without knowledge of the asymptotic variance. As it is generally unknown, we need an estimator. Given the equivalence $\boldsymbol{V}=\mathscr{I}_\theta^{-1}=\mathscr{H}_\theta^{-1}$ we can construct several feasible estimators of the asymptotic variance $\boldsymbol{V}$.

[2]Alternatively, assumption 3 could be replaced by a condition sufficient for asymptotic equicontinuity of the normalized score process $n^{-1/2}\sum_{i=1}^{n}\frac{\partial}{\partial\theta}\log f(X_i\mid\theta)$. This concept is defined in Section 18.6, with sufficient conditions in Section 18.7.

Expected Hessian Estimator. Recall the expected log density $\ell(\theta)$ and equation (10.2). Evaluated at the MLE, the latter is

$$\widehat{\mathscr{H}_\theta} = \mathscr{H}_\theta\left(\widehat{\theta}\right).$$

The expected Hessian estimator of the variance is

$$\widehat{\mathbf{V}}_0 = \widehat{\mathscr{H}_\theta}^{-1}.$$

This estimate can only be computed when $\mathscr{H}_\theta(\theta)$ is available as an explicit function of θ, which is not often.

Sample Hessian Estimator. This is the most common variance estimator. It is based on the formula for the expected Hessian and equals the second-derivative matrix of the log-likelihood function:

$$\widehat{\mathscr{H}_\theta} = \frac{1}{n}\sum_{i=1}^{n} -\frac{\partial^2}{\partial\theta\partial\theta'}\log f\left(X_i \mid \widehat{\theta}\right) = -\frac{1}{n}\frac{\partial^2}{\partial\theta\partial\theta'}\ell_n\left(\widehat{\theta}\right)$$

$$\widehat{\mathbf{V}}_1 = \widehat{\mathscr{H}_\theta}^{-1}.$$

The second-derivative matrix can be calculated analytically if the derivatives are known. Alternatively, it can be calculated using numerical derivatives.

Outer Product Estimator. This estimator is based on the formula for the Fisher information. It is

$$\widehat{\mathscr{I}_\theta} = \frac{1}{n}\sum_{i=1}^{n}\left(\frac{\partial}{\partial\theta}\log f\left(X_i \mid \widehat{\theta}\right)\right)\left(\frac{\partial}{\partial\theta}\log f\left(X_i \mid \widehat{\theta}\right)\right)'$$

$$\widehat{\mathbf{V}}_2 = \widehat{\mathscr{I}_\theta}^{-1}.$$

These three estimators can be shown to be consistent by using tools related to the proofs of consistency and asymptotic normality.

Theorem 10.11 Under the conditions of Theorem 10.9,

$$\widehat{\mathbf{V}}_0 \underset{p}{\longrightarrow} \mathbf{V}$$

$$\widehat{\mathbf{V}}_1 \underset{p}{\longrightarrow} \mathbf{V}$$

$$\widehat{\mathbf{V}}_2 \underset{p}{\longrightarrow} \mathbf{V}$$

where $\mathbf{V} = \mathscr{I}_\theta^{-1} = \mathscr{H}_\theta^{-1}$.

Asymptotic standard errors are constructed by taking the squares roots of the diagonal elements of $n^{-1}\widehat{\mathbf{V}}$. When θ is scalar, this is $s\left(\widehat{\theta}\right) = \sqrt{n^{-1}\widehat{V}}$.

Example: $f(x \mid \lambda) = \lambda^{-1} \exp(-x/\lambda)$. Recall that $\widehat{\lambda} = \overline{X}_n$, the first and second derivatives of the log density are $\frac{d}{d\lambda} \log f(x \mid \lambda) = -1/\lambda + x/\lambda^2$ and $\frac{d^2}{d\lambda^2} \log f(x \mid \lambda) = 1/\lambda^2 - 2x/\lambda^3$. We find

$$\widehat{\mathscr{H}}(\lambda) = \frac{1}{n} \sum_{i=1}^{n} -\frac{\partial^2}{\partial \lambda^2} \log f(X_i \mid \lambda) = \frac{1}{n} \sum_{i=1}^{n} \left(-\frac{1}{\lambda^2} + \frac{2X_i}{\lambda^3} \right) = -\frac{1}{\lambda^2} + \frac{2\overline{X}_n}{\lambda^3}$$

$$\widehat{\mathscr{H}} = -\frac{1}{\overline{X}_n^2} + \frac{2\overline{X}_n}{\overline{X}_n^3} = \frac{1}{\overline{X}_n^2}.$$

Hence

$$\widehat{V}_0 = \widehat{V}_1 = \overline{X}_n^2.$$

Also

$$\widehat{\mathscr{I}}(\lambda) = \frac{1}{n} \sum_{i=1}^{n} \left(\frac{\partial}{\partial \lambda} \log f(X_i \mid \lambda) \right)^2 = \frac{1}{n} \sum_{i=1}^{n} \left(\frac{X_i - \lambda}{\lambda^2} \right)^2$$

$$\widehat{\mathscr{I}} = \widehat{\mathscr{I}}\left(\widehat{\lambda}\right) = \frac{1}{n} \sum_{i=1}^{n} \left(\frac{X_i - \overline{X}}{\overline{X}_n^2} \right)^2 = \frac{\widehat{\sigma}^2}{\overline{X}^4}.$$

Hence

$$\widehat{V}_2 = \frac{\overline{X}^4}{\widehat{\sigma}^2}.$$

A standard error for $\widehat{\lambda}$ is $s\left(\widehat{\theta}\right) = n^{-1/2} \overline{X}_n$ if $\widehat{V}_0$ or $\widehat{V}_1$ is used, or $s\left(\widehat{\theta}\right) = n^{-1/2} \overline{X}_n^2 / \widehat{\sigma}$ if $\widehat{V}_2$ is used.

10.16 KULLBACK-LEIBLER DIVERGENCE

There is an interesting connection between MLE and the Kullback-Leibler divergence.

Definition 10.11 The **Kullback-Leibler divergence** between densities $f(x)$ and $g(x)$ is

$$\text{KLIC}(f, g) = \int f(x) \log \left(\frac{f(x)}{g(x)} \right) dx.$$

The Kullback-Leibler divergence is also known as the "Kullback-Leibler Information Criterion," and hence it is typically written using the acronym "KLIC." The KLIC distance is not symmetric; thus, $\text{KLIC}(f, g) \neq \text{KLIC}(g, f)$.

Theorem 10.12 Properties of KLIC

1. $\text{KLIC}(f, f) = 0$.
2. $\text{KLIC}(f, g) \geq 0$.
3. $f = \underset{g}{\operatorname{argmin}}\, \text{KLIC}(f, g)$.

Proof: Property 1 holds since $\log(f(x)/f(x))=0$. For property 2, let X be a random variable with density $f(x)$. Then since the logarithm is concave, using Jensen's inequality (Theorem 2.9), we have

$$-\mathrm{KLIC}(f,g)=\mathbb{E}\left[\log\left(\frac{g(X)}{f(X)}\right)\right]\leq\log\mathbb{E}\left[\frac{g(X)}{f(X)}\right]=\log\int g(x)dx=0$$

which is property 2. Property 3 follows from the first two properties. That is, $\mathrm{KLIC}(f,g)$ is nonnegative, and $\mathrm{KLIC}(f,f)=0$. So $\mathrm{KLIC}(f,g)$ is minimized over g by setting $g=f$. ■

Let $f_\theta=f(x\mid\theta)$ be a parametric family with $\theta\in\Theta$. From property 3 of Theorem 10.12, we deduce that θ_0 minimizes the KLIC divergence between f and f_θ.

Theorem 10.13 If $f(x)=f(x\mid\theta_0)$ for some $\theta\in\Theta_0$, then

$$\theta_0=\underset{\theta\in\Theta}{\operatorname{argmin}}\ \mathrm{KLIC}\left(f,f_\theta\right).$$

Theorem 10.13 simply points out that since the KLIC divergence is minimized by setting the densities equal, the KLIC divergence is minimized by setting the parameter equal to the true value.

10.17 APPROXIMATING MODELS

Suppose that $f_\theta=f(x\mid\theta)$ is a parametric family with $\theta\in\Theta$. Recall that a model is correctly specified if the true density is a member of the parametric family, and otherwise, the model is mis-specified. For example, consider the density $f(x)$ of hourly wages as displayed in Figure 2.7. The shape of $f(x)$ is similar to a log-normal density, so we may choose to use the log-normal density $f(x\mid\theta)$ as a parametric family for hourly wages (and indeed this is a common choice in applied labor economics). However, while $f(x)$ appears to be close to log-normal, there are differences. Consequently, it is correct to describe the log-normal parametric model as misspecified.

Despite being misspecified, a parametric model can be a good approximation to a given density function. Once again, the density of hourly wages in Figure 2.7 is reasonably close to a log-normal density, so it seems reasonable to use the log-normal as an approximating model.

Given the concept of an approximating model, a natural question is how to select its parameters. One solution is to minimize a measure of the divergence between densities. If we adopt minimization of Kullback-Leibler divergence, we arrive at the following criteria for selection of the parameters.

Definition 10.12 The **pseudo-true parameter** θ_0 for a model f_θ that best fits the true density f based on Kullback-Leibler divergence is

$$\theta_0=\underset{\theta\in\Theta}{\operatorname{argmin}}\ \mathrm{KLIC}\left(f,f_\theta\right).$$

A good property of this definition is that it corresponds to the true parameter value when the model is correctly specified. The name "pseudo-true parameter" refers to the fact that when f_θ is a misspecified parametric model, there is no true parameter, but there is a parameter value that produces the best-fitting density.

To further characterize the pseudo-true value, notice that we can write the KLIC divergence as

$$\begin{aligned}\mathrm{KLIC}\left(f, f_\theta\right) &= \int f(x)\log f(x)dx - \int f(x)\log f\left(x \mid \theta\right) dx \\ &= \int f(x)\log f(x)dx - \ell(\theta)\end{aligned}$$

where

$$\ell(\theta) = \int f(x)\log f\left(x \mid \theta\right) dx = \mathbb{E}\left[\log f(X \mid \theta)\right].$$

Since $\int f(x)\log f(x)dx$ does not depend on θ, θ_0 can be found by maximizing $\ell(\theta)$. Hence $\theta_0 = \underset{\theta\in\Theta}{\operatorname{argmax}}\, \ell(\theta)$. This property is shared by the true parameter value under correct specification.

Theorem 10.14 Under misspecification, the pseudo-true parameter satisfies $\theta_0 = \underset{\theta\in\Theta}{\operatorname{argmax}}\, \ell(\theta)$.

This theorem shows that the pseudo-true parameter, just like the true parameter, is an analog of the MLE that maximizes the sample analog of $\ell(\theta)$. The finite-sample analog of the pseudo-true parameter is thus the MLE. Consequently, the MLE is the natural analog estimator of the pseudo-true parameter.

For example, to estimate the parameters of a log-normal approximating model, we estimate the log-normal distribution, which is identical to fitting a normal model to the logarithm. This produces the best-fitting log-normal density function in the sense of minimizing the Kullback-Leibler divergence.

Thus, the MLE has twin roles. First, it is an estimator of the true parameter θ_0 when the model $f\left(x \mid \theta\right)$ is correctly specified. Second, it is an estimator of the pseudo-true parameter θ_0 otherwise—when the model is not correctly specified. The fact that the MLE is an estimator of the pseudo-true value means that MLE will produce an estimate of the best-fitting model in the class $f\left(x \mid \theta\right)$ whether or not the model is correct. If the model is correct, the MLE will produce an estimator of the true distribution, but otherwise, it will produce an approximation that produces the best fit as measured by the Kullback-Leibler divergence.

10.18 DISTRIBUTION OF THE MLE UNDER MISSPECIFICATION

We have seen that the MLE is an analog estimator of the pseudo-true value. Therefore, the MLE will be consistent for the pseudo-true value if the assumptions of Theorem 10.8 hold. The most important condition for this extension is that θ_0 is a unique maximum. Under misspecification, this is a strong assumption, as it is possible that multiple sets of parameters could be equally good approximations.

One important implication of misspecification concerns the information matrix equality. If we examine the proof of Theorem 10.5, the assumption of correct specification is used in equation (10.5) where the model density evaluated at the true parameter cancels the true density. Under misspecification, this cancellation does not occur. A consequence is that the information matrix equality fails.

Theorem 10.15 Under misspecification, $\mathscr{I}_\theta \neq \mathscr{H}_\theta$.

We can derive the asymptotic distribution of the MLE under misspecification using exactly the same steps as under correct specification. Examining our informal derivation of Theorem 10.9, the only place where the correct specification assumption is used is in the final equality. Omitting this step, we have the following asymptotic distribution.

Theorem 10.16 Assume the conditions of Theorem 10.9 hold excluding assumption 6. Then as $n \to \infty$,

$$\sqrt{n}\left(\widehat{\theta} - \theta_0\right) \underset{d}{\longrightarrow} \mathrm{N}\left(0, \mathscr{H}_\theta^{-1} \mathscr{I}_\theta \mathscr{H}_\theta^{-1}\right).$$

Thus the MLE is consistent for the pseudo-true value θ_0, converges at the conventional rate $n^{-1/2}$, and is asymptotically normally distributed. The difference between the misspecified and the correctly specified cases is that the asymptotic variance is $\mathscr{H}_\theta^{-1} \mathscr{I}_\theta \mathscr{H}_\theta^{-1}$ rather than $\mathscr{I}_\theta^{-1}$.

10.19 VARIANCE ESTIMATION UNDER MISSPECIFICATION

The conventional estimators of the asymptotic variance of the MLE use the assumption of correct specification. These variance estimators are inconsistent under misspecification. A consistent estimator needs to be an estimator of $\boldsymbol{V} = \mathscr{H}_\theta^{-1} \mathscr{I}_\theta \mathscr{H}_\theta^{-1}$. This requirement leads to the plug-in estimator

$$\widehat{\boldsymbol{V}} = \widehat{\mathscr{H}}^{-1} \widehat{\mathscr{I}} \widehat{\mathscr{H}}^{-1}.$$

This estimator uses both the Hessian and outer product variance estimators. It is consistent for the asymptotic variance under misspecification as well as correct specification.

Theorem 10.17 Under the conditions of Theorem 10.16, as $n \to \infty$, $\widehat{\boldsymbol{V}} \underset{p}{\longrightarrow} \boldsymbol{V}$.

Asymptotic standard errors for $\widehat{\theta}$ are constructed by taking the square roots of the diagonal elements of $n^{-1}\widehat{\boldsymbol{V}}$. When θ is scalar, this is $s\left(\widehat{\theta}\right) = \sqrt{n^{-1}\widehat{V}}$. These standard errors will be different than the conventional MLE standard errors. This covariance matrix estimator $\widehat{\boldsymbol{V}}$ is called a **robust covariance matrix estimator**. Similarly, standard errors constructed from $\widehat{\boldsymbol{V}}$ are called **robust standard errors**. As the word "robust" is used in a wide variety of situations, what is meant by "robust" is often confusing. The robustness for which $\widehat{\boldsymbol{V}}$ is relevant is misspecification of the parametric model. The covariance matrix estimator $\widehat{\boldsymbol{V}}$ is a valid estimator of the covariance matrix of the estimator $\widehat{\theta}$ whether or not the model is correctly specified. Thus, $\widehat{\boldsymbol{V}}$ and the associated standard errors are "model misspecification robust".

The theory of pseudo-true parameters, consistency of the MLE for the pseudo-true values, and robust covariance matrix estimation were developed by Halbert White (1982, 1984).

Example: $f(x \mid \lambda) = \lambda^{-1}\exp(-x/\lambda)$. We found earlier that the MLE is $\widehat{\lambda} = \overline{X}_n$, $\widehat{\mathscr{H}} = 1/\overline{X}_n^2$, and $\widehat{\mathscr{I}} = \widehat{\sigma}^2/\overline{X}^4$. It follows that the robust variance estimator is $\widehat{V} = \widehat{\mathscr{H}}^{-2}\widehat{\mathscr{I}} = \overline{X}_n^4\widehat{\sigma}^2/\overline{X}^4 = \widehat{\sigma}^2$.

The lesson is that in the exponential model, although the classical Hessian-based estimator for the asymptotic variance is $\overline{X}_n^2$, when we allow for potential misspecification, the estimator for the asymptotic variance is $\widehat{\sigma}^2$. The Hessian-based estimator exploits the knowledge that the variance of X is λ^2, while the robust estimator does not use this information.

Example: $X \sim \mathrm{N}(0, \sigma^2)$. We found earlier that the MLE for $\widehat{\sigma}^2$ is $\widehat{\sigma}^2 = n^{-1}\sum_{i=1}^n X_i^2$, $\frac{d}{d\sigma^2}\log f(x \mid \sigma^2) = \left(x^2 - \sigma^2\right)/2\sigma^4$, and $\frac{d^2}{d(\sigma^2)^2}\log f(x \mid \sigma^2) = \left(\sigma^2 - 2x^2\right)/2\sigma^6$. Hence

$$\widehat{\mathscr{H}} = \frac{1}{n}\sum_{i=1}^n \frac{-\widehat{\sigma}^2 + 2X_i^2}{2\widehat{\sigma}^6} = \frac{1}{2\widehat{\sigma}^4}$$

and

$$\widehat{\mathscr{I}} = \frac{1}{n}\sum_{i=1}^{n}\left(\frac{X_i^2 - \widehat{\sigma}^2}{2\widehat{\sigma}^4}\right)^2 = \frac{\widehat{v}^2}{4\widehat{\sigma}^8}$$

where

$$\widehat{v}^2 = \frac{1}{n}\sum_{i=1}^{n}\left(X_i^2 - \widehat{\sigma}^2\right)^2 = \widehat{\operatorname{var}\left[X_i^2\right]}.$$

It follows that the Hessian variance estimator is $\widehat{V}_1 = \widehat{\mathscr{H}}^{-1} = 2\widehat{\sigma}^4$, the outer product estimator is $\widehat{V}_2 = \widehat{\mathscr{I}}^{-1} = 4\widehat{\sigma}^8/\widehat{v}^2$, and the robust variance estimator is $\widehat{V} = \widehat{\mathscr{H}}^{-2}\widehat{\mathscr{I}} = \widehat{v}^2$.

The lesson is that the classical Hessian estimator $2\widehat{\sigma}^4$ uses the assumption that $\mathbb{E}\left[X^4\right] = 3\sigma^4$. The robust estimator does not use this assumption and instead uses the direct estimator of the variance of X^2.

These comparisons show a common finding in standard error calculation: In many cases, there is more than one choice. In this event, which should be used? There are different dimensions by which a choice can be made, including computation, convenience, robustness, and accuracy. The computation/convenience issue is that in many cases, it is easiest to use the formula automatically provided by an existing package, an estimator that is simple to program, and/or an estimator that is fast to compute. This reason is not compelling, however, if the variance estimator is an important component of your research. The robustness issue is that it is desirable to have variance estimators and standard errors that are valid under the broadest conditions. In the context of potential model misspecification, this is the robust variance estimator and associated standard errors. These provide variance estimators and standard errors that are valid in large samples, regardless of whether the model is correctly specified. The third issue, accuracy, refers to the desire to have an estimator of the asymptotic variance which itself is accurate for its target in the sense of having low variance. That is, an estimator $\widehat{\boldsymbol{V}}$ of $\boldsymbol{V}$ is random, and it is better to have a low-variance estimator than a high-variance estimator. While we do not know for certain which variance estimator has the lowest variance, it is typically the case that the Hessian-based estimator will have lower estimation variance than the robust estimator, because it uses the model information to simplify the formula before estimation. In the variance estimation example given above, the robust variance estimator is based on the sample fourth moment, while the Hessian estimator is based on the square of the sample variance. Fourth moments are more erratic and harder to estimate than second moments, so it is not hard to guess that the robust variance estimator will be less accurate, in the sense that it will have higher variance. Overall, we see an inherent trade-off. The robust estimator is valid under broader conditions, but the Hessian estimator is likely to be more accurate, at least when the model is correctly specified.

Contemporary econometrics tends to favor the robust approach. Since statistical models and assumptions are typically viewed as approximations rather than as literal truths, economists tend to prefer methods that are robust to a wide range of plausible assumptions. Therefore, my recommendation for practical applied econometrics is to use the robust estimator $\widehat{\boldsymbol{V}} = \widehat{\mathscr{H}}^{-1}\widehat{\mathscr{I}}\widehat{\mathscr{H}}^{-1}$.

10.20 TECHNICAL PROOFS*

Proof of Theorem 10.2 Take any $\theta \neq \theta_0$. Since the difference of logs is the log of the ratio,

$$\ell(\theta) - \ell(\theta_0) = \mathbb{E}\left[\log\left(f(X \mid \theta)\right) - \log(f(X \mid \theta_0)\right] = \mathbb{E}\left[\log\left(\frac{f(X \mid \theta)}{f(X \mid \theta_0)}\right)\right] < \log\left(\mathbb{E}\left[\frac{f(X \mid \theta)}{f(X \mid \theta_0)}\right]\right).$$

The inequality is Jensen's inequality (Theorem 2.9), since the logarithm is a concave function. The inequality is strict, since $f(X \mid \theta) \neq f(X \mid \theta_0)$ with positive probability. Let $f(x) = f(x \mid \theta_0)$ be the true density. The right-hand side of the above expression equals

$$\begin{aligned}
\log\left(\int \frac{f(x \mid \theta)}{f(x \mid \theta_0)} f(x)dx\right) &= \log\left(\int \frac{f(x \mid \theta)}{f(x \mid \theta_0)} f(x \mid \theta_0)dx\right) \\
&= \log\left(\int f(x \mid \theta)dx\right) \\
&= \log(1) \\
&= 0.
\end{aligned}$$

The third equality stems from the fact that any valid density integrates to 1. We have shown that for any $\theta \neq \theta_0$,

$$\ell(\theta) = \mathbb{E}\left[\log(f(X \mid \theta)\right] < \log\left(f(X \mid \theta_0)\right).$$

This proves that θ_0 maximizes $\ell(\theta)$, as claimed. ■

Proof of Theorem 10.3 We can write the likelihood for the transformed parameter as

$$L_n^*(\beta) = \max_{h(\theta)=\beta} L_n(\theta).$$

The MLE for β maximizes $L_n^*(\beta)$. Evaluating $L_n^*(\beta)$ at $h\left(\widehat{\theta}\right)$, we find

$$L_n^*\left(h\left(\widehat{\theta}\right)\right) = \max_{h(\theta)=h\left(\widehat{\theta}\right)} L_n(\theta) = L_n(\widehat{\theta}).$$

The final equality holds because $\theta = \widehat{\theta}$ satisfies $h(\theta) = h\left(\widehat{\theta}\right)$ and maximizes $L_n(\widehat{\theta})$, so it is the solution. This shows that $L_n^*\left(h\left(\widehat{\theta}\right)\right)$ achieves the maximized likelihood. Thus $h\left(\widehat{\theta}\right) = \widehat{\beta}$ is the MLE for β. ■

Proof of Theorem 10.5 The expected Hessian equals

$$\begin{aligned}
\mathscr{H}_\theta &= -\frac{\partial^2}{\partial\theta\partial\theta'}\mathbb{E}\left[\log f(X \mid \theta_0)\right] \\
&= -\left.\frac{\partial}{\partial\theta}\mathbb{E}\left[\frac{\frac{\partial}{\partial\theta'}f(X \mid \theta)}{f(X \mid \theta)}\right]\right|_{\theta=\theta_0} \\
&= -\left.\frac{\partial}{\partial\theta}\mathbb{E}\left[\frac{\frac{\partial}{\partial\theta'}f(X \mid \theta)}{f(X \mid \theta_0)}\right]\right|_{\theta=\theta_0} && (10.3) \\
&\quad + \mathbb{E}\left[\frac{\frac{\partial}{\partial\theta}f(X \mid \theta_0)\frac{\partial}{\partial\theta'}f(X \mid \theta_0)}{f(X \mid \theta_0)^2}\right]. && (10.4)
\end{aligned}$$

The second equality uses the Leibniz Rule to bring one of the derivatives inside the expectation. The third applies the product rule of differentiation. The outer derivative is applied first to the numerator of the second line (this is equation (10.3)) and applied second to the denominator of the second line (equation (10.4)).

Using the Leibniz Rule (Theorem A.22) to bring the derivative outside the expectation, equation (10.3) equals

$$
\begin{aligned}
-\frac{\partial^2}{\partial\theta\partial\theta'}\mathbb{E}\left[\frac{f(X\mid\theta)}{f(X\mid\theta_0)}\right]\Bigg|_{\theta=\theta_0} &= -\frac{\partial^2}{\partial\theta\partial\theta'}\int\frac{f(x\mid\theta)}{f(x\mid\theta_0)}f(x\mid\theta_0)dx\Bigg|_{\theta=\theta_0} \\
&= -\frac{\partial^2}{\partial\theta\partial\theta'}\int f(x\mid\theta)dx\Bigg|_{\theta=\theta_0} \qquad (10.5) \\
&= -\frac{\partial^2}{\partial\theta\partial\theta'}1 \\
&= 0.
\end{aligned}
$$

The first equality writes the expectation as an integral with respect to the true density $f(x\mid\theta_0)$ under correct specification. The densities cancel in the second line, due to the assumption of correct specification. The third equality uses the fact that $f(x\mid\theta)$ is a valid model so integrates to 1.

Using the definition of the efficient score S, equation (10.4) equals $\mathbb{E}\left[SS'\right]=\mathscr{I}_\theta$. Taken together, we have shown that $\mathscr{H}_\theta=\mathscr{I}_\theta$, as claimed. ■

Proof of Theorem 10.6 Write $x=(x_1,\ldots,x_n)'$ and $X=(X_1,\ldots,X_n)'$. The joint density of X is

$$f(x\mid\theta)=f(x_1\mid\theta)\times\cdots\times f(x_n\mid\theta).$$

The likelihood score is

$$S_n(\theta)=\frac{\partial}{\partial\theta}\log f(X\mid\theta).$$

Let $\widetilde{\theta}$ be an unbiased estimator of θ. Write it as $\widetilde{\theta}(X)$ to indicate that it is a function of X. Let $\mathbb{E}_\theta$ denote expectation with respect to the density $f(x\mid\theta)$, meaning $\mathbb{E}_\theta\left[g(X)\right]=\int g(x)f(x\mid\theta)dx$. Unbiasedness of $\widetilde{\theta}$ means that its expectation is θ for any value. Thus for all θ,

$$\theta=\mathbb{E}_\theta\left[\widetilde{\theta}(X)\right]=\int\widetilde{\theta}(x)f(x\mid\theta)dx.$$

The vector derivative of the left side is

$$\frac{\partial}{\partial\theta'}\theta=\boldsymbol{I}_m$$

and the vector derivative of the right side is

$$
\begin{aligned}
\frac{\partial}{\partial\theta'}\int\widetilde{\theta}(x)f(x\mid\theta)\,dx &= \int\widetilde{\theta}(x)\frac{\partial}{\partial\theta'}f(x\mid\theta)dx \\
&= \int\widetilde{\theta}(x)\frac{\partial}{\partial\theta'}\log f(x\mid\theta)f(x\mid\theta)dx.
\end{aligned}
$$

Evaluated at the true value, this is

$$\mathbb{E}\left[\widetilde{\theta}(X)\,S_n(\theta_0)'\right]=\mathbb{E}\left[\left(\widetilde{\theta}-\theta_0\right)S_n(\theta_0)'\right]$$

where the equality holds, since $\mathbb{E}\left[S_n(\theta_0)\right]=0$ by Theorem 10.4. Setting the two derivatives equal, we obtain

$$\boldsymbol{I}_m=\mathbb{E}\left[\left(\widetilde{\theta}-\theta_0\right)S_n(\theta_0)'\right].$$

Let $\boldsymbol{V} = \text{var}\left[\widetilde{\theta}\right]$. We have shown that the covariance matrix of stacked $\widetilde{\theta} - \theta_0$ and $S_n(\theta_0)$ is

$$\begin{bmatrix} \boldsymbol{V} & \boldsymbol{I}_m \\ \boldsymbol{I}_m & n\mathscr{I}_\theta \end{bmatrix}.$$

Since it is a covariance matrix, it is positive semi-definite and thus remains so if we pre- and post-multiply by the same matrix, for instance,

$$\begin{bmatrix} \boldsymbol{I}_m & -(n\mathscr{I}_\theta)^{-1} \end{bmatrix} \begin{bmatrix} \boldsymbol{V} & \boldsymbol{I}_m \\ \boldsymbol{I}_m & n\mathscr{I}_\theta \end{bmatrix} \begin{bmatrix} \boldsymbol{I}_m \\ -(n\mathscr{I}_\theta)^{-1} \end{bmatrix} = \boldsymbol{V} - (n\mathscr{I}_\theta)^{-1} \geq 0$$

which implies $\boldsymbol{V} \geq (n\mathscr{I}_\theta)^{-1}$, as claimed. ■

Proof of Theorem 10.7 The proof is similar to that of Theorem 10.6. Let $\widetilde{\beta} = \widetilde{\beta}(X)$ be an unbiased estimator of β. Unbiasedness implies

$$h(\theta) = \beta = \mathbb{E}_\theta\left[\widetilde{\beta}(X)\right] = \int \widetilde{\beta}(x) f(x \mid \theta) dx.$$

The vector derivative of the left side is

$$\frac{\partial}{\partial \theta'} h(\theta) = \boldsymbol{H}'$$

and the vector derivative of the right side is

$$\frac{\partial}{\partial \theta'} \int \widetilde{\beta}(x) f(x \mid \theta)\, dx = \int \widetilde{\beta}(x) \frac{\partial}{\partial \theta'} \log f(x \mid \theta) f(x \mid \theta) dx.$$

Evaluated at the true value, this is

$$\mathbb{E}\left[\widetilde{\beta} S_n(\theta_0)'\right] = \mathbb{E}\left[\left(\widetilde{\beta} - \beta_0\right) S_n(\theta_0)'\right].$$

Setting the two derivatives equal, we obtain

$$\boldsymbol{H}' = \mathbb{E}\left[\left(\widetilde{\beta} - \beta_0\right) S_n(\theta_0)'\right].$$

Post-multiply by $\mathscr{I}_\theta^{-1}\boldsymbol{H}$ to obtain

$$\boldsymbol{H}'\mathscr{I}_\theta^{-1}\boldsymbol{H} = \mathbb{E}\left[\left(\widetilde{\beta} - \beta_0\right) S_n(\theta_0)' \mathscr{I}_\theta^{-1}\boldsymbol{H}\right].$$

For simplicity, take the case of scalar β. Applying the expectation (Theorem 2.10) and Cauchy-Schwarz (Theorem 4.11) inequalities gives

$$\begin{aligned}
\boldsymbol{H}'\mathscr{I}_\theta^{-1}\boldsymbol{H} &= \left|\mathbb{E}\left[\left(\widetilde{\beta} - \beta_0\right) S_n(\theta_0)' \mathscr{I}_\theta^{-1}\boldsymbol{H}\right]\right| \\
&\leq \mathbb{E}\left|\left(\widetilde{\beta} - \beta_0\right) S_n(\theta_0)' \mathscr{I}_\theta^{-1}\boldsymbol{H}\right| \\
&\leq \left(\mathbb{E}\left[\left(\widetilde{\beta} - \beta_0\right)^2\right] \mathbb{E}\left[\left(S_n(\theta_0)' \mathscr{I}_\theta^{-1}\boldsymbol{H}\right)^2\right]\right)^{1/2} \\
&= \left(\text{var}\left[\widetilde{\beta}\right]\right)^{1/2} \left(\boldsymbol{H}'\mathscr{I}_\theta^{-1}\mathbb{E}\left[S_n(\theta_0) S_n(\theta_0)'\right]\mathscr{I}_\theta^{-1}\boldsymbol{H}\right)^{1/2} \\
&= \left(\text{var}\left[\widetilde{\beta}\right]\right)^{1/2} \left(n\boldsymbol{H}'\mathscr{I}_\theta^{-1}\boldsymbol{H}\right)^{1/2}
\end{aligned}$$

which implies

$$\text{var}\left[\widetilde{\beta}\right] \geq n^{-1}\boldsymbol{H}'\mathscr{I}_\theta^{-1}\boldsymbol{H}.$$

■

Proof of Theorem 10.8 The proof proceeds in three steps. First, we show that $\overline{\ell}_n(\theta) \underset{p}{\longrightarrow} \ell(\theta)$ uniformly in θ. Second, we show that $\ell(\widehat{\theta}) \underset{p}{\longrightarrow} \ell(\theta)$. Third, we show that this implies $\widehat{\theta} \underset{p}{\longrightarrow} \theta$.

We appeal to Theorem 18.2 of Chapter 18, setting $g(x, \theta) = \log f(x \mid \theta)$, whose assumptions (part (c)) are satisfied by those stated for Theorem 10.8. Theorem 18.2 implies

$$\sup_{\theta \in \Theta} \left\| \overline{\ell}_n(\theta) - \ell(\theta) \right\| \underset{p}{\longrightarrow} 0. \tag{10.6}$$

That is the first step.

Since θ_0 maximizes $\ell(\theta)$, $\ell(\theta_0) \geq \ell(\widehat{\theta})$. Hence

$$\begin{aligned}
0 &\leq \ell(\theta_0) - \ell(\widehat{\theta}) \\
&= \ell(\theta_0) - \overline{\ell}_n(\theta_0) + \overline{\ell}_n(\widehat{\theta}) - \ell(\widehat{\theta}) + \overline{\ell}_n(\theta_0) - \overline{\ell}_n(\widehat{\theta}) \\
&\leq \ell(\theta_0) - \overline{\ell}_n(\theta_0) + \overline{\ell}_n(\widehat{\theta}) - \ell(\widehat{\theta}) \\
&\leq 2 \sup_{\theta \in \Theta} \left\| \overline{\ell}_n(\theta) - \ell(\theta) \right\| \\
&\underset{p}{\longrightarrow} 0.
\end{aligned}$$

The second inequality uses the fact that $\widehat{\theta}$ maximizes $\overline{\ell}_n(\theta)$, so $\overline{\ell}_n(\theta_0) - \overline{\ell}_n(\widehat{\theta}) \leq 0$. The third inequality replaces the two pairwise comparisons by the supremum. The final convergence is given by equation (10.6). This completes the second step.

The preceeding argument is illustrated in Figure 10.7. The figure displays the expected log density $\ell(\theta)$ and the average log likelihood $\overline{\ell}_n(\theta)$. The distances between the two functions at the true value θ_0 and the MLE $\widehat{\theta}$ are marked by the two dashed lines. The sum of these two lengths is greater than the vertical distance between $\ell(\theta_0)$ and $\ell(\widehat{\theta})$, because the latter distance equals the sum of the two dashed lines plus the vertical height of the thick section of $\overline{\ell}_n(\theta)$ (between θ_0 and $\widehat{\theta}$), which is positive since $\overline{\ell}_n(\widehat{\theta}) \geq \overline{\ell}_n(\theta_0)$. Equation (10.6) implies that the sum of these components converges to 0. Hence $\ell(\widehat{\theta})$ converges to $\ell(\theta_0)$.

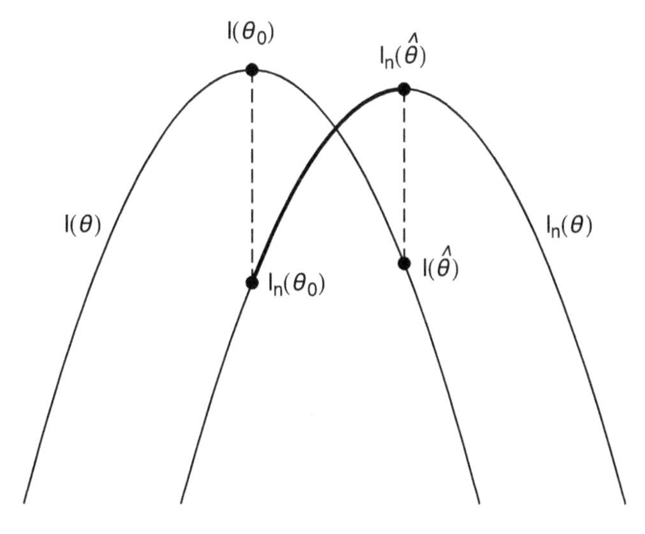

FIGURE 10.7 Consistency of MLE

In the third step of the proof, we show that $\widehat{\theta} \underset{p}{\longrightarrow} \theta$. Pick $\epsilon > 0$. The fifth assumption implies that there is a $\delta > 0$ such that $\|\theta_0 - \theta\| > \epsilon$ implies $\ell(\theta_0) - \ell(\theta) \geq \delta$. This means that $\|\theta_0 - \widehat{\theta}\| > \epsilon$ implies $\ell(\theta_0) - \ell(\widehat{\theta}) \geq \delta$. Hence

$$\mathbb{P}\left[\left\|\theta_0 - \widehat{\theta}\right\| > \epsilon\right] \leq \mathbb{P}\left[\ell(\theta_0) - \ell(\widehat{\theta}) \geq \delta\right].$$

The right-hand side converges to 0, since $\ell(\widehat{\theta}) \underset{p}{\longrightarrow} \ell(\theta)$. Thus the left-hand side converges to 0 as well. Since ϵ is arbitrary, we have that $\widehat{\theta} \underset{p}{\longrightarrow} \theta$, as stated.

To illustrate, again examine Figure 10.7. In it $\ell(\widehat{\theta})$ is marked on the graph of $\ell(\theta)$. Since $\ell(\widehat{\theta})$ converges to $\ell(\theta_0)$, $\ell(\widehat{\theta})$ must slide up the graph of $\ell(\theta)$ toward the maximum. The only way for $\widehat{\theta}$ to not converge to θ_0 would be if the function $\ell(\theta)$ were flat at the maximum, which is excluded by the assumption of a unique maximum. ■

10.21 EXERCISES

Exercise 10.1 Let X be distributed Poisson: $\pi(k) = \dfrac{\exp(-\theta)\theta^k}{k!}$ for nonnegative integer k and $\theta > 0$.

(a) Find the log-likelihood function $\ell_n(\theta)$.
(b) Find the MLE $\widehat{\theta}$ for θ.

Exercise 10.2 Let X be distributed as $\mathrm{N}(\mu, \sigma^2)$. The unknown parameters are μ and σ^2.

(a) Find the log-likelihood function $\ell_n(\mu, \sigma^2)$.
(b) Take the first-order condition with respect to μ, and show that the solution for $\widehat{\mu}$ does not depend on σ^2.
(c) Define the concentrated log-likelihood function $\ell_n(\widehat{\mu}, \sigma^2)$. Take the first-order condition for σ^2, and find the MLE $\widehat{\sigma}^2$ for σ^2.

Exercise 10.3 Let X be distributed Pareto with density $f(x) = \dfrac{\alpha}{x^{1+\alpha}}$ for $x \geq 1$ and $\alpha > 0$.

(a) Find the log-likelihood function $\ell_n(\alpha)$.
(b) Find the MLE $\widehat{\alpha}$ for α.

Exercise 10.4 Let X be distributed Cauchy with density $f(x) = \dfrac{1}{\pi(1 + (x - \theta)^2)}$ for $x \in \mathbb{R}$.

(a) Find the log-likelihood function $\ell_n(\theta)$.
(b) Find the first-order condition for the MLE $\widehat{\theta}$ for θ. You will not be able to solve for $\widehat{\theta}$.

Exercise 10.5 Let X be distributed double exponential with density $f(x) = \frac{1}{2}\exp(-|x - \theta|)$ for $x \in \mathbb{R}$.

(a) Find the log-likelihood function $\ell_n(\theta)$.
(b) Extra challenge: Find the MLE $\widehat{\theta}$ for θ. This is challenging as it does not involve simply solving the FOC due to the nondifferentiability of the density function.

Exercise 10.6 Let X be Bernoulli $\pi(X \mid p) = p^x(1-p)^{1-x}$.

(a) Calculate the information for p by taking the variance of the score.

(b) Calculate the information for p by taking the expectation of (minus) the second derivative. Did you obtain the same answer?

Exercise 10.7 Take the Pareto model $f(x) = \alpha x^{-1-\alpha}$, $x \geq 1$. Calculate the information for α using the second derivative.

Exercise 10.8 Find the Cramér-Rao lower bound for p in the Bernoulli model (use the results from Exercise 10.6). In Section 10.3, we showed that the MLE for p is $\widehat{p} = \overline{X}_n$. Compute var $\left[\widehat{p}\right]$. Compare var $\left[\widehat{p}\right]$ with the CRLB.

Exercise 10.9 Take the Pareto model, and recall the MLE $\widehat{\alpha}$ for α from Exercise 10.3. Show that $\widehat{\alpha} \xrightarrow[p]{} \alpha$ by using the WLLN and the CMT (Theorem 7.7).

Exercise 10.10 Take the model $f(x) = \theta \exp(-\theta x), x \geq 0, \theta > 0$.

(a) Find the Cramér-Rao lower bound for θ.

(b) Find the MLE $\widehat{\theta}$ for θ. Notice that it is a function of the sample mean. Use this formula and the Delta Method to find the asymptotic distribution for $\widehat{\theta}$.

(c) Find the asymptotic distribution for $\widehat{\theta}$ using the general formula for the asymptotic distribution of MLE. Do you find the same answer as in part (b)?

Exercise 10.11 Take the gamma model

$$f(x \mid \alpha, \beta) = \frac{\beta^\alpha}{\Gamma(\alpha)} x^{\alpha-1} e^{-\beta x}, \qquad x > 0.$$

Assume β is known, so the only parameter to estimate is α. Let $g(\alpha) = \log \Gamma(\alpha)$. Write your answers in terms of the derivatives $g'(\alpha)$ and $g''(\alpha)$. (You need not find the closed-form solution for these derivatives.)

(a) Calculate the information for α.

(b) Use the general formula for the asymptotic distribution of the MLE to find the asymptotic distribution of $\sqrt{n}(\widehat{\alpha} - \alpha)$, where $\widehat{\alpha}$ is the MLE.

(c) Letting V denote the asymptotic variance from part (b), propose an estimator $\widehat{V}$ for V.

Exercise 10.12 Take the Bernoulli model.

(a) Find the asymptotic variance of the MLE. (Hint: see Exercise 10.8.)

(b) Propose an estimator of the asymptotic variance V.

(c) Show that this estimator is consistent for V as $n \to \infty$.

(d) Propose a standard error $s(\widehat{p})$ for the MLE $\widehat{p}$. (Recall that the standard error is supposed to approximate the variance of $\widehat{p}$, not that of the variance of $\sqrt{n}(\widehat{p} - p)$. What would be a reasonable approximation of the variance of $\widehat{p}$ once you have a reasonable approximation of the variance of $\sqrt{n}(\widehat{p} - p)$ from part (b)?)

Exercise 10.13 Supose X has density $f(x)$ and $Y = \mu + \sigma X$. You have a random sample $\{Y_1, \ldots, Y_n\}$ from the distribution of Y.

(a) Find an expression for the density $f_Y(y)$ of Y.

(b) Suppose $f(x) = C\exp(-a(x))$ for some known differentiable function $a(x)$ and some C. Find C. (Since $a(x)$ is not specified, you cannot find an explicit answer but can express C in terms of an integral.)

(c) Given the density from part (b), find the log-likelihood function for the sample $\{Y_1, \ldots, Y_n\}$ as a function of the parameters (μ, σ).

(d) Find the pair of FOCs for the MLE $(\widehat{\mu}, \widehat{\sigma}^2)$. (You do not need to solve for the estimators.)

Exercise 10.14 Take the gamma density function. Assume that α is known.

(a) Find the MLE $\widehat{\beta}$ for β based on a random sample $\{X_1, \ldots, X_n\}$.

(b) Find the asymptotic distribution for $\sqrt{n}\left(\widehat{\beta} - \beta\right)$.

Exercise 10.15 Take the beta density function. Assume that β is known and equals $\beta = 1$.

(a) Find $f(x \mid \alpha) = f(x \mid \alpha, 1)$.

(b) Find the log-likelihood function $\ell_n(\alpha)$ for α from a random sample $\{X_1, \ldots, X_n\}$.

(c) Find the MLE $\widehat{\alpha}$.

Exercise 10.16 Let $g(x)$ be a density function of a random variable with mean μ and variance σ^2. Let X be a random variable with density function

$$f(x \mid \theta) = g(x)\left(1 + \theta\,(x - \mu)\right).$$

Assume that $g(x)$, μ, and σ^2 are known. The unknown parameter is θ. Assume that X has bounded support, so that $f(x \mid \theta) \geq 0$ for all x. (Basically, don't worry if $f(x \mid \theta) \geq 0$.)

(a) Verify that $\int_{-\infty}^{\infty} f(x \mid \theta) dx = 1$.

(b) Calculate $\mathbb{E}[X]$.

(c) Find the information $\mathscr{I}_\theta$ for θ. Write your expression as an expectation of some function of X.

(d) Find a simplified expression for $\mathscr{I}_\theta$ when $\theta = 0$.

(e) Given a random sample $\{X_1, \ldots, X_n\}$, write the log-likelihood function for θ.

(f) Find the first-order condition for the MLE $\widehat{\theta}$ for θ. You will not be able to solve for $\widehat{\theta}$.

(g) Using the known asymptotic distribution for maximum likelihood estimators, find the asymptotic distribution for $\sqrt{n}\left(\widehat{\theta} - \theta\right)$ as $n \to \infty$.

(h) How does the asymptotic distribution simplify when $\theta = 0$?

CHAPTER 11

METHOD OF MOMENTS

11.1 INTRODUCTION

Many popular estimation methods in economics are based on the method of moments. This chapter introduces the basic ideas and methods. We review estimation of multivariate means, moments, smooth functions, parametric models, and moment equations.

The method of moments allows for **semi-parametric** models. The term "semi-parametric" is used for estimation of finite-dimensional parameters when the distribution is not parametric. An example is estimation of the mean $\theta = \mathbb{E}[X]$ (which is a finite-dimensional parameter) when the distribution of X is unspecified. A distribution is called **nonparametric** if it cannot be described by a finite list of parameters.

To illustrate the methods, we use the same dataset as in Chapter 6, but take the subsample of married white male wage earners with 20 years of potential work experience. This subsample has $n = 529$ observations.

11.2 MULTIVARIATE MEANS

The multivariate expectation of the random vector X is $\mu = \mathbb{E}[X]$. The analog estimator is

$$\widehat{\mu} = \overline{X}_n = \frac{1}{n}\sum_{i=1}^{n} X_i.$$

The asymptotic distribution of $\widehat{\beta}$ is provided by Theorem 8.5:

$$\sqrt{n}\left(\widehat{\mu} - \mu\right) \underset{d}{\longrightarrow} \mathrm{N}(0, \Sigma)$$

where $\Sigma = \operatorname{var}[X]$.

The asymptotic covariance matrix Σ is estimated by the sample covariance matrix

$$\widehat{\Sigma} = \frac{1}{n-1}\sum_{i=1}^{n}\left(X_i - \overline{X}_n\right)\left(X_i - \overline{X}_n\right)'.$$

This estimator is consistent for Σ: $\widehat{\Sigma} \underset{p}{\longrightarrow} \Sigma$. Standard errors for the elements of $\widehat{\mu}$ are square roots of the diagonal elements of $n^{-1}\widehat{\Sigma}$.

Similarly, the expected value of any transformation $g(X)$ is $\theta = \mathbb{E}\left[g(X)\right]$. The analog estimator is

$$\widehat{\theta} = \frac{1}{n}\sum_{i=1}^{n} g(X_i).$$

The asymptotic distribution of $\widehat{\theta}$ is

$$\sqrt{n}\left(\widehat{\theta}-\theta\right) \underset{d}{\longrightarrow} \mathrm{N}\left(0, \boldsymbol{V}_{\theta}\right)$$

where $\boldsymbol{V}_{\theta}=\operatorname{var}\left[g(X)\right]$. The asymptotic covariance matrix is estimated by

$$\widehat{\boldsymbol{V}}_{\theta}=\frac{1}{n-1} \sum_{i=1}^{n}\left(g\left(X_{i}\right)-\widehat{\theta}\right)\left(g\left(X_{i}\right)-\widehat{\theta}\right)^{\prime}. \tag{11.1}$$

It is consistent for $\boldsymbol{V}_{\theta}$: $\widehat{\boldsymbol{V}}_{\theta} \underset{p}{\longrightarrow} \boldsymbol{V}_{\theta}$. Standard errors for the elements of $\widehat{\theta}$ are square roots of the diagonal elements of $n^{-1}\widehat{\boldsymbol{V}}_{\theta}$.

To illustrate, the sample mean and covariance matrix of wages and education are

$$\widehat{\mu}_{\text{wage}}=31.9$$

$$\widehat{\mu}_{\text{education}}=14.0$$

$$\widehat{\Sigma}=\begin{bmatrix} 834 & 35 \\ 35 & 7.2 \end{bmatrix}.$$

The standard errors for the two mean estimates are

$$s\left(\widehat{\mu}_{\text{wage}}\right)=\sqrt{834/529}=1.3$$

$$s\left(\widehat{\mu}_{\text{education}}\right)=\sqrt{7.2/529}=0.1.$$

11.3 MOMENTS

The mth moment of X is $\mu_{m}^{\prime}=\mathbb{E}\left[X^{m}\right]$. The sample analog estimator is

$$\widehat{\mu}_{m}^{\prime}=\frac{1}{n} \sum_{i=1}^{n} X_{i}^{m}.$$

The asymptotic distribution of $\widehat{\mu}_{m}^{\prime}$ is

$$\sqrt{n}\left(\widehat{\mu}_{m}^{\prime}-\mu_{m}^{\prime}\right) \underset{d}{\longrightarrow} \mathrm{N}\left(0, V_{m}\right)$$

where

$$V_{m}=\mathbb{E}\left[\left(X^{m}-\mu_{m}^{\prime}\right)^{2}\right]=\mathbb{E}\left[X^{2m}\right]-\left(\mathbb{E}\left[X^{m}\right]\right)^{2}=\mu_{2m}-\mu_{m}^{2}.$$

A consistent estimator of the asymptotic variance is

$$\widehat{V}_{m}=\frac{1}{n-1} \sum_{i=1}^{n}\left(X_{i}^{m}-\widehat{\mu}_{m}^{\prime}\right)^{2} \underset{p}{\longrightarrow} V_{m}.$$

A standard error for $\widehat{\mu}_{m}^{\prime}$ is $s\left(\widehat{\mu}_{m}^{\prime}\right)=\sqrt{n^{-1}\widehat{V}_{m}}$.

To illustrate, Table 11.1 displays the estimated moments of the hourly wage, expressed in units of \$100. Moments 1 through 4 are reported in the first column. Standard errors are reported in the second column.

Table 11.1
Moments of hourly wage (in $100)

Moment	$\widehat{\mu}'_m$	$s\left(\widehat{\mu}'_m\right)$	$\widehat{\mu}_m$	$s\left(\widehat{\mu}_m\right)$
1	0.32	0.01	0.32	0.01
2	0.19	0.02	0.08	0.01
3	0.19	0.03	0.07	0.02
4	0.25	0.06	0.10	0.02

11.4 SMOOTH FUNCTIONS

Many parameters of interest can be written in the form

$$\beta = h(\theta) = h\left(\mathbb{E}\left[g(X)\right]\right)$$

where X, g, and h are possibly vector-valued. When h is continuously differentiable, this is often called the **smooth function model**. The standard estimators take the plug-in form

$$\widehat{\beta} = h\left(\widehat{\theta}\right)$$

$$\widehat{\theta} = \frac{1}{n}\sum_{i=1}^{n} g(X_i).$$

The asymptotic distribution of $\widehat{\beta}$ is provided by Theorem 8.9:

$$\sqrt{n}\left(\widehat{\beta} - \beta\right) \underset{d}{\longrightarrow} \mathrm{N}\left(0, \boldsymbol{V}_\beta\right)$$

where $\boldsymbol{V}_\beta = \boldsymbol{H}'\boldsymbol{V}_\theta\boldsymbol{H}$, $\boldsymbol{V}_\theta = \operatorname{var}\left[g(X)\right]$, and $\boldsymbol{H} = \frac{\partial}{\partial\theta}h(\theta)'$.

The asymptotic covariance matrix $\boldsymbol{V}_\beta$ is estimated by replacing the components by sample estimators. The estimator of $\boldsymbol{V}_\theta$ is equation (11.1), that of $\boldsymbol{H}$ is

$$\widehat{\boldsymbol{H}} = \frac{\partial}{\partial\theta}h\left(\widehat{\theta}\right)',$$

and that of $\boldsymbol{V}_\beta$ is

$$\widehat{\boldsymbol{V}}_\beta = \widehat{\boldsymbol{H}}'\widehat{\boldsymbol{V}}_\theta\widehat{\boldsymbol{H}}.$$

This estimator is consistent: $\widehat{\boldsymbol{V}}_\beta \underset{p}{\longrightarrow} \boldsymbol{V}_\beta$.

We illustrate with three examples.

Example: Geometric mean. The geometric mean of a random variable X is

$$\beta = \exp\left(\mathbb{E}\left[\log(X)\right]\right).$$

In this case, $h(\theta) = \exp(\theta)$. The plug-in-estimator equals

$$\widehat{\beta} = \exp\left(\widehat{\theta}\right)$$

$$\widehat{\theta} = \frac{1}{n}\sum_{i=1}^{n} \log(X_i).$$

Since $H = \exp(\theta) = \beta$, the asymptotic variance of $\widehat{\beta}$ is

$$V_\beta = \beta^2 \sigma_\theta^2$$
$$\sigma_\theta^2 = \mathbb{E}\left[\log(X) - \mathbb{E}\left[\log(X)\right]\right]^2.$$

An estimator of the asymptotic variance is

$$\widehat{V}_\beta = \widehat{\beta}^2 \widehat{\sigma}_\theta^2$$
$$\widehat{\sigma}_\theta^2 = \frac{1}{n-1} \sum_{i=1}^{n} \left(\log(X_i) - \widehat{\theta}\right)^2.$$

In our sample, the estimated geometric mean of wages is $\widehat{\beta} = 24.9$. Its standard error is $s\left(\widehat{\beta}\right) = \sqrt{n^{-1}\widehat{\beta}^2\widehat{\sigma}_\theta^2} = 24.9 \times 0.66/\sqrt{529} = 0.7$.

Example: Ratio of means. Take a pair of random variables (Y, X) with means (μ_Y, μ_X). If $\mu_X > 0$, the ratio of their means is

$$\beta = \frac{\mu_Y}{\mu_X}.$$

In this case, $h(\mu_Y, \mu_X) = \mu_Y/\mu_X$. The plug-in-estimator equals

$$\widehat{\mu}_Y = \overline{Y}_n$$
$$\widehat{\mu}_X = \overline{X}_n$$
$$\widehat{\beta} = \frac{\widehat{\mu}_Y}{\widehat{\mu}_X}.$$

We calculate

$$\boldsymbol{H}(u) = \begin{pmatrix} \dfrac{\partial}{\partial u_1} \dfrac{u_1}{u_2} \\ \\ \dfrac{\partial}{\partial u_2} \dfrac{u_1}{u_2} \end{pmatrix} = \begin{pmatrix} \dfrac{1}{u_2} \\ \\ -\dfrac{u_1}{u_2^2} \end{pmatrix}.$$

Evaluated at the true values, this is

$$\boldsymbol{H} = \begin{pmatrix} \dfrac{1}{\mu_X} \\ \\ -\dfrac{\mu_Y}{\mu_X^2} \end{pmatrix}.$$

Notice that

$$\boldsymbol{V}_{Y,X} = \begin{pmatrix} \sigma_Y^2 & \sigma_{YX} \\ \sigma_{YX} & \sigma_X^2 \end{pmatrix}.$$

The asymptotic variance of $\widehat{\beta}$ is therefore

$$V_\beta = \boldsymbol{H}'\boldsymbol{V}_{Y,X}\boldsymbol{H} = \frac{\sigma_Y^2}{\mu_X^2} - 2\frac{\sigma_{YX}\mu_Y}{\mu_X^3} + \frac{\sigma_X^2\mu_Y^2}{\mu_X^4}.$$

An estimator is

$$\widehat{V}_\beta = \frac{s_Y^2}{\widehat{\mu}_X^2} - 2\frac{s_{XY}\widehat{\mu}_Y}{\widehat{\mu}_X^3} + \frac{s_X^2\widehat{\mu}_Y^2}{\widehat{\mu}_X^4}.$$

For example, let β be the ratio of the mean wage to mean years of education. The sample estimate is $\widehat{\beta}=31.9/14.0=2.23$. The estimated variance is

$$\widehat{V}_{\beta}=\frac{834}{14^2}-2\frac{35\times 31.9}{14^3}+\frac{7.2\times 31.9^2}{14^4}=3.6.$$

The standard error is

$$s\left(\widehat{\beta}\right)=\sqrt{\frac{\widehat{V}_{\beta}}{n}}=\sqrt{\frac{3.6}{529}}=0.08.$$

Example: Variance. The variance of X is $\sigma^2=\mathbb{E}\left[X^2\right]-(\mathbb{E}\left[X\right])^2$. We can set $g(X)=(X,X^2)'$ and $h(\mu_1,\mu_2)=\mu_2-\mu_1^2$. The plug-in-estimator of σ^2 equals

$$\widehat{\sigma}^2=\frac{1}{n}\sum_{i=1}^{n}X_i^2-\left(\frac{1}{n}\sum_{i=1}^{n}X_i\right)^2.$$

We calculate

$$\boldsymbol{H}(u)=\begin{pmatrix}\frac{\partial}{\partial u_1}\left(u_2-u_1^2\right)\\ \frac{\partial}{\partial u_2}\left(u_2-u_1^2\right)\end{pmatrix}=\begin{pmatrix}-2u_1\\1\end{pmatrix}.$$

Evaluated at the true values, this is

$$\boldsymbol{H}=\begin{pmatrix}-2\mathbb{E}\left[X\right]\\1\end{pmatrix}.$$

Notice that

$$\boldsymbol{V}_{\theta}=\begin{pmatrix}\operatorname{var}\left[X\right] & \operatorname{cov}\left(X,X^2\right)\\ \operatorname{cov}\left(X,X^2\right) & \operatorname{var}\left[X^2\right]\end{pmatrix}.$$

The asymptotic variance of $\widehat{\sigma}^2$ is therefore

$$V_{\sigma^2}=\boldsymbol{H}'\boldsymbol{V}_{\theta}\boldsymbol{H}=\operatorname{var}\left[X^2\right]-4\operatorname{cov}\left(X,X^2\right)\mathbb{E}\left[X\right]+4\operatorname{var}\left[X\right](\mathbb{E}\left[X\right])^2$$

which simplifies to

$$V_{\sigma^2}=\operatorname{var}\left[(X-\mathbb{E}\left[X\right])^2\right].$$

There are two ways to make this simplification. One is to expand out the second expression and verify that the two are the same. But it is hard to go in the reverse direction to see the simplification. The second way is to appeal to invariance. The variance σ^2 is invariant to the mean $\mathbb{E}\left[X\right]$. So is the sample variance $\widehat{\sigma}^2$. Consequently, the variance of the sample variance must be invariant to the mean $\mathbb{E}\left[X\right]$ and without loss of generality can be set to 0, which produces the simplified expression (if we are careful to re-center X at $X-\mathbb{E}\left[X\right]$).

The estimator of the variance is

$$\widehat{V}_{\sigma^2}=\frac{1}{n}\sum_{i=1}^{n}\left(\left(X_i-\overline{X}_n\right)^2-\widehat{\sigma}^2\right)^2.$$

The standard error is $s\left(\widehat{\sigma}^2\right)=\sqrt{n^{-1}\widehat{V}_{\sigma^2}}$.

Take, for example, the variance of the wage distribution. The point estimate is $\widehat{\sigma}^2 = 834$. The estimate of the variance of $\left(X - \overline{X}_n\right)^2$ is $\widehat{V}_{\sigma^2} = 3011^2$. The standard error for $\widehat{\sigma}^2$ is

$$s\left(\widehat{\sigma}^2\right) = \sqrt{\frac{\widehat{V}_{\sigma^2}}{n}} = \frac{3011}{\sqrt{529}} = 131.$$

11.5 CENTRAL MOMENTS

The mth central moment of X (for $m \geq 2$) is

$$\mu_m = \mathbb{E}\left[(X - \mathbb{E}[X])^m\right].$$

By the binomial theorem (Theorem 1.11), this can be written as

$$\begin{aligned}
\mu_m &= \sum_{k=0}^{m} \binom{m}{k} (-1)^{m-k} \mu'_k \mu^{m-k} \\
&= (-1)^m \mu^m + m(-1)^{m-1} \mu^m + \sum_{k=2}^{m} \binom{m}{k} (-1)^{m-k} \mu'_k \mu^{m-k} \\
&= (-1)^m (1-m)\mu^m + \sum_{k=2}^{m-1} \binom{m}{k} (-1)^{m-k} \mu'_k \mu^{m-k} + \mu'_m.
\end{aligned}$$

This is a nonlinear function of the uncentered moments $k = 1, \ldots, m$. The sample analog estimator is

$$\begin{aligned}
\widehat{\mu}_m &= \frac{1}{n} \sum_{i=1}^{n} \left(X_i - \overline{X}_n\right)^m \\
&= \sum_{k=0}^{m} \binom{m}{k} (-1)^{m-k} \widehat{\mu}'_k \widehat{\mu}^{m-k} \\
&= (-1)^m (1-m)\overline{X}_n^m + \sum_{k=2}^{m-1} \binom{m}{k} (-1)^{m-k} \widehat{\mu}'_k \widehat{\mu}^{m-k} + \widehat{\mu}'_m.
\end{aligned}$$

This estimator falls in the class of smooth function models. Since $\widehat{\mu}_m$ is invariant to μ, we can assume $\mu = 0$ for the distribution theory if we center X at μ. Set $\widetilde{X} = X - \mu$,

$$Z = \begin{pmatrix} \widetilde{X} \\ \widetilde{X}^2 \\ \vdots \\ \widetilde{X}^m \end{pmatrix}$$

and

$$\boldsymbol{V}_\theta = \begin{pmatrix} \operatorname{var}\left[\widetilde{X}\right] & \operatorname{cov}\left(\widetilde{X}, \widetilde{X}^2\right) & \cdots & \operatorname{cov}\left(\widetilde{X}, \widetilde{X}^m\right) \\ \operatorname{cov}\left(\widetilde{X}^2, \widetilde{X}\right) & \operatorname{var}\left[\widetilde{X}^2\right] & \cdots & \operatorname{cov}\left(X^2, \widetilde{X}^m\right) \\ \vdots & \vdots & \ddots & \vdots \\ \operatorname{cov}\left(\widetilde{X}^m, \widetilde{X}\right) & \operatorname{cov}\left(\widetilde{X}^m, \widetilde{X}^2\right) & \cdots & \operatorname{var}\left[\widetilde{X}^m\right] \end{pmatrix}.$$

Set

$$h(u) = \left[(-1)^m (1-m)u_1^m + \sum_{k=2}^{m-1} \binom{m}{k} (-1)^{m-k} u_k u_1^{m-k} + u_m \right].$$

It has derivative vector

$$\boldsymbol{H}(u) = \begin{pmatrix} (-1)^m (1-m) m u_1^{m-1} + \sum_{k=2}^{m-1} (-1)^{m-k} \binom{m}{k} (m-k) u_k u_1^{m-k-1} \\ (-1)^{m-2} \binom{m}{2} u_1^{m-2} \\ \vdots \\ (-1) \binom{m}{m-1} u_1 \\ 1 \end{pmatrix}.$$

Evaluated at the moments (and using $\mu = 0$) gives

$$\boldsymbol{H} = \begin{pmatrix} -m\mu_{m-1} \\ 0 \\ \vdots \\ 0 \\ 1 \end{pmatrix}.$$

Thus the asymptotic variance of $\widehat{\mu}_m$ is

$$\begin{aligned} V_m &= \boldsymbol{H}' \boldsymbol{V}_\theta \boldsymbol{H} \\ &= m^2 \mu_{m-1}^2 \operatorname{var}\left[\widetilde{X}\right] - 2m\mu_{m-1} \operatorname{cov}\left(\widetilde{X}^m, \widetilde{X}\right) + \operatorname{var}\left[\widetilde{X}^m\right] \\ &= m^2 \mu_{m-1}^2 \operatorname{var}[X] - 2m\mu_{m-1}\mu_{m+1} + \mu_{2m} - \mu_m^2 \end{aligned}$$

where we use $\operatorname{var}\left[\widetilde{X}^m\right] = \mu_{2m} - \mu_m^2$ and

$$\operatorname{cov}\left(\widetilde{X}^m, \widetilde{X}\right) = \mathbb{E}\left[\left((X - \mathbb{E}[X])^m - \mu_m\right)(X - \mathbb{E}[X])\right] = \mathbb{E}\left[(X - \mathbb{E}[X])^{m+1}\right] = \mu_{m+1}.$$

An estimator is

$$\widehat{V}_m = m^2 \widehat{\mu}_{m-1}^2 \widehat{\sigma}^2 - 2m\widehat{\mu}_{m-1}\widehat{\mu}_{m+1} + \widehat{\mu}_{2m} - \widehat{\mu}_m^2.$$

The standard error for $\widehat{\mu}_m$ is $s(\widehat{\mu}_m) = \sqrt{n^{-1}\widehat{V}_m}$.

To illustrate, Table 11.1 in Section 11.3 displays the estimated central moments of the hourly wage, expressed in units of \$100 for scaling. Central moments 1 through 4 are reported in the third column, with standard errors in the final column. The standard errors are reasonably small relative to the estimates, indicating precision.

11.6 BEST UNBIASED ESTIMATION

Up to this point, our only efficiency justification for the sample mean $\widehat{\mu} = \overline{X}_n$ as an estimator of the population mean is that it is BLUE—it has the lowest variance among linear unbiased estimators. This is a limited justification, as the restriction to linear estimators has no convincing justification. In this section, we find a much stronger efficiency justification, showing that the mean has the lowest variance among all unbiased estimators. This result is taken from Hansen (2022b).

Let $\{X_1, \ldots, X_n\}$ be a sample of i.i.d. draws from a distribution F with finite variance. Set $\mu = \mathbb{E}[X]$ and $\sigma^2 = \text{var}[X]$. The moment estimator of μ is the sample mean $\widehat{\mu} = \overline{X}_n$. It has the well-known properties $\mathbb{E}[\widehat{\mu}] = \mu$ and $\text{var}[\widehat{\mu}] = n^{-1}\sigma^2$. We are interested in whether any other unbiased estimator $\widetilde{\mu}$ has lower variance.

Theorem 11.1 If $\widetilde{\mu}$ is unbiased for μ for all distributions with a finite variance, then $\text{var}[\widetilde{\mu}] \geq n^{-1}\sigma^2$.

This is a strict improvement on the BLUE theorem, as the only restriction on the estimator is unbiasedness.

Recall that the sample mean $\widehat{\mu} = n^{-1}\sum_{i=1}^n X_i$ is unbiased and has variance $\text{var}[\widehat{\mu}] = \sigma^2/n$. Combined with Theorem 11.1, we deduce that no unbiased estimator (including both linear and nonlinear estimators) can have a lower variance than the sample mean.

Theorem 11.2 Assume $\mathbb{E}\left[X^2\right] < \infty$. The sample mean $\widehat{\mu} = n^{-1}\sum_{i=1}^n X_i$ has the lowest variance among all unbiased estimators of μ.

Theorem 11.2 states that the sample mean is the **best unbiased estimator** of the population expectation. In other words, the "L" in the BLUE label is unnecessary.

The key to the above result is unbiasedness. The assumption that an estimator is unbiased means that it is unbiased for all data distributions. This effectively rules out any estimator that exploits special information. This conclusion is quite important. It can be natural to think along the following lines: "If we only knew the correct parametric family $F(x \mid \beta)$ for the distribution of X, we could use this information to obtain a better estimator." The flaw with this reasoning is that any such "better estimator" would exploit this special information by producing an estimator that is biased when this information fails to be correct. It is impossible to improve estimation efficiency without compromising unbiasedness.

Let us complete this section with a proof of Theorem 11.1 under simplified assumptions. If desired, the argument is can be skipped without any loss of narrative flow.

For simplicity, assume that X has a distribution $F(x)$ with density[1] $f(x)$ which has bounded[2] support $\mathscr{X}$. Without loss of generality, assume that the true mean satisfies $\mathbb{E}[X] = 0$.

Define the auxiliary density function

$$f_\mu(x) = f(x)\left(1 + \frac{x\mu}{\sigma^2}\right). \tag{11.2}$$

Since X has bounded support $\mathscr{X}$, there is a set B such that $f_\mu(x) \geq 0$ for all $\mu \in B$ and $x \in \mathscr{X}$. You can check that $\int_{\mathscr{X}} f_\mu(x)dx = 1$, so $f_\mu(x)$ is a valid density function. Over $\mu \in B$, f_μ is a parametric family of density functions with associated distribution functions F_μ. Since $F_0 = F$, the parametric family F_μ is correctly specified with true parameter value $\mu_0 = 0$.

To illustrate, Figure 11.1(a) displays an example density $f(x) = (3/4)(1 - x^2)$ on $[-1, 1]$ with auxiliary density $f_\mu(x) = f(x)(1 + x)$. We can see how the auxiliary density is a tilted version of the original density $f(x)$.

Let $\mathbb{E}_\mu$ denote expectation with respect to the auxiliary density. Since $\int_{\mathscr{X}} xf(x)dx = 0$ and $\int_{\mathscr{X}} x^2 f(x)dx = \sigma^2$, we find

$$\mathbb{E}_\mu[X] = \int_{\mathscr{X}} xf_\mu(x)dx = \int_{\mathscr{X}} xf(x)dx + \int_{\mathscr{X}} x^2 f(x)dx\mu/\sigma^2 = \mu.$$

[1] The assumption of a density can be replaced by a derivation using the Radon-Nikodym derivative of the distribution function, which applies to all distributions.

[2] The assumption of bounded support can be replaced by a technical argument involving truncation and limits.

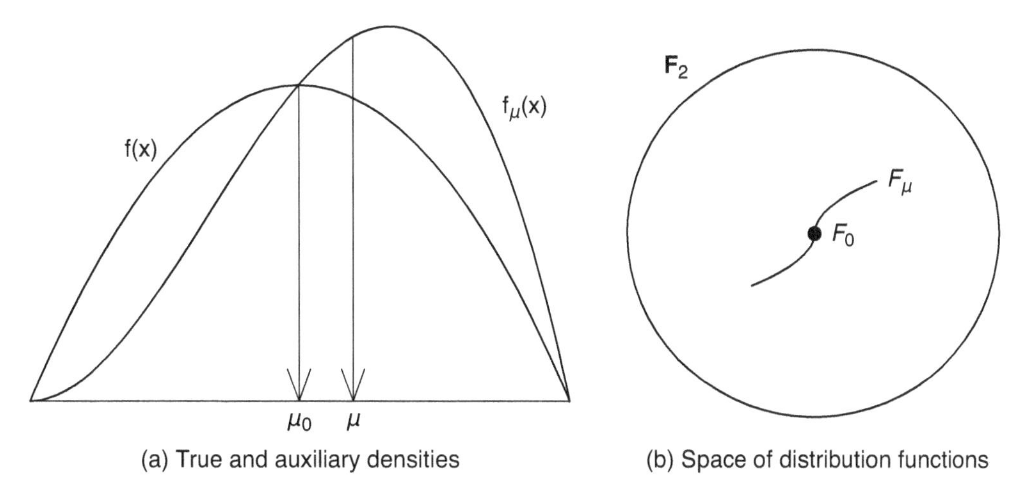

(a) True and auxiliary densities

(b) Space of distribution functions

FIGURE 11.1 Best unbiased estimation

Thus the auxiliary density has expectation μ. In Figure 11.1(a), the means of the two densities are indicated by the arrows to the x-axis. In this example, we can see how the auxiliary density has a larger expected value, since the density has been tilted to the right. You can also show that the auxiliary density has a finite variance.

To summarize, the auxiliary model F_μ is a regular parametric family over $\mu \in B$ with the properties that its expectation is μ, its variance is finite, the true value μ_0 lies in the interior of B, and the support of the density does not depend on μ. To visualize, Figure 11.1(b) displays $\mathbf{F}_2$, the space of distributions with a finite variance, by the large circle. The dot indicates the true distribution $F = F_0$. The curved line represents the distribution family F_μ. This family F_μ is a sliver in the general space of distributions $\mathbf{F}_2$ but includes the true distribution F.

Now, let $\widetilde{\mu}$ be an estimator for μ which (as stated in the assumptions) is unbiased for all distributions with a finite variance (the large circle in Figure 11.1(b)). This means that $\widetilde{\mu}$ is unbiased over the subset F_μ (the curve in Figure 11.1(b)) as it is a strict subset.

The likelihood score of the auxiliary model is

$$S = \frac{\partial}{\partial \mu}\left(\log f(X) + \log\left(1 + \frac{X\mu}{\sigma^2}\right)\right)\bigg|_{\mu=0} = \frac{X}{\sigma^2}.$$

Using Definition 10.6 and $\mathbb{E}\left[X^2\right] = \sigma^2$, we find that the model's information at $\mu = 0$ is

$$\mathscr{I}_0 = \mathbb{E}\left[S^2\right] = \frac{\mathbb{E}\left[X^2\right]}{\sigma^4} = \frac{1}{\sigma^2}.$$

Since $\widetilde{\mu}$ is unbiased in the model (11.2) and the latter is a regular parametric model, we can apply the Cramér-Rao Theorem (Theorem 10.6), which states that

$$\operatorname{var}\left[\widetilde{\mu}\right] \geq (n\mathscr{I}_0)^{-1} = \sigma^2/n.$$

This completes the proof.

It is rather surprising that the proof can be obtained by examining just a single auxiliary model. Why is that sufficient? The reason is that no Cramér-Rao bound can be larger than σ^2/n, as the latter is the variance of the sample mean $\overline{X}_n$. Thus σ^2/n is the best possible variance lower bound. The fact that the auxiliary model (11.2) has the Cramér-Rao bound σ^2/n is quite special, simplifying the argument.

11.7 PARAMETRIC MODELS

The classical method of moments often refers to the context where a random variable X has a parametric distribution with an unknown parameter, and the latter is estimated by solving moments of the distribution.

Let $f(x \mid \beta)$ be a parametric density with parameter vector $\beta \in \mathbb{R}^m$. The moments of the model are

$$\mu_k(\beta) = \int x^k f(x \mid \beta) dx.$$

These are mappings from the parameter space to $\mathbb{R}$.

Estimating the first m moments from the data gives

$$\widehat{\mu}_k = \frac{1}{n} \sum_{i=1}^{n} X_i^k$$

$k = 1, \ldots, m$. Take the m moment equations

$$\mu_1(\beta) = \widehat{\mu}_1$$
$$\vdots$$
$$\mu_m(\beta) = \widehat{\mu}_m.$$

These are m equations and m unknowns. Solve for the solution $\widehat{\beta}$. This is the method of moments estimator of β. In some cases, the solution can be found algebraically. In other cases, the solution must be found numerically.

In some cases, moments other than the first m integer moments may be selected. For example, if the density is symmetric, then there is no information in odd moments above 1, so only even moments above 1 should be used.

The method of moments estimator can be written as a function of the sample moments, thus $\widehat{\beta} = h(\widehat{\mu})$. Therefore, the asymptotic distribution is found by using Theorem 8.9:

$$\sqrt{n}\left(\widehat{\beta} - \beta\right) \underset{d}{\longrightarrow} \mathrm{N}\left(0, \boldsymbol{V}_\beta\right)$$

where $\boldsymbol{V}_\beta = \boldsymbol{H}' \boldsymbol{V}_\mu \boldsymbol{H}$. Here, $\boldsymbol{V}_\mu$ is the covariance matrix for the moments used, and $\boldsymbol{H}$ is the derivative of $\boldsymbol{h}(\mu)$.

11.8 EXAMPLES OF PARAMETRIC MODELS

Example: Exponential. The density is $f(x \mid \lambda) = \frac{1}{\lambda} \exp\left(-\frac{x}{\lambda}\right)$. In this model, $\lambda = \mathbb{E}[X]$. The expected value $\mathbb{E}[X]$ is estimated by the sample mean $\overline{X}_n$. The unknown λ is then estimated by matching the moment equation, for example,

$$\widehat{\lambda} = \overline{X}_n.$$

In this case, we have the simple solution $\widehat{\lambda} = \overline{X}_n$.

In this example, we have the immediate answer for the asymptotic distribution:

$$\sqrt{n}\left(\widehat{\lambda} - \lambda\right) \underset{d}{\longrightarrow} \mathrm{N}\left(0, \sigma_X^2\right).$$

Given the exponential assumption, we can write the variance as $\sigma_X^2 = \lambda^2$ if desired. Estimators for the asymptotic variance include $\widehat{\sigma}_X^2$ and $\widehat{\lambda}^2 = \overline{X}_n^2$ (the second is valid due to the parametric assumption).

Example: N $\left(\mu, \sigma^2\right)$. The mean and variance are μ and σ^2. The estimators are $\widehat{\mu} = \overline{X}_n$ and $\widehat{\sigma}^2$.

Example: Gamma(α, β). The first two central moments are

$$\mathbb{E}[X] = \frac{\alpha}{\beta}$$

$$\operatorname{var}[X] = \frac{\alpha}{\beta^2}.$$

Solving for the parameters in terms of the moments, we find

$$\alpha = \frac{(\mathbb{E}[X])^2}{\operatorname{var}[X]}$$

$$\beta = \frac{\mathbb{E}[X]}{\operatorname{var}[X]}.$$

The method of moments estimators are

$$\widehat{\alpha} = \frac{\overline{X}_n^2}{\widehat{\sigma}^2}$$

$$\widehat{\beta} = \frac{\overline{X}_n}{\widehat{\sigma}^2}.$$

To illustrate, consider estimation of the parameters using the wage observations. We have $\overline{X}_n = 31.9$ and $\widehat{\sigma}^2 = 834$, which implies $\widehat{\alpha} = 1.2$ and $\widehat{\beta} = 0.04$.

Now consider the asymptotic variance. Take $\widehat{\alpha}$. It can be written as

$$\widehat{\alpha} = h\left(\widehat{\mu}_1, \widehat{\mu}_2\right) = \frac{\widehat{\mu}_1^2}{\widehat{\mu}_2 - \widehat{\mu}_1^2}$$

$$\widehat{\mu}_1 = \frac{1}{n}\sum_{i=1}^n X_i$$

$$\widehat{\mu}_2 = \frac{1}{n}\sum_{i=1}^n X_i^2.$$

The asymptotic variance is

$$V_\alpha = \boldsymbol{H}' \boldsymbol{V} \boldsymbol{H}$$

where

$$\boldsymbol{V} = \begin{pmatrix} \operatorname{var}[X] & \operatorname{cov}\left(X, X^2\right) \\ \operatorname{cov}\left(X, X^2\right) & \operatorname{var}\left[X^2\right] \end{pmatrix}$$

and $\boldsymbol{H}$ is the vector of derivatives of $h\left(\mu_1, \mu_2\right)$:

$$\boldsymbol{H}(\mu_1,\mu_2) = \begin{pmatrix} \dfrac{\partial}{\partial\mu_1}\dfrac{\mu_1^2}{\mu_2-\mu_1^2} \\ \\ \dfrac{\partial}{\partial\mu_2}\dfrac{\mu_1^2}{\mu_2-\mu_1^2} \end{pmatrix} = \begin{pmatrix} \dfrac{2\mu_1\mu_2}{\left(\mu_2-\mu_1^2\right)^2} \\ \\ \dfrac{-2\mu_1^2}{\left(\mu_2-\mu_1^2\right)^2} \end{pmatrix} = \begin{pmatrix} \dfrac{2\mathbb{E}[X]\left(\text{var}[X]+(\mathbb{E}[X])^2\right)}{\text{var}[X]^2} \\ \\ \dfrac{-2(\mathbb{E}[X])^2}{\text{var}[X]^2} \end{pmatrix} = \begin{pmatrix} 2\beta(1+\alpha) \\ -2\beta^2 \end{pmatrix}.$$

Example: Log Normal. The density is $f(x\mid\theta,v^2) = \frac{1}{\sqrt{2\pi v^2}}x^{-1}\exp\left(-\left(\log x-\theta\right)^2/2v^2\right)$. Moments of X can be calculated and matched to find estimators of θ and v^2. A simpler (and preferred) method is to use the property that $\log X \sim \mathrm{N}(\theta, v^2)$. Hence the mean and variance are θ and v^2, so the moment estimators are the sample mean and variance of $\log X$. Thus

$$\widehat{\theta} = \frac{1}{n}\sum_{i=1}^n \log X_i$$

$$\widehat{v}^2 = \frac{1}{n-1}\sum_{i=1}^n \left(\log X_i - \widehat{\theta}\right)^2.$$

Using the wage data, we find $\widehat{\theta} = 3.2$ and $\widehat{v}^2 = 0.43$.

Example: Scaled student t. X has a scaled student t distribution if $T = (X-\theta)/v$ has a student t distribution with degree of freedom parameter r. If $r > 4$, the first four central moments of X are

$$\mu = \theta$$

$$\mu_2 = \frac{r}{r-2}v^2$$

$$\mu_3 = 0$$

$$\mu_4 = \frac{3r^2}{(r-2)(r-4)}v^4.$$

Inverting and using the fact that $\mu_4 \geq 3\mu_2$ (for the student t distribution), we can find that

$$\theta = \mu$$

$$v^2 = \frac{\mu_4\mu_2}{2\mu_4 - 3\mu_2^2}$$

$$r = \frac{4\mu_4 - 6\mu_2^2}{\mu_4 - 3\mu_2^2}.$$

To estimate the parameters (θ, v^2, r), we can use the first, second, and fourth moments. Let $\overline{X}_n$, $\widehat{\sigma}^2$, and $\widehat{\mu}_4$ be the sample mean, variance, and centered fourth moment. We have the moment equations

$$\widehat{\theta} = \overline{X}_n$$

$$\widehat{v}^2 = \frac{\widehat{\mu}_4\widehat{\sigma}^2}{2\widehat{\mu}_4 - 3\widehat{\sigma}^4}$$

$$\widehat{r} = \frac{4\widehat{\mu}_4 - 6\widehat{\sigma}^4}{\widehat{\mu}_4 - 3\widehat{\sigma}^4}.$$

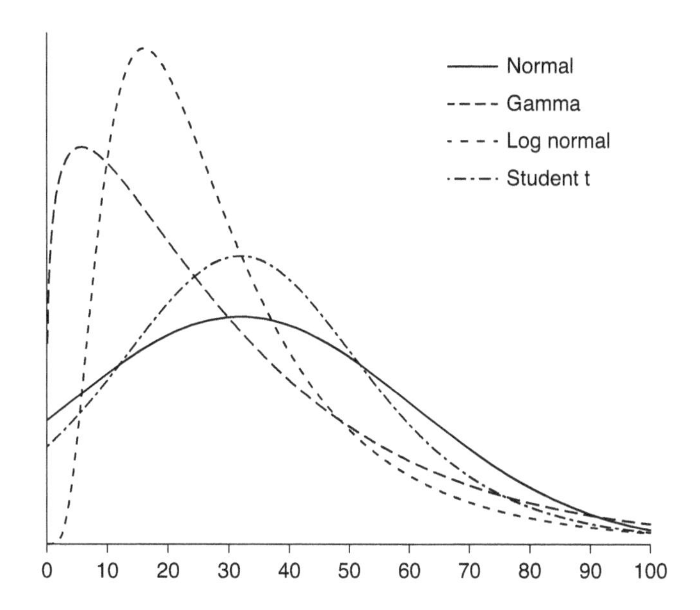

FIGURE 11.2 Parametric models fit to wage data using method of moments

Using the wage data, we find $\widehat{\theta} = 32$, $\widehat{v}^2 = 467$, and $\widehat{r} = 4.55$. The point estimate for the degree of freedom parameter is only slightly above the threshold for a finite fourth moment.

Figure 11.2 displays the density functions implied by method of moments estimates using four distributions. Displayed are the densities $\mathrm{N}(32, 834)$, $\mathrm{Gamma}(1.2, 0.04)$, $\mathrm{LogNormal}(3.2, 0.43)$, and $t(32, 467, 4.55)$. The normal and t densities spill over to the negative x axis. The gamma and log normal densities are highly skewed. The four estimated densities are, naturally, quite different from one another, as the four models are quite distinct. Thus, to reiterate, any fitted model should be viewed as misspecified. Some may be reasonable approximations, but some may not.

11.9 MOMENT EQUATIONS

In the previous examples, the parameter of interest can be written as a nonlinear function of population moments. This leads to the plug-in estimator obtained by using the nonlinear function of the corresponding sample moments.

In many econometric models, we cannot write the parameters as explicit functions of moments. However, we can write moments that are explicit functions of the parameters. We can still use the method of moments but need to numerically solve the moment equations to obtain estimators for the parameters.

This class of estimators takes the following form. We have a parameter $\beta \in \mathbb{R}^k$ and a $k \times 1$ function $m(x, \beta)$, which is a function of the random variable and the parameter. The model states that the expectation of the function is known and is normalized so that it equals 0. Thus

$$\mathbb{E}\left[m(X, \beta)\right] = 0.$$

The sample version of the moment function is

$$\overline{m}_n(\beta) = \frac{1}{n} \sum_{i=1}^{n} m(X_i, \beta).$$

The method of moments estimator $\widehat{\beta}$ is the value that solves the k nonlinear equations

$$\overline{m}_n\left(\widehat{\beta}\right)=0.$$

This includes all the explicit moment equations previously discussed. For example, the mean and variance satisfy

$$\mathbb{E}\begin{bmatrix}(X-\mu)^2-\sigma^2\\ (X-\mu)\end{bmatrix}=0.$$

We can set

$$m(x,\mu,\sigma^2)=\begin{pmatrix}(x-\mu)^2-\sigma^2\\ (x-\mu)\end{pmatrix}.$$

The population moment condition is

$$\mathbb{E}\left[m(X,\mu,\sigma^2)\right]=0.$$

The sample estimator of $\mathbb{E}\left[m(X,\mu,\sigma^2)\right]$ is

$$0=\overline{m}_n\left(\mu,\sigma^2\right)=\begin{pmatrix}\frac{1}{n}\sum_{i=1}^n (X_i-\mu)^2-\sigma^2\\ \frac{1}{n}\sum_{i=1}^n (X_i-\mu)\end{pmatrix}.$$

The estimators $\left(\widehat{\mu},\widehat{\sigma}^2\right)$ solve

$$0=\overline{m}_n\left(\widehat{\mu},\widehat{\sigma}^2\right).$$

These are identical to the sample mean and variance.

11.10 ASYMPTOTIC DISTRIBUTION FOR MOMENT EQUATIONS

An asymptotic theory for moment estimators can be derived using similar methods as we used for the MLE. I present the theory in this section without formal proofs. They are similar in nature to those for the MLE and so are omitted here for brevity.

A key issue for nonlinear estimation is identification. For moment estimation, this is the analog of the correct specification assumption for MLE.

Definition 11.1 The moment parameter β is **identified** in B if there is a unique $\beta_0 \in B$ that solves the moment equations

$$\mathbb{E}\left[m(X,\beta)\right]=0.$$

We first consider consistent estimation.

Theorem 11.3 Assume

1. X_i is i.i.d.
2. $\|m(x,\beta)\| \le M(x)$ and $\mathbb{E}\left[M(X)\right]<\infty$.
3. $m(X,\beta)$ is continuous in β with probability 1.
4. B is compact.
5. β_0 is the unique solution to $\mathbb{E}\left[m(X,\beta)\right]=0$.

Assumptions 1–4 imply that the moment function $\overline{m}_n(\beta)$ is uniformly consistent for its expectation. This along with assumption 5 can be used to show that the sample solution $\widehat{\beta}$ is consistent for the population solution β_0.

We next consider the asymptotic distribution. Define the moment matrices

$$\Omega = \operatorname{var}[m(X, \beta_0)]$$

$$\boldsymbol{Q} = \mathbb{E}\left[\frac{\partial}{\partial \beta} m(X, \beta_0)'\right].$$

Let $\mathscr{N}$ be a neighborhood of β_0.

Theorem 11.4 Assume the conditions of Theorem 11.3 hold, plus

1. $\mathbb{E}\left[\|m(X, \beta_0)\|^2\right] < \infty$.
2. $\frac{\partial}{\partial \beta}\mathbb{E}[m(x, \beta)]$ is continuous in $\beta \in \mathscr{N}$.
3. $m(X, \beta)$ is Lipschitz-continuous in $\mathscr{N}$.
4. $\boldsymbol{Q}$ is full rank.
5. β_0 is in the interior of B.

Then as $n \to \infty$,

$$\sqrt{n}\left(\widehat{\beta} - \beta_0\right) \underset{d}{\longrightarrow} \mathrm{N}(0, \boldsymbol{V})$$

where $\boldsymbol{V} = \boldsymbol{Q}^{-1\prime}\Omega \boldsymbol{Q}^{-1}$.

Assumption 1 states that the moment equation has a finite second moment. This is required to apply the central limit theory. Assumption 2 states that the derivative of the expected moment equation is continuous near β_0. This is a required technical condition. Assumption 3 states that $m(X, \beta)$ is Lipschitz-continuous. This is to ensure that the normalized moment $n^{-1/2}\sum_{i=1}^{n} m(X_i, \beta)$ is asymptotically equicontinuous. Alternatives to Assumption 3 are discussed in Sections 18.6 and 18.7. Assumption 4 states that the derivative matrix $\boldsymbol{Q}$ is full rank. This is an important identification condition. It means that the moment $\mathbb{E}[m(X, \beta)]$ changes with the parameter β. It is what ensures that the moment has information about the value of the parameter. Assumption 5 is necessary to justify a mean-value expansion. If the true parameter lies on the boundary of the parameter space, then the moment estimator will not have a standard asymptotic distribution.

The variance is estimated by the plug-in principle:

$$\widehat{\boldsymbol{V}} = \widehat{\boldsymbol{Q}}^{-1\prime}\widehat{\Omega}\widehat{\boldsymbol{Q}}^{-1}$$

$$\widehat{\Omega} = \frac{1}{n}\sum_{i=1}^{n} m(X_i, \widehat{\beta}) m(X_i, \widehat{\beta})'$$

$$\widehat{\boldsymbol{Q}} = \frac{1}{n}\sum_{i=1}^{n} \frac{\partial}{\partial \beta} m(X_i, \widehat{\beta})'.$$

11.11 EXAMPLE: EULER EQUATION

The following is a simplified version of a classic economic model that leads to a nonlinear moment equation.

Take a consumer who consumes C_t in period t and C_{t+1} in period $t+1$. Their utility is

$$U(C_t, C_{t+1}) = u(C_t) + \frac{1}{\beta} u(C_{t+1})$$

for some utility function $u(c)$ and some discount factor β. The consumer has the budget constraint

$$C_t + \frac{C_{t+1}}{R_{t+1}} \leq W_t$$

where W_t is their endowment, and R_{t+1} is the uncertain return on investment.

The expected utility from consumption of C_t in period t is

$$U^*(C_t) = \mathbb{E}\left[u(C_t) + \frac{1}{\beta} u((W_t - C_t) R_{t+1})\right]$$

where the budget constraint has been substituted for the second period. The first-order condition for selection of C_t is

$$0 = u'(C_t) - \mathbb{E}\left[\frac{R_{t+1}}{\beta} u'(C_{t+1})\right].$$

Suppose that the consumer's utility function is constant relative risk aversion: $u(c) = c^{1-\alpha}/(1-\alpha)$. Then $u'(c) = c^{-\alpha}$, and the above equation becomes

$$0 = C_t^{-\alpha} - \mathbb{E}\left[\frac{R_{t+1}}{\beta} C_{t+1}^{-\alpha}\right]$$

which can be rearranged as

$$\mathbb{E}\left[R_{t+1}\left(\frac{C_{t+1}}{C_t}\right)^{-\alpha} - \beta\right] = 0$$

since C_t is treated as known (as it is selected by the consumer at time t).

For simplicity, suppose that β is known. Then the equation is the expectation of a nonlinear equation of consumption growth C_{t+1}/C_t, the return on investment R_{t+1}, and the risk aversion parameter α.

We can put this in the method of moments framework by defining

$$m(R_{t+1}, C_{t+1}, C_t, \alpha) = R_{t+1}\left(\frac{C_{t+1}}{C_t}\right)^{-\alpha} - \beta.$$

The population parameter satisfies

$$\mathbb{E}\left[m(R_{t+1}, C_{t+1}, C_t, \alpha)\right] = 0.$$

The sample estimator of $\mathbb{E}[m(X, \alpha)]$ is

$$\overline{m}_n(\alpha) = \frac{1}{n}\sum_{t=1}^{n} R_{t+1}\left(\frac{C_{t+1}}{C_t}\right)^{-\alpha} - \beta.$$

The estimator $\widehat{\alpha}$ solves

$$\overline{m}_n(\widehat{\alpha}) = 0.$$

The solution is found numerically. The function $\overline{m}_n(\alpha)$ tends to be monotonic in α, so there is a unique solution $\widehat{\alpha}$.

The asymptotic distribution of the method of moments estimator is

$$\sqrt{n}(\widehat{\alpha} - \alpha) \underset{d}{\longrightarrow} \mathrm{N}(0, V)$$

$$V = \frac{\operatorname{var}\left[R_{t+1}\left(\frac{C_{t+1}}{C_t}\right)^{-\alpha} - \beta\right]}{\left(\mathbb{E}\left[R_{t+1}\left(\frac{C_{t+1}}{C_t}\right)^{-\alpha}\log\left(\frac{C_{t+1}}{C_t}\right)\right]\right)^2}.$$

The asymptotic variance can be estimated by

$$\widehat{V} = \frac{\frac{1}{n}\sum_{t=1}^{n}\left(R_{t+1}\left(\frac{C_{t+1}}{C_t}\right)^{-\widehat{\alpha}} - \frac{1}{n}\sum_{t=1}^{n}R_{t+1}\left(\frac{C_{t+1}}{C_t}\right)^{-\widehat{\alpha}}\right)^2}{\left(\frac{1}{n}\sum_{t=1}^{n}R_{t+1}\left(\frac{C_{t+1}}{C_t}\right)^{-\widehat{\alpha}}\log\left(\frac{C_{t+1}}{C_t}\right)\right)^2}.$$

11.12 EMPIRICAL DISTRIBUTION FUNCTION

Recall that the distribution function of a random variable X is $F(x) = \mathbb{P}[X \le x] = \mathbb{E}[\mathbb{1}\{X \le x\}]$. Given a sample $\{X_1, \ldots, X_n\}$ of observations from F, the method of moments estimator for $F(x)$ is the fraction of observations less than or equal to x:

$$F_n(x) = \frac{1}{n}\sum_{i=1}^{n}\mathbb{1}\{X_i \le x\}. \tag{11.3}$$

The function $F_n(x)$ is called the **empirical distribution function** (EDF):

For any sample, the EDF is a valid distribution function. (It is nondecreasing, right-continuous, and limits to 0 and 1.) It is the discrete distribution that puts probability mass $1/n$ on each observation. It is a nonparametric estimator, as it uses no prior information about the distribution function $F(x)$. Note that while $F(x)$ may be either discrete or continuous, $F_n(x)$ is by construction a step function.

The distribution function of a random vector $X \in \mathbb{R}^m$ is $F(x) = \mathbb{P}[X \le x] = \mathbb{E}[\mathbb{1}\{X \le x\}]$, where the inequality applies to all elements of the vector. The EDF for a sample $\{X_1, \ldots, X_n\}$ is

$$F_n(x) = \frac{1}{n}\sum_{i=1}^{n}\mathbb{1}\{X_i \le x\}.$$

As for scalar variables, the multivariate EDF is a valid distribution function, and it is the probability distribution that puts probability mass $1/n$ at each observation.

To illustrate, Figure 11.3(a) displays the empirical distribution function of the 20-observation sample from Table 6.1. You can see that it is a step function with each step of height 1/20. Each line segment takes the half-open form $[a, b)$. The steps occur at the sample values, which are marked on the x-axis with the crosses "×".

The EDF $F_n(x)$ is a consistent estimator of the distribution function $F(x)$. To see this, note that for any $x \in \mathbb{R}^m$, $\mathbb{1}\{X_i \le x\}$ is a bounded i.i.d. random variable with expectation $F(x)$. Thus by the WLLN (Theorem 7.4), $F_n(x) \underset{p}{\longrightarrow} F(x)$. It is also straightforward to derive the asymptotic distribution of the EDF.

Theorem 11.5 If $X_i \in \mathbb{R}^m$ are i.i.d., then for any $x \in \mathbb{R}^m$, as $n \to \infty$

$$\sqrt{n}\left(F_n(x) - F(x)\right) \underset{d}{\longrightarrow} \mathrm{N}\left(0, F(x)(1 - F(x))\right).$$

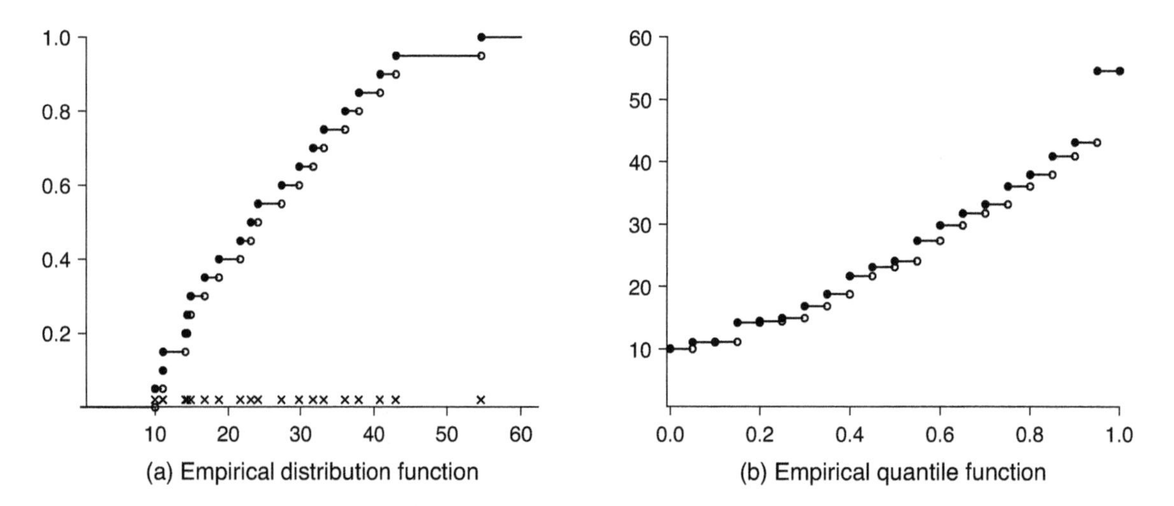

FIGURE 11.3 Empirical distribution and quantile functions

11.13 SAMPLE QUANTILES

Quantiles are a useful representation of a distribution. In Section 2.9, we defined the quantiles for a continuous distribution as the solution to $\alpha = F(q(\alpha))$. More generally, we can define the quantile of any distribution as follows.

Definition 11.2 For any $\alpha \in (0, 1]$, the αth quantile of a distribution $F(x)$ is $q(\alpha) = \inf\{x : F(x) \geq \alpha\}$.

When $F(x)$ is strictly increasing, then $q(\alpha)$ satisfies $F(q(\alpha)) = \alpha$ and is thus the "inverse" of the distribution function as defined previously.

One way to think about a quantile is that it is the point that splits the probabilty mass so that $100\alpha\%$ of the distribution is to the left of $q(\alpha)$, and $100(1 - \alpha)\%$ is to the right of $q(\alpha)$. Only univariate quantiles are defined; there is no multivariate version. A related concept is that of **percentiles**, which are expressed in terms of percentages. For any α, the αth quantile and 100th percentile are identical.

The empirical analog of $q(\alpha)$ given a univariate sample $\{X_1, \ldots, X_n\}$ is the **empirical quantile**, which is obtained by replacing $F(x)$ with the empirical distribution function $F_n(x)$. Thus

$$\widehat{q}(\alpha) = \inf\{x : F_n(x) \geq \alpha\}.$$

It turns out that this can be written as the jth-order statistic $X_{(j)}$ of the sample. (The jth ordered sample value; see Section 6.16.) Then $F_n(X_{(j)}) \geq j/n$ with equality if the sample values are unique. Set $j = \lceil n\alpha \rceil$, the value $n\alpha$ rounded up to the nearest integer (also known as the ceiling function). Thus $F_n(X_{(j)}) \geq j/n \geq \alpha$. For any $x < X_{(j)}$, $F_n(x) \leq (j - 1)/n < \alpha$. As asserted, we see that $X_{(j)}$ is the αth empirical quantile.

Theorem 11.6 $\widehat{q}(\alpha) = X_{(j)}$, where $j = \lceil n\alpha \rceil$.

To illustrate, consider estimation of the median wage from the dataset reported in Table 6.1. In this example, $n = 20$ and $\alpha = 0.5$. Thus $n\alpha = 10$ is an integer. The 10th-order statistic for the wage (the 10th smallest observed wage) is $wage_{(10)} = 23.08$. This is the empirical median $\widehat{q}(0.5) = 23.08$. To estimate the 0.66 quantile of this distribution, $n\alpha = 13.2$, so we round up to 14. The 14th-order statistic (and empirical 0.66 quantile) is $\widehat{q}(0.66) = 31.73$.

Figure 11.3(b) displays the empirical quantile function of the 20-observation sample from Figure 11.3(a). It is a step function with steps at 1/20, 2/20, and so forth. The height of the function at each step is the associated order statistic.

A useful property of quantiles and sample quantiles is that they are **equivariant to monotone transformations**. Specifically, let $h(\cdot): \mathbb{R} \to \mathbb{R}$ be nondecreasing, and set $W = h(Y)$. Let $q^y(\alpha)$ and $q^w(\alpha)$ be the quantile functions of Y and W. The equivariance property is $q^w(\alpha) = h(q^y(\alpha))$. That is, the quantiles of W are the transformations of the quantiles of Y. For example, the αth quantile of $\log(X)$ is the log of the αth quantile of X.

To illustrate with our empirical example, the log of the median wage is $\log(\widehat{q}(0.5)) = \log(23.08) = 3.14$. This equals the 10th-order statistic of the log(wage) observations. The two are identical because of the equivariance property.

The quantile estimator is consistent for $q(\alpha)$ when $F(x)$ is strictly increasing.

Theorem 11.7 If $F(x)$ is strictly increasing at $q(\alpha)$, then $\widehat{q}(\alpha) \underset{p}{\longrightarrow} q(\alpha)$ as $n \to \infty$.

Theorem 11.7 is a special case of Theorem 11.8 presented later in this section, so its proof is omitted. The assumption that $F(x)$ is strictly increasing at $q(\alpha)$ excludes discrete distributions and those with flat sections.

For most economists, the above information is sufficient to understand and work with quantiles. For completeness, we now give a few more details.

While Definition 11.2 is convenient because it defines quantiles uniquely, it may be more insightful to define the **quantile interval** as the set of solutions to $\alpha = F(q(\alpha))$. To handle this rigorously, it is useful to define the left-limit version of the probability function $F^+(q) = \mathbb{P}\left[X < q\right]$. We can then define the αth quantile interval as the set of numbers q that satisfy $F^+(q) \le \alpha \le F(q)$. This equals $[q(\alpha), q^+(\alpha)]$, where $q(\alpha)$ is from Definition 11.2, and $q^+(\alpha) = \sup\left\{x : F^+(x) \le \alpha\right\}$. We have the equality $q^+(\alpha) = q(\alpha)$ when $F(x)$ is strictly increasing (in both directions) at $q(\alpha)$.

We can similarly extend the definition of the empirical quantile. The empirical analog of the interval $[q(\alpha), q^+(\alpha)]$ is the empirical quantile interval $[\widehat{q}(\alpha), \widehat{q}^+(\alpha)]$, where $\widehat{q}(\alpha)$ is the empirical quantile defined earlier, and $\widehat{q}^+(\alpha) = \sup\left\{x : F_n^+(x) \le \alpha\right\}$, where $F_n^+(x) = \frac{1}{n}\sum_{i=1}^n \mathbb{1}\left(X_i < x\right)$. We can calculate that when $n\alpha$ is an integer, then $\widehat{q}^+(\alpha) = X_{(j+1)}$, where $j = \lceil n\alpha \rceil$ but otherwise $\widehat{q}^+(\alpha) = X_{(j)}$. Thus when $n\alpha$ is an integer, the empirical quantile interval is $[X_{(j)}, X_{(j+1)}]$, and otherwise it is the unique value $X_{(j)}$.

Various estimators for $q(\alpha)$ have been proposed and implemented in standard software. We discuss four of these estimators, using the labeling system expressed in the R documentation.

The Type 1 estimator is the empirical quantile, $\widehat{q}^1(\alpha) = \widehat{q}(\alpha)$.

The Type 2 estimator takes the midpoint of the empirical quantile interval $[\widehat{q}(\alpha), \widehat{q}^+(\alpha)]$. Thus the estimator is $\widehat{q}^2(\alpha) = (X_{(j)} + X_{(j+1)})/2$ when $n\alpha$ is an integer, and $X_{(j)}$ otherwise. This is the method implemented in Stata. Quantiles can be obtained by the `summarize, detail`, `xtile`, `pctile`, and `_pctile` commands.

The Type 5 estimator defines $m = n\alpha + 0.5$, $\ell = \text{int}(m)$ (integer part), and $r = m - \ell$ (remainder). It then sets $\widehat{q}^5(\alpha)$ as a weighted average of $X_{(\ell)}$ and $X_{(\ell+1)}$, using the interpolating weights $1 - r$ and r, respectively. This is the method implemented in MATLAB and can be obtained by using the `quantile` command.

The Type 7 estimator defines $m = n\alpha + 1 - \alpha$, $\ell = \text{int}(m)$, and $r = m - \ell$. It then sets $\widehat{q}^7(\alpha)$ as a weighted average of $X_{(\ell)}$ and $X_{(\ell+1)}$, using the interpolating weights $1 - r$ and r, respectively. This is the default method implemented in R and can be obtained by using the `quantile` command. The other methods (including Types 1, 2, and 5) can be obtained in R by specifying the Type as an option.

The Type 5 and Type 7 estimators may not be immediately intuitive. What they implement is to first smooth the empirical distribution function by interpolation, thus creating a strictly increasing estimator, and then they invert the interpolated EDF to obtain the corresponding quantile. The two methods differ in terms of how they implement interpolation. The estimates lie in the interval $[X_{(j-1)}, X_{(j+1)}]$, where $j = \lceil n\alpha \rceil$, but do not necessarily lie in the empirical quantile interval $[\widehat{q}(\alpha), \widehat{q}^+(\alpha)]$.

To illustrate, consider again estimation of the median wage from Table 6.1. The 10th-order and 11th-order statistics are 23.08 and 24.04, respectively, and $n\alpha = 10$ is an integer, so the empirical quantile interval for the median is $[23.08, 24.04]$. The point estimates are $\widehat{q}^1(0.5) = 23.08$ and $\widehat{q}^2(0.5) = \widehat{q}^5(0.5) = \widehat{q}^7(0.5) = 23.56$.

Consider the 0.66 quantile. The point estimates are $\widehat{q}^1(0.66) = \widehat{q}^2(0.66) = 31.73$, $\widehat{q}^5(0.66) = 31.15$, and $\widehat{q}^7(0.66) = 30.85$. Note that the latter two are smaller than the empirical quantile of 31.73.

The differences can be greatest at the extreme quantiles. Consider the 0.95 quantile. The empirical quantile is the 19th-order statistic $\widehat{q}^1(0.95) = 43.08$. Then $\widehat{q}^2(0.95) = \widehat{q}^5(0.95) = 48.85$ is the average of the 19th and 20th-order statistics, and $\widehat{q}^7(0.95) = 43.65$.

The differences between the methods diminish in large samples. However, it is useful to know that the packages implement distinct estimates when comparing results across packages.

Theorem 11.7 can be generalized to allow for interval-valued quantiles. To do so, we need a convergence concept for interval-valued parameters.

Definition 11.3 We say that a sequence of random variables Z_n **converges in probability** to the interval $[a, b]$ with $a \leq b$, as $n \to \infty$, denoted $Z_n \underset{p}{\longrightarrow} [a, b]$, if for all $\epsilon > 0$,

$$\mathbb{P}[a - \epsilon \leq Z_n \leq b + \epsilon] \underset{p}{\longrightarrow} 1.$$

This definition states that the variable Z_n lies within ϵ of the interval $[a, b]$ with probability approaching 1. The following result includes the quantile estimators described above (Types 1, 2, 5, and 7.)

Theorem 11.8 Let $\widehat{q}(\alpha)$ be any estimator satisfying $X_{(j-1)} \leq \widehat{q}(\alpha) \leq X_{(j+1)}$, where $j = \lceil n\alpha \rceil$. If X_i are i.i.d. and $0 < \alpha < 1$, then $\widehat{q}(\alpha) \underset{p}{\longrightarrow} [q(\alpha), q^+(\alpha)]$ as $n \to \infty$.

The proof is presented in Section 11.15. Theorem 11.8 applies to all distribution functions, including continuous and discrete distributions.

The following distribution theory applies to quantile estimators.

Theorem 11.9 If X has a density $f(x)$ which is positive at $q(\alpha)$, then

$$\sqrt{n}\left(\widehat{q}(\alpha) - q(\alpha)\right) \underset{d}{\longrightarrow} \mathrm{N}\left(0, \frac{F(q(\alpha))\left(1 - F(q(\alpha))\right)}{f\left(q(\alpha)\right)^2}\right)$$

as $n \to \infty$.

The proof is presented in Section 11.15. The result shows that precision of a quantile estimator depends on the distribution function and density at the point $q(\alpha)$. The formula indicates that the estimator is relatively imprecise in the tails of the distribution (where the density is low).

11.14 ROBUST VARIANCE ESTIMATION

The conventional estimator s of the standard deviation σ can be a poor measure of the spread of the distribution when the true distribution is non-normal. It can often be sensitive to the presence of a few large observations and thus overestimate the spread of the distribution. Consequently, some statisticians prefer a different estimator based on quantiles.

For a random variable X with quantile function q_α, we define the **interquartile range** as the difference between the 0.75 and 0.25 quantiles, $R = q_{.75} - q_{.25}$. A nonparametric estimator of R is the sample interquartile range $\widehat{R} = \widehat{q}_{.75} - \widehat{q}_{.25}$.

In general, R is proportional to σ, though the proportionality factor depends on the distribution. When $X \sim \mathrm{N}(\mu, \sigma^2)$, it turns out that $R \simeq 1.35\sigma$. Hence $\sigma \simeq R/1.35$. It follows that an estimator of σ (valid under normality) is $\widehat{R}/1.35$. While the estimator is not consistent for σ under non-normality, the deviation is small for a range of distributions. A further adjustment is to set

$$\widetilde{\sigma} = \min\left(s, \widehat{R}/1.35\right).$$

By taking the smaller of the conventional estimator and the interquartile estimator, $\widetilde{\sigma}$ is protected against being "too large".

The robust estimator $\widetilde{\sigma}$ is not commonly used in econometrics as a direct estimator of the standard deviation, but it is commonly used as a robust estimator for bandwidth selection. (See Section 17.9.)

11.15 TECHNICAL PROOFS*

Proof of Theorem 11.8 Fix $\epsilon > 0$. Set $\delta_1 = F(q(\alpha)) - F(q(\alpha) - \epsilon)$. Note that $\delta_1 > 0$ by the definition of $q(\alpha)$ and the assumption $\alpha = F(q(\alpha)) > 0$. The WLLN (Theorem 7.4) implies that

$$F_n(q(\alpha) - \epsilon) - F(q(\alpha) - \epsilon) = \frac{1}{n}\sum_{i=1}^{n} \mathbb{1}\left\{X_i \le q(\alpha) - \epsilon\right\} - \mathbb{E}\left[\mathbb{1}\left\{X \le q(\alpha) - \epsilon\right\}\right] \underset{p}{\longrightarrow} 0$$

which means that there is a $\overline{n}_1 < \infty$ such that for all $n \ge \overline{n}_1$,

$$\mathbb{P}\left(\left|F(q(\alpha) - \epsilon) - F_n(q(\alpha) - \epsilon)\right| > \delta_1/2\right) \le \epsilon.$$

Assume as well that $\overline{n}_1 > 2/\delta_1$. The inequality $\widehat{q}(\alpha) \ge X_{(j-1)}$ means that $\widehat{q}(\alpha) < q(\alpha) - \epsilon$ implies

$$F_n(q(\alpha) - \epsilon) \ge (j-1)/n \ge \alpha - 1/n.$$

Thus for all $n \ge \overline{n}_1$,

$$\begin{aligned}
\mathbb{P}\left[\widehat{q}(\alpha) < q_\alpha - \epsilon\right] &\le \mathbb{P}\left[F_n(q(\alpha) - \epsilon) \ge \alpha - 1/n\right] \\
&= \mathbb{P}\left[F_n(q(\alpha) - \epsilon) - F(q(\alpha) - \epsilon) \ge \delta_1 - 1/n\right] \\
&\le \mathbb{P}\left[\left|F_n(q(\alpha) - \epsilon) - F(q(\alpha) - \epsilon)\right| > \delta_1/2\right] \le \epsilon.
\end{aligned}$$

Now set $\delta_2 = F^+(q^+(\alpha) + \epsilon) - F^+(q^+(\alpha))$. Note that $\delta_2 > 0$ by the definition of $q^+(\alpha)$ and the assumption $\alpha = F^+(q^+(\alpha)) < 1$. The WLLN implies that

$$F_n^+(q(\alpha) + \epsilon) - F^+(q(\alpha) + \epsilon) = \frac{1}{n}\sum_{i=1}^{n} \mathbb{1}\left\{X_i < q(\alpha) + \epsilon\right\} - \mathbb{E}\left[\mathbb{1}\left\{X < q(\alpha) + \epsilon\right\}\right] \underset{p}{\longrightarrow} 0$$

which means that there is a $\overline{n}_2 < \infty$ such that for all $n \geq \overline{n}_2$,

$$\mathbb{P}\left[\left|F_n^+(q(\alpha)+\epsilon) - F^+(q(\alpha)+\epsilon)\right| > \delta_2/2\right] \leq \epsilon.$$

Again assume that $\overline{n}_2 > 2/\delta_2$. The inequality $\widehat{q}(\alpha) \leq X_{(j+1)}$ means that $\widehat{q}(\alpha) > q^+(\alpha) + \epsilon$ implies

$$F_n^+(q^+(\alpha)+\epsilon) \leq j/n \leq \alpha + 1/n.$$

Thus for all $n \geq \overline{n}_2$,

$$\begin{aligned}
\mathbb{P}\left[\widehat{q}(\alpha) > q^+(\alpha)+\epsilon\right] &\leq \mathbb{P}\left[F_n^+(q^+(\alpha)+\epsilon) \leq \alpha + 1/n\right] \\
&\leq \mathbb{P}\left[F^+(q^+(\alpha)+\epsilon) - F_n^+(q^+(\alpha)+\epsilon) > \delta_2/2\right] \\
&\leq \mathbb{P}\left[\left|F_n^+(q(\alpha)+\epsilon) - F^+(q(\alpha)+\epsilon)\right| > \delta_2/2\right] \leq \epsilon.
\end{aligned}$$

We have shown that for all $n \geq \max[\overline{n}_1, \overline{n}_2]$,

$$\mathbb{P}\left[q(\alpha) - \epsilon \leq \widehat{q}(\alpha) \leq q^+(\alpha) + \epsilon\right] \geq 1 - 2\epsilon$$

which establishes the result. ∎

Proof of Theorem 11.9 Fix x. By a Taylor expansion,

$$\sqrt{n}\left(F\left(q(\alpha)+n^{-1/2}x\right) - \alpha\right) = \sqrt{n}\left(F\left(q(\alpha)+n^{-1/2}x\right) - F\left(q(\alpha)\right)\right) = f(q(\alpha))x + O(n^{-1}).$$

Consider the random variable $U_{ni} = \mathbb{1}\left\{X_i \leq q(\alpha) + n^{-1/2}x\right\}$. It is bounded and has variance

$$\operatorname{var}\left[U_{ni}\right] = F\left(q(\alpha)+n^{-1/2}x\right) - F\left(q(\alpha)+n^{-1/2}x\right)^2 \to F\left(q(\alpha)\right) - F\left(q(\alpha)\right)^2.$$

Therefore by the CLT for heterogeneous random variables (Theorem 9.2),

$$\sqrt{n}\left(F_n\left(q(\alpha)+n^{-1/2}x\right) - F\left(q(\alpha)+n^{-1/2}x\right)\right) = \frac{1}{\sqrt{n}}\sum_{i=1}^{n} U_{ni} \underset{d}{\longrightarrow} Z \sim \mathrm{N}\left(0, F(q(\alpha))\left(1 - F(q(\alpha))\right)\right).$$

Then

$$\begin{aligned}
\mathbb{P}\left[\sqrt{n}\left(\widehat{q}(\alpha) - q(\alpha)\right) \leq x\right] &= \mathbb{P}\left[\widehat{q}(\alpha) \leq q(\alpha) + n^{-1/2}x\right] \\
&\leq \mathbb{P}\left[F_n\left(\widehat{q}(\alpha)\right) \leq F_n\left(q(\alpha)+n^{-1/2}x\right)\right] \\
&= \mathbb{P}\left[\alpha \leq F_n\left(q(\alpha)+n^{-1/2}x\right)\right] \\
&= \mathbb{P}\left[\sqrt{n}\left(\alpha - F\left(q(\alpha)+n^{-1/2}x\right)\right) \leq \sqrt{n}\left(F_n\left(q(\alpha)+n^{-1/2}x\right) - F\left(q(\alpha)+n^{-1/2}x\right)\right)\right] \\
&= \mathbb{P}\left[-f(q(\alpha))x + O(n^{-1}) \leq \sqrt{n}\left(F_n\left(q(\alpha)+n^{-1/2}x\right) - F\left(q(\alpha)+n^{-1/2}x\right)\right)\right] \\
&\to \mathbb{P}\left[-f(q(\alpha))x \leq Z\right] \\
&= \mathbb{P}\left[Z/f(q(\alpha)) \leq x\right].
\end{aligned} \tag{11.4}$$

Replacing the weak inequality $\leq$ in the probability statement in equation (11.4) with the strict inequality $<$, we find that the first two lines can be replaced by

$$\mathbb{P}\left[\sqrt{n}\left(\widehat{q}(\alpha) - q(\alpha)\right) < x\right] \geq \mathbb{P}\left[F_n\left(\widehat{q}(\alpha)\right) \leq F_n\left(q(\alpha)+n^{-1/2}x\right)\right]$$

which also converges to $\mathbb{P}\left[Z/f(q(\alpha)) \le x\right]$. We deduce that

$$\mathbb{P}\left[\sqrt{n}\left(\widehat{q}(\alpha) - q(\alpha)\right) \le x\right] \to \mathbb{P}\left[Z/f(q(\alpha)) \le x\right]$$

for all x.

Thus $\sqrt{n}\left(\widehat{q}(\alpha) - q(\alpha)\right)$ is asymptotically normal with variance $F(q(\alpha))\left(1 - F(q(\alpha))\right)/f(q(\alpha))^2$, as stated. ■

11.16 EXERCISES

Exercise 11.1 The coefficient of variation of X is $\text{cv} = 100 \times \sigma/\mu$, where $\sigma^2 = \text{var}[X]$ and $\mu = \mathbb{E}[X]$.

(a) Propose a plug-in moment estimator $\widehat{\text{cv}}$ for cv.
(b) Find the variance of the asymptotic distribution of $\sqrt{n}\,(\widehat{\text{cv}} - \text{cv})$.
(c) Propose an estimator of the asymptotic variance of $\widehat{\text{cv}}$.

Exercise 11.2 The skewness of a random variable is

$$\text{skew} = \frac{\mu_3}{\sigma^3}$$

where μ_3 is the third central moment.

(a) Propose a plug-in moment estimator $\widehat{\text{skew}}$ for skew.
(b) Find the variance of the asymptotic distribution of $\sqrt{n}\left(\widehat{\text{skew}} - \text{skew}\right)$.
(c) Propose an estimator of the asymptotic variance of $\widehat{\text{skew}}$.

Exercise 11.3 A Bernoulli random variable X is

$$\mathbb{P}[X = 0] = 1 - p$$
$$\mathbb{P}[X = 1] = p.$$

(a) Propose a moment estimator $\widehat{p}$ for p.
(b) Find the variance of the asymptotic distribution of $\sqrt{n}\,(\widehat{p} - p)$.
(c) Use your knowledge of the Bernoulli distribution to simplify the asymptotic variance.
(d) Propose an estimator of the asymptotic variance of $\widehat{p}$.

Exercise 11.4 Propose a moment estimator $\widehat{\lambda}$ for the parameter λ of a Poisson distribution (Section 3.6).

Exercise 11.5 Propose a moment estimator $\left(\widehat{p}, \widehat{r}\right)$ for the parameters (p, r) of the negative binomial distribution (Section 3.7).

Exercise 11.6 You travel to the planet *Estimation* as part of an economics expedition. Families are organized as pairs of individuals identified as *alphas* and *betas*. You are interested in their wage distributions. Let X_a, X_b denote the wages of a family pair. Let $\mu_a, \mu_b, \sigma_a^2, \sigma_b^2$, and σ_{ab} denote the means, variances, and covariance of the alpha and beta wages, respectively, in a family. You gather a random sample of n families and obtain the dataset

$\{(X_{a1}, X_{b1}), (X_{a2}, X_{b2}), \ldots, (X_{an}, X_{bn})\}$. Assume that the family pairs (X_{ai}, X_{bi}) are mutually independent and identically distributed across i.

(a) Write down an estimator $\widehat{\theta}$ for $\theta = \mu_b - \mu_a$, the wage difference between betas and alphas.

(b) Find var $[X_b - X_a]$. Write it in terms of the parameters given above.

(c) Find the asymptotic distribution for $\sqrt{n}\left(\widehat{\theta} - \theta\right)$ as $n \to \infty$. Write it in terms of the parameters given above.

(d) Construct an estimator of the asymptotic variance. Be explicit.

Exercise 11.7 You want to estimate the percentage of a population which has a wage below \$15 an hour. You consider three approaches.

(a) Estimate a normal distribution by MLE. Use the percentage of the fitted normal distribution below \$15.

(b) Estimate a log-normal distribution by MLE. Use the percentage of the fitted normal distribution below \$15.

(c) Estimate the empirical distribution function at \$15. Use the EDF estimate.
Which method do you advocate? Why?

Exercise 11.8 You believe X is distributed exponentially and want to estimate $\theta = \mathbb{P}[X > 5]$. You have the estimate $\overline{X}_n = 4.3$ with standard error $s\left(\overline{X}_n\right) = 1$. Find an estimate of θ and a standard error of estimation.

Exercise 11.9 You believe X is distributed Pareto$(\alpha, 1)$. You want to estimate α.

(a) Use the Pareto model to calculate $\mathbb{P}[X \leq x]$.

(b) You are given the EDF estimate $F_n(5) = 0.9$ from a sample of size $n = 100$. (That is, $F_n(x) = 0.9$ for $x = 5$.) Calculate a standard error for this estimate.

(c) Use the above information to form an estimator for α. Find $\widehat{\alpha}$ and a standard error for $\widehat{\alpha}$.

Exercise 11.10 Use Theorem 11.9 to find the asymptotic distribution of the sample median.

(a) Find the asymptotic distribution for general density $f(x)$.

(b) Find the asymptotic distribution when $f(x) \sim \mathrm{N}\left(\mu, \sigma^2\right)$.

(c) Find the asymptotic distribution when $f(x)$ is double exponential. Write your answer in terms of the variance σ^2 of X.

(d) Compare the answers from (b) and (c). In which case is the variance of the asymptotic distribution smaller? Do you have an intuitive explanation?

Exercise 11.11 Prove Theorem 11.5.

CHAPTER 12

NUMERICAL OPTIMIZATION

12.1 INTRODUCTION

The methods of moments estimator solves a set of nonlinear equations. The maximum likelihood estimator maximizes the log-likelihood function. In some cases, these functions can be solved algebraically to find the solution. In many cases, however, they cannot. When there is no algebraic solution, we turn to numerical methods.

Numerical optimization is a specialized branch of mathematics. Fortunately for us economists, experts in numerical optimization have constructed algorithms and implemented software, so we do not have to do so ourselves. Many popular estimation methods are coded into established econometrics software packages, such as Stata, so that the numerical optimization is done automatically without user intervention. Otherwise, numerical optimization can be performed in standard numerical packages, such as MATLAB and R.

In MATLAB, the most common packages for optimization are `fminunc`, `fminsearch`, and `fmincon`. `fminunc` is for unconstrained minimization of a differentiable criterion function using quasi-Newton and trust-region methods. `fmincon` is similar for constrained minimization. `fminsearch` is for minimization of a possibly nondifferentiable criterion function using the Nelder-Mead algorithm.

In R, the most common packages are `optimize`, `optim`, and `constrOptim`. `optimize` is for one-dimensional problems, `optim` is for unconstrained minimization in dimensions exceeding one, and `constrOptim` for constrained optimization. The latter two allow the user to select among several minimization methods.

It is not essential to understand the details of numerical optimization to use these packages, but it is helpful. Some basic knowledge will help you select the appropriate method and understand when to switch from one method to another. In practice, optimization is often an art, and knowledge of the underlying tools improves our use of them.

In this chapter, we discuss numerical differentiation, root finding, minimization in one dimension, minimization in multiple dimensions, and constrained optimization.

12.2 NUMERICAL FUNCTION EVALUATION AND DIFFERENTIATION

The goal in numerical optimization is to find a root, minimizer, or maximizer of a function $f(x)$ of a real or vector-valued input. In many estimation contexts, the function $f(x)$ takes the form of a sample moment $f(\theta)=\sum_{i=1}^{n} m(X_i,\theta)$, but this is not particularly important.

The first step is to code a function (or algorithm) to calculate the function $f(x)$ for any input x. This requires that the user write an algorithm to produce the output $y=f(x)$. This may seem obvious, but it is helpful to be explicit.

Some numerical optimization methods require calculation of derivatives of $f(x)$. The vector of first derivatives

$$g(x) = \frac{\partial}{\partial x}f(x)$$

is called the **gradient**. The matrix of second derivatives

$$\boldsymbol{H}(x) = \frac{\partial^2}{\partial x \partial x'}f(x)$$

is called the **Hessian**.

The preferred method to evaluate $g(x)$ and $\boldsymbol{H}(x)$ is by explicit algebraic algorithms. This requires that the user provides the algorithm. It may be helpful to know that many packages (including MATLAB, R, Mathematica, and Maple) have the ability to take symbolic derivatives of a function $f(x)$. These expressions cannot be used directly for numerical optimization, but they may be helpful in determining the correct formula. In MATLAB, use the function `diff`; in R, use `deriv`. In some cases, the symbolic answers will be helpful; in other cases they will not be helpful, as the result will not be computational.

Take, for example, the normal negative log likelihood

$$f\left(\mu, \sigma^2\right) = \frac{n}{2}\log(2\pi) + \frac{n}{2}\log\sigma^2 + \frac{\sum_{i=1}^{n}(X_i - \mu)^2}{2\sigma^2}.$$

We can calculate that the gradient is

$$g(\mu, \sigma^2) = \begin{pmatrix} \frac{\partial}{\partial \mu}f\left(\mu, \sigma^2\right) \\ \\ \frac{\partial}{\partial \sigma^2}f\left(\mu, \sigma^2\right) \end{pmatrix} = \begin{pmatrix} -\frac{1}{\sigma^2}\sum_{i=1}^{n}(X_i - \mu) \\ \\ \frac{n}{2\sigma^2} - \frac{1}{2\sigma^4}\sum_{i=1}^{n}(X_i - \mu)^2 \end{pmatrix}$$

and the Hessian is

$$\boldsymbol{H}(\mu, \sigma^2) = \begin{pmatrix} \frac{\partial^2}{\partial \mu^2}f\left(\mu, \sigma^2\right) & \frac{\partial^2}{\partial \mu \partial \sigma^2}f\left(\mu, \sigma^2\right) \\ \\ \frac{\partial^2}{\partial \sigma^2 \partial \mu}f\left(\mu, \sigma^2\right) & \frac{\partial}{\partial \left(\sigma^2\right)^2}f\left(\mu, \sigma^2\right) \end{pmatrix}$$

$$= \begin{pmatrix} \frac{n \;\; 2}{\sigma^2} & \frac{1}{\sigma 4}\sum_{i=1}^{n}(X_i - \mu) \\ \\ \frac{1}{\sigma^4}\sum_{i=1}^{n}(X_i - \mu) & -\frac{n}{2\sigma^4} + \frac{1}{\sigma^6}\sum_{i=1}^{n}(X_i - \mu)^2 \end{pmatrix}.$$

These can be directly calculated.

As another example, take the gamma distribution. It has negative log likelihood

$$f(\alpha, \beta) = n \log \Gamma(\alpha) + n\alpha \log(\beta) + (1 - \alpha)\sum_{i=1}^{n}\log(X_i) + \sum_{i=1}^{n}\frac{X_i}{\beta}.$$

The gradient is

$$g(\alpha,\beta)=\begin{pmatrix}\frac{\partial}{\partial\alpha}f(\alpha,\beta)\\ \\ \frac{\partial}{\partial\beta}f(\alpha,\beta)\end{pmatrix}=\begin{pmatrix}n\frac{d}{d\alpha}\log\Gamma(\alpha)+n\log(\beta)-\sum_{i=1}^{n}\log(X_i)\\ \\ \frac{n\alpha}{\beta}-\frac{1}{\beta^2}\sum_{i=1}^{n}X_i\end{pmatrix}.$$

The Hessian is

$$\boldsymbol{H}(\alpha,\beta)=\begin{pmatrix}\frac{\partial^2}{\partial\alpha^2}f(\alpha,\beta) & \frac{\partial^2}{\partial\alpha\partial\beta}f(\alpha,\beta)\\ \\ \frac{\partial^2}{\partial\beta\partial\alpha}f(\alpha,\beta) & \frac{\partial}{\partial\beta^2}f(\alpha,\beta)\end{pmatrix}=\begin{pmatrix}n\frac{d^2}{d\alpha^2}\log\Gamma(\alpha) & \frac{n}{\beta}\\ \\ \frac{n}{\beta} & -\frac{n\alpha}{\beta^2}+\frac{2}{\beta^3}\sum_{i=1}^{n}X_i\end{pmatrix}.$$

Notice that these expressions depend on $\log\Gamma(\alpha)$, its derivative, and second derivative. (The first and second derivatives of $\log\Gamma(\alpha)$ are known as the **digamma** and **trigamma** functions). None are available in closed form. However, all have been studied in numerical analysis, so algorithms are coded for numerical computation. This means that the above gradient and Hessian formulas are straightforward to code. The reason this is worth mentioning at this point is that just because a symbol can be written does not mean that it is numerically convenient for computation. It may be or may not be.

When explicit algebraic evalulation of derivatives are not available, we turn to numerical methods. To understand how this is done, it is useful to recall the definition

$$g(x)=\lim_{\epsilon\to 0}\frac{f(x+\epsilon)-f(x)}{\epsilon}.$$

A **numerical derivative** exploits this formula. A small value of ϵ is selected. The function f is evaluated at x and at $x+\epsilon$. We then have the numerical approximation

$$g(x)\simeq\frac{f(x+\epsilon)-f(x)}{\epsilon}.$$

This approximation will be good if $f(x)$ is smooth at x and ϵ is suitably small. The approximation will fail if $f(x)$ is discontinuous at x (in which case, the derivative is not defined) or if $f(x)$ is discontinuous near x and ϵ is too large. This calculation is also called a **discrete derivative**.

When x is a vector, numerical derivatives are a vector. In the two-dimensional case, the gradient equals

$$g(x)=\begin{pmatrix}\lim_{\epsilon\to 0}\frac{f(x_1+\epsilon,x_2)-f(x_1,x_2)}{\epsilon}\\ \\ \lim_{\epsilon\to 0}\frac{f(x_1,x_2+\epsilon)-f(x_1,x_2)}{\epsilon}\end{pmatrix}.$$

A numerical approximation is

$$g(x,\epsilon)=\begin{pmatrix}\frac{f(x_1+\epsilon,x_2)-f(x_1,x_2)}{\epsilon}\\ \\ \frac{f(x_1,x_2+\epsilon)-f(x_1,x_2)}{\epsilon}\end{pmatrix}.$$

In general, numerical computation of an m-dimensional gradient requires $m+1$ function evalutions.

Standard packages have functions to implement numerical (discrete) differentiation. In MATLAB, use `gradient`; in R, use `Deriv`.

An important practical challenge is selection of the step size ϵ. It may be tempting to set ϵ extremely small (e.g., near machine precision), but this choice can be unwise, because numerical evaluation of $f(x)$ may not be sufficiently accurate. Implementations in standard packages select ϵ automatically based on scaling.

The Hessian can be written as the first derivative of $g(x)$:

$$\boldsymbol{H}(x) = \frac{\partial}{\partial x'} g(x) = \begin{pmatrix} \frac{\partial}{\partial x_1} g(x) & \frac{\partial}{\partial x_2} g(x) \end{pmatrix}.$$

If the gradient can be calculated algebraically, then each derivative vector $\frac{\partial}{\partial x'} g_j(x)$ can be calculated by applying numerical derivatives to the gradient. Doing so, $\boldsymbol{H}(x)$ can be calculated using $m(m+1)$ function evaluations.

If the gradient is not calculated algebraically, then the Hessian can be calculated numerically. When $m = 2$, a numerical approximation to the Hessian is obtained by taking a numerical derivative of the numerical gradient approximation:

$$\boldsymbol{H}(x, \epsilon) = \begin{pmatrix} \dfrac{g_1(x_1+\epsilon, x_2, \epsilon) - g_1(x_1, x_2, \epsilon)}{\epsilon} & \dfrac{g_2(x_1+\epsilon, x_2, \epsilon) - g_2(x_1, x_2, \epsilon)}{\epsilon} \\ \dfrac{g_1(x_1, x_2+\epsilon, \epsilon) - g_1(x_1, x_2, \epsilon)}{\epsilon} & \dfrac{g_2(x_1, x_2+\epsilon, \epsilon) - g_2(x_1, x_2, \epsilon)}{\epsilon} \end{pmatrix}.$$

To numerically evaluate an $m \times m$ Hessian requires $2m(m+1)$ function evaluations. If m is large and $f(x)$ slow to compute, then this can be computationally costly.

Both algebraic and numerical derivative methods are useful. In many contexts, it is possible to calculate algebraic derivatives (perhaps with the aid of symbolic software) but the expressions are messy, and it is easy to make an error. It is therefore prudent to double-check your programming by comparing a proposed algebraic gradient or Hessian with a numerical calculation. The answers should be quite similar (up to numerical error) unless the programming is incorrect or the function highly non-smooth. Discrepancies typically indicate that a programming error has been made.

In general, if algebraic gradients and Hessians are available, they should be used instead of numerical calculations. Algebraic calculations tend to be much more accurate and computationally faster. However, in cases where numerical methods provide simple and quick results, numerical methods are sufficient.

12.3 ROOT FINDING

A **root** of a function $f(x)$ is a number x_0 such that $f(x_0) = 0$. The method of moments estimator, for example, is the root of a set of nonlinear equations.

We discuss three standard root-finding methods.

Grid Search. Grid search is a general-purpose but computationally inefficient method for finding a root.

Let $[a, b]$ be an interval that includes x_0. (If an appropriate choice is unknown, try an increasing sequence, such as $[-2^i, 2^i]$, until the function endpoints have opposite signs.) Call $[a, b]$ a **bracket** of x_0.

For some integer N, set $x_i = a + (b - a)i/N$ for $i = 0, \ldots, N$. This produces $N+1$ evenly spaced points on $[a, b]$. The set of points is called a **grid**. Calculate $f(x_i)$ at each point. Find the point x_i which minimizes $|f(x_i)|$. This point is the numerical approximation to x_0. The precision is $(b-a)/N$.

The primary advantages of grid search are that it is robust to functional form, and precision can be accurately stated. The primary disadvantage is computational cost; it is considerably most costly than other methods.

Newton's Method. Newton's method (or rule) is appropriate for monotonic differentiable functions $f(x)$. It makes a sequence of linear approximations to $f(x)$ to produce an iterative sequence x_i, $i = 1, 2, \ldots$, which converges to x_0.

Let x_1 be an initial guess for x_0. By a Taylor approximation of $f(x)$ about x_1, we have

$$f(x) \simeq f(x_1) + f'(x_1)(x - x_1).$$

Evaluated at $x = x_0$ and using the knowledge that $f(x_0) = 0$, we find that

$$x_0 \simeq x_1 - \frac{f(x_1)}{f'(x_1)}.$$

This suggests using the right-hand side as an updated guess for x_0. Write this as

$$x_2 = x_1 - \frac{f(x_1)}{f'(x_1)}.$$

In other words, given an initial guess x_1, we calculate an updated guess x_2. Repeating this argument, we obtain the sequence of approximations

$$f(x) \simeq f(x_i) + f'(x_i)(x - x_i)$$

and iteration rule

$$x_{i+1} = x_i - \frac{f(x_i)}{f'(x_i)}.$$

Newton's method repeats this rule until convergence is obtained. Convergence occurs when $|f(x_{i+1})|$ falls below a predetermined threshold. The method requires a starting value x_1 and either algebraic or numerical calculation of the derivative $f'(x)$.

The advantage of Newton's method is that convergence can be obtained in a small number of iterations when the function $f(x)$ is approximately linear. It has several disadvantages: (1) If $f(x)$ is non-monotonic, the iterations may not converge. (2) The method can be sensitive to starting value. (3) It requires that $f(x)$ is differentiable, and may require numerical calculation of the derivative.

To illustrate, consider the function $f(x) = x^{-1/2}\exp(x) - 2^{-1/2}\exp(2)$, which has root $x_0 = 2$. Figure 12.1(a) displays the function on the interval [1, 4] with the solid line. A Newton sequence is displayed by the labeled points and arrows. The starting point is $x_1 = 4$ and is marked as "1". The arrow from 1 shows the calculated derivative vector. The projected root is $x_2 = 3.1$ and is marked as "2". The arrow from 2 shows the derivative vector. The projected root is $x_3 = 2.4$ and is marked as "3". This continues to $x_4 = 2.06$ and $x_5 = 2.00$. The root is marked with the open circle. In this example, the algorithm converges in five iterations, which requires ten function evaluations (five function and five derivative calls).

Bisection. This method is appropriate for a function with a single root, and it does not require $f(x)$ to be differentiable.

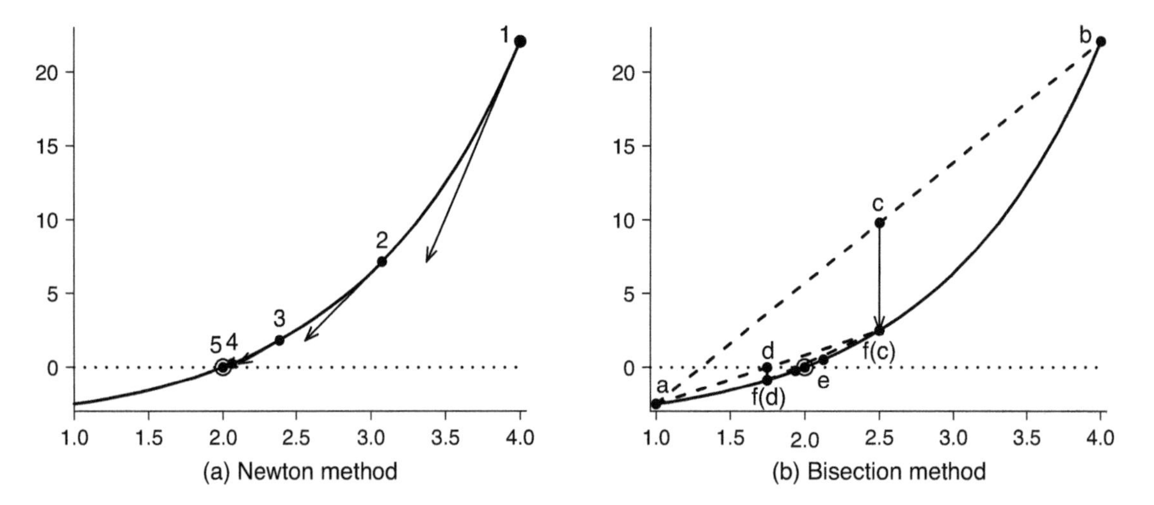

FIGURE 12.1 Root finding

Let $[a, b]$ be an interval that includes x_0, so that $f(a)$ and $f(b)$ have opposite signs. Let $c = (a+b)/2$ be the midpoint of the bracket. Calculate $f(c)$. If $f(c)$ has the same sign as $f(a)$, then reset the bracket as $[c, b]$; otherwise reset the bracket as $[a, c]$. The reset bracket has one-half its former length. By iteration, the bracket length is sequentially reduced until the desired precision is obtained. Precision is $(b-a)/2^N$, where N is the number of iterations.

The advantages of the bisection method is that it converges in a known number of steps, does not rely on a single starting value, does not use derivatives, and is robust to nondifferentiable and non-monotonic $f(x)$. A disadvantage is that it can take many iterations to converge.

Figure 12.1(b) illustrates the bisection method. The initial bracket is $[1, 4]$. The line segment joining these points is drawn in with midpoint $c = 2.5$ marked. The arrow marks $f(c)$. Since $f(c) > 0$, the bracket is reset as $[1, 2.5]$. The line segment joining these points is drawn with midpoint 1.75 and marked as "d". We find $f(d) < 0$, so the bracket is reset as $[1.75, 2.5]$. The following line segment is drawn with midpoint 2.12 and marked as "e". We find $f(e) > 0$, so the bracket is reset as $[1.75, 2.12]$. The subsequent brackets are $[1.94, 2.12]$, $[1.94, 2.03]$, $[1.98, 2.03]$, $[1.98, 2.01]$, and $[2.00, 2.01]$. (These brackets are not marked on the graph, because it becomes too cluttered.) Convergence within 0.01 accuracy occurs at the ninth function evaluation. Each iteration reduces the bracket by exactly one-half.

12.4 MINIMIZATION IN ONE DIMENSION

The **minimizer** of a function $f(x)$ is

$$x_0 = \underset{x}{\operatorname{argmin}} f(x).$$

The maximum likelihood estimator, for example, is the minimizer[1] of the negative log-likelihood function. Some vector-valued minimization algorithms use one-dimensional minimization as a subcomponent of their

[1] Most optimization software is written to minimize rather than maximize a function. We will follow this convention. There is no loss of generality, since the maximizer of $f(x)$ is the minimizer of $-f(x)$.

algorithms (often referred to as a **line search**). One-dimensional minimization is also widely used in applied econometrics for other purposes, such as the cross-validation selection of bandwidths.

We discuss four standard methods.

Grid Search. Grid search is a general-purpose but computationally inefficient method for finding a minimizer.

Let $[a, b]$ be an interval that includes x_0. For some integer N, create a grid of values x_i on the interval. Evaluate $f(x_i)$ at each point. Find the point x_i at which $f(x_i)$ is minimized. Precision is $(b-a)/N$.

Grid search has similar advantages and disadvantages as those for root finding.

Newton's Method. Newton's rule is appropriate for smooth (second-order differentiable) convex functions with a single local minimum. It makes a sequence of quadratic approximations to $f(x)$ to produce an iterative sequence x_i , $i=1,2,\ldots$ that converges to x_0.

Let x_1 be an initial guess. By a Taylor approximation of $f(x)$ about x_1,

$$f(x) \simeq f(x_1) + f'(x_1)(x-x_1) + \frac{1}{2} f''(x_1)(x-x_1)^2 .$$

The right side is a quadratic function of x. It is minimized at

$$x = x_1 - \frac{f'(x_1)}{f''(x_1)}.$$

This suggests the sequence of approximations

$$f(x) \simeq f(x_i) + f'(x_i)(x-x_i) + \frac{1}{2} f''(x_i)(x-x_i)^2$$

and iteration rule

$$x_{i+1} = x_i - \frac{f'(x_i)}{f''(x_i)}.$$

Newton's method iterates this rule until convergence. Convergence occurs when $\left|f'(x_{i+1})\right|$, $|x_{i+1}-x_i|$ and/or $\left|f(x_{i+1}) - f(x_i)\right|$ falls below a predetermined threshold. The method requires a starting value x_1 and algorithms to calculate the first and second derivatives.

If $f(x)$ is locally concave (if $f''(x) < 0$), a Newton iteration will effectively try to maximize $f(x)$ rather than minimize $f(x)$, and thus send the iterations in the wrong direction. A simple and useful modification is to constrain the estimated second derivative. One such modification is the iteration rule

$$x_{i+1} = x_i - \frac{f'(x_i)}{\left|f''(x_i)\right|}.$$

The primary advantage of Newton's method is that when the function $f(x)$ is close to quadratic, the method will converge in a small number of iterations. It can be a good choice when $f(x)$ is globally convex and $f''(x)$ is available algebraically. Its disadvantages are: (1) it relies on first and second derivatives, which may require numerical evaluation; (2) it is sensitive to departures from convexity; (3) it is sensitive to departures from a quadratic; and (4) it is sensitive to the starting value.

To illustrate, consider the function $f(x) = -\log(x) + x/2$, which has the minimizer $x_0 = 2$. The function is evaluated on the interval $[0.1, 4]$ and is displayed in Figure 12.2(a) by the solid line. The Newton iteration sequence is displayed by the marked points and arrows, using the starting value $x_1 = 0.1$. The derivatives are

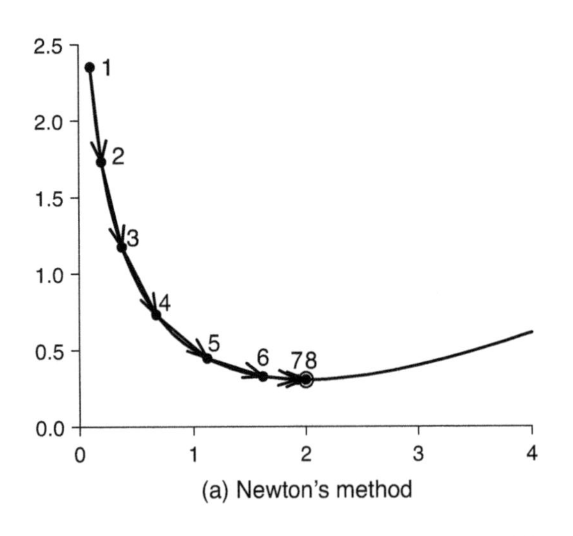

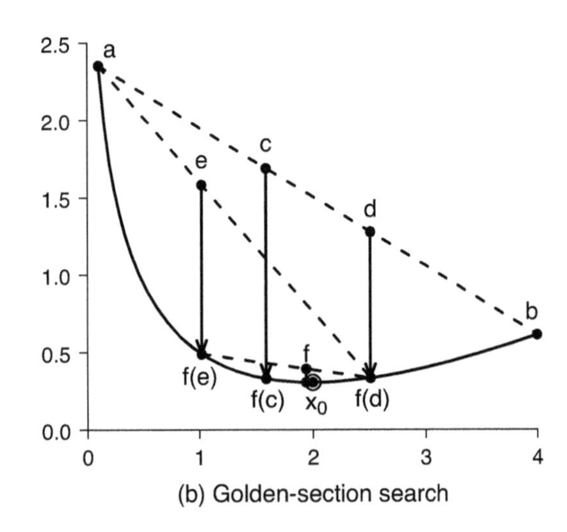

FIGURE 12.2 Minimization in one dimension

$f'(x) = -x^{-1} + 1/2$ and $f''(x) = x^{-2}$. Thus the updating rule is

$$x_{i+1} = 2x_i - \frac{x_i^2}{2}.$$

This produces the iteration sequence $x_2 = 0.195$, $x_3 = 0.37$, $x_4 = 0.67$, $x_5 = 1.12$, $x_6 = 1.61$, $x_7 = 1.92$, and $x_8 = 2.00$, which is the minimum, marked with the open circle in Figure 12.2(a). Since each iteration requires three function calls, this method requires $N = 24$ function evaluations.

Backtracking Algorithm. If the function $f(x)$ has a sharp minimum, a Newton iteration can overshoot the minimum. If this leads to an increase in the criterion, the algorithm may not converge. The backtracking algorithm is a simple adjustment that ensures that each iteration reduces the criterion.

The iteration sequence is

$$x_{i+1} = x_i - \alpha_i \frac{f'(x_i)}{f''(x_i)}$$

$$\alpha_i = \frac{1}{2^j}$$

where j is the smallest nonnegative integer satisfying

$$f\left(x_i - \frac{1}{2^j}\frac{f'(x_i)}{f''(x_i)}\right) < f(x_i).$$

The scalar α is called a **step-length**. Essentially, try $\alpha = 1$; if the criterion is not reduced, sequentially cut α in half until the criterion is reduced.

Golden-Section Search. Golden-section search is designed for unimodal functions $f(x)$.

Section search is a generalization of the bisection rule. The idea is to bracket the minimum, evaluate the function at two intermediate points, and move the bracket so that it brackets the lowest intermediate points. Golden-section search makes clever use of the "golden ratio" $\varphi = \left(1 + \sqrt{5}\right)/2 \simeq 1.62$ to select the intermediate points so that the relative spacing between the brackets and intermediate points are preserved across iterations, thus reducing the number of function evaluations.

1. Start with a bracket $[a, b]$ which contains x_0.
   (a) Compute $c = b - (b - a)/\varphi$.
   (b) Compute $d = a + (b - a)/\varphi$.
   (c) Compute $f(c)$ and $f(d)$.
   (d) These points (by construction) satisfy $a < c < d < b$.
2. Check if $f(a) > f(c)$ and $f(d) < f(b)$. These inequalities should hold if $[a, b]$ brackets x_0 and $f(x)$ is unimodal.
   (a) If these inequalities do not hold, increase the bracket $[a, b]$. If the bracket cannot be further increased, either try a different (reduced) bracket or switch to an alternative method. Return to step 1.
3. Check whether $f(c) < f(d)$ or $f(c) > f(d)$. If $f(c) < f(d)$:
   (a) Reset the bracket as $[a, d]$.
   (b) Rename d as b and drop the old b.
   (c) Rename c as d.
   (d) Compute a new $c = b - (b - a)/\varphi$.
   (e) Compute $f(c)$.
4. Otherwise, if $f(c) > f(d)$:
   (a) Reset the bracket as $[c, b]$.
   (b) Rename c as a and drop the old a.
   (c) Rename d as c.
   (d) Compute a new $d = a + (b - a)/\varphi$.
   (e) Compute $f(d)$.
5. Iterate steps 3 and 4 until convergence. Convergence obtains when $|b - a|$ is smaller than a predetermined threshold.

At each iteration, the bracket length is reduced by $100(1 - 1/\phi) \simeq 38\%$. Precision is $(b - a)\varphi^{-N}$, where N is the number of iterations.

The advantages of golden-section search are that it does not require derivatives and is relatively robust to the shape of $f(x)$. The main disadvantage is that it will take many iterations to converge if the initial bracket is large.

To illustrate, Figure 12.2(b) shows the golden-section iteration sequence starting with the bracket $[0.1, 4]$ marked with "a" and "b". Sketch a dotted line between $f(a)$ and $f(b)$, calculate the intermediate points $c = 1.6$ and $d = 2.5$, marked on the line, and compute $f(c)$ and $f(d)$ (marked with the arrows). Since $f(c) < f(d)$ (slightly), move the right endpoint to d, obtaining the bracket $[a, d] = [0.1, 2.5]$. Sketch a dotted line between $f(a)$, and $f(d)$, and calculate the new intermediate point $e = 1.0$, labeled "e", with an arrow to $f(e)$. Since $f(e) > f(c)$, move the left endpoint to e, obtaining the bracket $[e, d] = [1.0, 2.5]$. Sketche a dotted line between $f(e)$ and $f(d)$, and calculate the intermediate point $f = 1.9$. Since $f(c) > f(f)$, move the left endpoint to c, obtaining the bracket $[c, d] = [1.6, 2.5]$. Each iteration reduces the bracket length by 38%. By 16 iterations, the bracket is $[1.99, 2.00]$ with length less than 0.01. This search required $N = 16$ function evaluations.

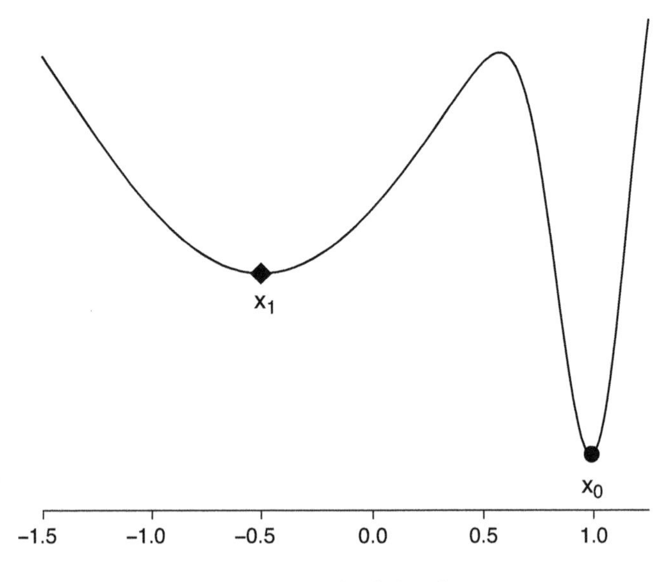

FIGURE 12.3 Multiple local minima

12.5 FAILURES OF MINIMIZATION

None of the minimization methods described in Section 12.4 are fail-proof. Each fails in one situation or another.

A grid search will fail when the grid is too coarse relative to the smoothness of $f(x)$. There is no way to know without increasing the number of grid points, though a plot of the function $f(x)$ will aid intuition. A grid search will also fail if the bracket $[a, b]$ does not contain x_0. If the numerical minimizer is one of the two endpoints, this is a signal that the initial bracket should be widened.

Newton's method can fail to converge if $f(x)$ is not convex or $f''(x)$ is not continuous. If the method fails to converge, it makes sense to switch to another method. If $f(x)$ has multiple local minima, Newton's method may converge to one of the local minima rather than the global minimum. The best way to investigate if this has occurred is to try iteration from multiple starting values. If you find multiple local minima x_0 and x_1, say, the global minimum is x_0 if $f(x_0) < f(x_1)$ and x_1 otherwise.

Golden-section search converges by construction, but it may converge to a local rather than a global minimum. If this is a concern, it is prudent to try different initial brackets.

The problem of multiple local minima is illustrated in Figure 12.3. Displayed is a function[2] $f(x)$ with a local minimum at $x_1 = -0.5$ and a global minimum at $x_0 \simeq 1$. The Newton and golden-section methods can converge to either x_1 or x_0, depending on the starting value and other idiosyncratic choices. If the iteration sequence converges to x_1 a user can mistakenly believe that the function has been minimized.

Minimization methods can be combined. A coarse grid search can be used to test out a criterion and get an initial sense of its shape. The Golden-section method is useful to start an iteration, as it avoids shape assumptions. Iteration can switch to Newton's method when the bracket has been reduced to a region where the function is convex.

[2]The function is $f(x) = -\phi(x + 1/2) - \phi(6x - 6)$.

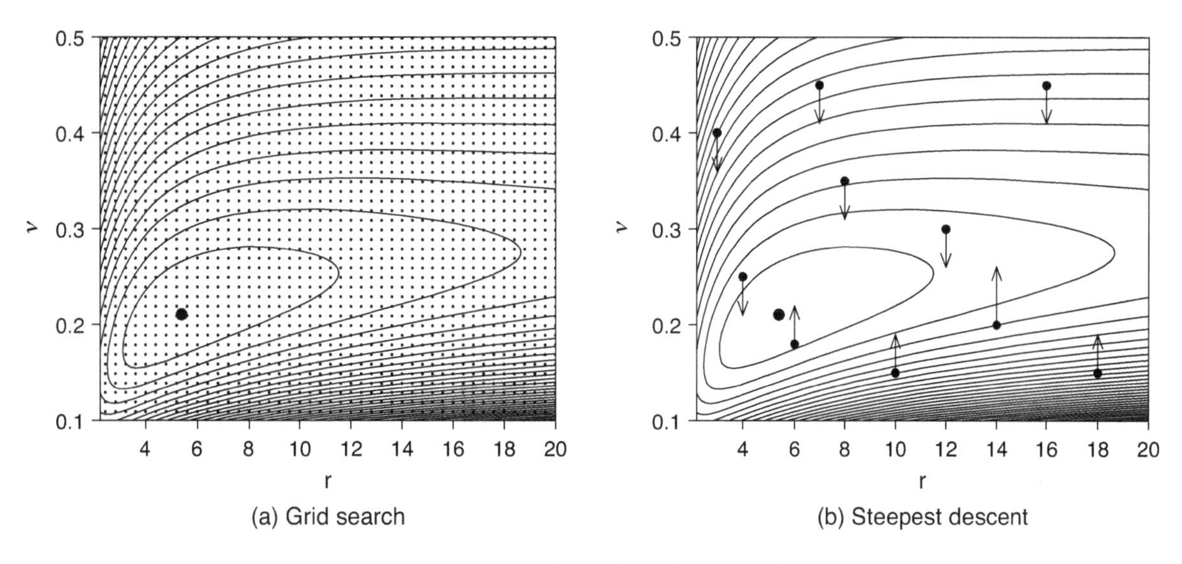

FIGURE 12.4 Grid search and steepest descent

12.6 MINIMIZATION IN MULTIPLE DIMENSIONS

To compute the MLE for a vector parameter, numerical minimization must be done in multiple dimensions. This is also true for other nonlinear estimation methods. Numerical minimization in high dimensions can be tricky. We discuss several popular methods. It is generally advised to not attempt to program these algorithms yourself, but instead use a well-established optimization package. The reason for understanding the algorithm is not to program it yourself, but to understand how it works as a guide for selection of an appropriate method.

The goal is to find the $m \times 1$ minimizer

$$x_0 = \underset{x}{\operatorname{argmin}} f(x).$$

Grid Search. Let $[a, b]$ be an m-dimensional bracket for x_0 in the sense that $a \leq x_0 \leq b$. For some integer G, create a grid of values on each axis. Taking all combinations, we obtain G^m gridpoints x_i. The function $f(x_i)$ is evaluated at each gridpoint, and the point is selected that minimizes $f(x_i)$. The computational cost of grid search increases exponentially with m. This method has the advantage of providing a full characterization of the function $f(x)$. The disadvantages are high computation cost if G is moderately large, and imprecision if the number of gridpoints G is small.

To illustrate, Figure 12.4(a) displays a scaled student t negative log-likelihood function as a function of the degrees-of-freedom parameter r and scale parameter v. The data are demeaned log wages for Hispanic women with 10 to 15 years of work experience ($n = 511$). This model nests the log normal distribution. By demeaning the series, we eliminate the location parameter, so we can focus on an easy-to-display two-dimensional problem. Figure 12.4(a) displays a contour plot of the negative log-likelihood function for $r \in [2.1, 20]$ and $v \in [.1, .7]$, with steps of 0.2 for r and 0.001 for v. The number of gridpoints in the two dimensions are 90 and 401, respectively, with a total of 36,090 gridpoints calculated. (A smaller array of 1,800 gridpoints is displayed in the figure by the dots.) The global minimum found by grid search is $\widehat{r} = 5.4$ and $\widehat{v} = .211$, and is shown at the center of the contour sets.

Steepest Descent. Also known as **gradient descent**, this method is appropriate for differentiable functions. The iteration sequence is

$$
\begin{aligned}
g_i &= \frac{\partial}{\partial x} f(x_i) \\
\alpha_i &= \underset{\alpha}{\operatorname{argmin}} f\left(x_i - \alpha g_i\right) \\
x_{i+1} &= x_i - \alpha_i g_i.
\end{aligned} \tag{12.1}
$$

The vector g is called the **direction**, and the scalar α is called the **step-length**. A motivation for the method is that for some $\alpha > 0, f\left(x_i - \alpha g_i\right) < f\left(x_i\right)$. Thus each iteration reduces $f(x)$. Gradient descent can be intrepreted as making a local linear approximation to $f(x)$. The minimization (12.1) can be replaced by the backtracking algorithm to produce an approximation that reduces (but only approximately minimizes) the criterion at each iteration.

The method requires a starting value x_1 and an algorithm to calculate the first derivative.

The method can work well when the function $f(x)$ is convex or "well behaved", but it can fail to converge in other cases. In some cases, it can be useful as a starting method, as it can swiftly move through the first iterations.

To illustrate, Figure 12.4(b) displays the gradients g_i of the scaled student t negative log likelihood from ten starting values, with each gradient normalized to have the same length, and each is displayed by an arrow. In this example, the gradients have the unusual property that they are nearly vertical, implying that a steepest-descent iteration will tend to move mostly the scale parameter v and only minimally r. Consequently (in this example), steepest descent will have difficulty finding the minimum.

Conjugate Gradient. A modification of gradient descent, this method is appropriate for differentiable functions and designed for convex functions. The iteration sequence is

$$
\begin{aligned}
g_i &= \frac{\partial}{\partial x} f(x_i) \\
\beta_i &= \frac{\left(g_i - g_{i-1}\right)' g_i}{g_{i-1}' g_{i-1}} \\
d_i &= -g_i + \beta_i d_{i-1} \\
\alpha_i &= \underset{\alpha}{\operatorname{argmin}} f\left(x_i + \alpha d_i\right) \\
x_{i+1} &= x_i + \alpha_i d_i
\end{aligned}
$$

with $\beta_1 = 0$. The sequence β_i introduces an acceleration. The algorithm requires first derivatives and a starting value.

It can converge with fewer iterations than steepest descent, but it can fail when $f(x)$ is not convex.

Newton's Method. This method is appropriate for smooth (second-order differentiable) convex functions with a single local minimum.

Let x_i be the current iteration. By a Taylor approximation of $f(x)$ about x_i,

$$f(x) \simeq f(x_i) + g_i'(x - x_i) + \frac{1}{2}(x - x_i)' \boldsymbol{H}_i (x - x_i) \tag{12.2}$$

$$g_i = \frac{\partial}{\partial x} f(x_i)$$

$$\boldsymbol{H}_i = \frac{\partial^2}{\partial x \partial x'} f(x_i).$$

The right side of equation (12.2) is a quadratic in the vector x. The quadratic is minimized (when $\boldsymbol{H}_i > 0$) at the unique solution

$$x = x_i - d_i$$

$$d_i = \boldsymbol{H}_i^{-1} g_i.$$

This behavior suggests the iteration rule

$$x_{i+1} = x_i - \alpha_i d_i$$

$$\alpha_i = \underset{\alpha}{\operatorname{argmin}} f(x_i + \alpha d_i). \tag{12.3}$$

The vector d is called the **direction**, and the scalar α is called the **step-length**. Newton's method iterates this equation until convergence. A simple (classic) Newton rule sets the step-length as $\alpha_i = 1$, but this is generally not recommended. The optimal step-length (12.3) can be replaced by the backtracking algorithm to produce an approximation that reduces (but only approximately minimizes) the criterion at each iteration.

The method requires a starting value x_1 and algorithms to calculate the first and second derivatives.

The Newton rule is sensitive to nonconvex functions $f(x)$. If the Hessian $\boldsymbol{H}$ is not positive definite, the updating rule will push the iteration in the wrong direction and can cause convergence failure. Nonconvexity can easily arise when $f(x)$ is complicated and multidimensional. Newton's method can be modified to eliminate this situation by forcing $\boldsymbol{H}$ to be positive semi-definite. One method is to bound or modify the eigenvalues of $\boldsymbol{H}$. A simple solution is to replace the eigenvalues by their absolute values. A method that implements this is as follows. Calculate $\lambda_{\min}(\boldsymbol{H})$. If it is positive, no modification is necessary. Otherwise, calculate the spectral decomposition

$$\boldsymbol{H} = \boldsymbol{Q}' \Lambda \boldsymbol{Q}$$

where $\boldsymbol{Q}$ is the matrix of eigenvectors of $\boldsymbol{H}$, and $\Lambda = \operatorname{diag}\{\lambda_1, \ldots, \lambda_m\}$ the associated eigenvalues. Define the function

$$c(\lambda) = \frac{1}{|\lambda|} \mathbb{1}\{\lambda \neq 0\}.$$

This inverts the absolute value of λ if λ is nonzero, otherwise it equals 0. Set

$$\Lambda^* = \operatorname{diag}\{c(\lambda_1), \ldots, c(\lambda_m)\}$$

$$\boldsymbol{H}^{*-1} = \boldsymbol{Q}' \Lambda^* \boldsymbol{Q}.$$

$\boldsymbol{H}^{*-1}$ can be defined for all Hessians $\boldsymbol{H}$ and is positive semi-definite. $\boldsymbol{H}^{*-1}$ can be used in place of $\boldsymbol{H}^{-1}$ in the Newton iteration sequence.

The advantage of Newton's method is that it can converge in a small number of iterations when $f(x)$ is convex. However, when $f(x)$ is not convex, the method can fail to converge. Computation cost can be high

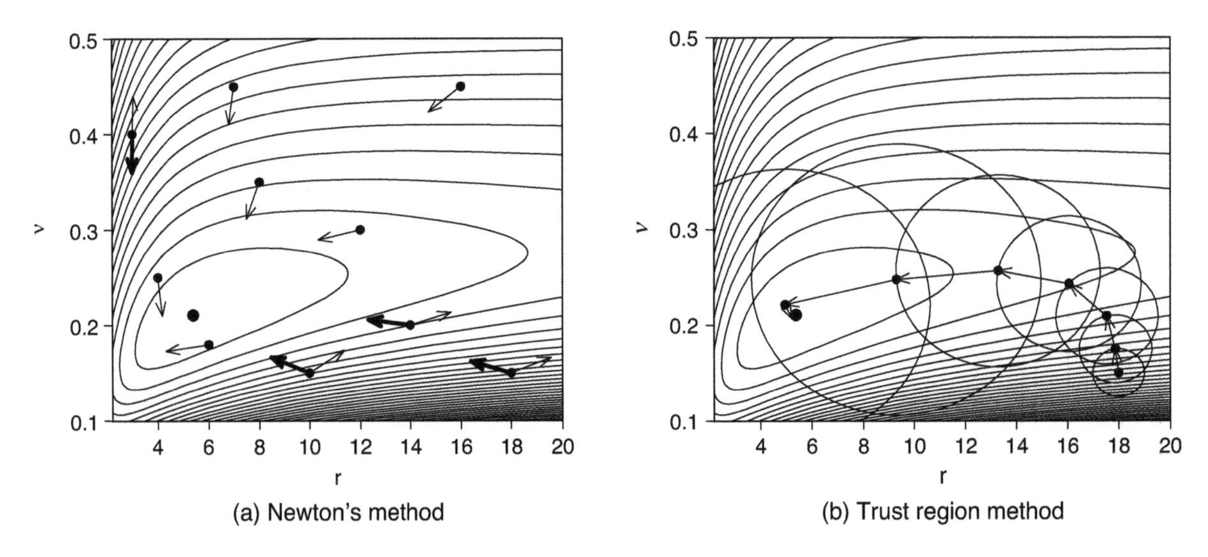

FIGURE 12.5 Newton's method and trust region method

when numerical second derivatives are used and m is large. Consequently, Newton's method is not typically used in higher-dimensional problems.

To illustrate, Figure 12.5(a) displays the Newton direction vectors d_i for the negatively scaled student t log likelihood from the same ten starting values as in Figure 12.4(b), with each gradient normalized to have the same length and displayed as an arrow. For four of the ten points, the Hessian is not positive definite, so the arrows (the light ones) point away from the global minimum. Making the eigenvalue correction described above, the modified Newton directions for these four starting values are displayed using heavy arrows. All four point toward the global maximum. What happens is that if the unmodified Hessian is used, Newton iteration will not converge for many starting values (in this example), but the modified Newton iteration converges for the starting values shown.

BFGS (Broyden-Fletcher-Goldfarb-Shanno). This method is appropriate for differentiable functions with a single local minimum. It is a common default method for differentiable functions.

BFGS is a popular member of the class of "Quasi-Newton" algorithms which replace the computationally costly second-derivative matrix $\boldsymbol{H}_i$ in the Newton method with an approximation $\boldsymbol{B}_i$. Additionally, BFGS is reasonably robust to non-convex functions $f(x)$. Let x_1 and $\boldsymbol{B}_1$ be starting values. The iteration sequence is defined as follows

$$
\begin{aligned}
g_i &= \frac{\partial}{\partial x} f(x_i) \\
d_i &= \boldsymbol{B}_i^{-1} g_i \\
\alpha_i &= \underset{\alpha}{\operatorname{argmin}} f(x_i - \alpha d_i) \\
x_{i+1} &= x_i - \alpha_i d_i.
\end{aligned}
$$

These steps are identical to Newton iteration when $\boldsymbol{B}_i$ is the matrix of second derivatives. The difference with the BFGS algorithm is how $\boldsymbol{B}_i$ is defined. It uses the iterative updating rule

$$h_{i+1} = g_{i+1} - g_i$$

$$\Delta x_{i+1} = x_{i+1} - x_i$$

$$\boldsymbol{B}_{i+1} = \boldsymbol{B}_i + \frac{h_{i+1}h'_{i+1}}{h'_{i+1}\Delta x_{i+1}} - \boldsymbol{B}_i \frac{\Delta x_{i+1}\Delta x'_{i+1}}{\Delta x'_{i+1}\boldsymbol{B}_i\Delta x_{i+1}}\boldsymbol{B}_i.$$

The method requires starting values x_1 and $\boldsymbol{B}_1$ and an algorithm to calculate the first derivative. One choice for $\boldsymbol{B}_1$ is the matrix of second derivatives (modified to be positive definite if necessary). Another choice sets $\boldsymbol{B}_1 = \boldsymbol{I}_m$.

The advantages of the BFGS method over the Newton method is that BFGS is less sensitive to non-convex $f(x)$, does not require $f(x)$ to be second differentiable, and each iteration is comptuationally less expensive. Disadvantages are that the initial iterations can be sensitive to the initial $\boldsymbol{B}_1$, and the method may fail to converge when $f(x)$ is not convex.

Trust-Region Method. This is a a modern modification of Newton's method and quasi-Newton search. The primary modification is that it uses a local (rather than global) quadratic approximation.

Recall that the Newton rule minimizes the Taylor series approximation

$$f(x) \simeq f(x_i) + g_i'(x - x_i) + \frac{1}{2}(x - x_i)' \boldsymbol{H}_i (x - x_i). \tag{12.4}$$

The BFGS algorithm is the same with $\boldsymbol{B}_i$ replacing $\boldsymbol{H}_i$. While the Taylor series approximation (12.4) only has local validity, Newton's method and BFGS iteration effectively treat (12.4) as a global approximation.

The trust-region method acknowledges that (12.4) is a local approximation. Assume that (12.4) using either $\boldsymbol{H}_i$ or $\boldsymbol{B}_i$ is a trustworthy approximation in a neighborhood of x_i

$$(x - x_i)' \boldsymbol{D} (x - x_i) \leq \Delta \tag{12.5}$$

where $\boldsymbol{D}$ is a diagonal scaling matrix, and Δ is a trust-region constant. The subsequent iteration is the value of x that minimizes (12.4) subject to the constraint (12.5) (implementation is discussed below). Imposing the constraint (12.5) is similar in concept to a step-length calculation for Newton's method but is quite different in practice.

The idea is that we "trust" the quadratic aproximation only in a neighborhood of where it is calculated. Imposing a trust-region constraint also prevents the algorithm from moving too quickly across iterations. This can be especially valuable when minimizing erratic functions.

The choice of Δ affects the speed of convergence. A small Δ leads to better quadratic approximations but small step sizes and consequently a large number of iterations. A large Δ can lead to larger steps and thus fewer iterations, but it also can lead to poorer quadratic approximations and consequently erratic iterations. Standard implementations of trust-region iteration modify Δ across iterations by increasing Δ if the decrease in $f(x)$ between two iterations is close to that predicted by the quadratic approximation (12.4), and decreasing Δ if the decrease in $f(x)$ between two iterations is small relative to that predicted by the quadratic approximation. We discuss this below.

Implementation of the minimization step (12.4)–(12.5) can be achieved as follows. First, solve for the standard Newton step $x_* = x_i - \boldsymbol{H}_i^{-1} g_i$. If x_* satisfies (12.5), then $x_{i+1} = x_*$. If not, write the minimization problem using the Lagrangian

$$\mathscr{L}(x, \alpha) = f(x_i) + g_i'(x - x_i) + \frac{1}{2}(x - x_i)' \boldsymbol{H}_i (x - x_i) + \frac{\alpha}{2}\left((x - x_i)' \boldsymbol{D} (x - x_i) - \Delta\right).$$

Given α, the first-order condition for x is

$$x(\alpha) = x_i - (\boldsymbol{H}_i + \alpha \boldsymbol{D})^{-1} g_i.$$

The Lagrange multiplier α is selected so that the constraint is exactly satisisfied, hence

$$\Delta = (x(\alpha) - x_i)' \boldsymbol{D} (x(\alpha) - x_i) = g_i' (\boldsymbol{H}_i + \alpha \boldsymbol{D})^{-1} \boldsymbol{D} (\boldsymbol{H}_i + \alpha \boldsymbol{D})^{-1} g_i.$$

The solution α can be found by one-dimensional root finding. An approximate solution can also be substituted.

The scaling matrix $\boldsymbol{D}$ should be selected so that the parameters are treated symmetrically by the constraint (12.5). A reasonable choice is to set the diagonal elements to be proportional to the inverse of the square of the range of each parameter.

It was mentioned above that the trust region Δ constant can be modified across iterations. Here is a standard recommendation, where we now index Δ_i by iteration:

1. After calculating the iteration $x_{i+1} = x_i - (\boldsymbol{H}_i + \alpha_i \boldsymbol{D})^{-1} g_i$, calculate the percentage improvement in the function relative to that predicted by the approximation (12.4). This is

$$\rho_i = \frac{f(x_i) - f(x_{i+1})}{-g_i' (x_{i+1} - x_i) - \frac{1}{2} (x_{i+1} - x_i)' \boldsymbol{H}_i (x_{i+1} - x_i)}.$$

2. If $\rho_i \geq 0.9$, then increase Δ_i, for example, $\Delta_{i+1} = 2\Delta_i$.
3. If $0.1 \leq \rho_i \geq 0.9$, then keep Δ_i constant.
4. If $\rho_i < 0.1$, then decrease Δ_i, for example, $\Delta_{i+1} = \frac{1}{2}\Delta_i$.

This modification increases Δ when the quadratic approximation is excellent and decreases Δ when the quadratic approximation is poor.

The trust-region method is a major improvement over previous-generation Newton methods. It is a standard implementation choice in contemporary optimization software.

To illustrate, Figure 12.5(b) displays a trust region convergence sequence using the negative scaled student t log likelihood from a difficult starting value ($r = 18$, $v = 0.15$), where the log likelihood is highly nonconvex. We set the scaling matrix $\boldsymbol{D}$ so that the trust regions have similar scales with respect to the displayed graph, and set the initial trust region constant so that the radius corresponds to one unit in r. The trust regions are shown by the circles, with arrows indicating the iteration steps. The iteration sequence moves in the direction of steepest descent, and the trust region constant doubles with each iteration for five iterations until the iteration is close to the global optimum. The first five iterations are constrained, with the minimum obtained on the boundary of the trust region. The sixth iteration is an interior solution, with the step selected by backtracking with $\alpha = 1/8$. The final two iterations are regular Newton steps. The global minimum is found at the eighth iteration.

Nelder-Mead Method. This method is appropriate for potentially nondifferentiable functions $f(x)$. It is a direct search method, also called the **downhill simplex** method.

Recall that m is the dimension of x. Let $\{x_1, \ldots, x_{m+1}\}$ be a set of linearly independent **test points** arranged in a **simplex**, meaning that none are in the interior of their convex hull. For example, when $m = 2$, the set $\{x_1, x_2, x_3\}$ should be the vertices of a triangle. The set is updated by replacing the highest point with a reflected point. This is achieved by applying the following operations.

1. **Order** the elements so that $f(x_1) \leq f(x_2) \leq \cdots \leq f(x_{m+1})$. The goal is to discard the highest point x_{m+1} and replace it by a better point.

2. Calculate $c = m^{-1} \sum_{i=1}^{m} x_i$, the **center point** of the best m points. For example, when $m = 3$, the center point is the midpoint of the side opposite the highest point.
3. **Reflect** x_{m+1} across c, by setting $x_r = 2c - x_{m+1}$. The idea is to go the opposite direction of the current highest point.
   If $f(x_1) \leq f(x_r) \leq f(x_m)$, then replace x_{m+1} with x_r. Retun to step 1 and repeat.
4. If $f(x_r) \leq f(x_1)$, then compute the **expanded** point $x_e = 2x_r - c$. Replace x_{m+1} with either x_e or x_r, depending on whether $f(x_e) \leq f(x_r)$ or conversely. The idea is to expand the simplex if it produces a greater improvement. Return to step 1 and repeat.
5. Compute the **contracted** point $x_c = (x_{m+1} + c)/2$. If $f(x_c) \leq f(x_{m+1})$, then replace x_{m+1} with x_c. Return to step 1 and repeat. The idea is that reflection did not result in an improvement, so reduce the size of the simplex.
6. **Shrink** all points but x_1 by the rule $x_i = (x_i + x_1)/2$. Return to step 1 and repeat. This step only arises in extreme cases.

The operations are repeated until a stopping rule is reached, at which point the lowest value x_1 is taken as the output.

To visualize the method, imagine placing a triangle on the side of a hill. You ask your student Sisyphus to move the triangle to the lowest point in the valley, but you cover his eyes, so he cannot see where the valley is. He can only feel the slope of the triangle. Sisyphus sequentially moves the triangle downhill by lifting the highest vertex and flipping the triangle over. If this results in a lower value than the other two vertices, he stretches the triangle so that it has double the length pointing downhill. If his flip results in an uphill move (whoops!), then he flips the triangle back and tries contracting the triangle by pushing the vertex toward the middle. If that also does not result in an improvement, then Sisyphus returns the triangle to its original position and then shrinks the higher two vertices toward the best vertex, so that the triangle is reduced. Sisyphus repeats this until the triangle is resting at the bottom and has been reduced to a small size, so that the lowest point can be determined.

To illustrate, Figure 12.6 displays Nelder-Mead searches starting with two different initial simplex sets. Panel (a) shows a search starting with a simplex in the lower right corner marked as "1". The first two iterations

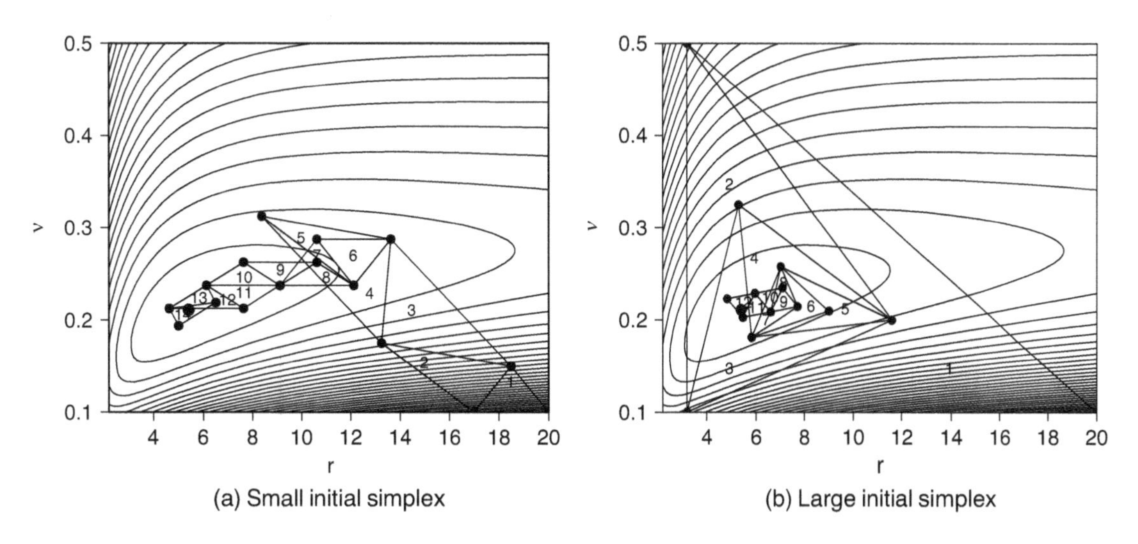

FIGURE 12.6 Nelder-mead search

are expansion steps followed by a reflection step, which yields the set "4". At that point, the algorithm makes a left turn, which requires two contraction steps (to "6"). This is followed by reflection and contraction steps (to "8"), and four reflection steps (to "12"). This is followed by a contraction and reflection step, leading to "14", which is a small set containing the global minimum. The simplex set is still quite wide and has not yet terminated, but it is difficult to display further iterations on the figure.

Figure 12.6(b) displays the Nelder-Mead search starting with a large simplex covering most of the relevant parameter space. The first eight iterations are all contraction steps (to "9"), followed by three reflection steps (to "12"). At this point, the simplex is a small set containing the global minimum.

Nelder-Mead is suitable for nondifferentiable or difficult-to-optimize functions. It is computationally slow, especially in high dimensions. It is not guarenteed to converge, and can converge to a local minimum rather than the global minimum.

12.7 CONSTRAINED OPTIMIZATION

In many cases, the function cannot literally be calculated for all values of θ unless the parameter space is appropriately restricted. Take, for example, the scaled student t log likelihood. The parameters must satisfy $s^2 > 0$ and $r > 2$, or the likelihood cannot be calculated. To enforce these constraints, we use constrained optimization methods. Most constraints can be written as either equality or inequality constraints on the parameters. We consider these two cases separately.

Equality Constraints. The general problem is

$$x_0 = \underset{x}{\operatorname{argmin}} f(x)$$

subject to

$$h(x) = 0.$$

In some cases, the constraints can be eliminated by substitution. In other cases, this cannot be done easily, or it is more convenient to write the problem in the above format. In this case, the standard approach to minimization is to use **Lagrange multiplier** methods. Define the Lagrangian

$$\mathscr{L}(x, \lambda) = f(x) + \lambda' h(x).$$

Let (x_0, λ_0) be the stationary points of $\mathscr{L}(x, \lambda)$ (they satisfy a saddlepoint condition). These can be found by analogs of Newton's method, BFGS, or Nelder-Mead.

Inequality Constraints. The general problem is

$$x_0 = \underset{x}{\operatorname{argmin}} f(x).$$

subject to

$$g(x) \geq 0.$$

It will be useful to write the inequalities separately as $g_j(x) \geq 0$ for $j = 1, \ldots, J$. The general formulation includes boundary constraints, such as $a \leq x \leq b$, by setting $g_1(x) = x - a$ and $g_2(x) = b - x$. It does not include open constraints, such as $\sigma^2 > 0$. When the latter is desired, the constraint $\sigma^2 \geq 0$ can be tried (but may deliver an error message if the boundary value is attempted) or the constraint $\sigma^2 \geq \epsilon$ can be used for some small $\epsilon > 0$.

The modern solution is the **interior point algorithm**. This works by first introducing slack coefficients w_j that satisfy

$$g_j(x) = w_j$$
$$w_j \geq 0.$$

The second step is to enforce the nonnegative constraints by a logarithmic barrier with a Lagrange multiplier constraint. For some small $\mu > 0$, consider the penalized criterion

$$f(x, w) = f(x) - \mu \sum_{j=1}^{J} \log w_j$$

subject to the constraint $g(x) = w$. This problem in turn can be written using the Lagrangian

$$\mathscr{L}(x, w, \lambda) = f(x) - \mu \sum_{j=1}^{J} \log w_j - \lambda' \left(g(x) - w\right).$$

The solution (x_0, w_0, λ_0) consists of the stationary points of $\mathscr{L}(x, w, \lambda)$. This problem is solved using the methods for equality constraints.

12.8 NESTED MINIMIZATION

When x is moderately high-dimensional, numerical minimization can be time consuming and tricky. If the dimension of numerical search can be reduced, this can greatly reduce computation time and improve accuracy. Dimension reduction can sometimes be achieved by using the principle of nested minimization. Write the criterion $f(x, y)$ as a function of two possibly vector-valued inputs. The **principle of nested minimization** states that the joint solution

$$\left(x_0, y_0\right) = \underset{x,y}{\operatorname{argmin}} f(x, y)$$

is identical to the nested solution

$$x_0 = \underset{x}{\operatorname{argmin}} \min_{y} f(x, y) \tag{12.6}$$
$$y_0 = \underset{y}{\operatorname{argmin}} f(x_0, y).$$

It is important to recognize that (12.6) is nested minimization, not sequential minimization. Thus the inner minimization in (12.6) over y is for any given x.

This result is valid for any partition of the variables. However, it is most useful when the partition is made so that the inner minimization has an algebraic solution. If the inner minimization is algebraic (or more generally, is computationally quick to compute), then the potentially difficult multivariate minimization over the pair (x, y) is reduced to minimization over the lower-dimensional x.

To see how this works, take the criterion $f(x, y)$ and imagine holding x fixed and minimizing over y. The solution is

$$y_0(x) = \underset{y}{\operatorname{argmin}} f(x, y).$$

Define the concentrated criterion

$$f^*(x) = \min_{y} f(x, y) = f(x, y_0(x)).$$

It has minimizer

$$x_0 = \underset{x}{\operatorname{argmin}} f^*(x). \tag{12.7}$$

We can use numerical methods to solve (12.7). Given the solution x_0, the solution for y is

$$y_0 = y_0(x_0).$$

For example, take the gamma distribution. It has the negative log likelihood

$$f(\alpha, \beta) = n \log \Gamma(\alpha) + n\alpha \log(\beta) + (1-\alpha) \sum_{i=1}^{n} \log(X_i) + \sum_{i=1}^{n} \frac{X_i}{\beta}.$$

The FOC for β is

$$0 = \frac{n\alpha}{\widehat{\beta}} - \sum_{i=1}^{n} \frac{X_i}{\widehat{\beta}^2}$$

which has solution

$$\widehat{\beta}(\alpha) = \frac{\overline{X}_n}{\alpha}.$$

This is a simple algebraic expression. Substitute this into the negative log likelihood to find the concentrated function:

$$f^*(\alpha) = f\left(\alpha, \widehat{\beta}(\alpha)\right) = f\left(\alpha, \frac{\overline{X}_n}{\alpha}\right) = n \log \Gamma(\alpha) + n\alpha \log\left(\frac{\overline{X}_n}{\alpha}\right) + (1-\alpha) \sum_{i=1}^{n} \log(X_i) + n\alpha.$$

The MLE $\widehat{\alpha}$ can be found by numerically minimizing $f^*(\alpha)$, which is a one-dimensional numerical search. Given $\widehat{\alpha}$, the MLE for β is $\widehat{\beta} = \overline{X}_n / \widehat{\alpha}$.

12.9 TIPS AND TRICKS

Numerical optimization is not always reliable. It requires careful attention and diligence. It is easy for errors to creep into computer code, so triple-checks are rarely sufficient to eliminate errors. Verify and validate your code by making validation and reasonability checks.

When you minimize a function, it is wise to be skeptical of your output. It can be valuable to try multiple methods and multiple starting values.

Optimizers can be sensitive to scaling. Often it is useful to scale your parameters so that the diagonal elements of the Hessian matrix are of similar magnitude. This is similar to scaling regressors so that variances are of similar magnitude.

The choice of parameterization matters. We can optimize over a variance σ^2, the standard deviation σ, or the precision $v = \sigma^{-2}$. While the variance or standard deviation may seem natural, what is better for optimization is convexity of the criterion. This may be difficult to know a priori but if an algorithm bogs down, it may be useful to consider alternative parameterizations.

Coefficient orthogonality can speed convergence and improve performance. A lack of orthogonality induces ridges in the criterion surface, which can be very difficult for the optimization to solve. Iterations need to navigate down a ridge, which can be painfully slow when curved. In a regression setting, we know that highly correlated regressors can often be rendered roughly orthogonal by taking differences. In nonlinear models, it can be more difficult to achieve orthogonality, but a rough guide is to aim to have coefficients that control different aspects of the problem. In the scaled student t example used in this chapter, part of the problem is that both the scale s^2 and degree-of-freedom parameter r affect the variance of the observations. An alternative parameterization is to write the model as a function of the variance σ^2 and degree-of-freedom parameter. This likelihood appears to be a more complicated function of the parameters, but the parameters are also more orthogonal. Reparameterization in this case leads to a likelihood that is simpler to optimize.

A possible alternative to constrained minimization is reparameterization. For example, if $\sigma^2 > 0$, we can reparameterize as $\theta = \log \sigma^2$. If $p \in [0, 1]$, we can reparameterize as $\theta = \log\left(p/(1-p)\right)$. This is generally not recommended, because the transformed criterion may be highly nonconvex and more difficult to optimize. It is generally better to use as convex a criterion as possible and impose constraints using the interior point algorithm and standard constrained optimization software.

12.10 EXERCISES

Exercise 12.1 Take the equation $f(x) = x^2 + x^3 - 1$. Consider the problem of finding the root in $[0, 1]$.

(a) Start with Newton's method. Find the derivative $f'(x)$ and the iteration rule $x_i \to x_{i+1}$.

(b) Starting with $x_1 = 1$, apply the Newton iteration to find x_2.

(c) Make a second Newton step to find x_3.

(d) Now try the bisection method. Calculate $f(0)$ and $f(1)$. Do they have opposite signs?

(e) Calculate two bisection iterations.

(f) Compare the Newton and bisection estimates for the root of $f(x)$.

Exercise 12.2 Take the equation $f(x) = x - 2x^2 + \frac{1}{4}x^4$. Consider the problem of finding the minimum over $x \geq 1$.

(a) For what values of x is $f(x)$ convex?

(b) Find the Newton iteration rule $x_i \to x_{i+1}$.

(c) Using the starting value $x_1 = 1$, calculate the Newton iteration x_2.

(d) Consider the Golden Section search. Start with the bracket $[a, b] = [1, 5]$. Calculate the intermediate points c and d.

(e) Calculate $f(x)$ at a, b, c, d. Does the function satisfy $f(a) > f(c)$ and $f(d) < f(b)$?

(f) Given these calculations, find the updated bracket.

Exercise 12.3 A parameter p lies in the interval $[0, 1]$. If you use the Golden Section search to find the minimum of the log-likelihood function with 0.01 accuracy, how many search iterations are required?

Exercise 12.4 Take the function $f(x, y) = -x^3 y + \frac{1}{2}y^2 x^2 + x^4 - 2x$. You want to find the joint minimizer (x_0, y_0) over $x \geq 0$, $y \geq 0$.

(a) Try nested minimization. Given x, find the minimizer of $f(x, y)$ over y. Write this solution as $y(x)$.

(b) Substitute $y(x)$ into $f(x, y)$. Find the minimizer x_0.

(c) Find y_0.

CHAPTER 13

HYPOTHESIS TESTING

13.1 INTRODUCTION

Economists make extensive use of hypothesis testing. Tests provide evidence concerning the validity of scientific hypotheses. By reporting hypothesis tests, we use statistical evidence to learn about the plausibility of models and assumptions.

Our maintained assumption is that some random vector X has a distribution $F(x)$. Interest focuses on a real-valued (scalar) parameter θ determined by $F \in \mathscr{F}$. The parameter space for θ is Θ. Hypothesis tests are constructed from a random sample $\{X_1, \ldots, X_n\}$ from the distribution F.

13.2 HYPOTHESES

A point hypothesis is the statement that θ equals a specific value θ_0, called the **hypothesized value**. This usage is a bit different from that in previous chapters, where we used the notation θ_0 to denote the true parameter value. In contrast, the hypothesized value is that implied by a theory or hypothesis.

A common example arises when θ measures the effect of a proposed policy. A typical question is whether the effect is 0. This can be answered by testing the hypothesis that the policy has no effect. The latter is equivalent to the statement that $\theta = 0$. This hypothesis is represented by the value $\theta_0 = 0$.

Another common example is when θ represents the difference in an average characteristic or choice between groups. The hypothesis that there is no average difference between the groups is the statement that $\theta = 0$. This hypothesis is $\theta = \theta_0$ with $\theta_0 = 0$.

We call the hypothesis to be tested the "null hypothesis".

Definition 13.1 The **null hypothesis**, written $\mathbb{H}_0 : \theta = \theta_0$, is the restriction $\theta = \theta_0$.

The complement of the null hypothesis (the collection of parameter values that do not satisfy the null hypothesis) is called the "alternative hypothesis".

Definition 13.2 The **alternative hypothesis**, written $\mathbb{H}_1 : \theta \neq \theta_0$, is the set $\{\theta \in \Theta : \theta \neq \theta_0\}$.

For simplicity, we often refer to these hypotheses as "the null" and "the alternative".

Alternative hypotheses can be **one-sided**: $\mathbb{H}_1 : \theta > \theta_0$ or $\mathbb{H}_1 : \theta < \theta_0$; or **two-sided**: $\mathbb{H}_1 : \theta \neq \theta_0$. A one-sided alternative arises when the null lies on the boundary of the parameter space (e.g., $\Theta = \{\theta \geq \theta_0\}$). This can be relevant in the context of a policy that is known to have a nonnegative effect. A two-sided alternative arises when the hypothesized value lies in the interior of the parameter space. Two-sided alternatives are more common in applications than one-sided alternatives, but the one-sided cases are

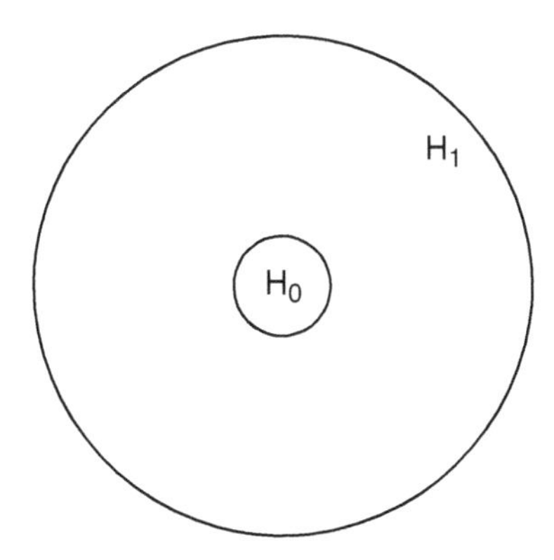

FIGURE 13.1 Null and alternative hypotheses

simpler to analyze. Figure 13.1 illustrates the division of the parameter space into null and alternative hypotheses.

In hypothesis testing, we assume that there is a true but unknown value of θ that either satisfies $\mathbb{H}_0$ or does not satisfy $\mathbb{H}_0$. The goal of testing is to assess whether $\mathbb{H}_0$ is true by examining the observed data.

We illustrate by two examples. The first is the question of the impact of an early childhood education program on adult wage earnings. We want to know whether participation in the program as a child increases wages years later as an adult. Write θ as the difference in average wages between the populations of individuals who participated in early childhood education versus individuals who did not. The null hypothesis is that the program has no effect on average wages, so $\theta_0 = 0$. The alternative hypothesis is that the program increases average wages, so the alternative is one-sided $\mathbb{H}_1 : \theta > 0$.

For our second example, suppose there are two bus routes from your home to the university: Bus 1 and Bus 2. You want to know which (on average) is faster. Write θ as the difference in mean travel times between Bus 1 and Bus 2. A reasonable starting point is the hypothesis that the two buses take the same amount of time. We set this as the null hypothesis, so $\theta_0 = 0$. The alternative hypothesis is that the two routes have different travel times, which is the two-sided alternative $\mathbb{H}_1 : \theta \neq 0$.

A hypothesis is a restriction on the distribution F. Let F_0 be the distribution of X under $\mathbb{H}_0$. The set of null distributions F_0 can be a singleton (a single distribution function), a parametric family, or a nonparametric family. The set is a singleton when $F(x \mid \theta)$ is parametric with θ fully determined by $\mathbb{H}_0$. In this case, F_0 is the single model $F(x \mid \theta_0)$. The set F_0 is a parametric family when there are remaining free parameters. For example, if the model is $\mathrm{N}(\theta, \sigma^2)$ and $\mathbb{H}_0 : \theta = 0$, then F_0 is the class of models $\mathrm{N}(0, \sigma^2)$. This is a class, because it varies across variance parameters σ^2. The set F_0 is nonparametric when F is nonparametric. For example, suppose that F is the class of random variables with a finite mean. This is a nonparametric family. Consider the hypothesis $\mathbb{H}_0 : \mathbb{E}[X] = 0$. In this case, the set F_0 is the class of mean-0 random variables.

For the construction of the theory of optimal testing, the case where F_0 is a singleton turns out to be special, so we introduce a pair of definitions to distinguish this case from the general case.

Definition 13.3 A hypothesis $\mathbb{H}$ is **simple** if the set $\{F \in \mathscr{F} : \mathbb{H} \text{ is true}\}$ is a single distribution.

Definition 13.4 A hypothesis $\mathbb{H}$ is **composite** if the set $\{F \in \mathscr{F} : \mathbb{H} \text{ is true}\}$ has multiple distributions.

In empirical practice, most hypotheses are composite.

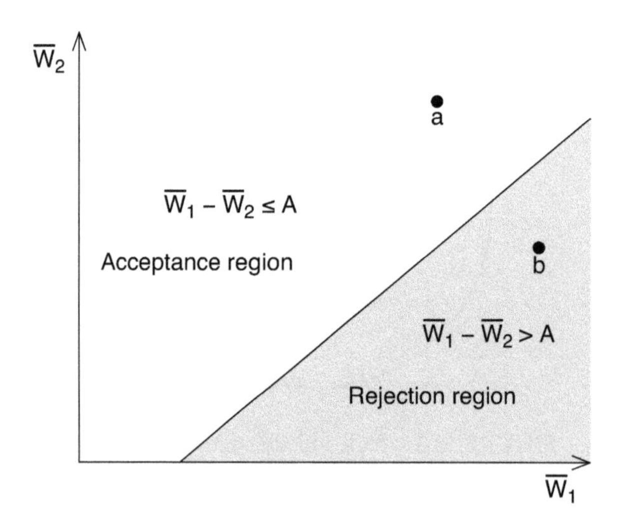

FIGURE 13.2 Acceptance and rejection regions: early childhood education example

13.3 ACCEPTANCE AND REJECTION

A hypothesis test is a **decision** is based on data. The decision either **accepts** the null hypothesis or **rejects** the null hypothesis in favor of the alternative hypothesis. We can describe these two decisions as "Accept $\mathbb{H}_0$" and "Reject $\mathbb{H}_0$".

The decision is based on the data, and so it is a mapping from the sample space to the decision set. One way to view a hypothesis test is as a division of the sample space into two regions S_0 and S_1. If the observed sample falls into S_0, we accept $\mathbb{H}_0$; while if the sample falls into S_1, we reject $\mathbb{H}_0$. The set S_0 is called the **acceptance region** and the set S_1 the **rejection** or **critical region**.

Take the early childhood education example from Section 13.3. To investigate, you find $2n$ adults who were raised in similar settings to one another, of whom n attended an early childhood education program. You interview the individuals and (among other questions) determine their current wage earnings. You decide to test the null hypothesis by comparing the average wages of the two groups. Let $\overline{W}_1$ be the average wage in the early childhood education group, and let $\overline{W}_2$ be the average wage in the remaining sample. You choose the following decision rule. If $\overline{W}_1$ exceeds $\overline{W}_2$ by some threshold (e.g., \$A/hour), you reject the null hypothesis that the average in the population is the same; if the difference between $\overline{W}_1$ and $\overline{W}_2$ is less than \$A/hour, you accept the null hypothesis. The acceptance region S_0 is the set $\left\{\overline{W}_1 - \overline{W}_2 \leq A\right\}$. The rejection region S_1 is the set $\left\{\overline{W}_1 - \overline{W}_2 > A\right\}$.

To illustrate, Figure 13.2 displays the acceptance and rejection regions for this example by a plot in $(\overline{W}_1, \overline{W}_2)$ space. The acceptance region is the light-shaded area, and the rejection region is the dark-shaded region. These regions are rules: they represent how a decision is determined by the data. For example, two observation pairs a and b are displayed. Point a satisfies $\overline{W}_1 - \overline{W}_2 \leq A$, so you "Accept $\mathbb{H}_0$". Point b satisfies $\overline{W}_1 - \overline{W}_2 > A$, so you "Reject $\mathbb{H}_0$".

Consider the bus route example. Suppose you investigate the hypotheses by conducting an experiment. You ride each bus once and record the time it takes to travel from home to the university. Let X_1 and X_2 be the two recorded travel times. You adopt the following decision rule: If the absolute difference in travel times is greater than B minutes, you will reject the hypothesis that the average travel times are the same; otherwise, you will accept the hypothesis. The acceptance region S_0 is the set $\{|X_1 - X_2| \leq B\}$. The rejection region S_1 is the set $\{|X_1 - X_2| > B\}$. These sets are displayed in Figure 13.3. Since the alternative hypothesis is two-sided,

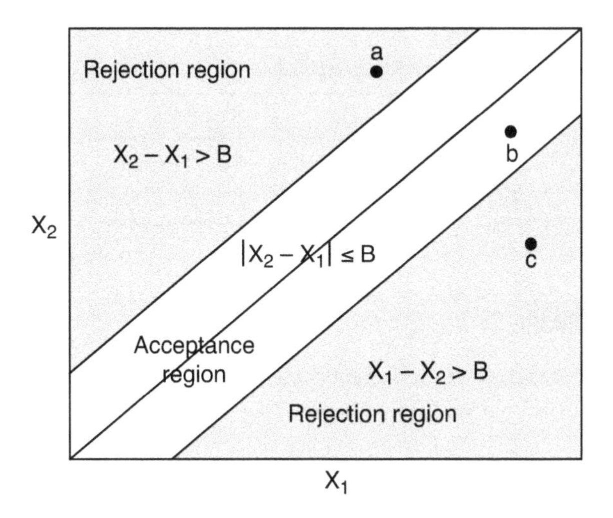

FIGURE 13.3 Acceptance and rejection regions: bus travel example

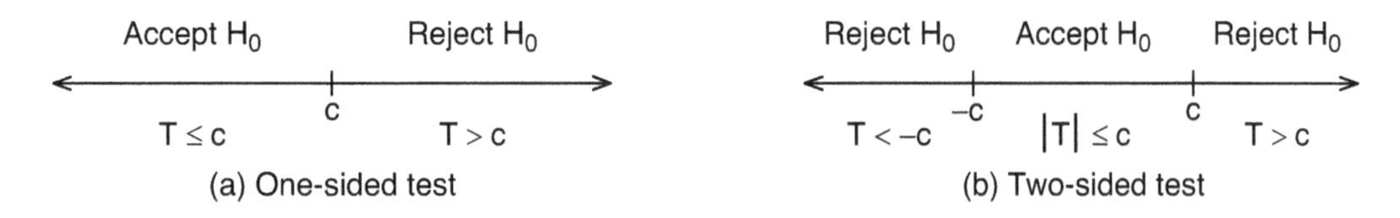

FIGURE 13.4 Acceptance and rejection regions for test statistic

the rejection region is the union of two disjoint sets. To illustrate how decisions are made, three observation pairs a, b, and c are displayed. Point a satisfies $X_2 - X_1 > B$, so you "Reject $\mathbb{H}_0$". Point c satisfies $X_1 - X_2 > B$, so you also "Reject $\mathbb{H}_0$". Point b satisfies $|X_1 - X_2| \leq B$, so you "Accept $\mathbb{H}_0$".

An alternative way to express a decision rule is to construct a real-valued function of the data called a **test statistic**

$$T = T(X_1, \ldots, X_n)$$

together with a **critical region** C. The statistic and the critical region are constructed so that $T \in C$ for all samples in S_1, and $T \notin C$ for all samples in S_0. For many tests, the critical region can be simplified to a **critical value** c, where $T \leq c$ for all samples in S_0 and $T > c$ for all samples in S_1. This is typical for most one-sided tests and most two-sided tests of multiple hypotheses. Two-sided tests of a scalar hypothesis typically take the form $|T| > c$, so the critical region is $C = \{x : |x| > c\}$.

The hypothesis test then can be written as the decision rule:

1. Accept $\mathbb{H}_0$ if $T \notin C$.
2. Reject $\mathbb{H}_0$ if $T \in C$.

In the education example, we set $T = \overline{W}_1 - \overline{W}_2$ and $c = A$. In the bus route example, we set $T = X_1 - X_2$ and $C = \{x < -B \text{ and } x > B\}$.

The acceptance and rejection regions are illustrated in Figure 13.4. Panel (a) illustrates tests where acceptance occurs for the rule $T \leq c$. Panel (b) illustrates tests where acceptance occurs for the rule $|T| \leq c$. These rules split the sample space by reducing an n-dimensional sample space to a single dimension, which is much easier to visualize.

Table 13.1
Hypothesis testing decisions

	Accept $\mathbb{H}_0$	Reject $\mathbb{H}_0$
$\mathbb{H}_0$ true	Correct Decision	Type I Error
$\mathbb{H}_1$ true	Type II Error	Correct Decision

13.4 TYPE I AND TYPE II ERRORS

A decision can be correct or incorrect. An incorrect decision is an error. There are two types of errors, which statisticians have given the bland names "Type I" and "Type II". A false rejection of the null hypothesis $\mathbb{H}_0$ (rejecting $\mathbb{H}_0$ when $\mathbb{H}_0$ is true) is a **Type I error**. A false acceptance of $\mathbb{H}_0$ (accepting $\mathbb{H}_0$ when $\mathbb{H}_1$ is true) is a **Type II error**. Given the two possible states of the world ($\mathbb{H}_0$ or $\mathbb{H}_1$) and the two possible decisions (Accept $\mathbb{H}_0$ or Reject $\mathbb{H}_0$), there are four possible pairings of states and decisions, as depicted in Table 13.1.

In the early education example, a Type I error is the incorrect decision that "early childhood education increases adult wages" when the truth is that the policy has no effect. A Type II error is the incorrect decision that "early childhood education does not increase adult wages" when the truth is that it does increase wages.

In our bus route example, a Type I error is the incorrect decision that the average travel times are different when the truth is that they are the same. A Type II error is the incorrect decision that the average travel times are the same when the truth is that they are different.

While both Type I and Type II errors are mistakes, they are different types of mistakes and lead to different consequences. They should not be viewed symmetrically. In the education example, a Type I error may lead to the decision to implement early childhood education. This means that resources will be devoted to a policy that does not accomplish the intended goal.[1] In contrast, a Type II error may lead to a decision to not implement early childhood education. This means that the benefits of the intervention will not be realized. Both errors lead to negative outcomes, but they are different types of errors and are not symmetric.

Again consider the bus route question. A Type I error may lead to the decision to exclusively take Bus 1. The cost is inconvenience due to avoiding Bus 2. A Type II error may lead to the decision to take whichever bus is more convenient. The cost is the differential travel time between the routes.

The importance or seriousness of the error depends on the (unknown) truth. In the bus route example, if both buses are equally convenient, then the cost of a Type I error may be negligible. If the difference in average travel times is small (e.g., 1 minute), then the cost of a Type II error may be small. But if the difference in convenience and/or travel times is greater, then a Type I and/or Type II error may have a more significant cost.

In an ideal world, a hypothesis test would make error-free decisions. To do so, it would need to be feasible to split the sample space so that samples in S_0 can only have occurred if $\mathbb{H}_0$ is true, and samples in S_1 can only have occurred if $\mathbb{H}_1$ is true. The implication is that we could exactly determine the truth or falsehood of $\mathbb{H}_0$ by examining the data. In the actual world, this is rarely feasible. The presence of randomness means that most decisions have a nontrivial probability of error. Rather than address the impossible goal of eliminating error, let us consider the contructive task of minimizing the probability of error.

Acknowledging that our test statistics are random variables, we can measure their accuracy by the probability that they make an accurate decision. To this end, we define the power function.

[1] We are abstracting from other potential benefits of an early childhood education program.

Definition 13.5 The **power function** of a hypothesis test is the probability of rejection:

$$\pi(F) = \mathbb{P}\left[\text{Reject } \mathbb{H}_0 \mid F\right] = \mathbb{P}\left[T \in C \mid F\right].$$

We can use the power function to calculate the probability of making an error. We have separate names for the probabilities of the two types of errors.

Definition 13.6 The **size** of a hypothesis test is the probability of a Type I error:

$$\mathbb{P}\left[\text{Reject } \mathbb{H}_0 \mid F_0\right] = \pi\left(F_0\right). \tag{13.1}$$

Definition 13.7 The **power** of a hypothesis test is the complement of the probability of a Type II error:

$$1 - \mathbb{P}\left[\text{Accept } \mathbb{H}_0 \mid \mathbb{H}_1\right] = \mathbb{P}\left[\text{Reject } \mathbb{H}_0 \mid F\right] = \pi(F)$$

for F in $\mathbb{H}_1$.

The size and power of a hypothesis test are both found from the power function. The size is the power function evaluated at the null hypothsis, the power is the power function evaluated under the alternative hypothesis.

To calculate the size of a test, observe that T is a random variable and so has a sampling distribution $G(x \mid F) = \mathbb{P}\left[T \leq x\right]$. In general, this depends on the population distribution F. The sampling distribution evaluated at the null distribution $G_0(x) = G(x \mid F_0)$ is called the **null sampling distribution**. It is the distribution of the statistic T when the null hypothesis is true.

13.5 ONE-SIDED TESTS

Let's focus on one-sided tests with rejection region $T > c$. In this case, the power function has the simple expression

$$\pi(F) = 1 - G(c \mid F) \tag{13.2}$$

and the size of the test is the power function evaluated at the null distribution F_0:

$$\pi(F_0) = 1 - G_0(c).$$

Since any distribution function is monotonically increasing, the expression (13.2) shows that the power function is monotonically decreasing in the critical value c. Thus the probability of Type I error is monotonically decreasing in c, and the probability of Type II error is monotonically increasing in c (the latter since the probability of Type II error is 1 minus the power function). Thus the choice of c induces a trade-off between the two error types. Decreasing the probability of one error necessarily increases the other.

Since both error probabilities cannot be simultaneously reduced, some sort of trade-off needs to be adopted. The **classical approach** initiated by Neyman and Pearson is to **control** the size of the test—meaning bounding the size, so that the probability of a Type I error is known—and then picking the test so as to maximize the power subject to this constraint. Today this remains the dominant approach in economics.

Definition 13.8 The **significance level** $\alpha \in (0, 1)$ is the probability selected by the researcher to be the maximal acceptable size of the hypothesis test.

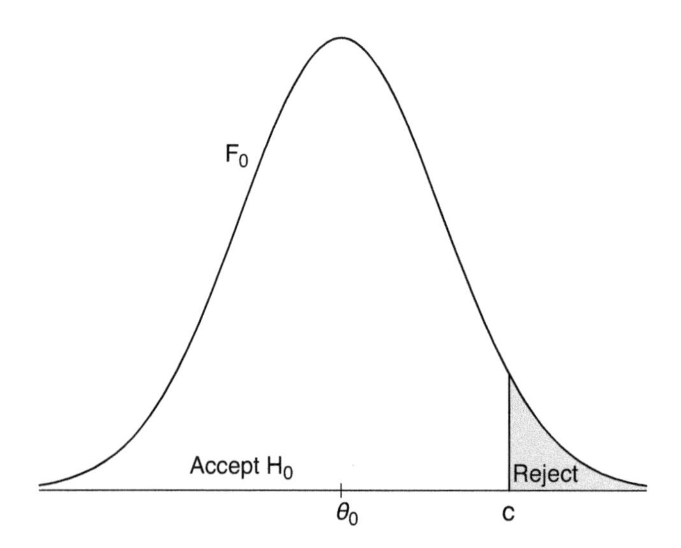

FIGURE 13.5 Null sampling distribution for a one-sided test

Traditional choices for significance levels are $\alpha = 0.10$, $\alpha = 0.05$, and $\alpha = 0.01$ with $\alpha = 0.05$ the most common, meaning that the researcher treats it as acceptable that there is a 1 in 20 chance of a false positive finding. This rule-of-thumb has guided empirical practice in medicine and the social sciences for many years. In much recent discussion, however, many scholars have been arguing that a much smaller significance level should be selected, because academic journals are publishing too many false positives. In particular, it has been suggested that researchers should set $\alpha = 0.005$, so that a false positive occurs once in 200 uses. In any event, there is no purely scientific basis for the choice of α.

Let's consider how the classical approach works for a one-sided test. You first select the significance level α. Following convention, let us use $\alpha = 0.05$. Second, you select a test statistic T and determine its null sampling distribution G_0. For example, if T is a sample mean from a normal population, it has the null sampling distribution $G_0(x) = \Phi(x/\sigma)$ for some σ. Third, you select the critical value c so that the size of the test is smaller than the signficance level:

$$1 - G_0(c) \leq \alpha. \tag{13.3}$$

Since G_0 is monotonically increasing, we can invert the function. The inverse G_0^{-1} of a distribution is its quantile function. We can write the solution to (13.3) as

$$c = G_0^{-1}(1 - \alpha). \tag{13.4}$$

This is the $1 - \alpha$ quantile of the null sampling distribution. For example, when $G_0(x) = \Phi(x/\sigma)$, the critical value (13.4) equals $c = \sigma Z_{1-\alpha}$ where $Z_{1-\alpha}$ is is the $1 - \alpha$ quantile of N(0, 1) (e.g., $Z_{.95} = 1.645$). Fourth, you calculate the statistic T on the dataset. Finally, you accept $\mathbb{H}_0$ if $T \leq c$ and reject $\mathbb{H}_0$ if $T > c$. This approach yields a test with size equal to α, as desired.

Theorem 13.1 If $c = G_0^{-1}(1 - \alpha)$, the size of a hypothesis test equals the significance level α:

$$\mathbb{P}\left[\text{Reject } \mathbb{H}_0 \mid F_0\right] = \alpha.$$

Figure 13.5 illustrates a null sampling distribution and critical region for a test with 5% size. The critical region $T > c$ is shaded and has 5% of the probability mass. The acceptance region $T \leq c$ is unshaded and has 95% of the probability mass.

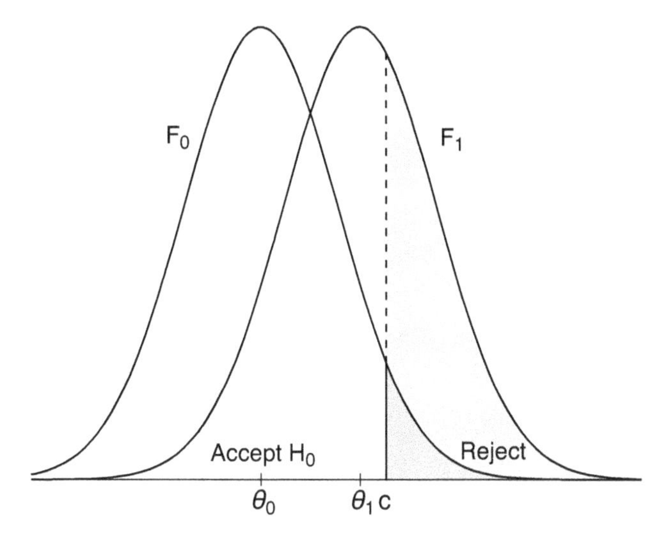

FIGURE 13.6 Sampling distribution under the alternative for a one-sided test

The power of the test is

$$\pi(F) = 1 - G\left(G_0^{-1}(1-\alpha) \mid F\right).$$

This depends on α and the distribution F. For illustration, again take the case of a normal sample mean where $G(x \mid F) = \Phi((x-\theta)/\sigma)$. In this case, the power equals

$$\begin{aligned}\pi(F) &= 1 - \Phi((\sigma Z_{1-\alpha} - \theta)/\sigma) \\ &= 1 - \Phi(Z_{1-\alpha} - \theta/\sigma).\end{aligned}$$

Given α, this is a monotonically increasing function of θ/σ. The power of the test is the probability of rejection when the alternative hypothesis is true. We see from the above expression that the power is increasing as θ increases and as σ decreases.

Figure 13.6 illustrates the sampling distribution under one value of the alternative. The density marked "F_0" is the null distribution. The density marked "F_1" is the distribution under the alternative hypothesis. The critical region is the shaded region. The dark shaded region is 5% of the probability mass of the critical region under the null. The full shaded region is the probability mass of the critical region under this alternative, and in this case equals 37%. Thus for this particular alternative θ_1, the power of the test equals 37%—the probability of rejecting the null is slightly higher than 1 in 3.

Figure 13.7 plots the normal power function as a function of θ/σ for $\alpha = 0.1, 0.05$, and 0.005. At $\theta/\sigma = 0$, the function equals the size α. The power function is monotonically increasing in θ/σ and asymptotes to 1. The three power functions are strictly ranked. This shows how increasing the size of the test (the probability of a Type I error) increases the power (the complement of the probability of a Type II error). In particular, you can see that the power of the test with size 0.005 is much lower than the other power functions.

13.6 TWO-SIDED TESTS

Consider the case of two-sided tests where the critical region is $|T| > c$. In this case, the power function equals

$$\pi(F) = 1 - G(c \mid F) + G(-c \mid F).$$

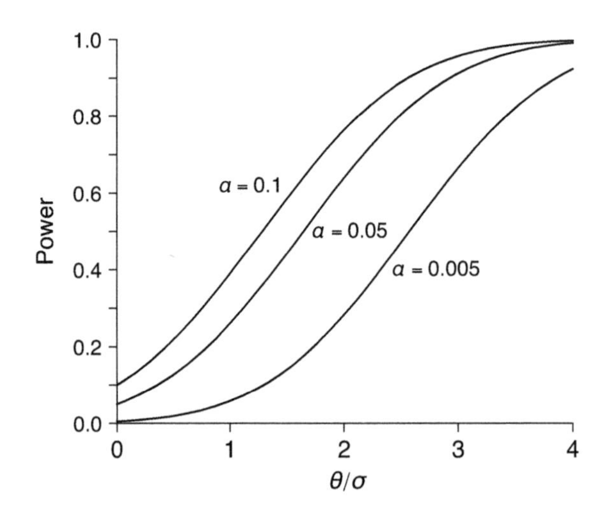

FIGURE 13.7 Power function

The size is

$$\pi(F_0) = 1 - G_0(c) + G_0(-c).$$

When the null sampling distribution is symmetric about 0 (as in the case of the normal distribution), then the size can be written as

$$\pi(F_0) = 2(1 - G_0(c)).$$

For a two-sided test, the critical value c is selected so that the size equals the significance level. Assuming G_0 is symmetric about 0, this is

$$2(1 - G_0(c)) \leq \alpha$$

which has solution

$$c = G_0^{-1}(1 - \alpha/2).$$

When $G_0(x) = \Phi(x/\sigma)$, the two-sided critical value (13.4) is $c = \sigma Z_{1-\alpha/2}$. For $\alpha = 0.05$, this implies $Z_{1-\alpha/2} = 1.96$. The test accepts $\mathbb{H}_0$ if $|T| \leq c$ and rejects $\mathbb{H}_0$ if $|T| > c$.

Figure 13.8 illustrates a null sampling distribution and critical region for a two-sided test with 5% size. The critical region is the union of the two tails of the sampling distribution. Each of the rejection subregions has 2.5% of the probability mass.

Figure 13.9 illustrates the sampling distribution under the alternative. The alternative shown has $\theta_1 > \theta_0$, so the sampling density is shifted to the right. Thus the distribution under the alternative has very little probability mass for the left rejection region $T < -c$ but instead has all probability mass in the right tail. In this case, the power of the test is 26%.

13.7 WHAT DOES "ACCEPT $\mathbb{H}_0$" MEAN ABOUT $\mathbb{H}_0$?

Spoiler alert: The decision "Accept $\mathbb{H}_0$" does not mean that $\mathbb{H}_0$ is true.

The classical approach prioritizes the control of Type I error. This means we require strong evidence to reject $\mathbb{H}_0$. The size of a test is set to be a small number, typically 0.05. Continuity of the power function means that power is small (close to 0.05) for alternatives close to the null hypothesis. Power can also be small for

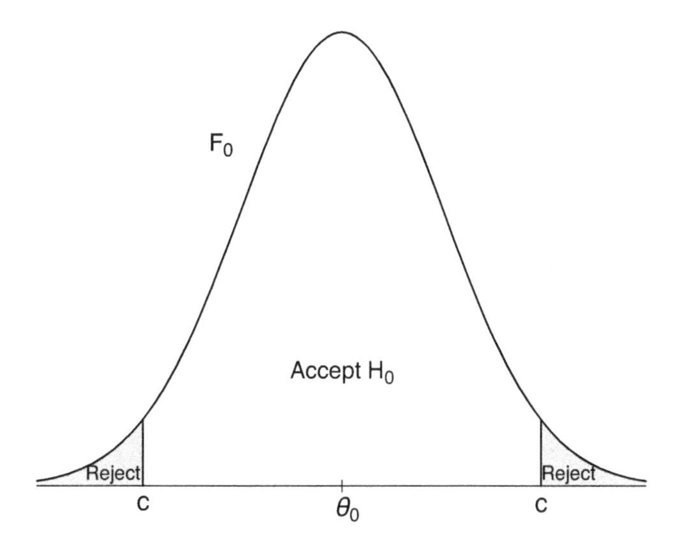

FIGURE 13.8 Null sampling distribution for a two-sided test

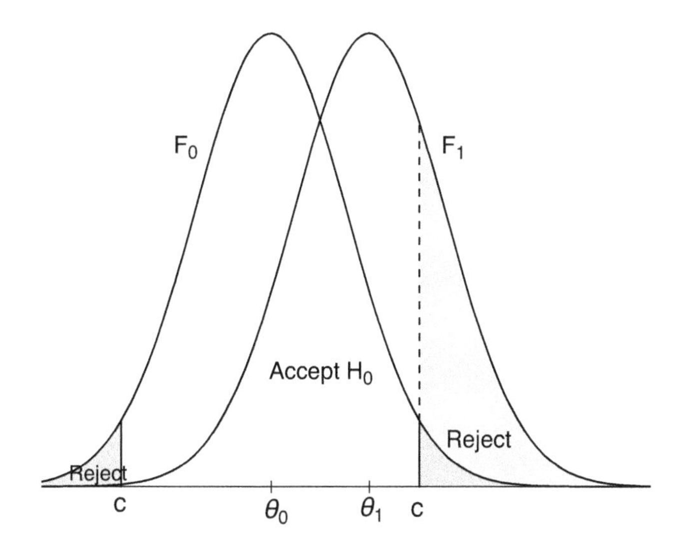

FIGURE 13.9 Sampling distribution under alternative for a two-sided test

alternatives far from the null hypothesis when the amount of information in the data is small. Suppose the power of a test is 25%. Then in 3 of 4 samples, the test does not reject, even though the null hypothesis is false and the alternative hypothesis is true. If the power is 50%, the likelihood of accepting the null hypothesis is equal to a coin flip. Even if the power is 80%, the probability is 20% that the null hypothesis is accepted. The bottom line is that failing to reject $\mathbb{H}_0$ does not meant that $\mathbb{H}_0$ is true.

It is more accurate to say "Fail to Reject $\mathbb{H}_0$" instead of "Accept $\mathbb{H}_0$". Some authors adopt this convention. The word choice is not important, as long as you are clear about what the statistical test means.

Take the early childhood education example. If we test the hypothesis $\theta = 0$, we are seeking evidence that the program had a positive effect on adult wages. Suppose we fail to reject the null hypothesis given the sample.

This could be because there is indeed no effect of the early childhood education program on adult wages. It could also be because the effect is small relative to the spread of wages in the population. Or it could be because the sample size was insufficiently large to be able to measure the effect sufficiently precisely. All of these are valid possibilities, and we do no know which possibility is true without more information. If the test statistic fails to reject $\theta = 0$, it is categorically incorrect to state "The evidence shows that early childhood education has no effect on wages". Instead you can state "The study failed to find evidence that there is an effect on wages." More information can be obtained by examining a confidence interval for θ, which is discussed in Chapter 14.

13.8 *t* TEST WITH NORMAL SAMPLING

We have talked about testing in general but have not been specific about how to design a test. The most common test statistic used in applications is a t test. We start by considering tests for the mean. Let $\mu = \mathbb{E}[X]$, and consider the hypothesis $\mathbb{H}_0 : \mu = \mu_0$. The t-statistic for $\mathbb{H}_0$ is

$$T = \frac{\overline{X}_n - \mu_0}{\sqrt{s^2/n}}$$

where $\overline{X}_n$ is the sample mean, and s^2 is the sample variance. The decision depends on the alternative. For the one-sided alternative $\mathbb{H}_1 : \mu > \mu_0$, the test rejects $\mathbb{H}_0$ in favor of $\mathbb{H}_1$ if $T > c$ for some c. For the one-sided alternative $\mathbb{H}_1 : \mu < \mu_0$, the test rejects $\mathbb{H}_0$ in favor of $\mathbb{H}_1$ if $T < c$. For the two-sided alternative $\mathbb{H}_1 : \mu \neq \mu_0$, the test rejects $\mathbb{H}_0$ in favor of $\mathbb{H}_1$ if $|T| > c$, or equivalently if $T^2 > c^2$.

If the variance σ^2 is known, it can be substituted in the formula for T. In this case some textbooks call T a "z-statistic".

There are alternative ways of writing the test. For example, we can define the statistic as $T = \overline{X}_n$ and reject in favor of $\mu > \mu_0$ if $T > \mu_0 + c\sqrt{s^2/n}$.

It is desirable to select the critical value to control the size, which requires that we know the null sampling distribution $G_0(x)$ of the statistic T. This is possible for the normal sampling model. That is, when $X \sim \mathrm{N}(\mu, \sigma^2)$, the exact distribution $G_0(x)$ of T is the student t with $n-1$ degress of freedom.

Let's consider the test against $\mathbb{H}_1 : \mu > \mu_0$. As described in Section 13.5, for a given significance level α, the goal is to find c so that

$$1 - G_0(c) \leq \alpha$$

which has the solution

$$c = G_0^{-1}(1-\alpha) = q_{1-\alpha}$$

which is the $1-\alpha$ quantile of the student t distribution with $n-1$ degrees of freedom. Similarly, the test against $\mathbb{H}_1 : \mu < \mu_0$ rejects if $T < q_\alpha$.

Now consider a two-sided test. As described in Section 13.6, the goal is to find c so that

$$2(1 - G_0(c)) \leq \alpha$$

which has the solution

$$c = G_0^{-1}(1-\alpha/2) = q_{1-\alpha/2}.$$

The test rejects if $|T| > q_{1-\alpha/2}$. Equivalently, the test rejects if $T^2 > q_{1-\alpha/2}^2$.

Theorem 13.2 In the normal sampling model $X \sim \mathrm{N}(\mu, \sigma^2)$:

1. The t test of $\mathbb{H}_0 : \mu = \mu_0$ against $\mathbb{H}_1 : \mu > \mu_0$ rejects if $T > q_{1-\alpha}$, where $q_{1-\alpha}$ is the $1-\alpha$ quantile of the t_{n-1} distribution.
2. The t test of $\mathbb{H}_0 : \mu = \mu_0$ against $\mathbb{H}_1 : \mu < \mu_0$ rejects if $T < q_{\alpha}$.
3. The t test of $\mathbb{H}_0 : \mu = \mu_0$ against $\mathbb{H}_1 : \mu \neq \mu_0$ rejects if $|T| > q_{1-\alpha/2}$.

   These tests have exact size α.

13.9 ASYMPTOTIC t TEST

When the null sampling distribution is unknown, we typically use asymptotic tests. These tests are based on an asymptotic (large sample) approximation to the probability of a Type I error.

Again consider tests for the mean. The t-statistic for $\mathbb{H}_0 : \mu = \mu_0$ against $\mathbb{H}_1 : \mu > \mu_0$ is

$$T = \frac{\overline{X}_n - \mu_0}{\sqrt{s^2/n}}.$$

The statistic can also be constructed with $\widehat{\sigma}^2$ instead of s^2. The test rejects $\mathbb{H}_0$ if $T > c$. This has size

$$\mathbb{P}\left[\text{Reject } \mathbb{H}_0 \mid F_0\right] = \mathbb{P}\left[T > c \mid F_0\right]$$

which is generally unknown. However, since T is asymptotically standard normal, as $n \to \infty$,

$$\mathbb{P}\left[T > c \mid F_0\right] \to \mathbb{P}\left[\mathrm{N}(0,1) > c\right] = 1 - \Phi(c)$$

which suggests using the critical value $c = Z_{1-\alpha}$ from the normal distribution. For n reasonably large (e.g., $n \geq 60$), this is essentially the same as using student-t quantiles.

The test "Reject $\mathbb{H}_0$ if $T > Z_{1-\alpha}$" does not control the exact size of the test. Instead it controls the asymptotic (large sample) size of the test.

Definition 13.9 The **asymptotic size** of a test is the limiting probability of a Type I error as $n \to \infty$:

$$\alpha = \limsup_{n \to \infty} \mathbb{P}\left[\text{Reject } \mathbb{H}_0 \mid F_0\right].$$

Theorem 13.3 If X has finite mean μ and variance σ^2, then

1. The asymptotic t test of $\mathbb{H}_0 : \mu = \mu_0$ against $\mathbb{H}_1 : \mu > \mu_0$ rejects if $T > Z_{1-\alpha}$, where $Z_{1-\alpha}$ is the $1-\alpha$ quantile of the standard normal distribution.
2. The asymptotic t test of $\mathbb{H}_0 : \mu = \mu_0$ against $\mathbb{H}_1 : \mu < \mu_0$ rejects if $T < Z_{\alpha}$.
3. The asymptotic t test of $\mathbb{H}_0 : \mu = \mu_0$ against $\mathbb{H}_1 : \mu \neq \mu_0$ rejects if $|T| > Z_{1-\alpha/2}$.

   These tests have asymptotic size α.

The same testing method applies to any real-valued hypothesis for which we can calculate a t-ratio. Let θ be a parameter of interest, $\widehat{\theta}$ its estimator, and $s\left(\widehat{\theta}\right)$ its standard error. Under standard conditions,

$$\frac{\widehat{\theta} - \theta}{s\left(\widehat{\theta}\right)} \xrightarrow[d]{} \mathrm{N}(0,1). \tag{13.5}$$

Consequently, under $\mathbb{H}_0 : \theta = \theta_0$,

$$T = \frac{\widehat{\theta} - \theta_0}{s\left(\widehat{\theta}\right)} \underset{d}{\longrightarrow} \mathrm{N}(0,1)$$

which implies that this t-statistic T can be used identically as for the sample mean. Specifically, tests of $\mathbb{H}_0$ compare T with quantiles of the standard normal distribution.

Theorem 13.4 If (13.5) holds, then for $T = \left(\widehat{\theta} - \theta_0\right)/s\left(\widehat{\theta}\right)$,

1. The asymptotic t test of $\mathbb{H}_0 : \theta = \theta_0$ against $\mathbb{H}_1 : \theta > \theta_0$ rejects if $T > Z_{1-\alpha}$, where $Z_{1-\alpha}$ is the $1-\alpha$ quantile of the standard normal distribution.
2. The asymptotic t test of $\mathbb{H}_0 : \theta = \theta_0$ against $\mathbb{H}_1 : \theta < \theta_0$ rejects if $T < Z_{\alpha}$.
3. The asymptotic t test of $\mathbb{H}_0 : \theta = \theta_0$ against $\mathbb{H}_1 : \theta \neq \theta_0$ rejects if $|T| > Z_{1-\alpha/2}$.

   These tests have asymptotic size α.

Since these tests have asymptotic size α but not exact size α, they do not completely control the size of the test. Thus the actual probability of a Type I error could be higher than the significance level α. The theory states that the probability converges to α as $n \to \infty$, but the discrepancy is unknown in any given application. Thus asymptotic tests adopt additional error (unknown finite sample size) in return for broad applicability and convenience.

13.10 LIKELIHOOD RATIO TEST FOR SIMPLE HYPOTHESES

Another important test statistic is the likelihood ratio. In this section, we consider the case of simple hypotheses. Recall that the likelihood informs us about which values of the parameter are most likely to be compatible with the observations. The maximum likelihood estimator $\widehat{\theta}$ is motivated as the value most likely to have generated the data. It therefore seems reasonable to use the likelihood function to assess the likelihood of specific hypotheses concerning θ.

Take the case of simple null and alternative hypotheses. The likelihood at the null and alternative are $L_n(\theta_0)$ and $L_n(\theta_1)$, respectively. If $L_n(\theta_1)$ is sufficiently larger than $L_n(\theta_0)$, this is evidence that θ_1 is more likely than θ_0. We can formalize this by defining a test statistic as the ratio of the two likelihood functions:

$$\frac{L_n(\theta_1)}{L_n(\theta_0)}.$$

A hypothesis test accepts $\mathbb{H}_0$ if $L_n(\theta_1)/L_n(\theta_0) \leq c$ for some c and rejects $\mathbb{H}_0$ if $L_n(\theta_1)/L_n(\theta_0) > c$. Since we typically work with the log-likelihood function, it is more convenient to take the logarithm of the above statistic. This results in the difference in the log likelihoods. For historical reasons (and to simplify how critical values are often calculated), it is convenient to multiply this difference by 2. We therefore define the **likelihood ratio statistic** of two simple hypotheses as

$$\mathrm{LR}_n = 2\left(\ell_n(\theta_1) - \ell_n(\theta_0)\right).$$

The test accepts $\mathbb{H}_0$ if $\mathrm{LR}_n \leq c$ for some c and rejects $\mathbb{H}_0$ if $\mathrm{LR}_n > c$.

Example: $X \sim \mathrm{N}(\theta, \sigma^2)$ with σ^2 known. The log-likelihood function is

$$\ell_n(\theta) = -\frac{n}{2}\log\left(2\pi\sigma^2\right) - \frac{1}{2\sigma^2}\sum_{i=1}^{n}(X_i - \theta)^2.$$

The likelihood ratio statistic for $\mathbb{H}_0 : \theta = \theta_0$ against $\mathbb{H}_1 : \theta = \theta_1 > \theta_0$ is

$$\begin{aligned}
\mathrm{LR}_n &= 2\left(\ell_n(\theta_1) - \ell_n(\theta_0)\right) \\
&= \frac{1}{\sigma^2}\sum_{i=1}^{n}\left((X_i - \theta_0)^2 - (X_i - \theta_1)^2\right) \\
&= \frac{2n}{\sigma^2}\overline{X}_n(\theta_1 - \theta_0) + \frac{n}{\sigma^2}\left(\theta_0^2 - \theta_1^2\right).
\end{aligned}$$

The test rejects $\mathbb{H}_0$ in favor of $\mathbb{H}_1$ if $\mathrm{LR}_n > c$ for some c, which is the same as rejecting if

$$T = \sqrt{n}\left(\frac{\overline{X}_n - \theta_0}{\sigma}\right) > b$$

for some b. Set $b = Z_{1-\alpha}$, so that

$$\mathbb{P}\left[T > Z_\alpha \mid \theta_0\right] = 1 - \Phi\left(Z_{1-\alpha}\right) = \alpha.$$

We find that the likelihood ratio test of simple hypotheses for the normal model with known variance is identical to a t-test with known variance.

13.11 NEYMAN-PEARSON LEMMA

I mentioned in Section 13.5 that the classical approach to hypothesis testing controls the size and picks the test to maximize the power subject to this constraint. We now explore the issue of selecting a test to maximize power. There is a clean solution in the special case of testing simple hypotheses.

Write the joint density of the observations as $f(x \mid \theta)$ for $x \in \mathbb{R}^{nm}$. The likelihood function is $L_n(\theta) = f(X \mid \theta)$. We are testing the simple null $\mathbb{H}_0 : \theta = \theta_0$ against the simple alternative $\mathbb{H}_1 : \theta = \theta_1$ given a fixed significance level α. The likelihood ratio test rejects $\mathbb{H}_0$ if $L_n(\theta_1)/L_n(\theta_0) > c$, where c is selected so that

$$\mathbb{P}\left[\frac{L_n(\theta_1)}{L_n(\theta_0)} > c \,\middle|\, \theta_0\right] = \alpha.$$

Let $\psi_a(x) = \mathbb{1}\left\{f(x \mid \theta_1) > cf(x \mid \theta_0)\right\}$ be the likelihood ratio test function. That is, $\psi_a(X) = 1$ when the likelihood ratio test rejects, and $\psi_a(X) = 0$ otherwise. Let $\psi_b(x)$ be the test function for any other test with size α. Since both tests have size α

$$\mathbb{P}\left[\psi_a(X) = 1 \mid \theta_0\right] = \mathbb{P}\left[\psi_b(X) = 1 \mid \theta_0\right] = \alpha.$$

We can write this as

$$\int \psi_a(x) f(x \mid \theta_0)\,dx = \int \psi_b(x) f(x \mid \theta_0)\,dx = \alpha. \tag{13.6}$$

The power of the likelihood ratio test is

$$\begin{aligned}
\mathbb{P}\left[\frac{L_n(\theta_1)}{L_n(\theta_0)} > c \,\middle|\, \theta_1\right] &= \int \psi_a(x) f(x \mid \theta_1)\, dx \\
&= \int \psi_a(x) f(x \mid \theta_1)\, dx - c\left(\int \psi_a(x) f(x \mid \theta_0)\, dx - \int \psi_b(x) f(x \mid \theta_0)\, dx\right) \\
&= \int \psi_a(x) \left(f(x \mid \theta_1) - cf(x \mid \theta_0)\right) dx + c\int \psi_b(x) f(x \mid \theta_0)\, dx \\
&\geq \int \psi_b(x) \left(f(x \mid \theta_1) - cf(x \mid \theta_0)\right) dx + c\int \psi_b(x) f(x \mid \theta_0)\, dx \\
&= \int \psi_b(x) f(x \mid \theta_1)\, dx \\
&= \pi_b(\theta_1).
\end{aligned}$$

The second equality uses equation (13.6). The inequality in line four holds because if $f(x \mid \theta_1) - cf(x \mid \theta_0) > 0$, then $\psi_a(x) = 1 \geq \psi_b(x)$. If $f(x \mid \theta_1) - cf(x \mid \theta_0) < 0$, then $\psi_a(x) = 0 \geq -\psi_b(x)$. The final expression (line six) is the power of the test ψ_b. This shows that the power of the likelihood ratio test is greater than the power of the test ψ_b, which means that the likelihood ratio test has higher power than any other test with the same size.

Theorem 13.5 Neyman-Pearson Lemma. Among all tests of a simple null hypothesis against a simple alternative hypothesis with size α, the likelihood ratio test has the greatest power.

The Neyman-Pearson Lemma is a foundational result in testing theory.

In Section 13.11, we found that in the normal sampling model with known variance, the likelihood ratio test of simple hypotheses is identical to a t test using a known variance. The Neyman-Pearson Lemma shows that the latter is the most powerful test of this hypothesis in this model.

13.12 LIKELIHOOD RATIO TEST AGAINST COMPOSITE ALTERNATIVES

We now consider two-sided alternatives $\mathbb{H}_1 : \theta \neq \theta_0$. The log likelihood under $\mathbb{H}_1$ is the unrestricted maximum. Let $\widehat{\theta}$ be the MLE that maximizes $\ell_n(\theta)$. Then the maximized likelihood is $\ell_n(\widehat{\theta})$. The likelihood ratio statistic of $\mathbb{H}_0 : \theta = \theta_0$ against $\mathbb{H}_1 : \theta \neq \theta_0$ is twice the difference in the maximized log-likelihoods:

$$\mathrm{LR}_n = 2\left(\ell_n(\widehat{\theta}) - \ell_n(\theta_0)\right).$$

The test accepts $\mathbb{H}_0$ if $\mathrm{LR}_n \leq c$ for some c and rejects $\mathbb{H}_0$ if $\mathrm{LR}_n > c$.

To illustrate, Figure 13.10 displays an exponential log-likelihood function along with a hypothesized value θ_0 and the MLE $\widehat{\theta}$. The likelihood ratio statistic is twice the difference between the log-likelihood function at the two values. The likelihood ratio test rejects the null hypothesis if this difference is sufficiently large.

The test against one-sided alternatives is a bit less intuitive, combining a comparison of the log-likelihoods with an inequality check on the coefficient estimates. Let $\widehat{\theta}_+$ be the maximizer of $\ell_n(\theta)$ over $\theta \geq \theta_0$. In general,

$$\widehat{\theta}_+ = \begin{cases} \widehat{\theta} & \text{if } \widehat{\theta} > \theta_0 \\ \theta_0 & \text{if } \widehat{\theta} \leq \theta_0. \end{cases}$$

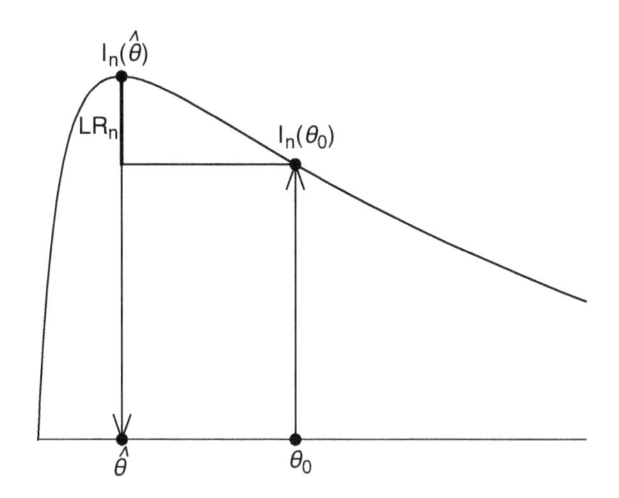

FIGURE 13.10 Likelihood ratio

The maximized log likelihood is defined similarly, so the likelihood ratio statistic is

$$\begin{aligned}\mathrm{LR}_n^+ &= 2\left(\ell_n\left(\widehat{\theta}_+\right) - \ell_n(\theta_0)\right)\\ &= \begin{cases} 2\left(\ell_n\left(\widehat{\theta}\right) - \ell_n(\theta_0)\right) & \text{if } \widehat{\theta} > \theta_0 \\ 0 & \text{if } \widehat{\theta} \le \theta_0 \end{cases}\\ &= \mathrm{LR}_n \mathbb{1}\left\{\widehat{\theta} > \theta_0\right\}.\end{aligned}$$

The test rejects if $\mathrm{LR}_n^* > c$, thus if $\mathrm{LR}_n > c$ and $\widehat{\theta} > \theta_0$.

Example: $X \sim \mathrm{N}(\theta, \sigma^2)$ with σ^2 known. We found before that the likelihood ratio test of $\mathbb{H}_0 : \theta = \theta_0$ against $\mathbb{H}_1 : \theta = \theta_1 > \theta_0$ rejects for $T > Z_{1-\alpha}$, where $T = \sqrt{n}\left(\overline{X}_n - \theta_0\right)/\sigma$. This does not depend on the specific alternative θ_1. Thus this t-ratio is also the test against the one-sided alternative $\mathbb{H}_1 : \theta > \theta_0$. Furthermore, the Neyman-Pearson Lemma shows that this test is the most powerful test against $\mathbb{H}_1 : \theta = \theta_1$ for every $\theta_1 > \theta_0$. Thus the t-ratio test is the uniformly most powerful test against the one-sided alternative $\mathbb{H}_1 : \theta > \theta_0$.

13.13 LIKELIHOOD RATIO AND *t* TESTS

An asymptotic approximation shows that the likelihood ratio of a simple null against a composite alternative is similar to a t test. This can be seen by taking a Taylor series expansion. Let $\theta \in \mathbb{R}^m$. The likelihood ratio statistic is

$$\mathrm{LR}_n = 2\left(\ell_n\left(\widehat{\theta}\right) - \ell_n(\theta_0)\right).$$

Take a second-order Taylor series approximation to $\ell_n(\theta_0)$ about the MLE $\widehat{\theta}$:

$$\ell_n(\theta_0) \simeq \ell_n\left(\widehat{\theta}\right) + \frac{\partial}{\partial\theta}\ell_n\left(\widehat{\theta}\right)'\left(\widehat{\theta} - \theta_0\right) + \frac{1}{2}\left(\widehat{\theta} - \theta_0\right)'\frac{\partial^2}{\partial\theta\partial\theta'}\ell_n\left(\widehat{\theta}\right)\left(\widehat{\theta} - \theta_0\right).$$

The first-order condition for the MLE is $\frac{\partial}{\partial\theta}\ell_n\left(\widehat{\theta}\right) = 0$, so the second term on the right-hand side is 0. The second derivative of the log likelihood is the negative inverse of the Hessian estimator $\widehat{\boldsymbol{V}}$ of the asymptotic variance of $\widehat{\theta}$. Thus the above expression can be written as

$$\ell_n(\theta_0) \simeq \ell_n(\widehat{\theta}) - \frac{1}{2}(\widehat{\theta} - \theta_0)' \widehat{\boldsymbol{V}}^{-1}(\widehat{\theta} - \theta_0).$$

Inserting this expression into the formula for the likelihood ratio statistic, we have

$$\mathrm{LR}_n \simeq (\widehat{\theta} - \theta_0)' \widehat{\boldsymbol{V}}^{-1}(\widehat{\theta} - \theta_0).$$

As $n \to \infty$, this converges in distribution to a χ^2_m distribution, where m is the dimension of θ.

Thus the critical value should be the $1 - \alpha$ quantile of the χ^2_m distribution for asymptotically correct size.

Theorem 13.6 For simple null hypotheses, under $\mathbb{H}_0$,

$$\mathrm{LR}_n = (\widehat{\theta} - \theta_0)' \widehat{\boldsymbol{V}}^{-1}(\widehat{\theta} - \theta_0) + o_p(1) \underset{d}{\longrightarrow} \chi^2_m.$$

The test "Reject $\mathbb{H}_0$ if $\mathrm{LR}_n > q_{1-\alpha}$", the latter being the $1-\alpha$ quantile of the χ^2_m distribution, has asymptotic size α.

Furthermore, in the one-dimensional case, we find $(\widehat{\theta} - \theta_0)' \widehat{\boldsymbol{V}}^{-1}(\widehat{\theta} - \theta_0) = T^2$, the square of the conventional t-ratio. Hence we find that the likelihood ratio and t tests are asymptotically equivalent tests.

Theorem 13.7 As $n \to \infty$, the tests "Reject $\mathbb{H}_0$ if $\mathrm{LR}_n > c$" and "Reject $\mathbb{H}_0$ if $|T| > c$" are asymptotically equivalent tests.

In practice, this theorem does not mean that the likelihood ratio and t tests will give the same answer in a given application. In small and moderate samples, or when the likelihood function is highly nonquadratic, the likelihood ratio and t statistics can be meaningfully different from one another. Thus, tests based on one statistic can differ from tests based on the other. When they do differ, there are two general pieces of advice. First, trust the likelihood ratio over the t test. It has advantages (including invariance to reparameterizations) and tends to work better in finite samples. Second, be skeptical of the asymptotic normal approximation. If the two tests are different, then the large sample approximation theory is not providing a good approximation. It follows that the distributional approximation is unlikely to be accurate.

13.14 STATISTICAL SIGNIFICANCE

When θ represents an important magnitude or effect, it is often of interest to test the hypothesis of no effect, thus $\mathbb{H}_0 : \theta = 0$. If the test rejects the null hypothesis, it is common to describe the finding as "the effect is statistically significant". Often asterisks are appended, indicating the degree of rejection. It is common to see * for a rejection at the 10% significance level, ** for a rejection at the 5% level, and *** for a rejection at the 1% level. I am not a fan of this practice. It focuses attention on statistical significance rather than on interpretation and meaning. It is also common to see phrases such as "the estimate is mildly significant" if it rejects at the 10%, "is statistically significant" if it rejects at the 5% level, and "is highly significant" if it rejects at the 1% level. This information can be useful when these phrases are used in exposition and focus attention on the key parameters of interest.

While statistical significance can be important when judging the relevance of a proposed policy or the scientific merit of a proposed theory, it is not relevant for all parameters and coefficients. Blindly testing that coefficients equal 0 is rarely insightful.

Furthermore, statistical significance is frequently less important than interpretations of magnitudes and statements concerning precision. Chapter 14 introduces confidence intervals, which focus on the assessment of precision. There is a close relationship between statistical tests and confidence intervals, but the two have different uses and interpretations. Confidence intervals are generally better tools than hypothesis tests for the assessment of an economic model.

13.15 p-VALUE

Suppose we use a test that has the form: "Reject $\mathbb{H}_0$ when $T > c$". How should we report our result? Should we report only "Accept" or "Reject"? Should we report the value of T and the critical value c? Or, should we report the value of T and the null distribution of T?

A simple choice is to report the **p-value**, which is

$$p = 1 - G_0(T)$$

where $G_0(x)$ is the null sampling distribution. Since $G_0(x)$ is monotonically increasing, p is a monotonically decreasing function of T. Furthermore, $G_0(c) = \alpha$. Thus the decision "Reject $\mathbb{H}_0$ when $T > c$" is identical to "Reject $\mathbb{H}_0$ when $p < \alpha$". Thus a simple reporting rule is: Report p. Given p, a user can interpret the test using any desired critical value. The p-value p transforms T to an easily interpretable universal scale.

Reporting p-values is especially useful when T has a complicated or unusual distribution. The p-value scale makes interpretation immediately convenient for all users.

Reporting p-values also removes the necessity for adding labels such as "mildly significant". Users are free to interpret $p = 0.09$ as they see fit.

Reporting p-values also allows inference to be continuous rather than discrete. Suppose that based on a 5% level, you find that one statistic "Rejects" and another statistic "Accepts". At first reading, that seems clear. But then on further examination, you find that the first statistic has the p-value 0.049 and the second statistic has the p-value 0.051. These are essentially the same. The fact that one crosses the 0.05 boundary and the other does not is partly an artifact of the 0.05 boundary. If you had selected $\alpha = 0.052$, both would would be "significant". If you had selected $\alpha = 0.048$, neither would be. It seems more sensible to treat these two results identically. Both are (essentially) 0.05, both are suggestive of a statistically significant effect, but neither is strongly persuasive. Looking at the p-value, we are led to treat the two statistics symmetrically rather than differently.

A p-value can be interpreted as the "marginal significance value". That is, it tells you how small α would need to be to reject the null hypothesis. Suppose, for example, that $p = 0.11$. This tells you that if $\alpha = 0.12$, you would have "rejected" the null hypothesis, but not if $\alpha = 0.10$. In contrast, suppose that $p = 0.005$. This tells you that you would have rejected the null for α as small as 0.006. Thus the p-value can be interpreted as the "degree of evidence" against the null hypothesis. The smaller the p-value is, the stronger the evidence.

p-values can be misused. A common misinterpretation is that they are "the probability that the null hypothesis is true". This is incorrect. Some writers have made a big deal about the fact that p-values are not Bayesian probabilities. This is true but irrelevant. A p-value is a transformation of a test statistic and has exactly the same information; it is the statistic written on the $[0, 1]$ scale. It should not be interpreted as a probability.

Some of the recent literature has attacked the overuse of p-values and the excessive use of the 0.05 threshold. These criticisms should not be interpreted as attacks on the p-value transformation. Instead, they are valid attacks on the overuse of testing and a recommendation for a much smaller significance threshold. Hypothesis testing should be applied judiciously and appropriately.

13.16 COMPOSITE NULL HYPOTHESIS

Our discussion up to this point has been implicitly assuming that the sampling distribution G_0 is unique, which occurs when the null hypothesis is simple. Most commonly, the null hypothesis is composite, which introduces some complications.

A parametric example is the normal sampling model $X \sim \mathrm{N}(\mu, \sigma^2)$. The null hypothesis $\mathbb{H}_0 : \mu = \mu_0$ does not specify the variance σ^2, so the null is composite. A nonparametric example is $X \sim F$ with null hypothesis $\mathbb{H}_0 : \mathbb{E}[X] = \mu_0$. This null hypothesis does not restrict F other than restricting its mean.

In this section, we focus on parametric examples where likelihood analysis applies. Assume that X has a known parametric distribution with vector-valued parameter $\beta \in \mathbb{R}^k$. (In the case of the normal distribution, $\beta = (\mu, \sigma^2)$.) Let $\ell_n(\beta)$ be the log-likelihood function. For some real-valued parameter $\theta = h(\beta)$, the null and alternative hypotheses are $\mathbb{H}_0 : \theta = \theta_0$ and $\mathbb{H}_1 : \theta \neq \theta_0$, respectively.

First, consider estimation of β under $\mathbb{H}_1$. The latter is straightforward. It is the unconstrained MLE

$$\widehat{\beta} = \underset{\beta}{\operatorname{argmax}}\, \ell_n(\beta).$$

Second, consider estimation under $\mathbb{H}_0$. This is MLE subject to the contraint $h(\beta) = \theta_0$:

$$\widetilde{\beta} = \underset{h(\beta)=\theta_0}{\operatorname{argmax}}\, \ell_n(\beta).$$

We use the notation $\widetilde{\beta}$ to refer to constrained estimation. By construction, the constrained estimator satisfies the null hypothesis $h\left(\widetilde{\beta}\right) = \theta_0$.

The likelihood ratio statistic for $\mathbb{H}_0$ against $\mathbb{H}_1$ is twice the difference in the log-likelihood function evaluated at the two estimators, or

$$\mathrm{LR}_n = 2\left(\ell_n\left(\widehat{\beta}\right) - \ell_n\left(\widetilde{\beta}\right)\right).$$

This likelihood ratio statistic is necessarily nonnegative, since the unrestricted maximum cannot be smaller than the restricted maximum. The test rejects $\mathbb{H}_0$ in favor of $\mathbb{H}_1$ if $\mathrm{LR}_n > c$ for some c.

When testing simple hypotheses in parametric models, the likelihood ratio is the most powerful test. When the hypotheses are composite, it is (in general) unknown how to construct the most powerful test. Regardless, it is still a standard default choice to use the likelihood ratio to test composite hypotheses in parametric models. Although there are known counterexamples where other tests have better power, in general, the evidence suggests that it is difficult to obtain a test with better power than the likelihood ratio.

Construction of the likelihood ratio is generally straightforward. We estimate the model under the null and alternative and take twice the difference in the maximized log-likelihood ratios.

In some cases, however, we can obtain specific simplications and insight. The primary example is the normal sampling model $X \sim \mathrm{N}(\mu, \sigma^2)$ under a test of $\mathbb{H}_0 : \mu = \mu_0$ against $\mathbb{H}_1 : \mu \neq \mu_0$. The log-likelihood function is

$$\ell_n(\beta) = -\frac{n}{2}\log\left(2\pi\sigma^2\right) - \frac{1}{2\sigma^2}\sum_{i=1}^{n}(X_i - \mu)^2.$$

The unconstrained MLE is $\widehat{\beta}=(\overline{X}_n,\widehat{\sigma}^2)$ with maximized log likelihood

$$\ell_n\left(\widehat{\beta}\right)=-\frac{n}{2}\log\left(2\pi\right)-\frac{n}{2}\log\left(\widehat{\sigma}^2\right)-\frac{n}{2}.$$

The constrained estimator sets $\mu=\mu_0$ and then maximizes the likelihood over σ^2 alone. This yields $\widetilde{\beta}=(\mu_0,\widetilde{\sigma}^2)$, where

$$\widetilde{\sigma}^2=\frac{1}{n}\sum_{i=1}^{n}\left(X_i-\mu_0\right)^2.$$

The maximized log likelihood is

$$\ell_n\left(\widetilde{\beta}\right)=-\frac{n}{2}\log\left(2\pi\right)-\frac{n}{2}\log\left(\widetilde{\sigma}^2\right)-\frac{n}{2}.$$

The likelihood ratio statistic is

$$\begin{aligned}\mathrm{LR}_n&=2\left(\ell_n\left(\widehat{\beta}\right)-\ell_n\left(\widetilde{\beta}\right)\right)\\&=-n\log\left(\widehat{\sigma}^2\right)+n\log\left(\widetilde{\sigma}^2\right)\\&=n\log\left(\frac{\widetilde{\sigma}^2}{\widehat{\sigma}^2}\right).\end{aligned}$$

The test rejects if $\mathrm{LR}_n>c$, which is the same as

$$n\left(\frac{\widetilde{\sigma}^2-\widehat{\sigma}^2}{\widehat{\sigma}^2}\right)>b^2$$

for some b^2. We can write

$$n\left(\frac{\widetilde{\sigma}^2-\widehat{\sigma}^2}{\widehat{\sigma}^2}\right)=\frac{\sum_{i=1}^{n}\left(X_i-\mu_0\right)^2-\sum_{i=1}^{n}\left(X_i-\overline{X}_n\right)^2}{\widehat{\sigma}^2}=\frac{n\left(\overline{X}_n-\mu_0\right)^2}{\widehat{\sigma}^2}=T^2$$

where

$$T=\frac{\overline{X}_n-\mu_0}{\sqrt{\widehat{\sigma}^2/n}}$$

is the t-ratio for the sample mean centered at the hypothesized value. Rejection if $T^2>b^2$ is the same as rejection if $|T|>b$. To summarize, we have shown that the likelihood ratio test is equivalent to the absolute t-ratio.

Since the t-ratio has an exact student t distribution with $n-1$ degrees of freedom, we deduce that the probability of a Type I error is

$$\mathbb{P}\left[|T|>b\mid\mu_0\right]=2\mathbb{P}\left[t_{n-1}>b\right]=2\left(1-G_{n-1}\left(b\right)\right)$$

where $G_{n-1}(x)$ is the student t distribution function. To achieve a test of size α, we set

$$2\left(1-G_{n-1}\left(b\right)\right)=\alpha$$

or

$$G_{n-1}\left(b\right)=1-\alpha/2.$$

Thus b equals the $1-\alpha/2$ quantile of the t_{n-1} distribution.

Theorem 13.8 In the normal sampling model $X\sim\mathrm{N}(\mu,\sigma^2)$, the likelihood ratio test of $\mathbb{H}_0:\mu=\mu_0$ against $\mathbb{H}_1:\mu\neq\mu_0$ rejects if $|T|>q$, where q is the $1-\alpha/2$ quantile of the t_{n-1} distribution. This test has exact size α.

13.17 ASYMPTOTIC UNIFORMITY

When the null hypothesis is composite, the null sampling distribution $G_0(x)$ may vary with the null distribution F_0. In this case, it is not clear how to construct an appropriate critical value or assess the size of the test. A classical approach is to define the uniform size of a test. This is the largest rejection probability across all distributions that satisfy the null hypothesis. Let $\mathscr{F}$ be the class of distributions in the model, and let $\mathscr{F}_0$ be the class of distributions satisfying $\mathbb{H}_0$.

Definition 13.10 The **uniform size** of a hypothesis test is

$$\alpha = \sup_{F_0 \in \mathscr{F}_0} \mathbb{P}\left[\text{Reject } \mathbb{H}_0 \mid F_0\right].$$

Many classical authors simply call this the "size of the test". A difficulty with this concept is that it is challenging to calculate the uniform size in reasonable applications.

In practice, most econometric tests are asymptotic tests. While the size of a test may converge pointwise, it does not necessarily converge uniformly across the distributions in the null hypothesis. Consequently, it is useful to define the **uniform asymptotic size** of a test as

$$\alpha = \limsup_{n\to\infty} \sup_{F_0 \in \mathscr{F}_0} \mathbb{P}\left[\text{Reject } \mathbb{H}_0 \mid F_0\right].$$

This is a stricter concept than the asymptotic size.

In Section 9.4, we showed that uniform convergence in distribution can fail. Thus the uniform asymptotic size may be excessively high. Specifically, if $\mathscr{F}$ denotes the set of distributions with a finite variance, and $\mathscr{F}_0$ is the subset satisfying $\mathbb{H}_0 : \theta = \theta_0$ where $\theta = \mathbb{E}[X]$, then the uniform asymptotic size of any test based on a t-ratio is 1. It is not possible to control the size of the test.

For a test to control size, we need to restrict the class of distributions. Let $\mathscr{F}$ denote the set of distributions such that for some $r > 2$, $B < \infty$, and $\delta > 0$, $\mathbb{E}|X|^r \leq B$ and $\text{var}[X] \geq \delta$. Let $\mathscr{F}_0$ be the subset satisfying $\mathbb{H}_0 : \theta = \theta_0$. Then

$$\limsup_{n\to\infty} \sup_{F_0 \in \mathscr{F}_0} \mathbb{P}\left[|T| > Z_{1-\alpha/2} \mid \mathscr{F}_0\right] = \alpha.$$

Thus a test based on the t-ratio has uniform asymptotic size α. The message from this result is that uniform size control is possible in an asymptotic sense if stronger assumptions are used.

13.18 SUMMARY

To summarize this chapter on hypothesis testing:

1. A hypothesis is a statement about the population. A hypothesis test assesses whether the hypothesis is true or not true based on the data.
2. Classical testing makes the decision either "Accept $\mathbb{H}_0$" or "Reject $\mathbb{H}_0$". There are two possible errors: Type I and Type II. Classical testing attempts to control the probability of Type I error.
3. In many cases, a sensible test statistic is a t-ratio. The null is accepted for small values of the t ratio. The null is rejected for large values. The critical value is selected based on either the student t distribution (in the normal sampling model) or the normal distribution (in other cases).

4. The likelihood ratio statistic is a generally appropriate test statistic. The null is rejected for large values of the likelihood ratio. Critical values are based on the chi-square distribution. In the one-dimensional case, it is equivalent to the t statistic.
5. The Neyman-Pearson Lemma (Theorem 13.5) shows that in certain restricted settings, the likelihood ratio is the most powerful test statistic.
6. Testing can be reported simply by a p-value.

13.19 EXERCISES

For all exercises concerning developing a test, assume that you have an i.i.d. sample of size n from the given distribution.

Exercise 13.1 Take the Bernoulli model with probability parameter p. We want a test for $\mathbb{H}_0 : p = 0.05$ against $\mathbb{H}_1 : p \neq 0.05$.

(a) Develop a test based on the sample mean $\overline{X}_n$.
(b) Find the likelihood ratio statistic.

Exercise 13.2 Take the Poisson model with parameter λ. We want a test for $\mathbb{H}_0 : \lambda = 1$ against $\mathbb{H}_1 : \lambda \neq 1$.

(a) Develop a test based on the sample mean $\overline{X}_n$.
(b) Find the likelihood ratio statistic.

Exercise 13.3 Take the exponential model with parameter λ. We want a test for $\mathbb{H}_0 : \lambda = 1$ against $\mathbb{H}_1 : \lambda \neq 1$.

(a) Develop a test based on the sample mean $\overline{X}_n$.
(b) Find the likelihood ratio statistic.

Exercise 13.4 Take the Pareto model with parameter α. We want a test for $\mathbb{H}_0 : \alpha = 4$ against $\mathbb{H}_1 : \lambda \neq 4$.

(a) Find the likelihood ratio statistic.

Exercise 13.5 Take the model $X \sim \mathrm{N}(\mu, \sigma^2)$. Propose a test for $\mathbb{H}_0 : \mu = 1$ against $\mathbb{H}_1 : \mu \neq 1$.

Exercise 13.6 Test the following hypotheses against two-sided alternatives using the given information. In the following, $\overline{X}_n$ is the sample mean, $s\left(\overline{X}_n\right)$ is its standard error, s^2 is the sample variance, n the sample size, and $\mu = \mathbb{E}[X]$.

(a) $\mathbb{H}_0 : \mu = 0$, $\overline{X}_n = 1.2$, $s\left(\overline{X}_n\right) = 0.4$.
(b) $\mathbb{H}_0 : \mu = 0$, $\overline{X}_n = -1.6$, $s\left(\overline{X}_n\right) = 0.9$.
(c) $\mathbb{H}_0 : \mu = 0$, $\overline{X}_n = -3.5$, $s^2 = 36$, $n = 100$.
(d) $\mathbb{H}_0 : \mu = 1$, $\overline{X}_n = 0.4$, $s^2 = 100$, $n = 1000$.

Exercise 13.7 In a likelihood model with parameter λ, a colleague tests $\mathbb{H}_0 : \lambda = 1$ against $\mathbb{H}_1 : \lambda \neq 1$. They claim to find a negative likelihood ratio statistic, $\mathrm{LR} = -3.4$. What do you think? What do you conclude?

Exercise 13.8 You teach a section of undergraduate statistics for economists with 100 students. You give the students an assignment: They are to find a creative variable (e.g., snowfall in Wisconsin), calculate the correlation of their selected variable with stock price returns, and test the hypothesis that the correlation is 0. Assuming that each of the 100 students selects a variable that is truly unrelated to stock price returns, how many of the 100 students do you expect to obtain a p-value that is significant at the 5% level? Consequently, how should we interpret their results?

Exercise 13.9 Take the model $X \sim \mathrm{N}(\mu, 1)$. Consider testing $\mathbb{H}_0 : \mu \in \{0, 1\}$ against $\mathbb{H}_1 : \mu \notin \{0, 1\}$. Consider the test statistic

$$T = \min\{|\sqrt{n}\overline{X}_n|, |\sqrt{n}(\overline{X}_n - 1)|\}.$$

Let the critical value be the $1-\alpha$ quantile of the random variable $\min\{|Z|, |Z - \sqrt{n}|\}$, where $Z \sim \mathrm{N}(0, 1)$. Show that $\mathbb{P}[T > c \mid \mu = 0] = \mathbb{P}[T > c \mid \mu = 1] = \alpha$. Conclude that the size of the test $\phi_n = \mathbb{1}\{T > c\}$ is α.

Hint: Use the fact that Z and $-Z$ have the same distribution.

This is an example where the null distribution is the same under different points in a composite null. The test $\phi_n = 1(T > c)$ is called a **similar test** because $\inf_{\theta_0 \in \Theta_0} \mathbb{P}[T > c \mid \theta = \theta_0] = \sup_{\theta_0 \in \Theta_0} \mathbb{P}[T > c \mid \theta = \theta_0]$.

Exercise 13.10 The government implements a new banking policy. You want to assess whether the policy has had an impact. You gather information on 10 affected banks (so your sample size is $n = 10$). You conduct a t-test for the effectiveness of the policy and obtain a p-value of 0.20. Since the test is insignificant, what should you conclude? In particular, should you write "The policy has no effect"?

Exercise 13.11 You have two samples (Madison and Ann Arbor) of monthly rents paid by n individuals in each sample. You want to test the hypothesis that the average rent in the two cities is the same. Construct an appropriate test.

Exercise 13.12 You have two samples (mathematics and literature) of size n of the length of a Ph.D. thesis measured by the number of characters. You believe the Pareto model fits the distribution of "length" well. You want to test the hypothesis that the Pareto parameter is the same in the two disciplines. Construct an appropriate test.

Exercise 13.13 You design a statistical test of some hypothesis $\mathbb{H}_0$ which has asymptotic size 5%, but you are unsure of the approximation in finite samples. You run a simulation experiment on your computer to check whether the asymptotic distribution is a good approximation. You generate data that satisfies $\mathbb{H}_0$. On each simulated sample, you compute the test. Out of $B = 50$ independent trials, you find 5 rejections and 45 acceptances.

(a) Based on the $B = 50$ simulation trials, what is your estimate $\widehat{p}$ of p, the probability of rejection?

(b) Find the asymptotic distribution for $\sqrt{B}\left(\widehat{p} - p\right)$ as $B \to \infty$.

(c) Test the hypothesis that $p = 0.05$ against $p \neq 0.05$. Does the simulation evidence support or reject the hypothesis that the size is 5%?

Hint: $\mathbb{P}[|N(0, 1)| \geq 1.96] = 0.05$.

Hint: There may be more than one method to implement a test. That is okay. It is sufficient to describe one method.

CHAPTER 14

CONFIDENCE INTERVALS

14.1 INTRODUCTION

Confidence intervals are used as a tool to report estimation uncertainty.

14.2 DEFINITIONS

Definition 14.1 An **interval estimator** of a real-valued parameter θ is an interval $C = [L, U]$ where L and U are statistics.

The endpoints L and U are statistics, meaning that they are functions of the data, and hence are random. The goal of an interval estimator is to cover (include) the true value θ.

Definition 14.2 The **coverage probability** of an interval estimator $C = [L, U]$ is the probability that the random interval contains the true θ: $\mathbb{P}[L \leq \theta \leq U] = \mathbb{P}[\theta \in C]$.

The coverage probability in general depends on the distribution F.

Definition 14.3 A $1-\alpha$ **confidence interval** for θ is an interval estimator $C = [L, U]$ that has coverage probability $1-\alpha$.

A confidence interval C is a pair of statistics written as a range $[L, U]$ that includes the true value with a prespecified probability. Due to randomness, we rarely seek a confidence interval with 100% coverage, as this would typically need to be the entire parameter space. Instead we seek an interval that includes the true value with reasonably high probability. When we produce a confidence interval, the goal is to produce a range that conveys the uncertainty in the estimation of θ. The interval C shows a range of values that are reasonably plausible, given the data and information. Confidence intervals are reported to indicate the degree of precision of our estimates; the endpoints can be used as reasonable ranges of plausible values.

The value α is similar to the significance level in hypothesis testing, but plays a different role. In hypothesis testing, the significance level is often set to be a very small number to prevent the occurrance of false positives. In confidence interval construction, however, it is often of interest to report the likely range of plausible values of the parameter, not the extreme cases. Standard choices are $\alpha = 0.05$ and 0.10, corresponding to 95% and 90% confidence, respectively.

Since a confidence interval only has interpretation in connection with the coverage probability, it is important that the latter be reported when reporting confidence intervals.

When the finite sample distribution is unknown, we can approximate the coverage probability by its asymptotic limit.

Definition 14.4 The **asymptotic coverage probability** of interval estimator C is $\liminf_{n\to\infty} \mathbb{P}[\theta \in C]$.

Definition 14.5 An $1-\alpha$ **asymptotic confidence interval** for θ is an interval estimator $C = [L, U]$ with asymptotic coverage probability $1-\alpha$.

14.3 SIMPLE CONFIDENCE INTERVALS

Suppose we have a parameter θ, an estimator $\widehat{\theta}$, and a standard error $s\left(\widehat{\theta}\right)$. These can be used to describe three basic confidence intervals.

The default 95% confidence interval for θ is the simple rule

$$C = \left[\widehat{\theta} - 2s\left(\widehat{\theta}\right), \quad \widehat{\theta} + 2s\left(\widehat{\theta}\right)\right]. \tag{14.1}$$

This interval is centered at the parameter estimator $\widehat{\theta}$ and adds plus and minus twice the standard error.

A normal-based $1-\alpha$ confidence interval is

$$C = \left[\widehat{\theta} - Z_{1-\alpha/2}s\left(\widehat{\theta}\right), \quad \widehat{\theta} + Z_{1-\alpha/2}s\left(\widehat{\theta}\right)\right] \tag{14.2}$$

where $Z_{1-\alpha/2}$ is the $1-\alpha/2$ quantile of the standard normal distribution.

A student-based $1-\alpha$ confidence interval is

$$C = \left[\widehat{\theta} - q_{1-\alpha/2}s\left(\widehat{\theta}\right), \quad \widehat{\theta} + q_{1-\alpha/2}s\left(\widehat{\theta}\right)\right] \tag{14.3}$$

where $q_{1-\alpha/2}$ is the $1-\alpha/2$ quantile of the student t distribution with some degree of freedom r.

The default interval (14.1) is similar to the intervals (14.2) and (14.3) for $\alpha = 0.05$. The interval (14.2) is slightly smaller, since $Z_{0.975} = 1.96$. The student-based interval (14.3) is identical for $r = 59$, larger for $r < 59$, and slightly smaller for $r > 59$. Thus (14.1) is a simplified approximation to (14.2) and (14.3).

The normal-based interval (14.2) is exact when the t-ratio

$$T = \frac{\widehat{\theta} - \theta_0}{s\left(\widehat{\theta}\right)}$$

has an exact standard normal distribution. It is an asymptotic approximation when T has an asymptotic standard normal distribution.

The student-based interval (14.3) is exact when the t-ratio has an exact student t distribution. It is also valid when T has an asymptotic standard normal distribution, since (14.3) is wider than (14.2).

We discuss motivations for (14.2) and (14.3) in the following sections and then describe other confidence intervals.

14.4 CONFIDENCE INTERVALS FOR THE SAMPLE MEAN UNDER NORMAL SAMPLING

The student-based interval (14.3) is designed for the mean in the normal sampling model. The degrees of freedom for the critical value is $n-1$.

Let $X \sim \mathrm{N}(\mu, \sigma^2)$ with estimators $\widehat{\mu} = \overline{X}_n$ for μ and s^2 for σ^2. The standard error for $\widehat{\mu}$ is $s\left(\widehat{\mu}\right) = s/n^{1/2}$. The interval (14.3) equals

$$C = \left[\widehat{\mu} - q_{1-\alpha/2}s/n^{1/2}, \quad \widehat{\mu} + q_{1-\alpha/2}s/n^{1/2}\right].$$

To calculate the coverage probability, let μ be the true value. Then

$$\mathbb{P}\left[\mu \in C\right] = \mathbb{P}\left[\widehat{\mu} - q_{1-\alpha/2}s/n^{1/2} \leq \mu \leq \widehat{\mu} + q_{1-\alpha/2}s/n^{1/2}\right].$$

The first step is to subtract $\widehat{\mu}$ from the three components of the equation. This yields

$$\mathbb{P}\left[-q_{1-\alpha/2}s/n^{1/2} \leq \mu - \widehat{\mu} \leq q_{1-\alpha/2}s/n^{1/2}\right].$$

Multiply by -1 and switch the inequalities to obtain

$$\mathbb{P}\left[q_{1-\alpha/2}s/n^{1/2} \geq \widehat{\mu} - \mu \geq -q_{1-\alpha/2}s/n^{1/2}\right].$$

Rewrite as

$$\mathbb{P}\left[-q_{1-\alpha/2}s/n^{1/2} \leq \widehat{\mu} - \mu \leq q_{1-\alpha/2}s/n^{1/2}\right].$$

Divide by $s/n^{1/2}$ and use the fact that the middle term is the t-ratio

$$\mathbb{P}\left[-q_{1-\alpha/2} \leq T \leq q_{1-\alpha/2}\right].$$

Use the fact that $-q_{1-\alpha/2} = q_{\alpha/2}$ and then evaluate the probability

$$\mathbb{P}\left[q_{\alpha/2} \leq T \leq q_{1-\alpha/2}\right] = G(q_{1-\alpha/2}) - G(q_{\alpha/2}) = 1 - \frac{\alpha}{2} - \frac{\alpha}{2} = 1 - \alpha.$$

Here, $G(x)$ is the distribution function of the student t with $n-1$ degrees of freedom

We have calculated that the interval C is a $1-\alpha$ confidence interval.

14.5 CONFIDENCE INTERVALS FOR THE SAMPLE MEAN UNDER NON-NORMAL SAMPLING

Let $X \sim F$ with mean μ, variance σ^2, with estimators $\widehat{\mu} = \overline{X}_n$ for μ and s^2 for σ^2. The standard error for $\widehat{\mu}$ is $s(\widehat{\mu}) = s/n^{1/2}$. The interval (14.2) equals

$$C = \left[\widehat{\mu} - Z_{1-\alpha/2}s/n^{1/2}, \quad \widehat{\mu} + Z_{1-\alpha/2}s/n^{1/2}\right].$$

Using the same steps as in Section 14.4, the coverage probability is

$$\begin{aligned}
\mathbb{P}\left[\mu \in C\right] &= \mathbb{P}\left[\widehat{\mu} - Z_{1-\alpha/2}s/n^{1/2} \leq \mu \leq \widehat{\mu} + Z_{1-\alpha/2}s/n^{1/2}\right] \\
&= \mathbb{P}\left[-Z_{1-\alpha/2}s/n^{1/2} \leq \mu - \widehat{\mu} \leq Z_{1-\alpha/2}s/n^{1/2}\right] \\
&= \mathbb{P}\left[Z_{1-\alpha/2}s/n^{1/2} \geq \widehat{\mu} - \mu \geq -Z_{1-\alpha/2}s/n^{1/2}\right] \\
&= \mathbb{P}\left[-Z_{1-\alpha/2}s/n^{1/2} \leq \widehat{\mu} - \mu \leq Z_{1-\alpha/2}s/n^{1/2}\right] \\
&= \mathbb{P}\left[Z_{\alpha/2} \leq T \leq Z_{1-\alpha/2}\right] \\
&= G_n\left(Z_{1-\alpha/2} \mid F\right) - G_n\left(Z_{\alpha/2} \mid F\right).
\end{aligned}$$

Here, $G_n(x \mid F)$ is the sampling distribution of T.

By the CLT (Theorem 8.3), for any F with a finite variance, $G_n(x \mid F) \longrightarrow \Phi(x)$. Thus

$$\begin{aligned}
\mathbb{P}\left[\mu \in C\right] &= G_n\left(Z_{1-\alpha/2} \mid F\right) - G_n\left(Z_{\alpha/2} \mid F\right) \\
&\to \Phi(Z_{1-\alpha/2}) - \Phi(Z_{\alpha/2}) \\
&= 1 - \frac{\alpha}{2} - \frac{\alpha}{2} = 1 - \alpha.
\end{aligned}$$

Thus the interval C is a $1-\alpha$ asymptotic confidence interval.

14.6 CONFIDENCE INTERVALS FOR ESTIMATED PARAMETERS

Suppose we have a parameter θ, an estimator $\widehat{\theta}$, and a standard error $s\left(\widehat{\theta}\right)$. Assume that they satisfy CLT

$$T=\frac{\widehat{\theta}-\theta}{s\left(\widehat{\theta}\right)} \underset{d}{\longrightarrow} \mathrm{N}(0,1).$$

The interval (14.2) equals

$$C=\left[\widehat{\theta}-Z_{1-\alpha/2}s\left(\widehat{\theta}\right), \quad \widehat{\theta}+Z_{1-\alpha/2}s\left(\widehat{\theta}\right)\right].$$

Using the same steps as in Sections 14.3–14.5, the coverage probability is

$$\begin{aligned}
\mathbb{P}\left[\theta\in C\right] &= \mathbb{P}\left[\widehat{\theta}-Z_{1-\alpha/2}s\left(\widehat{\theta}\right)\le\theta\le\widehat{\theta}+Z_{1-\alpha/2}s\left(\widehat{\theta}\right)\right]\\
&= \mathbb{P}\left[-Z_{1-\alpha/2}\le\frac{\widehat{\theta}-\theta}{s\left(\widehat{\theta}\right)}\le Z_{1-\alpha/2}\right]\\
&= \mathbb{P}\left[Z_{\alpha/2}\le T\le Z_{1-\alpha/2}\right]\\
&= G_n\left(Z_{1-\alpha/2}\mid F\right)-G_n\left(Z_{\alpha/2}\mid F\right)\\
&\to \Phi(Z_{1-\alpha/2})-\Phi(Z_{\alpha/2})\\
&= 1-\frac{\alpha}{2}-\frac{\alpha}{2}=1-\alpha.
\end{aligned}$$

In the fourth line, $G_n(x\mid F)$ is the sampling distribution of T. This shows that the interval C is a $1-\alpha$ asymptotic confidence interval.

14.7 CONFIDENCE INTERVAL FOR THE VARIANCE

Let $X\sim\mathrm{N}(\mu,\sigma^2)$ with mean μ, variance σ^2 with estimators $\widehat{\mu}=\overline{X}_n$ for μ and s^2 for σ^2. The variance estimator has the exact distribution

$$\frac{(n-1)\,s^2}{\sigma^2}\sim\chi^2_{n-1}.$$

Let $G(x)$ denote the χ^2_{n-1} distribution, and set $q_{\alpha/2}$ and $q_{1-\alpha/2}$ to be the $\alpha/2$ and $1-\alpha/2$ quantiles of this distribution. The distribution implies that

$$\mathbb{P}\left[q_{\alpha/2}\le\frac{(n-1)\,s^2}{\sigma^2}\le q_{1-\alpha/2}\right]=1-\alpha.$$

Rewriting the inequalities, we find

$$\mathbb{P}\left[\frac{(n-1)\,s^2}{q_{1-\alpha/2}}\le\sigma^2\le\frac{(n-1)\,s^2}{q_{\alpha/2}}\right]=1-\alpha.$$

Set

$$C=\left[\frac{(n-1)\,s^2}{q_{1-\alpha/2}},\quad\frac{(n-1)\,s^2}{q_{\alpha/2}}\right].$$

This is an exact $1-\alpha$ confidence interval for σ^2.

The interval C is asymmetric about the point estimator s^2, unlike the intervals (14.1)–(14.3). It also respects the natural boundary $\sigma^2\ge 0$, as the lower endpoint is nonnegative.

14.8 CONFIDENCE INTERVALS BY TEST INVERSION

The intervals (14.1)–(14.3) are not the only possible confidence intervals. Why should we use these specific intervals? It turns out that a universally appropriate way to construct a confidence interval is by **test inversion**. The essential idea is to select a sensible test statistic and find the collection of parameter values which are not rejected by the hypothesis test. This produces a set that typically satisfies the properties of a confidence interval.

For a parameter $\theta \in \Theta$, let $T(\theta)$ be a test statistic and c a critical value with the property that to test $\mathbb{H}_0 : \theta = \theta_0$ against $\mathbb{H}_1 : \theta \neq \theta_0$ at level α, we reject if $T(\theta_0) > c$. Define the set

$$C = \{\theta \in \Theta : T(\theta) \leq c\}. \tag{14.4}$$

The set C is the set of parameters θ which are "accepted" by the test "reject if $T(\theta) > c$". The complement of C is the set of parameters which are "rejected" by the same test. This set C is a sensible choice for an interval estimator if the test $T(\theta)$ is a sensible test statistic.

Theorem 14.1 If $T(\theta)$ has exact size α for all $\theta \in \Theta$, then C is a $1-\alpha$ confidence interval for θ.

Theorem 14.2 If $T(\theta)$ has asymptotic size α for all $\theta \in \Theta$, then C is a $1-\alpha$ asymptotic confidence interval for θ.

To prove these results, let θ_0 be the true value of θ. For Theorem 14.1, notice that

$$\begin{aligned}\mathbb{P}\left[\theta_0 \in C\right] &= \mathbb{P}\left[T(\theta_0) \leq c\right] \\ &= 1 - \mathbb{P}\left[T(\theta_0) > c\right] \\ &\geq 1 - \alpha\end{aligned}$$

where the final equality holds if $T(\theta_0)$ has exact size α. Thus C is a confidence interval.

For Theorem 14.2, apply limits to the second line above. This has the limit on the third line if $T(\theta_0)$ has asymptotic size α. Thus C is an asymptotic confidence interval.

We can now see that the intervals (14.2) and (14.3) are identical to (14.4) for the t-statistic

$$T(\theta) = \frac{\widehat{\theta} - \theta}{s\left(\widehat{\theta}\right)}$$

with critical value $c = Z_{1-\alpha/2}$ for (14.2) and $c = q_{1-\alpha/2}$ (14.3).

Thus the conventional "simple" confidence intervals correspond to inversion of t-statistics.

The other major test statistic we have studied is the likelihood ratio statistic. The likelihood ratio can also be "inverted" to obtain a test statistic. Recall the general case. Suppose the full parameter vector is β, and we want a confidence interval for function $\theta = h(\beta)$. The likelihood ratio statistic for a test of $\mathbb{H}_0 : \theta = \theta_0$ against $\mathbb{H}_1 : \theta \neq \theta_0$ is $\mathrm{LR}_n(\theta_0)$, where

$$\mathrm{LR}_n(\theta) = 2\left(\max_{\beta} \ell_n(\beta) - \max_{h(\beta)=\theta} \ell_n\left(\widetilde{\beta}\right)\right).$$

This is twice the difference between the log-likelihood function maximized without restriction and the log-likelihood function maximized subject to the restriction that $h(\beta) = \theta$. The hypothesis $\mathbb{H}_0 : \theta = \theta_0$ is rejected at the asymptotic significance level α if $\mathrm{LR}_n(\theta) > q_{1-\alpha}$, where the latter is the $1-\alpha$ quantile of the χ_1^2 distribution. It may be helpful to know that since $\chi_1^2 = Z^2$, $q_{1-\alpha} = Z_{1-\alpha/2}^2$. Thus, for example, $q_{0.95} = 1.96^2 = 3.84$.

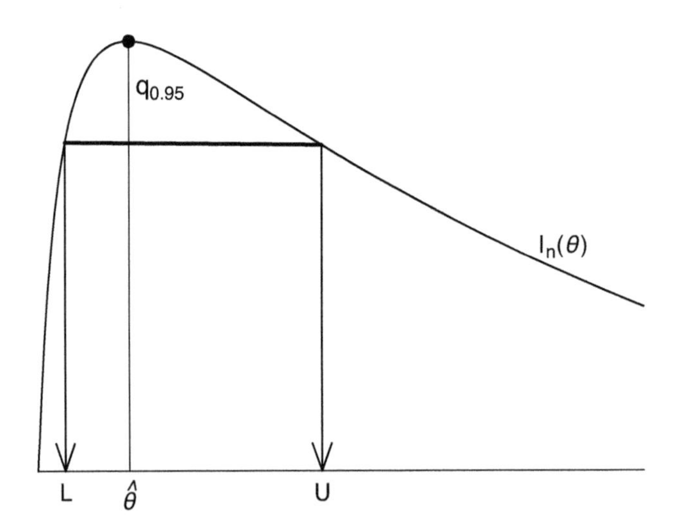

FIGURE 14.1 Confidence interval by test inversion

The likelihood test-inversion interval estimator for θ is the set of values that are not rejected by this test, which is the set

$$C = \left\{ \theta \in \Theta : \mathrm{LR}_n(\theta) \leq q_{1-\alpha} \right\}.$$

This is a level set of the likelihood surface. The set C is an asymptotic confidence interval under standard conditions.

To illustrate, Figure 14.1 displays a log-likelihood function for an exponential distribution. The highest point of the likelihood coincides with the MLE. Subtracting the chi-square critical value from this point yields the horizontal line. The value of the parameter θ for which the log likelihood exceeds this value are parameter values that are not rejected by the likelihood ratio test. This set is compact, because the log likelihood is single peaked, and the set is asymmetric, because the log-likelihood function is asymmetric about the MLE. The test-inversion asymptotic confidence interval is $[L, U]$, The interval includes more values of θ to the right of the MLE than to the left, because of the asymmetry of the log-likelihood function.

14.9 USE OF CONFIDENCE INTERVALS

In applied economic practice, it is common to report parameter estimates $\widehat{\theta}$ and associated standard errors $s\left(\widehat{\theta}\right)$. Good papers also report standard errors for calculations (estimators) made from the estimated model parameters. The routine approximation is that the parameter estimators are asymptotically normal, so the normal intervals (14.2) are asymptotic confidence intervals. The student t intervals (14.3) are also valid, since they are slightly larger than the normal intervals and are exact for the normal sampling model.

Most papers do not explicitly report confidence intervals. Instead, most implicitly use the rule-of-thumb interval (14.1) when discussing parameter estimates and associated effects. The rule-of-thumb interval does not need to be explicitly reported, since it is easy to visually calculate from reported parameter estimates and standard errors.

Confidence intervals should be interpreted relative to the meaning and interpretation of the parameter θ. The upper and lower endpoints can be used to assess plausible ranges. Wide confidence intervals (relative to the meaning of the parameter) indicate that the parameter estimate is imprecise and therefore does not give useful information. When confidence intervals are wide, they include many values. This can lead to an "acceptance" of a default null hypothesis (such as the hypothesis that a policy effect is 0). In this context, it is a mistake to deduce that the empirical study provides evidence in favor of the null hypothesis of no policy effect. Instead, the correct deduction is that the sample was insufficiently informative to shed light on the question. Understanding why the sample was insufficiently informative may lead to a substantive finding (for example, that the underlying population is highly heterogeneous), or it may not. In any event, this analysis indicates that when testing a policy effect, it is important to examine both the p-value and the width of the confidence interval relative to the meaning of the policy effect parameter.

Narrow confidence intervals (again relative to the meaning of the parameter) indicate that the parameter estimate is precise and can be taken seriously. Narrow confidence intervals (small standard errors) are quite common in contemporary applied economics studies which use large datasets.

14.10 UNIFORM CONFIDENCE INTERVALS

Our definitions of coverage probability and asymptotic coverage probability are pointwise in the parameter space. These are simple concepts but have their limitations. A stronger concept is uniform coverage, where the uniformity is over a class of distributions. This is similar to the concept of uniform size of a hypothesis test.

Let $\mathscr{F}$ be a class of distributions.

Definition 14.6 The **uniform coverage probability** of an interval estimator C is $\inf_{F\in\mathscr{F}} \mathbb{P}[\theta \in C]$.

Definition 14.7 A $1-\alpha$ **uniform confidence interval** for θ is an interval estimator C with uniform coverage probability $1-\alpha$.

Definition 14.8 The **uniform asymptotic coverage probability** of interval estimator C is $\liminf_{n\to\infty} \inf_{F\in\mathscr{F}} \mathbb{P}[\theta \in C]$.

Definition 14.9 An $1-\alpha$ **uniform asymptotic confidence interval** for θ is an interval estimator C with uniform asymptotic coverage probability $1-\alpha$.

The uniform coverage probability is the worst-case coverage among distributions in the class $\mathscr{F}$.

In the case of the normal sampling model, the coverage probability of the student interval (14.3) is exact for all distributions. Thus interval (14.3) is a uniform confidence interval.

Asymptotic intervals can be strengthed to uniform asymptotic intervals by the same method as for uniform tests. The moment conditions need to be strengthed, so that the central limit theory applies uniformly.

14.11 EXERCISES

Exercise 14.1 You have the point estimate $\widehat{\theta} = 2.45$ and standard error $s\left(\widehat{\theta}\right) = 0.14$.

(a) Calculate a 95% asymptotic confidence interval.
(b) Calculate a 90% asymptotic confidence interval.

Exercise 14.2 You have the point estimate $\widehat{\theta}=-1.73$ and standard error $s\left(\widehat{\theta}\right)=0.84$.

(a) Calculate a 95% asymptotic confidence interval.

(b) Calculate a 90% asymptotic confidence interval.

Exercise 14.3 You have two independent samples with estimates and standard errors $\widehat{\theta}_1=1.4$, $s\left(\widehat{\theta}_1\right)=0.2$, $\widehat{\theta}_2=0.7$, $s\left(\widehat{\theta}_2\right)=0.3$. You are interested in the difference $\beta=\theta_1-\theta_2$.

(a) Find $\widehat{\beta}$.

(b) Find a standard error $s\left(\widehat{\beta}\right)$.

(c) Calculate a 95% asymptotic confidence interval for β.

Exercise 14.4 You have the point estimate $\widehat{\theta}=0.45$ and standard error $s\left(\widehat{\theta}\right)=0.28$. You are interested in $\beta=\exp(\theta)$.

(a) Find $\widehat{\beta}$.

(b) Use the Delta Method to find a standard error $s\left(\widehat{\beta}\right)$.

(c) Use your answers to parts (a) and (b) to calculate a 95% asymptotic confidence interval for β.

(d) Calculate a 95% asymptotic confidence interval $[L, U]$ for the orginal parameter θ. Calculate a 95% asymptotic confidence interval for β as $[\exp(L), \exp(U)]$. Can you explain why this is a valid choice? Compare this interval with your answer in (c).

Exercise 14.5 To estimate a variance σ^2, you have an estimate $\widehat{\sigma}^2=10$ with standard error $s\left(\widehat{\sigma}^2\right)=7$.

(a) Construct the standard 95% asymptotic confidence interval for σ^2.

(b) Is there a problem in (a)? Explain the difficulty.

(c) Can a similar problem arise for estimation of a probability p?

Exercise 14.6 A confidence interval for the mean of a variable X is $[L, U]$. You decide to rescale your data, so you set $Y=X/1000$. Find the confidence interval for the mean of Y.

Exercise 14.7 A friend suggests the following confidence interval for θ. They draw a random number $U\sim U[0,1]$ and set

$$C=\begin{cases}\mathbb{R} & \text{if} \quad U\leq 0.95\\ \varnothing & \text{if} \quad U>0.05\end{cases}$$

(a) What is the coverage probability of C?

(b) Is C a good choice for a confidence interval? Explain.

Exercise 14.8 A colleague reports a 95% confidence interval $[L, U]=[0.1, 3.4]$ for θ and also states "The t-statistic for $\theta=0$ is insignificant at the 5% level". How do you interpret this? What could explain this situation?

Exercise 14.9 Use the Bernoulli model with probability parameter p. Let $\widehat{p}=\overline{X}_n$ from a random sample of size n.

(a) Find $s\left(\widehat{p}\right)$ and a default 95% confidence interval for p.

(b) Given n, what is the widest possible length of a default 95% interval?

(c) Find n such that the length is less than 0.02.

Exercise 14.10 Let $C = [L, U]$ be a $1-\alpha$ confidence interval for θ. Consider $\beta = h(\theta)$, where $h(\theta)$ is monotonically increasing. Set $C_\beta = [h(L), h(U)]$. Evaluate the converage probability of C_β for β. Is C_β a $1-\alpha$ confidence interval?

Exercise 14.11 If $C = [L, U]$ is a $1-\alpha$ confidence interval for σ^2, find a confidence interval for the standard deviation σ.

Exercise 14.12 You work in a government agency supervising job training programs. A research paper examines the effect of a specific job training program on hourly wages. The reported estimate of the effect (in U.S. dollars) is $\widehat{\theta} = \$0.50$ with a standard error of 1.20. Your supervisor concludes: "The effect is statistically insignificant. The program has no effect and should be canceled." Is your supervisor correct? What do you think is the correct interpretation?

CHAPTER 15

SHRINKAGE ESTIMATION

15.1 INTRODUCTION

This chapter introduces the Stein-Rule shrinkage estimators. These are biased estimators that have lower MSE than the MLE. Throughout this chapter, assume that we have a parameter of interest $\theta \in \mathbb{R}^K$ and an initial estimator $\widehat{\theta}$ for θ. For example, the estimator $\widehat{\theta}$ could be a vector of sample means or a MLE. Assume that $\widehat{\theta}$ is unbiased for θ and has covariance matrix $\boldsymbol{V}$. James-Stein shrinkage theory is thoroughly covered in Lehmann and Casella (1998). See also Wasserman (2006) and Efron (2010).

15.2 MEAN SQUARED ERROR

In this section, we define weighted MSE as our measure of estimation accuracy and illustrate some of its features.

For a scalar estimator $\widetilde{\theta}$ of a parameter θ, a common measure of efficiency is MSE

$$\operatorname{mse}\left[\widetilde{\theta}\right]=\mathbb{E}\left[\left(\widetilde{\theta}-\theta\right)^2\right].$$

For vector-valued ($K \times 1$) estimators $\widetilde{\theta}$, there is no simple extension. Unweighted mean squared error is

$$\begin{aligned}\operatorname{mse}\left[\widetilde{\theta}\right]&=\sum_{j=1}^{K}\mathbb{E}\left[\left(\widehat{\theta}_j-\theta_j\right)^2\right]\\&=\mathbb{E}\left[\left(\widetilde{\theta}-\theta\right)'\left(\widetilde{\theta}-\theta\right)\right].\end{aligned}$$

This is generally unsatisfactory, as it is not invariant to re-scalings of individual parameters. It is therefore useful to define weighted **mean squared error (MSE)**

$$\operatorname{mse}\left[\widetilde{\theta}\right]=\mathbb{E}\left[\left(\widetilde{\theta}-\theta\right)'\boldsymbol{W}\left(\widetilde{\theta}-\theta\right)\right]$$

where $\boldsymbol{W}$ is a weight matrix.

A particularly convenient choice for the weight matrix is $\boldsymbol{W}=\boldsymbol{V}^{-1}$, where $\boldsymbol{V}$ is the covariance matrix for the initial estimator $\widehat{\theta}$. By setting the weight matrix equal to this choice, the weighted MSE is invariant to linear rotations of the parameters. This MSE is

$$\operatorname{mse}\left[\widetilde{\theta}\right]=\mathbb{E}\left[\left(\widetilde{\theta}-\theta\right)'\boldsymbol{V}^{-1}\left(\widetilde{\theta}-\theta\right)\right].$$

For simplicity, we focus on this definition in this chapter.

We can write the MSE as

$$\begin{aligned}\operatorname{mse}\left[\widetilde{\theta}\right] &= \mathbb{E}\left[\operatorname{tr}\left(\left(\widetilde{\theta}-\theta\right)' V^{-1}\left(\widetilde{\theta}-\theta\right)\right)\right] \\ &= \mathbb{E}\left[\operatorname{tr}\left(V^{-1}\left(\widetilde{\theta}-\theta\right)\left(\widetilde{\theta}-\theta\right)'\right)\right] \\ &= \operatorname{tr}\left(V^{-1}\mathbb{E}\left[\left(\widetilde{\theta}-\theta\right)\left(\widetilde{\theta}-\theta\right)'\right]\right) \\ &= \operatorname{bias}\left[\widetilde{\theta}\right]' V^{-1} \operatorname{bias}\left[\widetilde{\theta}\right] + \operatorname{tr}\left(V^{-1}\operatorname{var}\left[\widetilde{\theta}\right]\right).\end{aligned}$$

The first equality uses the fact that a scalar is a 1×1 matrix and thus equals the trace of that matrix. The second equality uses the property $\operatorname{tr}(\boldsymbol{AB}) = \operatorname{tr}(\boldsymbol{BA})$. The third equality uses the fact that the trace is a linear operator, so we can exchange expectation with the trace. (For more information on the trace operator, see Section A.11 in the Appendix.) The final line sets $\operatorname{bias}\left[\widetilde{\theta}\right] = \mathbb{E}\left[\widetilde{\theta}-\theta\right]$.

Now consider the MSE of the initial estimator $\widehat{\theta}$, which is assumed to be unbiased and has variance $\boldsymbol{V}$. Its weighted MSE is $\operatorname{mse}\left[\widehat{\theta}\right] = \operatorname{tr}\left(V^{-1}\operatorname{var}\left[\widehat{\theta}\right]\right) = \operatorname{tr}\left(V^{-1}V\right) = K$.

Theorem 15.1 $\operatorname{mse}\left[\widehat{\theta}\right] = K$.

We are interested in finding an estimator with reduced MSE. This means finding an estimator whose weighted MSE is less than K.

15.3 SHRINKAGE

We focus on shrinkage estimators, which shrink the initial estimator toward the zero vector. The simplest shrinkage estimator takes the form $\widetilde{\theta} = (1-w)\widehat{\theta}$ for some shrinkage weight $w \in [0, 1]$. Setting $w = 0$, we obtain $\widetilde{\theta} = \widehat{\theta}$ (no shrinkage) and setting $w = 1$, we obtain $\widetilde{\theta} = 0$ (full shrinkage). It is straightforward to calculate the MSE of this estimator. Recall that $\widehat{\theta} \sim (\theta, \boldsymbol{V})$. The shrinkage estimator $\widetilde{\theta}$ has bias

$$\operatorname{bias}\left[\widetilde{\theta}\right] = \mathbb{E}\left[\widetilde{\theta}\right] - \theta = \mathbb{E}\left[(1-w)\widehat{\theta}\right] - \theta = -w\theta, \tag{15.1}$$

and variance

$$\operatorname{var}\left[\widetilde{\theta}\right] = \operatorname{var}\left[(1-w)\widehat{\theta}\right] = (1-w)^2 \boldsymbol{V}. \tag{15.2}$$

Its MSE equals

$$\begin{aligned}\operatorname{mse}\left[\widetilde{\theta}\right] &= \operatorname{bias}\left[\widetilde{\theta}\right]' V^{-1} \operatorname{bias}\left[\widetilde{\theta}\right] + \operatorname{tr}\left(V^{-1}\operatorname{var}\left[\widetilde{\theta}\right]\right) \\ &= w\theta' V^{-1} w\theta + \operatorname{tr}\left(V^{-1}V(1-w)^2\right) \\ &= w^2\theta' V^{-1}\theta + (1-w)^2 \operatorname{tr}(\boldsymbol{I}_K) \\ &= w^2\lambda + (1-w)^2 K \end{aligned} \tag{15.3}$$

where $\lambda = \theta' V^{-1}\theta$. We deduce the following.

Theorem 15.2 If $\widehat{\theta} \sim (\theta, \boldsymbol{V})$ and $\widetilde{\theta} = (1-w)\widehat{\theta}$, then

1. $\operatorname{mse}\left[\widetilde{\theta}\right] < \operatorname{mse}\left[\widehat{\theta}\right]$ if $0 < w < 2K/(K+\lambda)$.
2. $\operatorname{mse}\left[\widetilde{\theta}\right]$ is minimized by the shrinkage weight $w_0 = K/(K+\lambda)$.
3. The minimized MSE is $\operatorname{mse}\left[\widetilde{\theta}\right] = K\lambda/(K+\lambda)$.

Part 1 of the theorem shows that the shrinkage estimator has reduced MSE for a range of values of the shrinkage weight w. Part 2 of the theorem shows that the MSE-minimizing shrinkage weight is a simple function of K and λ. The latter is a measure of the magnitude of θ relative to the estimation variance. When λ is large (the coefficients are large), then the optimal shrinkage weight w_0 is small; when λ is small (the coefficients are small), then the optimal shrinkage weight w_0 is large. Part 3 calculates the associated optimal MSE. This can be substantially less than the MSE of the sample mean $\widehat{\theta}$. For example, if $\lambda = K$, then $\operatorname{mse}\left[\widetilde{\theta}\right] = K/2$, one-half of the MSE of $\widehat{\theta}$.

The optimal shrinkage weight is infeasible, since we do not know λ. Therefore this result is tantilizing but not a receipe for empirical practice.

Let's consider a feasible version of the shrinkage estimator. A plug-in estimator for λ is $\widehat{\lambda} = \widehat{\theta}' \boldsymbol{V}^{-1} \widehat{\theta}$, which is biased for λ, as it has expectation $\lambda + K$. An unbiased estimator is $\widetilde{\lambda} = \widehat{\theta}' \boldsymbol{V}^{-1} \widehat{\theta} - K$. If this estimator is plugged into the formula for the optimal weight, we obtain

$$\widehat{w} = \frac{K}{K + \widetilde{\lambda}} = \frac{K}{\widehat{\theta}' \boldsymbol{V}^{-1} \widehat{\theta}}. \tag{15.4}$$

Replacing the numerator K with a free parameter c (which we call the **shrinkage coefficient**), we obtain a feasible shrinkage estimator

$$\widetilde{\theta} = \left(1 - \frac{c}{\widehat{\theta}' \boldsymbol{V}^{-1} \widehat{\theta}}\right) \widehat{\theta}. \tag{15.5}$$

This class of estimators is known as **Stein-Rule estimators.**

15.4 JAMES-STEIN SHRINKAGE ESTIMATOR

Take any estimator that satisfies $\widehat{\theta} \sim \mathrm{N}(\theta, \boldsymbol{V})$. This includes the sample mean in the normal sampling model. James and Stein (1961) made the following discovery.

Theorem 15.3 (James-Stein) Assume $K > 2$. For $\widetilde{\theta}$ defined in equation (15.5) and $\widehat{\theta} \sim \mathrm{N}(\theta, \boldsymbol{V})$:

1. $\operatorname{mse}\left[\widetilde{\theta}\right] = \operatorname{mse}\left[\widehat{\theta}\right] - c\,(2(K-2) - c)\, J_K$, where

$$J_K = \mathbb{E}\left[\frac{1}{\widehat{\theta}' \boldsymbol{V}^{-1} \widehat{\theta}}\right] > 0.$$

2. $\operatorname{mse}\left[\widetilde{\theta}\right] < \operatorname{mse}\left[\widehat{\theta}\right]$, if $0 < c < 2(K-2)$.
3. $\operatorname{mse}\left[\widetilde{\theta}\right]$ is minimized by $c = K - 2$ and equals $\operatorname{mse}\left[\widetilde{\theta}\right] = K - (K-2)^2 J_K$.

The proof of part 1 is presented in Section 15.9.

Part 1 of the theorem gives an expression for the MSE of the shrinkage estimator (15.5). The expression depends on the number of parameters K, the shrinkage constant c, and the expectation J_K. (An explicit formula for J_K is given in Section 15.5.)

Part 2 of the theorem shows that the MSE of $\widetilde{\theta}$ is strictly less than $\widehat{\theta}$ for a range of values of the shrinkage coefficient. This is a strict inequality, meaning that the Stein Rule estimator has strictly smaller MSE. The inequality holds for all values of the parameter θ, meaning that this is a uniform strict inequality. Thus the Stein Rule estimator uniformly dominates the estimator $\widehat{\theta}$. Part 2 follows from part 1, the fact $\operatorname{mse}\left[\widehat{\theta}\right] = K$, and $J_K > 0$.

The Stein Rule estimator depends on the choice of shrinkage parameter c. Part 3 shows that the MSE-minimizing choice is $c = K - 2$. This follows from part 1 and minimizing over c and leads to the explicit feasible estimator

$$\widetilde{\theta}_{\mathrm{JS}} = \left(1 - \frac{K-2}{\widehat{\theta}' \boldsymbol{V}^{-1} \widehat{\theta}}\right) \widehat{\theta}. \tag{15.6}$$

This is called the **James-Stein estimator**.

Theorem 15.3 stunned the world of statistics. In the normal sampling model, $\widehat{\theta}$ is Cramér-Rao efficient. Theorem 15.3 shows that the shrinkage estimator $\widetilde{\theta}$ dominates the MLE $\widehat{\theta}$. This result is stunning, because it had been previously assumed that it would be impossible to find an estimator that dominates a Cramér-Rao efficient MLE.

Theorem 15.3 critically depends on the condition $K > 2$, which means that shrinkage achieves uniform MSE reductions in dimensions three or higher.

In practice, $\boldsymbol{V}$ is unknown, so we substitute an estimator $\widehat{\boldsymbol{V}}$, leading to

$$\widetilde{\theta}_{\mathrm{JS}} = \left(1 - \frac{K-2}{\widehat{\theta}' \widehat{\boldsymbol{V}}^{-1} \widehat{\theta}}\right) \widehat{\theta}.$$

The substitution of $\widehat{\boldsymbol{V}}$ for $\boldsymbol{V}$ can be justified by finite sample or asymptotic arguments, but we do not do so here.

The proof of Theorem 15.3 uses a clever result known as Stein's Lemma. It is a simple yet famous application of integration by parts.

Theorem 15.4 **Stein's Lemma**. If $X \sim \mathrm{N}(\theta, \boldsymbol{V})$ and $g(x) : \mathbb{R}^K \to \mathbb{R}^K$ is absolutely continuous, then

$$\mathbb{E}\left[g(X)' \boldsymbol{V}^{-1} (X - \theta)\right] = \mathbb{E}\left[\operatorname{tr}\left(\frac{\partial}{\partial x} g(X)'\right)\right].$$

The proof is presented in Section 15.9.

15.5 NUMERICAL CALCULATION

We numerically calculate the MSE of $\widetilde{\theta}_{\mathrm{JS}}$ (15.6) and plot the results in Figure 15.1. As shown below, J_K is a function of $\lambda = \theta' \boldsymbol{V}^{-1} \theta$.

We plot $\operatorname{mse}\left[\widetilde{\theta}\right]/K$ as a function of λ/K for $K = 4$, 6, 12, and 48. The plots are uniformly below 1 (the normalized MSE of the MLE) and are substantially so for small and moderate values of λ. The MSE falls as K increases, demonstrating that the MSE reductions are more substantial when K is large. The MSE increases with λ, appearing to asymptote to K, which is the MSE of $\widehat{\theta}$.

The plots require numerical evaluation of J_K. A computational formula requires a technical calculation.

From Theorem 5.23 the random variable $\widehat{\theta}' \boldsymbol{V}^{-1} \widehat{\theta}$ is distributed as $\chi^2_K(\lambda)$, a non-central chi-square random variable with degree of freedom K and non-centrality parameter λ, with density defined in (3.4). Using this definition, we can calculate an explicit formula for J_K.

Theorem 15.5 For $K > 2$

$$J_K = \sum_{i=0}^{\infty} \frac{e^{-\lambda/2}}{i!} \frac{(\lambda/2)^i}{K + 2i - 2}. \tag{15.7}$$

The proof is presented in Section 15.9.

This sum is a convergent series. For the computations reported in Figure 15.1, J_K was computed by evaluating the first 200 terms in the series.

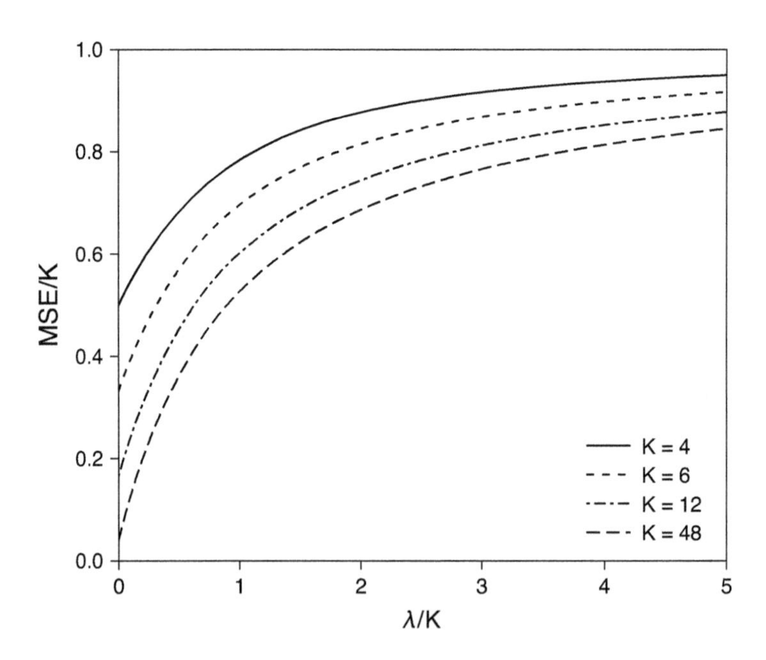

FIGURE 15.1 MSE of james-stein estimator

15.6 INTERPRETATION OF THE STEIN EFFECT

The James-Stein Theorem (Theorem 15.3) may appear to conflict with our previous efficiency theory. The sample mean $\widehat{\theta}$ is the maximum likelihood estimator. It is unbiased. It is minimum variance unbiased. It is Cramér-Rao efficient. How can it be that the James-Stein shrinkage estimator achieves uniformly smaller MSE?

Part of the answer is that the previous theory has caveats. The Cramér-Rao Theorem (Theorem 10.6) restricts attention to unbiased estimators and thus precludes consideration of shrinkage estimators. The James-Stein estimator has reduced MSE, but is not Cramér-Rao efficient, since it is biased. Therefore the James-Stein Theorem (Theorem 15.3) does not conflict with the Cramér-Rao Theorem. Instead, they are complementary. On the one hand, the Cramér-Rao Theorem describes the best possible variance when unbiasedness is required. On the other hand, the James-Stein Theorem shows that if unbiasedness is relaxed, there are lower MSE estimators than the MLE.

The MSE improvements achieved by the James-Stein estimator are greatest when λ is small. This occurs when the parameters θ are small in magnitude relative to the estimation variance $\boldsymbol{V}$. Thus the user needs to choose the centering point wisely.

15.7 POSITIVE-PART ESTIMATOR

The simple James-Stein estimator has the odd property that it can "over-shrink". When $\widehat{\theta}'\boldsymbol{V}^{-1}\widehat{\theta} < K-2$, then $\widetilde{\theta}$ has the opposite sign from $\widehat{\theta}$. This behavior does not make sense and suggests that further improvements can be made to the estimator. The standard solution is to use "positive-part" trimming by bounding the shrinkage weight (15.4) to be less than 1. This estimator can be written as

$$\widetilde{\theta}^{+}=\begin{cases}\widetilde{\theta}, & \widehat{\theta}'\boldsymbol{V}^{-1}\widehat{\theta}\geq K-2\\ 0, & \widehat{\theta}'\boldsymbol{V}^{-1}\widehat{\theta}<K-2\end{cases}$$
$$=\left(1-\frac{K-2}{\widehat{\theta}'\boldsymbol{V}^{-1}\widehat{\theta}}\right)_{+}\widehat{\theta}$$

where $(a)_{+}=\max[a,0]$ is the "positive-part" function. Alternatively, it can be written as

$$\widetilde{\theta}^{+}=\widehat{\theta}-\left(\frac{K-2}{\widehat{\theta}'\boldsymbol{V}^{-1}\widehat{\theta}}\right)_{1}\widehat{\theta}$$

where $(a)_{1}=\min[a,1]$.

The positive-part estimator simultaneously performs "selection" as well as "shrinkage". If $\widehat{\theta}'\boldsymbol{V}^{-1}\widehat{\theta}$ is sufficiently small, $\widetilde{\theta}^{+}$ "selects" 0. When $\widehat{\theta}'\boldsymbol{V}^{-1}\widehat{\theta}$ is of moderate size, $\widetilde{\theta}^{+}$ shrinks $\widehat{\theta}$ toward 0. When $\widehat{\theta}'\boldsymbol{V}^{-1}\widehat{\theta}$ is very large, $\widetilde{\theta}^{+}$ is close to the original estimator $\widehat{\theta}$.

We now show that the positive-part estimator has uniformly lower MSE than the unadjusted James-Stein estimator.

Theorem 15.6 Under the assumptions of Theorem 15.3,

$$\operatorname{mse}\left[\widetilde{\theta}^{+}\right]<\operatorname{mse}\left[\widetilde{\theta}\right]. \tag{15.8}$$

Theorem 15.7 provides an expression for the MSE.

Theorem 15.7 Under the assumptions of Theorem 15.3,

$$\operatorname{mse}\left[\widetilde{\theta}^{+}\right]=\operatorname{mse}\left[\widetilde{\theta}\right]-2KF_{K}\left(K-2,\lambda\right)+KF_{K+2}(K-2,\lambda)+\lambda F_{K+4}(K-2,\lambda)$$
$$+(K-2)^{2}\sum_{i=0}^{\infty}\frac{e^{-\lambda/2}}{i!}\left(\frac{\lambda}{2}\right)^{i}\frac{F_{K+2i-2}\left(K-2\right)}{K+2i-2}$$

where $F_r(x)$ and $F_r(x,\lambda)$ are the distribution functions of the central chi-square and non-central chi-square, respectively.

The proofs of Theorems 15.6 and 15.7 are provided in Section 15.9.

To illustrate the improvement in MSE, Figure 15.2 plots the MSE of the unadjusted ("JS" in the figure) and positive-part (JS+) James-Stein estimators for $K=4$ and $K=12$. For $K=4$, the positive-part estimator has meaningfully reduced MSE relative to the unadjusted estimator, especially for small values of λ. For $K=12$, the difference between the estimators is smaller.

In summary, the positive-part transformation is an important improvement over the unadjusted James-Stein estimator. It is more reasonable and reduces the MSE. The broader message is that imposing boundary conditions can often help regularize estimators and improve their performance.

15.8 SUMMARY

The James-Stein estimator is a specific shrinkage estimator that has well-studied efficiency properties. It is not the only shrinkage or shrinkage-type estimator. Methods described as "model selection" and "machine learning" all involve shrinkage techniques. These methods are becoming increasingly popular in applied statistics and econometrics. At the core, all involve an essential bias-variance trade-off. By reducing the parameter

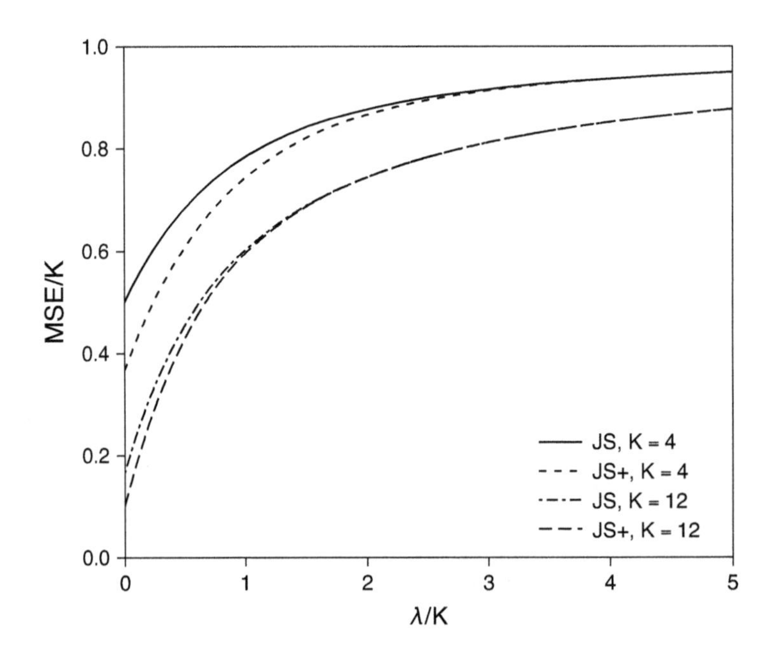

FIGURE 15.2 MSE of positive-part james-stein estimator

space or shrinking toward a preselected point, we can reduce estimation variance. Bias is introduced, but the trade-off may be beneficial.

15.9 TECHNICAL PROOFS*

Proof of Theorem 15.3, part 1 Using the definition of $\widetilde{\theta}$ and expanding the quadratic gives

$$\begin{aligned}
\operatorname{mse}\left[\widetilde{\theta}\right] &= \mathbb{E}\left[\left(\widetilde{\theta}-\theta\right)' \boldsymbol{V}^{-1}\left(\widetilde{\theta}-\theta\right)\right] \\
&= \mathbb{E}\left[\left(\widehat{\theta}-\theta-\widehat{\theta}\frac{c}{\widehat{\theta}'\boldsymbol{V}^{-1}\widehat{\theta}}\right)' \boldsymbol{V}^{-1}\left(\widehat{\theta}-\theta-\widehat{\theta}\frac{c}{\widehat{\theta}'\boldsymbol{V}^{-1}\widehat{\theta}}\right)\right] \\
&= \mathbb{E}\left[\left(\widehat{\theta}-\theta\right)' \boldsymbol{V}^{-1}\left(\widehat{\theta}-\theta\right)\right] + c^2\mathbb{E}\left[\frac{1}{\widehat{\theta}'\boldsymbol{V}^{-1}\widehat{\theta}}\right] - 2c\mathbb{E}\left[\frac{\widehat{\theta}'\boldsymbol{V}^{-1}\left(\widehat{\theta}-\theta\right)}{\widehat{\theta}'\boldsymbol{V}^{-1}\widehat{\theta}}\right].
\end{aligned} \tag{15.9}$$

Take the first term in (15.9). It equals $\operatorname{mse}\left[\widehat{\theta}\right]=K$. The second term in (15.9) is $c^2 J_K$. The third term equals $-2c$ multiplied by $\mathbb{E}\left[g\left(\widehat{\theta}\right)' \boldsymbol{V}^{-1}\left(\widehat{\theta}-\theta\right)\right]$, where we define $g(x)=x\left(x'\boldsymbol{V}^{-1}x\right)^{-1}$. Using the rules of matrix differentiation, we have

$$\operatorname{tr}\left(\frac{\partial}{\partial x} g(x)'\right) = \operatorname{tr}\left(\boldsymbol{I}_K\left(x'\boldsymbol{V}^{-1}x\right)^{-1} - 2\boldsymbol{V}^{-1}xx'\left(x'\boldsymbol{V}^{-1}x\right)^{-2}\right) = \frac{K-2}{x'\boldsymbol{V}^{-1}x}. \tag{15.10}$$

Applying Stein's Lemma, we find

$$\mathbb{E}\left[g\left(\widehat{\theta}\right)' \boldsymbol{V}^{-1}\left(\widehat{\theta}-\theta\right)\right] = \mathbb{E}\left[\operatorname{tr}\left(\frac{\partial}{\partial x} g\left(\widehat{\theta}\right)'\right)\right]$$
$$= \mathbb{E}\left[\frac{K-2}{\widehat{\theta}' \boldsymbol{V}^{-1} \widehat{\theta}}\right]$$
$$= (K-2) J_K.$$

The second line is equation (15.10), and the final line uses the definition of J_K. We have shown that the third term in equation (15.9) is $-2c(K-2)J_K$. Summing the three terms, we find that equation (15.9) equals

$$\operatorname{mse}\left[\widetilde{\theta}\right] = \operatorname{mse}\left[\widehat{\theta}\right] + c^2 J_K - 2c(K-2)J_K = \operatorname{mse}\left[\widehat{\theta}\right] - c\left(2(K-2) - c\right) J_K$$

which is part 1 of the theorem. ■

Proof of Theorem 15.4 For simplicity, assume $\boldsymbol{V} = \boldsymbol{I}_K$.

Since the multivariate normal density is $\phi(x) = (2\pi)^{-K/2} \exp\left(-x'x/2\right)$, then

$$\frac{\partial}{\partial x} \phi(x-\theta) = -(x-\theta)\,\phi(x-\theta).$$

By integration by parts

$$\mathbb{E}\left[g(X)'(X-\theta)\right] = \int g(x)'(x-\theta)\,\phi(x-\theta)\,dx$$
$$= -\int g(x)' \frac{\partial}{\partial x} \phi(x-\theta) dx$$
$$= \int \operatorname{tr}\left(\frac{\partial}{\partial x} g(x)'\right) \phi(x-\theta)\,dx$$
$$= \mathbb{E}\left[\operatorname{tr}\left(\frac{\partial}{\partial x} g(X)'\right)\right].$$

This is the stated result. ■

Proof of Theorem 15.5 Using equation (3.4) and integrating term-by-term, we find

$$J_K = \int_0^\infty x^{-1} f_K(x,\lambda) dx = \sum_{i=0}^\infty \frac{e^{-\lambda/2}}{i!} \left(\frac{\lambda}{2}\right)^i \int_0^\infty x^{-1} f_{K+2i}(x) dx.$$

The chi-square density satisfies $x^{-1} f_r(x) = (r-2)^{-1} f_{r-2}(x)$ for $r > 2$. Thus

$$\int_0^\infty x^{-1} f_r(x) dx = \frac{1}{r-2} \int_0^\infty f_{r-2}(x) dx = \frac{1}{r-2}.$$

Making this substitution, we find (15.7). ■

Proof of Theorem 15.6 It will be convenient to denote $Q_K = \widehat{\theta}' \boldsymbol{V}^{-1}\widehat{\theta}$. Observe that

$$\begin{aligned}
\left(\widetilde{\theta}-\theta\right)' \boldsymbol{V}^{-1}\left(\widetilde{\theta}-\theta\right)-\left(\widetilde{\theta}^{+}-\theta\right)' \boldsymbol{V}^{-1}\left(\widetilde{\theta}^{+}-\theta\right) &= \widetilde{\theta}' \boldsymbol{V}^{-1}\widetilde{\theta}-\widetilde{\theta}^{+\prime} \boldsymbol{V}^{-1}\widetilde{\theta}^{+}+2\theta' \boldsymbol{V}^{-1}\left(\widetilde{\theta}-\widetilde{\theta}^{+}\right) \\
&= \left(\widetilde{\theta}' \boldsymbol{V}^{-1}\widetilde{\theta}-\widetilde{\theta}^{+\prime} \boldsymbol{V}^{-1}\widetilde{\theta}^{+}+2\theta' \boldsymbol{V}^{-1}\left(\widetilde{\theta}-\widetilde{\theta}^{+}\right)\right)\mathbb{1}\left\{Q_K<K-2\right\} \\
&= \left(\widetilde{\theta}' \boldsymbol{V}^{-1}\widetilde{\theta}+2\theta' \boldsymbol{V}^{-1}\widetilde{\theta}\right)\mathbb{1}\left\{Q_K<K-2\right\} \\
&\geq 2\theta' \boldsymbol{V}^{-1}\widetilde{\theta}\mathbb{1}\left\{Q_K<K-2\right\} \\
&= 2\theta' \boldsymbol{V}^{-1}\widehat{\theta}\left(1-\frac{(K-2)}{Q_K}\right)\mathbb{1}\left\{Q_K<K-2\right\} \\
&\geq 0.
\end{aligned}$$

The second equality holds because the expression is identically 0 on the event $Q_K \geq K-2$. The third equality holds because $\widetilde{\theta}^{+}=0$ on the event $Q_K<K-2$. The first inequality is strict unless $\widetilde{\theta}=0$, which is a probability 0 event. The following equality substitutes the definition of $\widetilde{\theta}$. The final inequality is the fact that $(K-2)/Q_K>1$ on the event $Q_K \geq K-2$.

Taking expectations, we find $\operatorname{mse}\left[\widetilde{\theta}\right]-\operatorname{mse}\left[\widetilde{\theta}^{+}\right]>0$, as claimed. ■

Proof of Theorem 15.7 It will be convenient to denote $Q_K=\widehat{\theta}' \boldsymbol{V}^{-1}\widehat{\theta} \sim \chi_K^2(\lambda)$. The estimator can be written as

$$\widetilde{\theta}^{+}=\widehat{\theta}-\frac{(K-2)}{Q_K}\widehat{\theta}-h(\widehat{\theta})$$

where

$$h(x)=x\left(1-\frac{(K-2)}{x' \boldsymbol{V}^{-1} x}\right)\mathbb{1}\left\{x' \boldsymbol{V}^{-1} x<K-2\right\}.$$

Observe that

$$\begin{aligned}
\operatorname{tr}\left(\frac{\partial}{\partial x} h(x)'\right) &= \operatorname{tr}\left(\boldsymbol{I}_K\left(1-\frac{K-2}{x' \boldsymbol{V}^{-1} x}\right)-2\left(\frac{K-2}{\left(x' \boldsymbol{V}^{-1} x\right)^2}\right)\boldsymbol{V}^{-1} x x'\right)\mathbb{1}\left\{x' \boldsymbol{V}^{-1} x<K-2\right\} \\
&= \left(K-\frac{(K-2)^2}{x' \boldsymbol{V}^{-1} x}\right)\mathbb{1}\left\{x' \boldsymbol{V}^{-1} x<K-2\right\}.
\end{aligned}$$

Using these expressions, the definition of $\widetilde{\theta}$, and expanding the quadratic, we have

$$\begin{aligned}
\operatorname{mse}\left[\widetilde{\theta}^{+}\right] &= \mathbb{E}\left[\left(\widetilde{\theta}-\theta\right)' \boldsymbol{V}^{-1}\left(\widetilde{\theta}-\theta\right)\right] \\
&\quad+\mathbb{E}\left[h(\widehat{\theta})' \boldsymbol{V}^{-1} h(\widehat{\theta})\right] \\
&\quad+\mathbb{E}\left[\frac{2(K-2)}{Q_K}\widehat{\theta}' \boldsymbol{V}^{-1} h(\widehat{\theta})\right] \\
&\quad-2\mathbb{E}\left[h(\widehat{\theta})' \boldsymbol{V}^{-1}\left(\widehat{\theta}-\theta\right)\right].
\end{aligned}$$

Evaluating the four terms, we have

$$\mathbb{E}\left[\left(\widetilde{\theta}-\theta\right)' \boldsymbol{V}^{-1}\left(\widetilde{\theta}-\theta\right)\right]=\operatorname{mse}\left[\widetilde{\theta}\right]$$

$$\mathbb{E}\left[h(\widehat{\theta})' \boldsymbol{V}^{-1} h(\widehat{\theta})\right]=\mathbb{E}\left[\left(Q_K-2(K-2)+\frac{(K-2)^2}{Q_K}\right) \mathbb{1}\left\{Q_K<K-2\right\}\right]$$

$$\mathbb{E}\left[\frac{2(K-2)}{Q_K} \widehat{\theta}' \boldsymbol{V}^{-1} h(\widehat{\theta})\right]=\mathbb{E}\left[\left(2(K-2)-2 \frac{(K-2)^2}{Q_K}\right) \mathbb{1}\left\{Q_K<K-2\right\}\right]$$

$$\mathbb{E}\left[h(\widehat{\theta})' \boldsymbol{V}^{-1}\left(\widehat{\theta}-\theta\right)\right]=\mathbb{E}\left[\left(K-\frac{(K-2)^2}{Q_K}\right) \mathbb{1}\left\{Q_K<K-2\right\}\right]$$

the last using Stein's Lemma. Summing, we find

$$\begin{aligned}
\operatorname{mse}\left[\widetilde{\theta}^{+}\right] & =\operatorname{mse}\left[\widetilde{\theta}\right]-\mathbb{E}\left[\left(2K-Q_K-\frac{(K-2)^2}{Q_K}\right) \mathbb{1}\left\{Q_K<K-2\right\}\right] \\
& =\operatorname{mse}\left[\widetilde{\theta}\right]-2K F_K(K-2, \lambda)+\int_0^{K-2} x f_K(x, \lambda) d x+(K-2)^2 \int_0^{K-2} x^{-1} f_K(x, \lambda) d x . \qquad (15.11)
\end{aligned}$$

By manipulating the definition of the non-central chi-square density (3.4) we can show the identity

$$f_K(x, \lambda)=\frac{K}{x} f_{K+2}(x, \lambda)+\frac{\lambda}{x} f_{K+4}(x, \lambda). \qquad (15.12)$$

This allows us to calculate that

$$\begin{aligned}
\int_0^{K-2} x f_K(x, \lambda) d x & =K \int_0^{K-2} f_{K+2}(x, \lambda) d x+\lambda \int_0^{K-2} f_{K+4}(x, \lambda) d x \\
& =K F_{K+2}(K-2, \lambda)+\lambda F_{K+4}(K-2, \lambda).
\end{aligned}$$

Writing out the non-central chi-square density (3.4) and applying (15.12) with $\lambda=0$, we calculate that

$$\begin{aligned}
\int_0^{K-2} x^{-1} f_K(x, \lambda) d x & =\sum_{i=0}^{\infty} \frac{e^{-\lambda / 2}}{i!}\left(\frac{\lambda}{2}\right)^i \int_0^{K-2} x^{-1} f_{K+2i}(x) d x \\
& =\sum_{i=0}^{\infty} \frac{e^{-\lambda / 2}}{i!}\left(\frac{\lambda}{2}\right)^i \frac{1}{K+2i-2} \int_0^{K-2} f_{K+2i-2}(x) d x \\
& =\sum_{i=0}^{\infty} \frac{e^{-\lambda / 2}}{i!}\left(\frac{\lambda}{2}\right)^i \frac{F_{K+2i-2}(K-2)}{K+2i-2}.
\end{aligned}$$

Together, these results show that

$$\begin{aligned}
\operatorname{mse}\left[\widetilde{\theta}^{+}\right]= & \operatorname{mse}\left[\widetilde{\theta}\right]-2K F_K(K-2, \lambda)+K F_{K+2}(K-2, \lambda)+\lambda F_{K+4}(K-2, \lambda) \\
& +(K-2)^2 \sum_{i=0}^{\infty} \frac{e^{-\lambda / 2}}{i!}\left(\frac{\lambda}{2}\right)^i \frac{F_{K+2i-2}(K-2)}{K+2i-2}
\end{aligned}$$

as claimed. ■

15.10 EXERCISES

Exercise 15.1 Let $\overline{X}_n$ be the sample mean from a random sample. Consider the estimators $\widehat{\theta} = \overline{X}_n$, $\widetilde{\theta} = \overline{X}_n - c$, and $\overline{\theta} = c\overline{X}_n$ for $0 < c < 1$.

(a) Calculate the bias and variance of the three estimators.

(b) Compare the estimators based on MSE.

Exercise 15.2 Let $\widehat{\theta}$ and $\widetilde{\theta}$ be two estimators. Suppose $\widehat{\theta}$ is unbiased. Suppose $\widetilde{\theta}$ is biased but has variance that is 0.09 less than $\widehat{\theta}$.

(a) If $\widetilde{\theta}$ has a bias of -0.1, which estimator is preferred based on MSE?

(b) Find the level of the bias of $\widetilde{\theta}$ at which the two estimators have the same MSE.

Exercise 15.3 For scalar $X \sim \mathrm{N}(\theta, \sigma^2)$, use Stein's Lemma (Theorem 15.4) to calculate the following:

(a) $\mathbb{E}\left[g(X)(X-\theta)\right]$ for scalar continuous $g(x)$.
(b) $\mathbb{E}\left[X^3(X-\theta)\right]$.
(c) $\mathbb{E}\left[\sin(X)(X-\theta)\right]$.
(d) $\mathbb{E}\left[\exp(tX)(X-\theta)\right]$.
(e) $\mathbb{E}\left[\left(\frac{X}{1+X^2}\right)(X-\theta)\right]$.

Exercise 15.4 Assume $\widehat{\theta} \sim \mathrm{N}(0, V)$. Calculate the following:

(a) λ.
(b) J_K in Theorem 15.5.
(c) $\operatorname{mse}\left[\widehat{\theta}\right]$.
(d) $\operatorname{mse}\left[\widetilde{\theta}_{\mathrm{JS}}\right]$.

Exercise 15.5 Bock (1975, Theorem A) established that if $X \sim \mathrm{N}(\theta, \boldsymbol{I}_K)$, then for any scalar function $g(x)$,

$$\mathbb{E}\left[Xh\left(X'X\right)\right] = \theta\mathbb{E}\left[h\left(Q_{K+2}\right)\right]$$

where $Q_{K+2} \sim \chi^2_{K+2}(\lambda)$. Assume $\widehat{\theta} \sim \mathrm{N}(\theta, \boldsymbol{I}_K)$ and $\widetilde{\theta}_{\mathrm{JS}} = \left(1 - \frac{K-2}{\widehat{\theta}'\widehat{\theta}}\right)\widehat{\theta}$.

(a) Use Bock's result to calculate $\mathbb{E}\left[\frac{1}{\widehat{\theta}'\widehat{\theta}}\widehat{\theta}\right]$. You can represent your answer using the function J_K.
(b) Calculate the bias of $\widetilde{\theta}_{\mathrm{JS}}$.
(c) Describe the bias. Is the estimator biased downward, upward, or toward 0?

CHAPTER 16

BAYESIAN METHODS

16.1 INTRODUCTION

The statistical methods reviewed up to this point are called **classical** or **frequentist**. An alternative class of statistical methods is called **Bayesian**. Frequentist statistical theory takes the probability distribution and parameters as fixed unknowns. Bayesian statistical theory treats the parameters of the probability model as random variables and conducts inference on the parameters conditional on the observed sample. There are advantages and disadvantages to each approach.

Bayesian methods can be briefly summarized as follows. The user selects a probability model $f(x \mid \theta)$, as in Chapter 10. This implies a joint density $L_n(x \mid \theta)$ for the sample. The user also specifies a prior distribution $\pi(\theta)$ for the parameters. Together these choices imply a joint distribution $L_n(x \mid \theta)\, \pi(\theta)$ for the random variables and parameters. The conditional distribution of the parameters θ given the observed data X (calculated by Bayes Rule) is called the "posterior distribution" $\pi(\theta \mid X)$. In Bayes analysis, the posterior summarizes the information about the parameter given the data, model, and prior. The standard Bayes estimator of θ is the posterior mean $\widehat{\theta} = \int_{\Theta} \theta \pi(\theta \mid X)\, d\theta$. Interval estimators are constructed as credible sets, whose coverage probabilities are calculated based on the posterior. The standard Bayes credible set is constructed from the principle of highest posterior density. Bayes tests select hypotheses based on which has the greatest posterior probability of truth. An important class of prior distributions are conjugate priors, which are distributional families matched with specific likelihoods such that the posterior and prior both are members of the same distributional family.

One advantage of Bayesian methods is that they provide a unified, coherent approach for estimation and inference. The difficulties of complicated sampling distributions are avoided. A second advantage is that Bayesian methods induce shrinkage similar to James-Stein estimation. This can be sensibly employed to improve precision relative to MLE.

One possible disadvantage of Bayesian methods is that they can be computationally burdensome. Once you move beyond simple textbook models, closed-form posteriors and estimators cease to exist, and numerical methods must be employed. The good news is that Bayesian econometrics has made great advances in computational implementation, so this disadvantage should not be viewed as a barrier. A second disadvantage of Bayesian methods is that the results are by construction dependent on the choice of prior distribution. Most Bayesian applications select their prior based on mathematical convenience, which implies that the results are a by-product of this arbitrary selection. When the data have little information about the parameters, the Bayes posterior will be similar to the prior, meaning that inferential statements are about the prior, not about the actual world. A third disadvantage of Bayesian methods is that they are built for parametric models. To a first approximation, Bayesian methods are an analog of MLE (not an analog of the method of moments). In contrast, most econometric models are either semi-parametric or nonparametric. There are extensions of Bayesian methods to handle nonparametric models, but they are not the norm. A fourth disadvantage is that it is difficult to robustify Bayesian inference to allow unmodeled dependence (clustered or time series) among the observations, or to allow for misspecification. Dependence (modeled and unmodeled) is not discussed in this textbook, but it is an important topic in *Econometrics*.

A good introduction to Bayesian econometric methods is Koop, Poirier, and Tobias (2007). For theoretical properties, see Lehmann and Casella (1998) and van der Vaart (1998).

16.2 BAYESIAN PROBABILITY MODEL

Bayesian methods are applied in contexts of complete probability models, which means that the user has a probability model $f(x \mid \theta)$ for a random variable X with parameter space $\theta \in \Theta$. In Bayesian analysis, we assume that the model is correctly specified in the sense that the true distribution $f(x)$ is a member of this parametric family.

A key feature of Bayesian inference is that the parameter θ is treated as random. A way to think about this is that anything unknown (e.g., a parameter) is random until the uncertainty is resolved. To do so, it is necessary that the user specify a probability density $\pi(\theta)$ for θ called a **prior**. This is the distribution of θ before the data are observed. The prior should have support equal to Θ. Otherwise, if the support of $\pi(\theta)$ is a strict subset of Θ, this is equivalent to restricting θ to that subset.

The choice of prior distribution is a critically important step in Bayesian analysis. More than one approach can be used for the selection of a prior, including the subjectivist, the objectivist, and the shrinkage approaches. The **subjectivist** approach states that the prior should reflect the user's prior beliefs about the parameter. In effect, the prior should summarize the current state of knowledge. In this case, the goal of estimation and inference is to use the current dataset to improve the state of knowledge. This approach is useful for personal use, business decisions, and policy decisions, when the state of knowledge is well articulated. However it is less clear if it has a role in scientific discourse, where the researcher and reader may have different prior beliefs. The **objectivist** approach states that the prior should be non-informative about the parameter. The goal of estimation and inference is to learn what this dataset tells us about the world. This is a scientific approach. A challenge is that there are multiple ways to describe a prior as non-informative. The **shrinkage** approach uses the prior as a tool to regularize estimation and improve estimation precision. For shrinkage, the prior is selected similar to the restrictions used in Stein-Rule estimation. An advantage of Bayesian shrinkage is that it can be flexibly applied outside the narrow confines of Stein Rule estimation theory. A disadvantage is that there are no guidelines for selecting the degree of shrinkage.

The joint density of X and θ is

$$f(x, \theta) = f(x \mid \theta)\, \pi(\theta).$$

This is a standard factorization of a joint density into the product of a conditional $f(x \mid \theta)$ and marginal $\pi(\theta)$. The implied **marginal density** of X is obtained by integrating over θ:

$$m(x) = \int_{\Theta} f(x, \theta)\, d\theta.$$

As an example, suppose that a model is $X \sim \mathrm{N}(\theta, 1)$ with prior $\theta \sim \mathrm{N}(0, 1)$. From Theorem 5.17, we can calculate that the joint distribution of (X, θ) is bivariate normal with mean vector $(0, 0)'$ and covariance matrix $\begin{bmatrix} 2 & 1 \\ 1 & 1 \end{bmatrix}$. The marginal density of X is $X \sim \mathrm{N}(0, 2)$.

The joint distribution for n independent draws $X=(X_1,\ldots,X_n)$ from $f(x\mid\theta)$ is

$$L_n(x\mid\theta)=\prod_{i=1}^{n}f(x_i\mid\theta)$$

where $x=(x_1,\ldots,x_n)$. The joint density of X and θ is

$$f(x,\theta)=L_n(x\mid\theta)\,\pi(\theta).$$

The marginal density of X is

$$m(x)=\int_\Theta f(x,\theta)\,d\theta=\int_\Theta L_n(x\mid\theta)\,\pi(\theta)d\theta.$$

In simple examples, the marginal density can be straightforward to evaluate. However, in many applications, computation of the marginal density requires some sort of numerical method. This is often an important computational hurdle in applications.

16.3 POSTERIOR DENSITY

The user has a random sample of size n. The maintained assumption is that the observations are i.i.d. draws from the model $f(x\mid\theta)$ for some θ. The joint conditional density $L_n(x\mid\theta)$ evaluated at the observations X is the likelihood function $L_n(X\mid\theta)=L_n(\theta)$. The marginal density $m(x)$ evaluated at the observations X is called the **marginal likelihood** $m(X)$.

Bayes Rule states that the conditional density of the parameters given $X=x$ is

$$\pi(\theta\mid x)=\frac{f(x,\theta)}{\int_\Theta f(x,\theta)\,d\theta}=\frac{L_n(x\mid\theta)\,\pi(\theta)}{m(x)}.$$

This density is conditional on $X=x$. As a result, many authors describe Bayesian analysis as treating the data as fixed. This is not quite right, but if it helps the intuition, that's okay. The correct statement is that Bayesian analysis is conditional on the data.

Evaluated at the observations X, this conditional density is called the **posterior density** of θ, or more simply, the **posterior**:

$$\pi(\theta\mid X)=\frac{f(X,\theta)}{\int_\Theta f(X,\theta)\,d\theta}=\frac{L_n(X\mid\theta)\,\pi(\theta)}{m(X)}.$$

The labels "prior" and "posterior" convey that the prior $\pi(\theta)$ is the distribution for θ before observing the data, and the posterior $\pi(\theta\mid X)$ is the distribution for θ after observing the data.

Take our earlier example of $X\sim\mathrm{N}(\theta,1)$ and $\theta\sim\mathrm{N}(0,1)$. We can calculate from Theorem 5.17 that the posterior density is $\theta\sim\mathrm{N}\left(\frac{X}{2},\frac{1}{2}\right)$.

16.4 BAYESIAN ESTIMATION

Recall that the the maximum likelihood estimator maximizes the likelihood $L(X\mid\theta)$ over $\theta\in\Theta$. In contrast, the standard Bayes estimator is the mean of the posterior density:

$$\widehat{\theta}_{\text{Bayes}}=\int_\Theta\theta\,\pi(\theta\mid X)\,d\theta=\frac{\int_\Theta\theta\,L_n(X\mid\theta)\,\pi(\theta)d\theta}{m(X)}.$$

This is called the **Bayes estimator** or the **posterior mean**.

One way to justify this choice of estimator is as follows. Let $\ell(T,\theta)$ denote the loss associated with using the estimator $T=T(X)$ when θ is the truth. You can think of this as the cost (lost profit, lost utility) induced by using T instead of θ. The **Bayes Risk** of an estimator T is the expected loss calculated using the posterior density $\pi(\theta \mid X)$. This is

$$R(T \mid X)=\int_{\Theta} \ell(T,\theta)\, \pi(\theta \mid X)\, d\theta.$$

This is the average loss weighted by the posterior.

Given a loss function $\ell(T,\theta)$ the optimal Bayes estimator is the one that minimizes the Bayes Risk. In principle, this means that a user can tailor a specialized estimator for a situation where a loss function is well specified by the economic problem. In practice, however, it is can be difficult to find the estimator T that minimizes the Bayes Risk outside a few well-known loss functions.

The most common choice for the loss is quadratic:

$$\ell(T,\theta)=(T-\theta)'(T-\theta).$$

In this case, the Bayes Risk is

$$\begin{aligned}
R(T \mid X) &= \int_{\Theta} (T-\theta)'(T-\theta)\, \pi(\theta \mid X)\, d\theta. \\
&= T'T\int_{\Theta} \pi(\theta \mid X)\, d\theta - 2T'\int_{\Theta} \theta\pi(\theta \mid X)\, d\theta + \int_{\Theta} \theta'\theta\pi(\theta \mid X)\, d\theta \\
&= T'T - 2T'\widehat{\theta}_{\text{Bayes}} + \widehat{\mu}_2
\end{aligned}$$

where $\widehat{\theta}_{\text{Bayes}}$ is the posterior mean, and $\widehat{\mu}_2=\int_{\Theta} \theta'\theta\pi(\theta \mid X)\, d\theta$. This risk is linear-quadratic in T. The FOC for minimization is $0=2T-2\widehat{\theta}_{\text{Bayes}}$, which implies $T=\widehat{\theta}_{\text{Bayes}}$. Thus the posterior mean minimizes the Bayes Risk under quadratic loss.

As another example, for scalar θ, consider absolute loss $\ell(T,\theta)=|T-\theta|$. In this case, the Bayes Risk is

$$R(T \mid X)=\int_{\Theta} |T-\theta|\, \pi(\theta \mid X)\, d\theta.$$

which is minimized by

$$\widetilde{\theta}_{\text{Bayes}}=\text{median}(\theta \mid X),$$

the median of the posterior density. Thus the posterior median is the Bayes estimator under absolute loss.

These examples show that the Bayes estimator is constructed from the posterior given a choice of loss function. In practice, the most common choice is the posterior mean, with the posterior median being the second most common choice.

16.5 PARAMETRIC PRIORS

It is typical to specify the prior density $\pi(\theta \mid \alpha)$ as a parametric distribution on Θ. A parametric prior depends on a parameter vector α, which is used to control the shape (centering and spread) of the prior. The prior density is typically selected from a standard parametric family.

Choices for the prior parametric family will be partly determined by the parameter space Θ.

1. Probabilities (such as p in a Bernoulli model) lie in $[0,1]$, so they need a distribution constrained to that interval. The beta distribution is a natural choice.

2. Variances lie in $\mathbb{R}_+$ so they need a distribution on the positive real line. The gamma or inverse-gamma are natural choices.
3. The mean μ from a normal distribution is unconstrained, so it needs an unconstrained distribution. The normal is a typical choice.

Given a parametric prior $\pi(\theta \mid \alpha)$, the user needs to select the prior parameters α. When selecting the prior parameters, it is useful to think about centering and spread. Take the mean μ from a normal distribution, and consider the prior $\mathrm{N}(\overline{\mu}, v)$. The prior is centered at $\overline{\mu}$, and the spread is controlled by v. These parameters control the shape of the prior.

The choice of prior parameters may differ among objectivists, subjectivists, and shrinkage Bayesians. An objectivist wants the prior to be relatively uninformative, so that the posterior reflects the information in the data about the parameters. Hence an objectivist will set the centering parameters so that the prior is situated at the relative center of the parameter space and will set the spread parameters so that the prior is diffusely spread throughout the parameter space. An example is the normal prior $\mathrm{N}(0, v)$, with v set relatively large. In contrast, a subjectivist will set the centering parameters to match the user's prior knowledge and beliefs about the likely values of the parameters, guided by prior studies and information. The spread parameters will be set depending on the strength of the knowledge about those parameters. If good information is available concerning a parameter, then the spread parameter can be set relatively small, while if the information is vague, then the spread parameter can be set relatively large similarly to an objectivist prior. A shrinkage Bayesian will center the prior on "default" parameter values and will select the spread parameters to control the degree of shrinkage. The spread parameters may be set to counterbalance limited sample information. The Bayes estimator will shrink the imprecise MLE toward the default parameter values, reducing estimation variance.

16.6 NORMAL-GAMMA DISTRIBUTION

The following hierarchical distribution is widely used in the Bayesian analysis of the normal sampling model:

$$X \mid \nu \sim \mathrm{N}(\mu, 1/(\lambda\nu))$$
$$\nu \sim \mathrm{gamma}(\alpha, \beta)$$

with parameters $(\mu, \lambda, \alpha, \beta)$. We write it as $(X, \nu) \sim \mathrm{NormalGamma}(\mu, \lambda, \alpha, \beta)$.

The joint density of (X, ν) is

$$f(x, \nu \mid \mu, \lambda, \alpha, \beta) = \frac{\lambda^{1/2}\beta^{\alpha}}{\sqrt{2\pi}} \nu^{\alpha-1/2} \exp(-\nu\beta) \exp\left(-\frac{\lambda\nu(x-\mu)^2}{2}\right).$$

The distribution has the following moments:

$$\mathbb{E}[X] = \mu$$
$$\mathbb{E}[\nu] = \frac{\alpha}{\beta}.$$

The marginal distribution of X is a scaled student t. To see this, define $Z = \sqrt{\lambda\nu}(X - \mu)$, which is scaled so that $Z \mid \nu \sim \mathrm{N}(0, 1)$. Since this is independent of ν, then Z and ν are independent. This implies that Z and $Q = 2\beta\nu \sim \chi^2_{2\alpha}$ are independent. Hence $Z/\sqrt{Q/2\alpha}$ is distributed student t with 2α degrees of freedom.

We can write X as

$$X - \mu = \sqrt{\frac{\beta}{\lambda\alpha}} \frac{Z}{\sqrt{Q/2\alpha}}$$

which is scaled student t with scale parameter $\beta/\lambda\alpha$ and degrees of freedom 2α.

16.7 CONJUGATE PRIOR

We say that the prior $\pi(\theta)$ is **conjugate** to the likelihood $L(X \mid \theta)$ if the prior and posterior $\pi(\theta \mid X)$ are members of the same parametric family. This property is convenient for estimation and inference, because in this case, the posterior mean and other statistics of interest are simple calculations. It is therefore useful to know how to calculate conjugate priors (when they exist).

For the following discussion, it will be useful to define what it means for a function to be proportional to a density. We say that $g(\theta)$ is **proportional** to a density $f(\theta)$, written $g(\theta) \propto f(\theta)$, if $g(\theta) = cf(\theta)$ for some $c > 0$. When discussing likelihoods and posteriors, the constant c may depend on the data X, but it should not depend on θ. The constant c can be calculated, but its value is not important for the purpose of selecting a conjugate prior.

For example,

1. $g(p) = p^x (1-p)^y$ is proportional to the beta$(1+x, 1+y)$ density.
2. $g(\theta) = \theta^a \exp(-\theta b)$ is proportional to the gamma$(1+a, b)$ density.
3. $g(\mu) = \exp\left(-a(\mu - b)^2\right)$ is proportional to the $\mathrm{N}(b, 1/2a)$ density.

The formula

$$\pi(\theta \mid X) = \frac{L_n(X \mid \theta)\, \pi(\theta \mid \alpha)}{m(X)}$$

shows that the posterior is proportional to the product of the likelihood and prior. We deduce that the posterior and prior will be members of the same parametric family when two conditions hold:

1. The likelihood viewed as a function of the parameters θ is proportional to $\pi(\theta \mid \alpha)$ for some α.
2. The product $\pi(\theta \mid \alpha_1)\pi(\theta \mid \alpha_2)$ is proportional to $\pi(\theta \mid \alpha)$ for some α.

These conditions imply that the posterior is proportional to $\pi(\theta \mid \alpha)$ and consequently, the prior is conjugate to the likelihood.

It may be helpful to give examples of parametric families that satisfy the second condition—that products of the density are members of the same family.

Example: Beta

$$\begin{aligned}
\text{beta}(\alpha_1, \beta_1) \times \text{beta}(\alpha_2, \beta_2) &\propto x^{\alpha_1 - 1} (1-x)^{\beta_1 - 1} x^{\alpha_2 - 1} (1-x)^{\beta_2 - 1} \\
&= x^{\alpha_1 + \alpha_2 - 2} (1-x)^{\beta_1 + \beta_2 - 2} \\
&\propto \text{beta}(\alpha_1 + \alpha_2 - 1, \beta_1 + \beta_2 - 1).
\end{aligned}$$

Example: Gamma

$$\begin{aligned}\text{gamma}(\alpha_1,\beta_1)\times\text{gamma}(\alpha_2,\beta_2) &\propto x^{\alpha_1-1}\exp(-x\beta_1)\,x^{\alpha_2-1}\exp(-x\beta_2)\\ &= x^{\alpha_1+\alpha_2-2}\exp(-x(\beta_1+\beta_2))\\ &\propto \text{gamma}(\alpha_1+\alpha_2-1,\beta_1+\beta_2)\end{aligned}$$

if $\alpha_1+\alpha_2>1$.

Example: Normal

$$\text{N}(\mu_1,1/\nu_1)\times\text{N}(\mu_2,1/\nu_2)\propto\text{N}(\overline{\mu},1/\overline{\nu}) \tag{16.1}$$

$$\overline{\mu}=\frac{\nu_1\mu_1+\nu_2\mu_2}{\nu_1+\nu_2}$$

$$\overline{\nu}=\nu_1+\nu_2.$$

Here we have parameterized the normal distribution in terms of the precision $\nu=1/\sigma^2$ rather than the variance. This is commonly done in Bayesian analysis for algebraic convenience. The derivation of this result is algebraic. See Exercise 16.1.

Example: Normal-Gamma

$$\begin{aligned}&\text{NormalGamma}(\mu_1,\lambda_1,\alpha_1,\beta_1)\times\text{NormalGamma}(\mu_2,\lambda_2,\alpha_2,\beta_2)\\ &=\text{N}(\mu_1,1/(\lambda_1\nu))\times\text{N}(\mu_2,1/(\lambda_2\nu))\times\text{gamma}(\alpha_1,\beta_1)\times\text{gamma}(\alpha_2,\beta_2)\\ &\propto\text{N}\left(\frac{\lambda_1\mu_1+\lambda_2\mu_2}{\lambda_1+\lambda_2},\frac{1}{(\lambda_1+\lambda_2)\nu}\right)\text{gamma}(\alpha_1+\alpha_2-1,\beta_1+\beta_2)\\ &=\text{NormalGamma}\left(\frac{\lambda_1\mu_1+\lambda_2\mu_2}{\lambda_1+\lambda_2},\lambda_1+\lambda_2,\alpha_1+\alpha_2-1,\beta_1+\beta_2\right).\end{aligned}$$

16.8 BERNOULLI SAMPLING

The probability mass function is $f(x\mid p)=p^x(1-p)^{1-x}$.

Consider a random sample with n observations. Let $S_n=\sum_{i=1}^n X_i$. The likelihood is

$$L_n(X\mid p)=p^{S_n}(1-p)^{n-S_n}.$$

This likelihood is proportional to a $p\sim\text{beta}(1+S_n,1+n-S_n)$ density. Since multiples of beta densities are beta densities, this shows that the conjugate prior for the Bernoulli likelihood is the beta density.

Given the beta prior

$$\pi(p\mid\alpha,\beta)=\frac{p^{\alpha-1}(1-p)^{\beta-1}}{B(\alpha,\beta)}$$

the product of the likelihood and prior is

$$\begin{aligned}L_n(X\mid p)\,\pi(p\mid\alpha,\beta)&\propto p^{S_n}(1-p)^{n-S_n}p^{\alpha-1}(1-p)^{\beta-1}\\ &=p^{S_n+\alpha-1}(1-p)^{n-S_n+\beta-1}\\ &\propto\text{beta}(p\mid S_n+\alpha,n-S_n+\beta).\end{aligned}$$

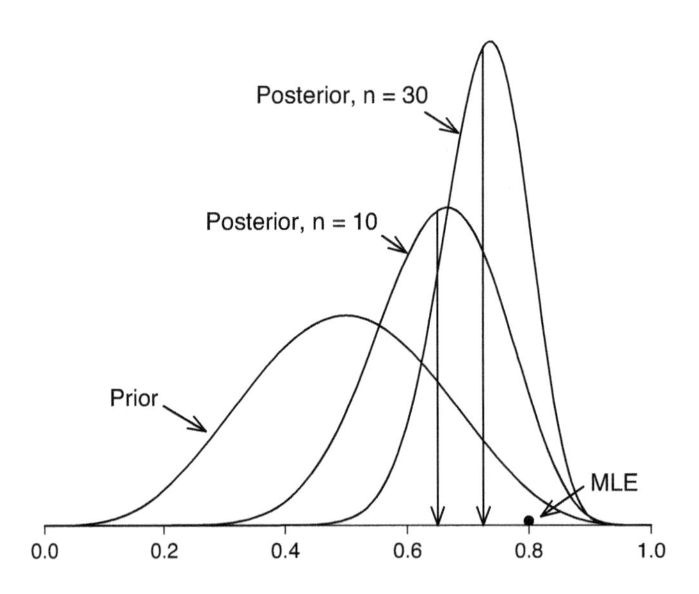

FIGURE 16.1 Posterior for Bernoulli model

It follows that the posterior density for p is

$$\begin{aligned}\pi\left(p \mid X\right) &= \text{beta}\left(x \mid S_n + \alpha, n - S_n + \beta\right) \\ &= \frac{p^{S_n+\alpha-1}\left(1-p\right)^{n-S_n+\beta-1}}{B\left(S_n+\alpha, n-S_n+\beta\right)}.\end{aligned}$$

Since the mean of beta(α, β) is $\alpha/(\alpha+\beta)$, the posterior mean (the Bayes estimator for p) is

$$\begin{aligned}\widehat{p}_{\text{Bayes}} &= \int_0^1 p\,\pi\left(p \mid X\right) dp \\ &= \frac{S_n+\alpha}{n+\alpha+\beta}.\end{aligned}$$

In contrast, the MLE for p is

$$\widehat{p}_{\text{mle}} = \frac{S_n}{n}.$$

To illustrate, Figure 16.1 displays the prior beta (5, 5) and two posteriors, the latter calculated with samples of size $n=10$ and $n=30$, respectively, and MLE $\widehat{p}=0.8$ in each. The prior is centered at 0.5 with most probability mass in the center, indicating a prior belief that p is close to 0.5. Another way of viewing this is that the Bayes estimator will shrink the MLE toward 0.5. The posterior densities are skewed, with posterior means $\widehat{p}=0.65$ and $\widehat{p}=0.725$ marked by the arrows from the posterior densities to the x-axis. The posterior densities are a mix of the information in the prior and the MLE, with the latter receiving a greater weight with the larger sample size.

16.9 NORMAL SAMPLING

As mentioned earlier, in Bayesian analysis, it is typical to rewrite the model in terms of the **precision** $\nu = \sigma^{-2}$ rather than the variance. The likelihood written as a function of μ and ν is

$$L(X \mid \mu, \nu) = \frac{\nu^{n/2}}{(2\pi)^{n/2}} \exp\left(-\frac{\sum_{i=1}^{n}(X_i - \mu)^2}{2/\nu}\right)$$
$$= \frac{\nu^{n/2}}{(2\pi)^{n/2}} \exp\left(-\nu \frac{n\widehat{\sigma}^2_{\text{mle}}}{2}\right) \exp\left(-n\nu \frac{\left(\overline{X}_n - \mu\right)^2}{2}\right) \tag{16.2}$$

where $\widehat{\sigma}^2_{\text{mle}} = n^{-1}\sum_{i=1}^{n}\left(X_i - \overline{X}_n\right)^2$.

We consider three cases: (a) estimation of μ with known ν; (b) estimation of ν with known $\mu = 0$; and (c) estimation of μ and ν.

Estimation of μ with known ν. The likelihood as a function of μ is proportional to $\text{N}\left(\overline{X}_n, 1/(n\nu)\right)$. It follows that the natural conjugate prior is the Normal distribution. Let the prior be $\pi(\mu) = \text{N}(\overline{\mu}, 1/\overline{\nu})$. For this prior, $\overline{\mu}$ controls location and $\overline{\nu}$ controls the spread of the prior.

The product of the likelihood and prior is

$$L_n(X \mid \mu)\,\pi(\mu \mid \overline{\mu}, \overline{\nu}) = \text{N}\left(\overline{X}_n, 1/(n\nu)\right) \times \text{N}(\overline{\mu}, 1/\overline{\nu})$$
$$\propto \text{N}\left(\frac{n\nu\overline{X}_n + \overline{\nu}\,\overline{\mu}}{n\nu + \overline{\nu}}, \frac{1}{n\nu + \overline{\nu}}\right)$$
$$= \pi(\mu \mid X).$$

Thus the posterior density is $\text{N}\left(\frac{n\nu\overline{X}_n + \overline{\nu}}{n\nu + \overline{\nu}}, \frac{1}{n\nu + \overline{\nu}}\right)$. The posterior mean (the Bayes estimator) for the mean μ is

$$\widehat{\mu}_{\text{Bayes}} = \frac{n\nu\overline{X}_n + \overline{\nu}\,\overline{\mu}}{n\nu + \overline{\nu}}.$$

This estimator is a weighted average of the sample mean $\overline{X}_n$ and the prior mean $\overline{\mu}$. Greater weight is placed on the sample mean when $n\nu$ is large (when estimation variance is small) or when $\overline{\nu}$ is small (when the prior variance is large) and vice versa. For fixed priors, as the sample size n increases, the posterior mean converges to the sample mean $\overline{X}_n$.

One way of interpreting the prior spread parameter $\overline{\nu}$ is in terms of "number of observations relative to variance". Suppose that instead of using a prior we added $N = \overline{\nu}/\nu$ observations to the sample, each taking the value $\overline{\mu}$. The sample mean on this augmented sample is $\widehat{\mu}_{\text{Bayes}}$. Thus one way to imagine selecting $\overline{\nu}$ relative to ν is to think of prior knowledge in terms of the number of observations.

Estimation of ν with known $\mu = 0$. Set $\widetilde{\sigma}^2_{\text{mle}} = n^{-1}\sum_{i=1}^{n} X_i^2$. The likelihood equals

$$L_n(X \mid \nu) = \frac{\nu^{n/2}}{(2\pi)^{n/2}} \exp\left(-\nu \frac{n\widetilde{\sigma}^2_{\text{mle}}}{2}\right).$$

which is proportional to gamma $\left(1 + n/2, \sum_{i=1}^{n} X_i^2/2\right)$. It follows that the natural conjugate prior is the gamma distribution. Let the prior be $\pi(\nu) = \text{gamma}(\alpha, \beta)$. It is helpful to recall that the mean of this distribution is

$\overline{\nu} = \alpha/\beta$, which controls the location. We can alternatively control the location through its inverse $\overline{\sigma}^2 = 1/\overline{\nu} = \beta/\alpha$. The spread of the prior for ν increases (and the spread for σ^2 decreases) as α and β increase.

The product of the likelihood and prior is

$$\begin{aligned} L_n(X \mid \nu)\,\pi(\nu \mid \alpha, \beta) &= \text{gamma}\left(1 + \frac{n}{2}, \frac{n\widetilde{\sigma}^2_{\text{mle}}}{2}\right) \times \text{gamma}(\alpha, \beta) \\ &\propto \text{gamma}\left(\frac{n}{2} + \alpha, \frac{n\widetilde{\sigma}^2_{\text{mle}}}{2} + \beta\right) \\ &= \pi(\nu \mid X). \end{aligned}$$

Thus the posterior density is gamma $\left(n/2 + \alpha, \frac{1}{2}n\widetilde{\sigma}^2_{\text{mle}} + \beta\right) = \chi^2_{n+\alpha} / \left(n\widetilde{\sigma}^2_{\text{mle}} + 2\beta\right)$. The posterior mean (the Bayes estimator) for the precision ν is

$$\widehat{\nu}_{\text{Bayes}} = \frac{\frac{n}{2} + \alpha}{\frac{n}{2}\widetilde{\sigma}^2_{\text{mle}} + \beta} = \frac{n + 2\alpha}{n\widetilde{\sigma}^2_{\text{mle}} + 2\alpha\overline{\sigma}^2}.$$

where the second equality uses the relationship $\beta = \alpha\overline{\sigma}^2$. An estimator for the variance σ^2 is the inverse of this estimator

$$\widehat{\sigma}^2_{\text{Bayes}} = \frac{n\widetilde{\sigma}^2_{\text{mle}} + 2\alpha\overline{\sigma}^2}{n + 2\alpha}.$$

This is a weighted average of the MLE $\widetilde{\sigma}^2_{\text{mle}}$ and prior value $\overline{\sigma}^2$, with greater weight on the MLE when n is large and/or α is small, the latter corresponding to a prior that is dispersed in σ^2.

This prior parameter α can also be interpreted in terms of the number of observations. Suppose we add 2α observations to the sample, each taking the value $\overline{\sigma}$. The MLE on this augmented sample equals $\widehat{\sigma}^2_{\text{Bayes}}$. Thus the prior parameter α can be interpreted in terms of confidence about σ^2 as measured by number of observations.

Estimation of μ and ν. The likelihood (16.2) as a function of (μ, ν) is proportional to

$$(\mu, \nu) \sim \text{NormalGamma}\left(\overline{X}_n, n, (n+1)/2, n\widehat{\sigma}^2_{\text{mle}}/2\right).$$

It follows that the natural conjugate prior is the NormalGamma distribution.

Let the prior be NormalGamma $(\overline{\mu}, \lambda, \alpha, \beta)$. The parameters $\overline{\mu}$ and λ control the location and spread of the prior for the mean μ. The parameters α and β control the location and spread of the prior for ν. As discussed in Section 16.8, it is useful to define $\overline{\sigma}^2 = \beta/\alpha$ as the location of the prior variance. We can interpret λ in terms of the number of observations controlled by the prior for estimation of μ, and 2α as the number of observations controlled by the prior for estimation of ν.

The product of the likelihood and prior is

$$\begin{aligned} L_n(X \mid \mu, \nu)\,\pi(\mu, \nu \mid \overline{\mu}, \lambda, \alpha, \beta) &= \text{NormalGamma}\left(\overline{X}_n, n, \frac{n+1}{2}, \frac{n\widehat{\sigma}^2_{\text{mle}}}{2}\right) \times \text{NormalGamma}(\overline{\mu}, \lambda, \alpha, \beta) \\ &\propto \text{NormalGamma}\left(\frac{n\overline{X}_n + \lambda\overline{\mu}}{n + \lambda}, n + \lambda, \frac{n-1}{2} + \alpha, \frac{n\widehat{\sigma}^2_{\text{mle}}}{2} + \beta\right) \\ &= \pi(\mu, \nu \mid X). \end{aligned}$$

Thus the posterior density is NormalGamma.

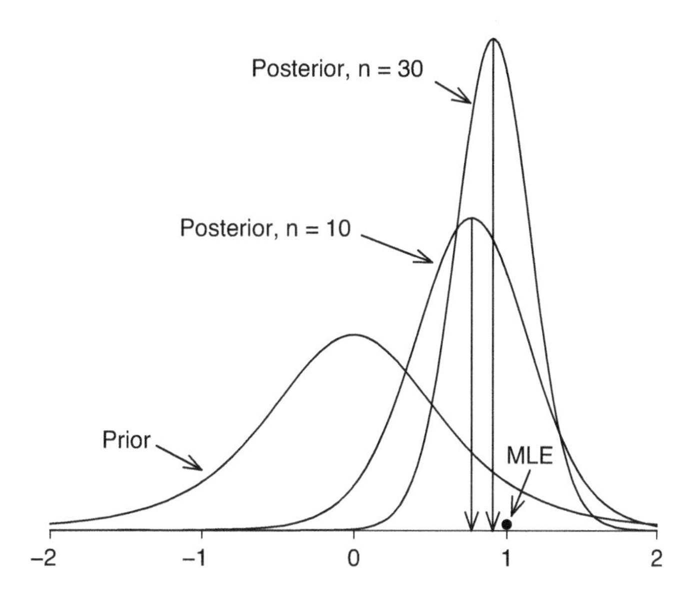

FIGURE 16.2 Posterior for mean in normal model

The posterior mean for μ is

$$\widehat{\mu}_{\text{Bayes}} = \frac{n\overline{X}_n + \lambda\overline{\mu}}{n+\lambda}.$$

This is a simple weighted average of the sample mean and the prior mean with weights n and λ, respectively. Thus the prior parameter λ can be interpreted in terms of the number of observations controlled by the prior.

The posterior mean for ν is

$$\widehat{\nu}_{\text{Bayes}} = \frac{n-1+2\alpha}{n\widehat{\sigma}^2_{\text{mle}} + 2\beta}.$$

An estimator for the variance σ^2 is

$$\widehat{\sigma}^2_{\text{Bayes}} = \frac{(n-1)s^2 + 2\alpha\overline{\sigma}^2}{n-1+2\alpha}$$

where we have used the relationship $\beta = \alpha\overline{\sigma}^2$ and set $s^2 = (n-1)^{-1}\sum_{i=1}^{n}\left(X_i - \overline{X}_n\right)^2$. We see that the Bayes estimator is a weighted average of the bias-corrected variance estimator s^2 and the prior value $\overline{\sigma}^2$, with greater weight on s^2 when n is large and/or α is small.

It is useful to know the marginal posterior densities. By the properties of the NormalGamma distribution, the marginal posterior density for μ is scaled student t with mean $\widehat{\mu}_{\text{Bayes}}$, degrees of freedom $n-1+2\alpha$, and scale parameter $1/\left((n+\lambda)\,\widehat{\nu}_{\text{Bayes}}\right)$. The marginal posterior for ν is

$$\text{gamma}\left((n-1)/2+\alpha, \frac{1}{2}n\widehat{\sigma}^2_{\text{mle}} + \beta\right) = \chi^2_{n-1+2\alpha}/\left(n\widehat{\sigma}^2_{\text{mle}} + 2\beta\right).$$

When the prior sets $\lambda = \alpha = \beta = 0$, these correspond to the exact distributions of the sample mean and variance estimators in the normal sampling model.

To illustrate, Figure 16.2 displays the marginal prior for μ and two marginal posteriors, the latter calculated with samples of size $n = 10$ and $n = 30$, respectively. The MLE are $\widehat{\mu} = 1$ and $1/\widehat{\sigma}^2 = 0.5$. The prior is NormalGamma(0,3,2,2). The marginal priors and posteriors are student t. The posterior means are marked

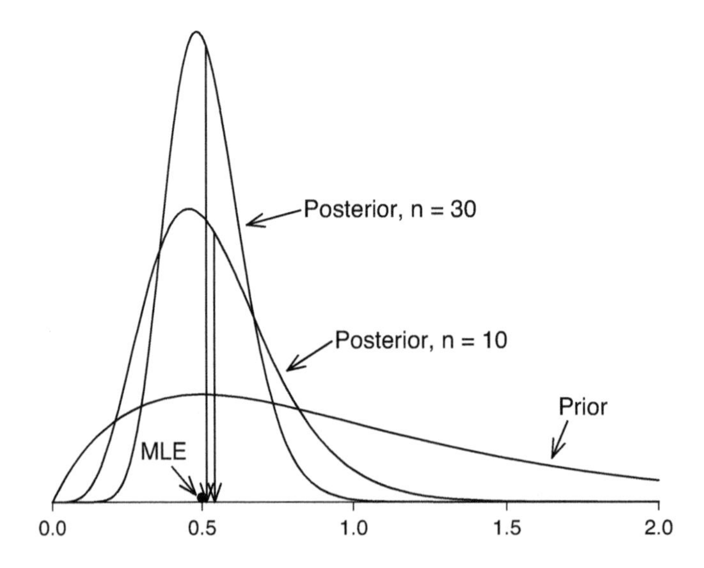

FIGURE 16.3 Posterior for precision in normal model

by arrows from the posterior densities to the x-axis, showing how the posterior mean approaches the MLE as n increases.

Figure 16.3 displays the marginal prior for ν and two marginal posteriors, calculated with the same sample sizes, MLE, and prior as in Figure 16.2. The marginal priors and posteriors are gamma distributions. The prior is diffuse, so the posterior means are close to the MLE.

16.10 CREDIBLE SETS

Recall that an interval estimator for a real-valued parameter θ is an interval $C = [L, U]$, which is a function of the observations. For interval estimation, Bayesians select interval estimates C as credible regions which treat parameters as random and are conditional on the observed sample. In contrast, frequentists select interval estimates as confidence region which treat the parameters as fixed and the observations as random.

Definition 16.1 A $1-\eta$ **credible interval** for θ is an interval estimator $C = [L, U]$ such that

$$\mathbb{P}\left[\theta \in C \mid X\right] = 1 - \eta.$$

The credible interval C is a function of the data, and the probability calculation treats C as fixed. The randomness is therefore in the parameter θ, not C. The probability stated in the credible interval is calculated from the posterior for θ:

$$\mathbb{P}\left[\theta \in C \mid X\right] = \int_C \pi\left(\theta \mid X\right) d\theta.$$

The most common approach in Bayesian statistics is to select C by the principle of highest posterior density (HPD). A property of HPD intervals is that they have the smallest length among all $1-\eta$ credible intervals.

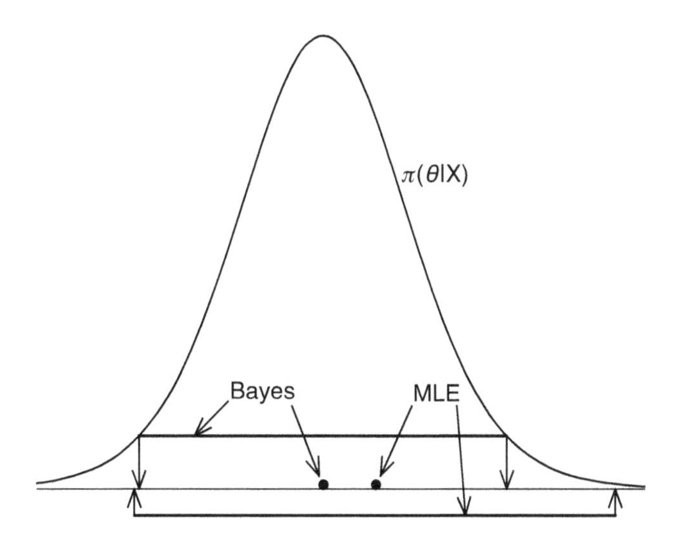

FIGURE 16.4 Credible set for normal mean

Definition 16.2 Let $\pi\ (\theta \mid X)$ be a posterior density. If C is an interval such that

$$\int_C \pi\ (\theta \mid X)\ d\theta = 1 - \eta$$

and for all $\theta_1, \theta_2 \in C$

$$\pi\ (\theta_1 \mid X) \geq \pi\ (\theta_2 \mid X)$$

then the interval C is called a **highest posterior density (HPD)** interval of probability $1 - \alpha$.

HPD intervals are constructed as level sets of the (marginal) posterior. Given a posterior density $\pi\ (\theta \mid X)$, a $1 - \eta$ HPD interval $C = [L, U]$ satisfies $\pi\ (L \mid X) = \pi\ (U \mid X)$ and $\int_L^U \pi\ (\theta \mid X)\ d\theta = 1 - \eta$. In most cases, there is no algebraic formula for the credible interval. However, it can be found numerically by solving the following equation. Let $F(\theta), f(\theta)$, and $Q(\eta)$ denote the posterior distribution, density, and quantile functions, respectively. The lower endpoint L solves the equation

$$f\ (L) - f\ (Q\ (1 - \eta + F(L))) = 0.$$

Given L, the upper endpoint is $U = Q\ (1 - \eta + F(L))$.

Example: Normal mean. Consider the normal sampling model with unknown mean μ and precision v. The marginal posterior for μ is scaled student t with mean $\widehat{\mu}_{\text{Bayes}}$, degrees of freedom $n - 1 + 2\alpha$, and scale parameter $1/\left((n + \lambda)\widehat{v}_{\text{Bayes}}\right)$. Since the student t is a symmetric distribution, a HPD credible interval will be symmetric about $\widehat{\mu}_{\text{Bayes}}$. Thus a $1 - \eta$ credible interval equals

$$C = \left[\widehat{\mu}_{\text{Bayes}} - \frac{q_{1-\eta/2}}{\sqrt{(n + \lambda)\ \widehat{v}_{\text{Bayes}}}}, \qquad \widehat{\mu}_{\text{Bayes}} + \frac{q_{1-\eta/2}}{\sqrt{(n + \lambda)\ \widehat{v}_{\text{Bayes}}}}\right]$$

where $q_{1-\eta/2}$ is the $1 - \eta/2$ quantile of the student t density with $n - 1 - 2\alpha$ degrees of freedom. When the prior coefficients λ and α are small, this is similar to the $1 - \eta$ classical (frequentist) confidence interval under normal sampling.

To illustrate, Figure 16.4 displays the posterior density for the mean from Figure 16.2 with $n = 10$. The 95% Bayes credible interval is marked by the horizontal line leading to the arrows to the x-axis. The area under

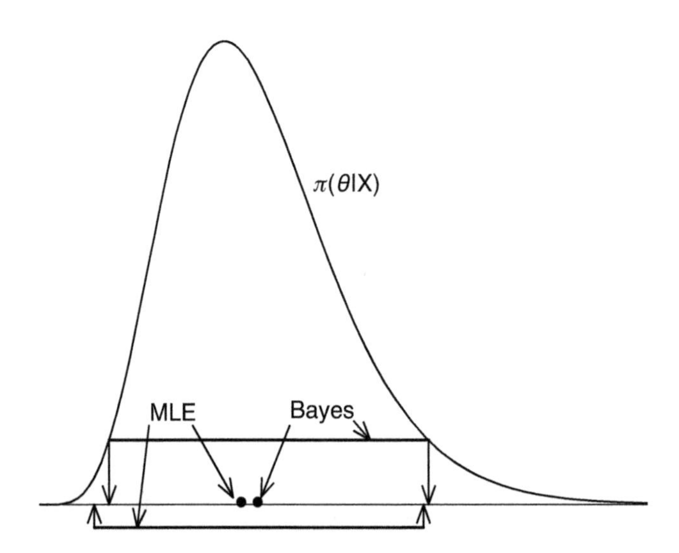

FIGURE 16.5 Credible set for normal precision

the posterior between these arrows is 95%, and the heights of the posterior density at the two endpoints are equal, so this is the HPD credible interval. The Bayes estimator (the posterior mean) is marked. Also displayed is the MLE and the 95% classical confidence interval. The latter is marked by the arrows underneath the x-axis. The Bayes credible interval has shorter length than the classical confidence interval, because the former utilizes the information in the prior to sharpen inference on the mean.

Example: Normal Variance. Consider the same model but now construct a credible interval for v. Its marginal posterior is $\chi^2_{n-1+2\alpha}/\left(n\widehat{\sigma}^2_{\text{mle}}+2\beta\right)$. Since the posterior is asymmetric, the credible set will be asymmetric as well.

Figure 16.5 displays the posterior density for the precision from Figure 16.3 with $n=10$. The 95% Bayes credible interval is marked by the horizontal line leading to the arrows to the x-axis. In contrast to the previous example, this posterior is asymmetric, so the credible set is asymmetric about the posterior mean (the latter is marked as "Bayes"). Also displayed is the MLE and the 95% classical confidence interval. The latter is marked by the arrows underneath the x-axis. The Bayes interval has shorter length than the classical interval, because the former utilizes the information in the prior. However, the difference is not large in this example, because the prior is relatively diffuse reflecting low prior confidence in knowledge about the precision.

If a confidence interval is desired for a transformation of the precision, this transformation can be applied directly to the endpoints of the credible set. In this example, the endpoints are $L=0.17$ and $U=0.96$. Thus the 95% HPD credible interval for the variance σ^2 is $[1/.96, 1/.17]=[1.04, 5.88]$ and for the standard deviation is $[1/\sqrt{.96}, 1/\sqrt{.17}]=[1.02, 2.43]$.

16.11 BAYESIAN HYPOTHESIS TESTING

As a general rule, Bayesian econometricians make less use of hypothesis testing than frequentist econometricians, and when they do so, it takes a very different form. In contrast to the Neyman-Pearson approach to hypothesis testing, Bayesian tests attempt to calculate which model has the highest probability of truth.

Bayesian hypothesis tests treat models symmetrically rather than labeling one the "null" and the other the "alternative".

Assume that we have a set of J models, or hypotheses, denoted by $\mathbb{H}_j$ for $j = 1, \ldots, J$. Each model $\mathbb{H}_j$ consists of a parametric density $f_j(x \mid \theta_j)$ with vector-valued parameter θ_j, likelihood function $L_j(X \mid \theta_j)$, prior density $\pi_j(\theta_j)$, marginal likelihood $m_j(X)$, and posterior density $\pi_j(\theta_j \mid X)$. In addition, we specify the prior probability that each model is true,

$$\pi_j = \mathbb{P}\left[\mathbb{H}_j\right]$$

with

$$\sum_{j=1}^{J} \pi_j = 1.$$

By Bayes Rule, the posterior probability that model $\mathbb{H}_j$ is the true model is

$$\pi_j(X) = \mathbb{P}\left[\mathbb{H}_j \mid X\right] = \frac{\pi_j m_j(X)}{\sum_{i=1}^{J} \pi_i m_i(X)}.$$

A Bayes test selects the model with the highest posterior probability. Since the denominator of this expression is common across models, this is equivalent to selecting the model with the highest value of $\pi_j m_j(X)$. When all models are given equal prior probability (which is common for an agnostic Bayesian hypothesis test), this is the same as selecting the model with the highest marginal likelihood $m_j(X)$.

When comparing two models, say $\mathbb{H}_1$ versus $\mathbb{H}_2$, we select $\mathbb{H}_2$ if

$$\pi_1 m_1(X) < \pi_2 m_2(X)$$

or equivalently if

$$1 < \frac{\pi_2}{\pi_1} \frac{m_2(X)}{m_1(X)}.$$

We use the following terms. The **prior odds** for $\mathbb{H}_2$ versus $\mathbb{H}_1$ are π_2/π_1. The **posterior odds** for $\mathbb{H}_2$ versus $\mathbb{H}_1$ are $\pi_2(X)/\pi_1(X)$. The **Bayes Factor** for $\mathbb{H}_2$ versus $\mathbb{H}_1$ is $m_2(X)/m_1(X)$. Thus we select $\mathbb{H}_2$ over $\mathbb{H}_1$ if the posterior odds exceed 1, or equivalently, if the prior odds multiplied by the Bayes Factor exceeds 1. When models are given equal prior probability, we select $\mathbb{H}_2$ over $\mathbb{H}_1$ when the Bayes Factor exceeds 1.

16.12 SAMPLING PROPERTIES IN THE NORMAL MODEL

It is useful to evaluate the sampling properties of the Bayes estimator and compare them with classical estimators. In the normal model, the Bayes estimator of μ is

$$\widehat{\mu}_{\text{Bayes}} = \frac{n\overline{X}_n + \lambda\overline{\mu}}{n + \lambda}.$$

We now evalulate its bias, variance, and exact sampling distribution.

Theorem 16.1 In the normal sampling model

1. $\mathbb{E}\left[\widehat{\mu}_{\text{Bayes}}\right] = \dfrac{n\mu + \lambda\overline{\mu}}{n + \lambda}.$

2. $\text{bias}\left[\widehat{\mu}_{\text{Bayes}}\right] = \frac{\lambda}{n+\lambda}(\overline{\mu} - \mu)$.
3. $\text{var}\left[\widehat{\mu}_{\text{Bayes}}\right] = \frac{\sigma^2}{n+\lambda}$.
4. $\widehat{\mu}_{\text{Bayes}} \sim \text{N}\left(\frac{n\mu + \lambda\overline{\mu}}{n+\lambda}, \frac{\sigma^2}{n+\lambda}\right)$.

Thus the Bayes estimator has reduced variance (relative to the sample mean) when $\lambda > 0$ but has bias when $\lambda > 0$ and $\overline{\mu} \neq \mu$. Its sampling distribution is normal.

16.13 ASYMPTOTIC DISTRIBUTION

If the MLE is asymptotically normally distributed, the corresponding Bayes estimator behaves similarly to the MLE for large samples. That is, if the priors are fixed with support containing the true parameter, then the posterior distribution converges to a normal distribution, and the standardized posterior mean converges in distribution to a normal random vector.

I state the results for the general case of a vector-valued estimator. Let θ be $k \times 1$. Define the rescaled parameter $\zeta = \sqrt{n}\left(\widehat{\theta}_{\text{mle}} - \theta\right)$ and the recentered posterior density

$$\pi^*(\zeta \mid X) = \pi\left(\widehat{\theta}_{\text{mle}} - n^{-1/2}\zeta \mid X\right). \tag{16.3}$$

Theorem 16.2 Bernstein–von Mises. Assume the conditions of Theorem 10.9 hold, θ_0 is in the interior of the support of the prior $\pi(\theta)$, and the prior is continuous in a neighborhood of θ_0. As $n \to \infty$,

$$\pi^*(\zeta \mid X) \underset{p}{\longrightarrow} \frac{\det\left(\mathscr{I}_\theta\right)^{k/2}}{(2\pi)^{k/2}} \exp\left(-\frac{1}{2}\zeta' \mathscr{I}_\theta \zeta\right).$$

The theorem states that the posterior density converges to a normal density. Examining the posterior densities in Figures 16.1–16.3, this seems credible as in each case, the densities become more normal-shaped as the sample size increases. The Bernstein–von Mises Theorem states that this holds for all parametric models which satisfy the conditions for asymptotic normality of the MLE. The critical condition is that the prior must have positive support at the true parameter value. Otherwise if the prior places no support at the true value, the posterior will place zero support there as well. Consequently it is prudent to always use a prior which has positive support on the entire relevant parameter space.

Theorem 16.2 is a statement about the asymptotic shape of the posterior density. We can also obtain the asymptotic distribution of the Bayes estimator.

Theorem 16.3 Assume the conditions of Theorem 16.2 hold. As $n \to \infty$,

$$\sqrt{n}\left(\widehat{\theta}_{\text{Bayes}} - \theta_0\right) \underset{d}{\longrightarrow} \text{N}\left(0, \mathscr{I}_\theta^{-1}\right).$$

Theorem 16.3 shows that the Bayes estimator has the same asymptotic distribution as the MLE. In fact, the proof of the theorem shows that the Bayes estimator is asymptotically equivalent to the MLE in the sense that

$$\sqrt{n}\left(\widehat{\theta}_{\text{Bayes}} - \widehat{\theta}_{\text{mle}}\right) \underset{p}{\longrightarrow} 0.$$

Bayesians typically do not use Theorem 16.3 for inference. Instead, they form credible sets as previously described. One role of Theorem 16.3 is to show that the distinction between classical and Bayesian estimation diminishes in large samples. The intuition is that when sample information is strong, it will dominate prior information. One message is that if a large difference exists between a Bayes estimator and the MLE, it is a cause for concern and investigation. One possibility is that the sample has low information for the parameters. Another is that the prior is highly informative. In either event, it is useful to know the source before making a strong conclusion.

The proofs of the two theorems are provided in Section 16.14.

16.14 TECHNICAL PROOFS*

Proof of Theorem 16.2* Using the ratio of recentered densities (16.3), we see that

$$\frac{\pi^*(\zeta \mid X)}{\pi^*(0 \mid X)} = \frac{\pi\left(\widehat{\theta}_{\mathrm{mle}} - n^{-1/2}\zeta \mid X\right)}{\pi\left(\widehat{\theta}_{\mathrm{mle}} \mid X\right)} = \frac{L_n\left(X \mid \widehat{\theta}_{\mathrm{mle}} - n^{-1/2}\zeta\right)}{L_n\left(X \mid \widehat{\theta}_{\mathrm{mle}}\right)} \frac{\pi\left(\widehat{\theta}_{\mathrm{mle}} - n^{-1/2}\zeta\right)}{\pi\left(\widehat{\theta}_{\mathrm{mle}}\right)}.$$

For fixed ζ, the prior satisfies

$$\frac{\pi\left(\widehat{\theta}_{\mathrm{mle}} - n^{-1/2}\zeta\right)}{\pi\left(\widehat{\theta}_{\mathrm{mle}}\right)} \underset{p}{\longrightarrow} 1.$$

By a Taylor series expansion and the first-order condition for the MLE,

$$\begin{aligned}\ell_n\left(\widehat{\theta}_{\mathrm{mle}} - n^{-1/2}\zeta\right) - \ell_n\left(\widehat{\theta}_{\mathrm{mle}}\right) &= -n^{-1/2}\frac{\partial}{\partial\theta}\ell_n\left(\widehat{\theta}_{\mathrm{mle}}\right)'\zeta + \frac{1}{2n}\zeta'\frac{\partial^2}{\partial\theta\partial\theta'}\ell_n\left(\theta^*\right)\zeta \\ &= \frac{1}{2n}\zeta'\frac{\partial^2}{\partial\theta\partial\theta'}\ell_n\left(\theta^*\right)\zeta \\ &\underset{p}{\longrightarrow} -\frac{1}{2}\zeta'\mathscr{I}_\theta\zeta \end{aligned} \tag{16.4}$$

where θ^* is intermediate between $\widehat{\theta}_{\mathrm{mle}}$ and $\widehat{\theta}_{\mathrm{mle}} - n^{-1/2}\zeta$. Therefore the ratio of likelihoods satisfies

$$\frac{L_n\left(X \mid \widehat{\theta}_{\mathrm{mle}} - n^{-1/2}\zeta\right)}{L_n\left(X \mid \widehat{\theta}_{\mathrm{mle}}\right)} = \exp\left(\ell_n\left(\widehat{\theta}_{\mathrm{mle}} - n^{-1/2}\zeta\right) - \ell_n\left(\widehat{\theta}_{\mathrm{mle}}\right)\right) \underset{p}{\longrightarrow} \exp\left(-\frac{1}{2}\zeta'\mathscr{I}_\theta\zeta\right).$$

These results show that

$$\frac{\pi^*(\zeta \mid X)}{\pi^*(0 \mid X)} \underset{p}{\longrightarrow} \exp\left(-\frac{1}{2}\zeta'\mathscr{I}_\theta\zeta\right).$$

Thus the posterior density $\pi^*(\zeta \mid X)$ is asymptotically proportional to $\exp\left(-\frac{1}{2}\zeta'\mathscr{I}_\theta\zeta\right)$, which is the multivariate normal distribution, as stated. ■

Proof of Theorem 16.3* A technical demonstration shows that for any $\eta > 0$, it is possible to find a compact set C symmetric about 0 such that the posterior $\pi^*(\zeta \mid X)$ concentrates in C with probability exceeding $1 - \eta$. Consequently, it is sufficient to treat the posterior as if it were truncated on this set. For a full proof of this technical detail, see section 10.3 of van der Vaart (1998).

Theorem 16.2 shows that the recentered posterior density converges pointwise to a normal density. We extend this result to uniform convergence over C. Since $\frac{1}{n}\frac{\partial^2}{\partial\theta\partial\theta'}\ell_n(\theta)$ converges uniformly in probability to its expectation in a neighborhood of θ_0 under the assumptions of Theorem 10.9, the convergence in (16.4) is uniform for $\zeta \in C$. This implies that the remaining convergence statements are uniform as well, and the recentered posterior density converges uniformly to a normal density $\phi_\theta(\zeta)$.

The difference between the posterior mean and the MLE equals

$$\begin{aligned}
\sqrt{n}\left(\widehat{\theta}_{\text{Bayes}} - \widehat{\theta}_{\text{mle}}\right) &= \sqrt{n}\int \left(\theta - \widehat{\theta}_{\text{mle}}\right)\pi(\theta \mid X)\, d\theta - \widehat{\theta}_{\text{mle}} \\
&= \int \zeta\pi\left(\widehat{\theta}_{\text{mle}} - n^{-1/2}\zeta \mid X\right) d\zeta \\
&= \int \zeta\pi^*(\zeta \mid X)\, d\zeta \\
&= \widehat{m},
\end{aligned}$$

the mean of the recentered posterior density. We calculate that

$$\begin{aligned}
|\widehat{m}| &= \left|\int_C \zeta\pi^*(\zeta \mid X)\, d\zeta\right| \\
&= \left|\int_C \zeta\phi_\theta(\zeta)\, d\zeta + \int_C \zeta\left(\pi^*(\zeta \mid X) - \phi_\theta(\zeta)\right) d\zeta\right| \\
&= \left|\int_C \zeta\left(\pi^*(\zeta \mid X) - \phi_\theta(\zeta)\right) d\zeta\right| \\
&\le \int_C |\zeta|\left|\pi^*(\zeta \mid X) - \phi_\theta(\zeta)\right| d\zeta \\
&\le \int_C |\zeta|\, d\zeta \sup_{\zeta\in C}\left|\pi^*(\zeta \mid X) - \phi_\theta(\zeta)\right| \\
&= o_p(1).
\end{aligned}$$

The third equality uses the fact that the mean of the normal density $\phi_\theta(\zeta)$ is 0 and the final $o_p(1)$ holds because the posterior converges uniformly to the normal density and the set C is compact.

We have showed that

$$\sqrt{n}\left(\widehat{\theta}_{\text{Bayes}} - \widehat{\theta}_{\text{mle}}\right) = \widehat{m} \underset{p}{\longrightarrow} 0.$$

This shows that $\widehat{\theta}_{\text{Bayes}}$ and $\widehat{\theta}_{\text{mle}}$ are asymptotically equivalent. Since the latter has the asymptotic distribution $\mathrm{N}\left(0, \mathscr{I}_\theta^{-1}\right)$ under Theorem 10.9, so does $\widehat{\theta}_{\text{Bayes}}$. ■

16.15 EXERCISES

Exercise 16.1 Show equation (16.1). This is the same as showing that

$$\sqrt{v_1 v_2}\phi\left(v_1(x-\mu_1)\right)\phi\left(v_2(x-\mu_2)\right) = c\sqrt{\overline{v}}\phi\left(\overline{v}\left(x-\overline{\mu}\right)\right)$$

where c can depend on the parameters but not on x.

Exercise 16.2 Let p be the probability of your textbook author Bruce making a free throw shot.

(a) Consider the prior $\pi(p) = \text{beta}(1, 1)$. (Beta with $\alpha = 1$ and $\beta = 1$.) Write and sketch the prior density. Does this prior seem to you to be a good choice for representing uncertainty about p?

(b) Bruce attempts one free throw shot and misses. Write the likelihood and posterior densities. Find the posterior mean $\widehat{p}_{\text{Bayes}}$ and the MLE $\widehat{p}_{\text{mle}}$.

(c) Bruce attempts a second free throw shot and misses. Find the posterior density and mean given the two shots. Find the MLE.

(d) Bruce attempts a third free throw shot and makes the basket. Find the estimators $\widehat{p}_{\text{Bayes}}$ and $\widehat{p}_{\text{mle}}$.

(e) Compare the sequence of estimators. In this context, is the Bayes or MLE more sensible?

Exercise 16.3 Consider the exponential density $f(x \mid \lambda) = \frac{1}{\lambda}\exp\left(-\frac{x}{\lambda}\right)$ for $\lambda > 0$.

(a) Find the conjugate prior $\pi(\lambda)$.

(b) Find the posterior $\pi(\lambda \mid X)$ given a random sample.

(c) Find the posterior mean $\widehat{\lambda}_{\text{Bayes}}$.

Exercise 16.4 Consider the Poisson density $f(x \mid \lambda) = \dfrac{e^{-\lambda}\lambda^x}{x!}$ for $\lambda > 0$.

(a) Find the conjugate prior $\pi(\lambda)$.

(b) Find the posterior $\pi(\lambda \mid X)$ given a random sample.

(c) Find the posterior mean $\widehat{\lambda}_{\text{Bayes}}$.

Exercise 16.5 Consider the Poisson density $f(x \mid \lambda) = \dfrac{e^{-\lambda}\lambda^x}{x!}$ for $\lambda > 0$.

(a) Find the conjugate prior $\pi(\lambda)$.

(b) Find the posterior $\pi(\lambda \mid X)$ given a random sample.

(c) Find the posterior mean $\widehat{\lambda}_{\text{Bayes}}$.

Exercise 16.6 Consider the Pareto density $f(x \mid \alpha) = \dfrac{\alpha}{x^{\alpha+1}}$ for $x > 1$ and $\alpha > 0$.

(a) Find the conjugate prior $\pi(\alpha)$. (This may be tricky.)

(b) Find the posterior $\pi(\alpha \mid X)$ given a random sample.

(c) Find the posterior mean $\widehat{\alpha}_{\text{Bayes}}$.

Exercise 16.7 Consider the $U[0, \theta]$ density $f(x \mid \theta) = 1/\theta$ for $0 \leq x \leq \theta$.

(a) Find the conjugate prior $\pi(\theta)$. (This may be tricky.)

(b) Find the posterior $\pi(\theta \mid X)$ given a random sample.

(c) Find the posterior mean $\widehat{\theta}_{\text{Bayes}}$.

CHAPTER 17

NONPARAMETRIC DENSITY ESTIMATION

17.1 INTRODUCTION

Sometimes it is useful to estimate the density function of a continuously distributed variable. As a general rule, density functions can take any shape. As a result, density functions are inherently **nonparametric**, which means that the density function cannot be described by a finite set of parameters. The most common method for estimating density functions is by using **kernel smoothing** estimators, which are related to the kernel regression estimators explored in Chapter 19 of *Econometrics*.

There are many excellent monographs written on nonparametric density estimation, including Silverman (1986) and Scott (1992). The methods are also covered in detail in Pagan and Ullah (1999) and Li and Racine (2007).

In this chapter, we focus on univariate density estimation. The setting is a real-valued random variable X for which we have n independent observations. The maintained assumption is that X has a continuous density $f(x)$. The goal is to estimate $f(x)$ either at a single point x or at a set of points in the interior of the support of X. For purposes of presentation, we focus on estimation at a single point x.

17.2 HISTOGRAM DENSITY ESTIMATION

To make things concrete, let us again use the March 2009 Current Population Survey (see Section 6.3), but in this application, we use the subsample of Asian women, which has $n = 1,149$ observations. Our goal is to estimate the density $f(x)$ of hourly wages for this group.

A simple and familiar density estimator is a histogram. We divide the range of $f(x)$ into B bins of width w and then count the number of observations n_j in each bin. The histogram estimator of $f(x)$ for x in the jth bin is

$$\widehat{f}(x) = \frac{n_j}{nw}. \tag{17.1}$$

The histogram is the plot of these heights, displayed as rectangles. The scaling is set so that the sum of the area of the rectangles is $\sum_{j=1}^{B} wn_j/nw = 1$, and therefore the histogram estimator is a valid density.

Figure 17.1(a) displays the histogram of the sample described above, using bins of width \$10. For example, the first bar shows that 189 of the 1,149 individuals had wages in the range [0, 10), so the height of the histogram is $189/(1149 \times 10) = 0.016$.

The histogram in Figure 17.1(a) is a rather crude estimate. For example, it in uninformative as to whether \$11 or \$19 wages are more prevalent. To address this shortcoming, Figure 17.1(b) displays a histogram calculated with bins of width \$1. In contrast with part (a), this appears quite noisy, and we might guess that the

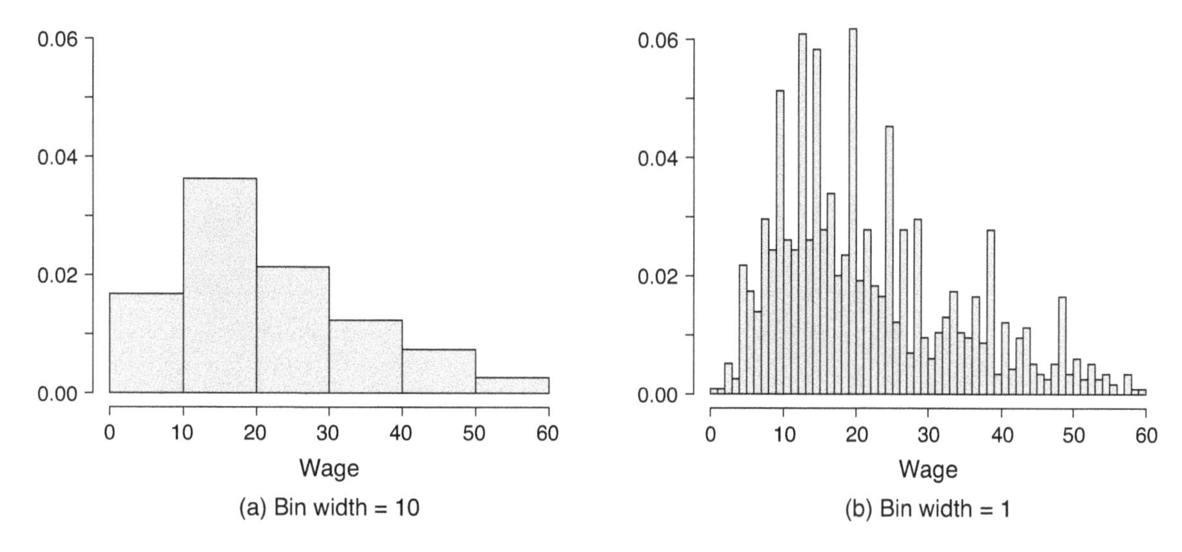

FIGURE 17.1 Histogram estimate of wage density for asian women

peaks and valleys are sampling noise. We would like an estimator which avoids the two extremes shown in Figure 17.1. For this, we consider smoother estimators in the following section.

17.3 KERNEL DENSITY ESTIMATOR

Continuing the wage density example from Section 17.2, suppose we want to estimate the density at $x = \$13$. Consider the histogram density estimate in Figure 17.1(a). It is based on the frequency of observations in the interval $[10, 20)$, which is a skewed window about $x = 13$. It seems more sensible to center the window at $x = 13$, for example, to use $[8, 18)$ instead of $[10, 20)$. It also seems sensible to give more weight to observations close to $x = 13$ and less to those at the edge of the window.

These considerations give rise to what is called the **kernel density estimator** of $f(x)$:

$$\widehat{f}(x) = \frac{1}{nh}\sum_{i=1}^{n} K\left(\frac{X_i - x}{h}\right) \tag{17.2}$$

where $K(u)$ is a weighting function known as a **kernel function**, and $h > 0$ is a scalar known as a **bandwidth.** The estimator (17.2) is the sample average of the "kernel smooths" $h^{-1}K\left(\frac{X_i - x}{h}\right)$. The estimator (17.2) was first proposed by Rosenblatt (1956) and Parzen (1962), and is often called the "Rosenblatt" or "Rosenblatt-Parzen" kernel density estimator.

Kernel density estimators (17.2) can be constructed with any kernel satisfying the following definition.

Definition 17.1 A (second-order) **kernel function** $K(u)$ satisfies

1. $0 \le K(u) \le \overline{K} < \infty$.
2. $K(u) = K(-u)$.
3. $\int_{-\infty}^{\infty} K(u)du = 1$.
4. $\int_{-\infty}^{\infty} |u|^r K(u)du < \infty$ for all positive integers r.

Table 17.1
Common normalized second-order kernels

Kernel	Formula	R_K	C_K
Rectangular	$K(u)=\begin{cases}\frac{1}{2\sqrt{3}} & \text{if } \lvert u\rvert<\sqrt{3}\\ 0 & \text{otherwise}\end{cases}$	$\frac{1}{2\sqrt{3}}$	1.064
Gaussian	$K(u)=\frac{1}{\sqrt{2\pi}}\exp\left(-\frac{u^2}{2}\right)$	$\frac{1}{2\sqrt{\pi}}$	1.059
Epanechnikov	$K(u)=\begin{cases}\frac{3}{4\sqrt{5}}\left(1-\frac{u^2}{5}\right) & \text{if } \lvert u\rvert<\sqrt{5}\\ 0 & \text{otherwise}\end{cases}$	$\frac{3\sqrt{5}}{25}$	1.049
Triangular	$K(u)=\begin{cases}\frac{1}{\sqrt{6}}\left(1-\frac{\lvert u\rvert}{\sqrt{6}}\right) & \text{if } \lvert u\rvert<\sqrt{6}\\ 0 & \text{otherwise}\end{cases}$	$\frac{\sqrt{6}}{9}$	1.052
Biweight	$K(u)=\begin{cases}\frac{15}{16\sqrt{7}}\left(1-\frac{u^2}{7}\right)^2 & \text{if } \lvert u\rvert<\sqrt{7}\\ 0 & \text{otherwise}\end{cases}$	$\frac{5\sqrt{7}}{49}$	1.050

Essentially, a kernel function is a bounded probability density function that is symmetric about 0. Assumption 4 in Definition 17.1 is not essential for most results but is a convenient simplification and does not exclude any kernel function used in standard empirical practice.

Furthermore, some of the mathematical expressions are simplified if we restrict attention to kernels whose variance is normalized to unity.

Definition 17.2 A **normalized kernel function** satisfies $\int_{-\infty}^{\infty} u^2K(u)du = 1$.

A large number of functions satisfy Definition 17.1, and many are programmed as options in statistical packages. Table 17.1 lists the most important functions: the **Rectangular**, **Gaussian**, **Epanechnikov**, **Triangular**, and **Biweight** (Quartic) kernels. These kernel functions are displayed (for $u \geq 0$) in Figure 17.2. In practice, it is unnecessary to consider kernels beyond these five.

The histogram density estimator (17.1) equals the kernel density estimator (17.2) at the bin midpoints (e.g., $x = 5$ or $x = 15$ in Figure 17.1(a)) when $K(u)$ is a uniform density function. This is known as the Rectangular kernel. The kernel density estimator generalizes the histogram estimator in two important ways. First, the window is centered at the point x rather than by bins, and second, the observations are weighted by the kernel function. Thus, the estimator (17.2) can be viewed as a smoothed histogram. $\widehat{f}(x)$ counts the frequency of observations X_i that are close to x.

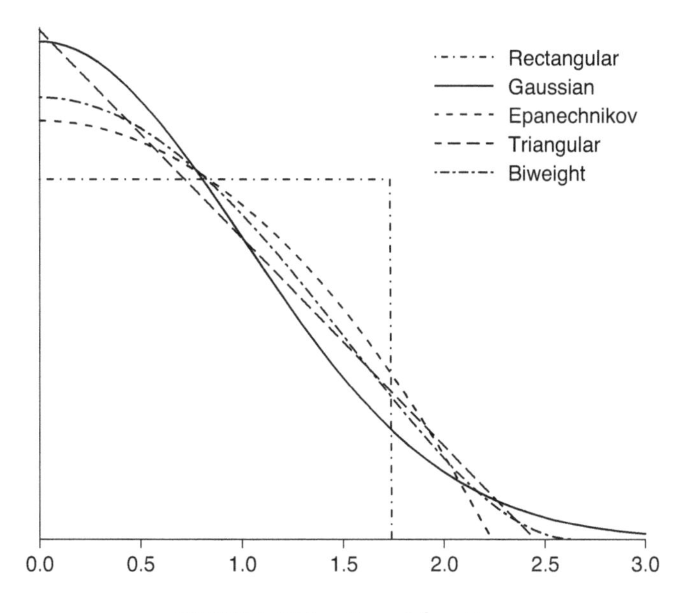

FIGURE 17.2 Kernel functions

I advise against the Rectangular kernel, as it produces discontinuous density estimates. Better choices are the Epanechnikov, Biweight, and Gaussian kernels, which give more weight to observations X_i near the point of evaluation x. In most practical applications, these three kernels provide very similar density estimates, with the Gaussian somewhat smoother. In practice, the Gaussian kernel is a convenient choice, because it produces a density estimator which possesses derivatives of all orders. The Triangular kernel is not typically used for density estimation but is used for other purposes in nonparametric estimation.

The kernel $K(u)$ weights observations based on the distance between X_i and x. The bandwidth h determines what is meant by "close". Consequently, the estimator (17.2) critically depends on the bandwidth h.

Definition 17.3 A **bandwidth** or **tuning parameter** $h > 0$ is a real number used to control the degree of smoothing of a nonparametric estimator.

Typically, larger values of a bandwidth h result in smoother estimators, and smaller values of h result in less-smooth estimators.

Kernel estimators are invariant to rescaling the kernel function and bandwidth. That is, the estimator (17.2) using a kernel $K(u)$ and bandwidth h is equal for any $b > 0$ to a kernel density estimator using the kernel $K_b(u) = K(u/b)/b$ with bandwith h/b.

Kernel density estimators are also invariant to data rescaling. That is, let $Y = cX$ for some $c > 0$. Then the density of Y is $f_Y(y) = f_X(y/c)/c$. If $\widehat{f}_X(x)$ is the estimator (17.2) using the observations X_i and bandwidth h, and $\widehat{f}_Y(y)$ is the estimator using the scaled observations Y_i with bandwidth ch, then $\widehat{f}_Y(y) = \widehat{f}_X(y/c)/c$.

The kernel density estimator (17.2) is a valid density function. Specifically, it is nonnegative and integrates to 1. To see the latter point,

$$\int_{-\infty}^{\infty} \widehat{f}(x)dx = \frac{1}{nh}\sum_{i=1}^{n}\int_{-\infty}^{\infty} K\left(\frac{X_i - x}{h}\right) dx = \frac{1}{n}\sum_{i=1}^{n}\int_{-\infty}^{\infty} K(u)\, du = 1$$

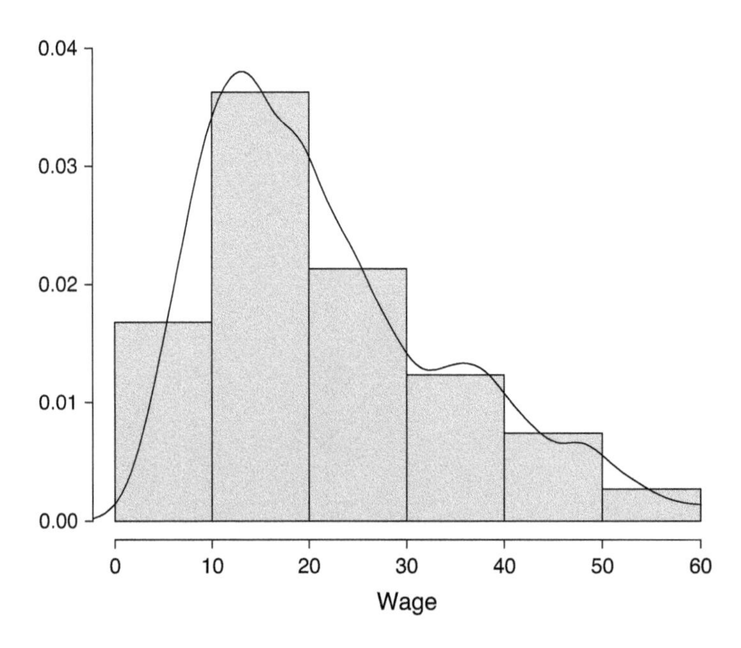

FIGURE 17.3 Kernel density estimator of wage density for asian women

where the second equality makes the change of variables $u = (X_i - x)/h$, and the final uses assumption 3 in Definition 17.1.

To illustrate, Figure 17.3 displays the histogram estimator along with the kernel density estimator using the Gaussian kernel with the bandwidth $h = 2.14$. (Bandwidth selection is discussed in Section 17.9.) You can see that the density estimator is a smoothed version of the histogram. The density appears to be single-peaked and asymmetric, with its mode about $x = \$13$.

17.4 BIAS OF DENSITY ESTIMATOR

In this section, we discuss how to approximate the bias of the density estimator. Since the kernel density estimator (17.2) is an average of i.i.d. observations, its expectation is

$$\mathbb{E}\left[\widehat{f}(x)\right] = \mathbb{E}\left[\frac{1}{nh}\sum_{i=1}^{n} K\left(\frac{X_i - x}{h}\right)\right] = \mathbb{E}\left[\frac{1}{h}K\left(\frac{X - x}{h}\right)\right].$$

At this point, you may feel unsure whether we can proceed further, as $K((X_i - x)/h)$ is a nonlinear function of the random variable X_i. To make progress, let us write the expectation as an explicit integral:

$$\int_{-\infty}^{\infty} \frac{1}{h}K\left(\frac{v - x}{h}\right) f(v)dv.$$

The next step is a trick. Make the change of variables $u = (v - x)/h$, so that the integral equals

$$\int_{-\infty}^{\infty} K(u) f(x + hu)du = f(x) + \int_{-\infty}^{\infty} K(u)\left(f(x + hu) - f(x)\right) du \tag{17.3}$$

where the final equality uses assumption 3 in Definition 17.1.

Expression (17.3) shows that the expected value of $\widehat{f}(x)$ is a weighted average of the function $f(u)$ about the point $u=x$. When $f(x)$ is linear, then $\widehat{f}(x)$ will be unbiased for $f(x)$. In general, however, $\widehat{f}(x)$ is a biased estimator.

As h decreases to 0, the bias term in (17.3) tends to 0:

$$\mathbb{E}\left[\widehat{f}(x)\right]=f(x)+o(1).$$

Intuitively, (17.3) is an average of $f(u)$ in a local window about x. If the window is sufficiently small, then this average should be close to $f(x)$.

Under a stronger smoothness condition, we can provide an improved characterization of the bias. Make a second-order Taylor series expansion of $f(x+hu)$ so that

$$f(x+hu)=f(x)+f'(x)hu+\frac{1}{2}f''(x)h^2u^2+o(h^2).$$

Substituting, we find that (17.3) equals

$$\begin{aligned}
&=f(x)+\int_{-\infty}^{\infty}K(u)\left(f'(x)hu+\frac{1}{2}f''(x)h^2u^2\right)du+o(h^2)\\
&=f(x)+f'(x)h\int_{-\infty}^{\infty}uK(u)\,du+\frac{1}{2}f''(x)h^2\int_{-\infty}^{\infty}u^2K(u)\,du+o(h^2)\\
&=f(x)+\frac{1}{2}f''(x)h^2+o(h^2).
\end{aligned}$$

The final equality uses $\int_{-\infty}^{\infty}uK(u)\,du=0$ and $\int_{-\infty}^{\infty}u^2K(u)\,du=1$. We have shown that (17.3) simplifies to

$$\mathbb{E}\left[\widehat{f}(x)\right]=f(x)+\frac{1}{2}f''(x)h^2+o\left(h^2\right).$$

This result is revealing. It shows that the approximate bias of $\widehat{f}(x)$ is $\frac{1}{2}f''(x)h^2$, which is consistent with our earlier finding that the bias decreases as h tends to 0, but is a more constructive characterization. We see that the bias depends on the underlying curvature of $f(x)$ through its second derivative. If $f''(x)<0$ (as it typical at the mode), then the bias is negative, meaning that $\widehat{f}(x)$ is typically less than the true $f(x)$. If $f''(x)>0$ (as may occur in the tails), then the bias is positive, meaning that $\widehat{f}(x)$ is typically greater than the true $f(x)$. This is smoothing bias.

Let us summarize our findings. Let $\mathcal{N}$ be a neighborhood of x.

Theorem 17.1 If $f(x)$ is continuous in $\mathcal{N}$, then as $h\to 0$,

$$\mathbb{E}\left[\widehat{f}(x)\right]=f(x)+o(1). \tag{17.4}$$

If $f''(x)$ is continuous in $\mathcal{N}$, then as $h\to 0$,

$$\mathbb{E}\left[\widehat{f}(x)\right]=f(x)+\frac{1}{2}f''(x)h^2+o\left(h^2\right). \tag{17.5}$$

A formal proof is presented in Section 17.16.

The asymptotic unbiasedness result (17.4) holds under the minimal assumption that $f(x)$ is continuous. The asymptotic expansion (17.5) holds under the stronger assumption that the second derivative is continuous. These are examples of what are often called **smoothness** assumptions. They are interpreted as meaning that

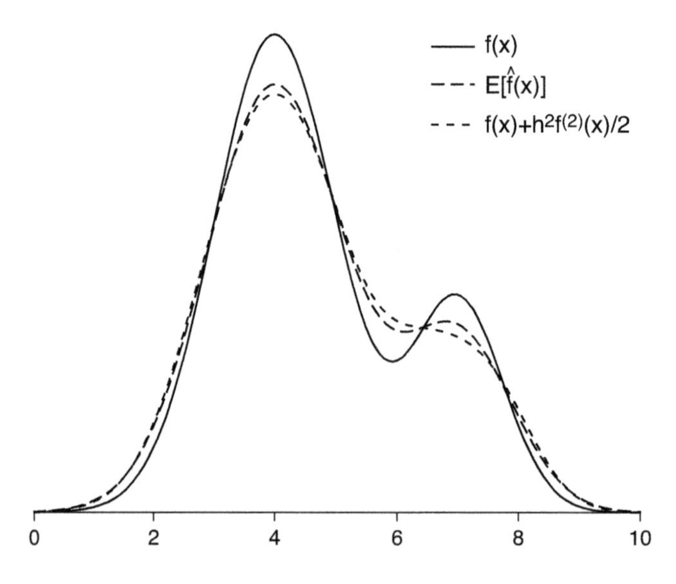

FIGURE 17.4 Smoothing bias

the density is not too variable. It is a common feature of nonparametric theory to use smoothness assumptions to obtain asymptotic approximations.

To illustrate the bias of the kernel density estimator, Figure 17.4 displays the density

$$f(x) = \frac{3}{4}\phi(x-4) + \frac{1}{3}\phi\left(\frac{x-7}{3/4}\right)$$

with the solid line. You can see that the density is bimodal, with local peaks at 4 and 7. Now imagine estimating this density using a Gaussian kernel and a bandwidth of $h = 0.5$ (which turns out to be the reference rule (see Section 17.9) for a sample size $n = 200$). The expectation $\mathbb{E}\left[\widehat{f}(x)\right]$ of this estimator is plotted using the long dashes. You can see that it has the same general shape as $f(x)$, with the same local peaks, but the peak and valley are attenuated. The expectation is a smoothed version of the actual density $f(x)$. The asymptotic approximation $f(x) + f''(x)h^2/2$ is displayed using the short dashes. It is similar to the expectation $\mathbb{E}\left[\widehat{f}(x)\right]$ but is not identical. The difference between $f(x)$ and $\mathbb{E}\left[\widehat{f}(x)\right]$ is the bias of the estimator.

17.5 VARIANCE OF DENSITY ESTIMATOR

Since $\widehat{f}(x)$ is a sample average of the kernel smooths and the latter are i.i.d., the exact variance of $\widehat{f}(x)$ is

$$\operatorname{var}\left[\widehat{f}(x)\right] = \frac{1}{n^2h^2}\operatorname{var}\left[\sum_{i=1}^{n} K\left(\frac{X_i - x}{h}\right)\right] = \frac{1}{nh^2}\operatorname{var}\left[K\left(\frac{X-x}{h}\right)\right].$$

This can be approximated by calculations similar to those used for the bias.

Theorem 17.2 The exact variance of $\widehat{f}(x)$ is

$$V_{\widehat{f}} = \operatorname{var}\left[\widehat{f}(x)\right] = \frac{1}{nh^2}\operatorname{var}\left[K\left(\frac{X-x}{h}\right)\right]. \tag{17.6}$$

If $f(x)$ is continuous in $\mathcal{N}$, then as $h \to 0$ and $nh \to \infty$,

$$V_{\widehat{f}} = \frac{f(x)R_K}{nh} + o\left(\frac{1}{nh}\right) \tag{17.7}$$

where

$$R_K = \int_{-\infty}^{\infty} K(u)^2 du \tag{17.8}$$

is known as the **roughness** of the kernel $K(u)$.

The proof is presented in Section 17.16.

Equation (17.7) shows that the asymptotic variance of $\widehat{f}(x)$ is inversely proportional to nh, which can be viewed as the effective sample size. The variance is proportional to the height of the density $f(x)$ and the kernel roughness R_K. The values of R_K for the kernel functions are displayed in Table 17.1.

17.6 VARIANCE ESTIMATION AND STANDARD ERRORS

The expressions (17.6) and (17.7) can be used to motivate estimators of the variance $V_{\widehat{f}}$. An estimator based on the finite sample formula (17.6) is the scaled sample variance of the kernel smooths $h^{-1}K\left(\frac{X_i-x}{h}\right)$:

$$\widehat{V}_{\widehat{f}}(x) = \frac{1}{n-1}\left(\frac{1}{nh^2}\sum_{i=1}^{n} K\left(\frac{X_i-x}{h}\right)^2 - \widehat{f}(x)^2\right).$$

An estimator based on the asymptotic formula (17.7) is

$$\widehat{V}_{\widehat{f}}(x) = \frac{\widehat{f}(x)R_K}{nh}. \tag{17.9}$$

Using either estimator, the standard error for $\widehat{f}(x)$ is $\widehat{V}_{\widehat{f}}(x)^{1/2}$.

17.7 INTEGRATED MEAN SQUARED ERROR OF DENSITY ESTIMATOR

A useful measure of precision of a density estimator is its **integrated mean squared error** (IMSE):

$$\text{IMSE} = \int_{-\infty}^{\infty} \mathbb{E}\left[\left(\widehat{f}(x) - f(x)\right)^2\right] dx.$$

It is the average precision of $\widehat{f}(x)$ over all values of x. Using Theorems 17.1 and 17.2, we can calculate that it equals

$$\text{IMSE} = \frac{1}{4}R\left(f''\right)h^4 + \frac{R_K}{nh} + o\left(h^4\right) + o\left(\frac{1}{nh}\right)$$

where

$$R\left(f''\right)=\int_{-\infty}^{\infty}\left(f''(x)\right)^2 dx$$

is called the **roughness** of the second derivative $f''(x)$. The leading term,

$$\text{AIMSE}=\frac{1}{4}R\left(f''\right)h^4+\frac{R_K}{nh} \tag{17.10}$$

is called the **asymptotic integrated mean squared error**. The AIMSE is an asymptotic approximation to the IMSE. In nonparametric theory, it is common to use AIMSE to assess precision.

The AIMSE (17.10) shows that $\widehat{f}(x)$ is less accurate when $R\left(f''\right)$ is large, meaning that accuracy deteriorates with increased curvature in $f(x)$. The expression also shows that the first term (the squared bias) of the AIMSE is increasing in h, but the second term (the variance) is decreasing in h. Thus the choice of h affects (17.10) with a tradeoff between bias and variance.

We can calculate the bandwidth h which minimizes the AIMSE by solving the first-order condition (See Exercise 17.2). The solution is

$$h_0=\left(\frac{R_K}{R\left(f''\right)}\right)^{1/5} n^{-1/5}. \tag{17.11}$$

This bandwidth takes the form $h_0=cn^{-1/5}$, so it satisfies the intriguing rate $h_0\sim n^{-1/5}$.

A common error is to interpret $h_0\sim n^{-1/5}$ as meaning that a user can set $h=n^{-1/5}$. This is incorrect and can be a huge mistake in an application. The constant c is critically important as well.

When $h\sim n^{-1/5}$, then $\text{AIMSE}\sim n^{-4/5}$, which means that the density estimator converges at the rate $n^{-2/5}$. This rate is slower than the standard $n^{-1/2}$ parametric rate, which is a common finding in nonparametric analysis. An interpretation is that nonparametric estimation problems are harder than parametric problems, so more observations are required to obtain accurate estimates.

Theorem 17.3 summarizes our findings.

Theorem 17.3 If $f''(x)$ is uniformly continuous, then

$$\text{IMSE}=\frac{1}{4}R\left(f''\right)h^4+\frac{R_K}{nh}+o\left(h^4\right)+o\left((nh)^{-1}\right).$$

The leading terms (the AIMSE) are minimized by the bandwidth h_0 in (17.11).

17.8 OPTIMAL KERNEL

Expression (17.10) shows that the choice of kernel function affects the AIMSE only through R_K. Thus the kernel with the smallest R_K will have the smallest AIMSE. As shown by Hodges and Lehmann (1956), R_K is minimized by the Epanechnikov kernel. Thus the density estimation with the Epanechnikov kernel is AIMSE efficient. This observation led Epanechnikov (1969) to recommend this kernel for density estimation.

Theorem 17.4 AIMSE is minimized by the Epanechnikov kernel.

Theorem 17.4 is proved later in this section.

It is also interesting to calculate the efficiency loss obtained by using a different kernel. Inserting the optimal bandwidth (17.11) into the AIMSE (17.10) and doing a little algebra, we find that for any kernel, the optimal AIMSE is

$$\text{AIMSE}_0(K) = \frac{5}{4} R\left(f''\right)^{1/5} R_K^{4/5} n^{-4/5}.$$

The square root of the ratio of the optimal AIMSE of the Gaussian kernel to the Epanechnikov kernel is

$$\left(\frac{\text{AIMSE}_0(\text{Gaussian})}{\text{AIMSE}_0(\text{Epanechnikov})}\right)^{1/2} = \left(\frac{R_K(\text{Gaussian})}{R_K(\text{Epanechnikov})}\right)^{2/5} = \left(\frac{1/2\sqrt{\pi}}{3\sqrt{5}/25}\right)^{2/5} \simeq 1.02.$$

Thus the efficiency loss[1] from using the Gaussian kernel relative to the Epanechnikov is only 2%. This is not particularly large. Therefore from an efficiency viewpoint, the Epanechnikov is optimal, and the Gaussian is near-optimal.

The Gaussian kernel has other advantages over the Epanechnikov. The Gaussian kernel possesses derivatives of all orders (is infinitely smooth), so kernel density estimates with the Gaussian kernel will also have derivatives of all orders. This is not the case with the Epanechnikov kernel, as its first derivative is discontinuous at the boundary of its support. Consequently, estimates calculated using the Gaussian kernel are smoother and particularly well suited for estimation of density derivatives. Another feature is that the density estimator $\widehat{f}(x)$ with the Gaussian kernel is nonzero for all x, which can be a useful if the inverse $\widehat{f}(x)^{-1}$ is desired. These considerations lead to the practical recommendation to use the Gaussian kernel. A compromise is the Biweight kernel, which has nearly the efficiency of the Epanechnikov, yet is fourth-order differentiable, so it is useful for the estimation of low-order density derivatives.

Let us now prove Theorem 17.4. To do so, we use the calculus of variations. Construct the Lagrangian

$$\mathscr{L}(K, \lambda_1, \lambda_2) = \int_{-\infty}^{\infty} K(u)^2 du - \lambda_1 \left(\int_{-\infty}^{\infty} K(u) du - 1\right) - \lambda_2 \left(\int_{-\infty}^{\infty} u^2 K(u) du - 1\right).$$

The first term is R_K. The constraints are that the kernel integrates to 1, and the second moment is 1. Taking the derivative with respect to $K(u)$ and setting to 0, we obtain

$$\frac{d}{dK(u)} \mathscr{L}(K, \lambda_1, \lambda_2) = \left(2K(u) - \lambda_1 - \lambda_2 u^2\right) \mathbb{1}\left\{K(u) \geq 0\right\} = 0.$$

Solving for $K(u)$, we find the solution

$$K(u) = \frac{1}{2}\left(\lambda_1 + \lambda_2 u^2\right) \mathbb{1}\left\{\lambda_1 + \lambda_2 u^2 \geq 0\right\}$$

which is a truncated quadratic. The constants λ_1 and λ_2 may be found by seting $\int_{-\infty}^{\infty} K(u) du = 1$ and $\int_{-\infty}^{\infty} u^2 K(u) du = 1$. After some algebra, we find that the solution is the Epanechnikov kernel as listed in Table 17.1.

17.9 REFERENCE BANDWIDTH

The density estimator (17.2) depends critically on the bandwidth h. Without a specific rule to select h, the method is incomplete. Consequently, an important component of nonparametric estimation methods are data-dependent bandwidth selection rules.

[1] Measured by the square root of AIMSE.

A simple bandwidth selection rule proposed by Silverman (1986) has come to be known as the **reference bandwidth** or **Silverman's Rule-of-Thumb**. It uses the bandwidth (17.11) calculated under the simplifying assumption that the true density $f(x)$ is normal, with a few variations. The rule produces a reasonable bandwidth for many estimation contexts.

The Silverman rule is

$$h_r = \sigma C_K n^{-1/5} \tag{17.12}$$

where σ is the standard deviation of the distribution of X, and

$$C_K = \left(\frac{8\sqrt{\pi} R_K}{3}\right)^{1/5}.$$

The constant C_K is determined by the kernel. Its values are recorded in Table 17.1.

The Silverman rule is simple to derive. Using a change of variables, you can calculate that when $f(x) = \sigma^{-1}\phi(x/\sigma)$, then $R(f'') = \sigma^{-5}R(\phi'')$. A technical calculation (see Theorem 17.5 below) shows that $R(\phi'') = 3/8\sqrt{\pi}$. Using these two results, we obtain the reference estimate $R(f'') = \sigma^{-5}3/8\sqrt{\pi}$. Inserted into the Silverman rule (17.11), we obtain equation (17.12).

For the Gaussian kernel, $R_K = 1/2\sqrt{\pi}$, so the constant C_K is

$$C_K = \left(\frac{8\sqrt{\pi}}{3}\frac{1}{2\sqrt{\pi}}\right)^{1/5} = \left(\frac{4}{3}\right)^{1/5} \simeq 1.059. \tag{17.13}$$

Thus the Silverman rule (17.12) is often written as

$$h_r = \sigma 1.06 n^{-1/5}. \tag{17.14}$$

It turns out that the constant (17.13) is remarkably robust to the choice of kernel. Notice that C_K depends on the kernel only through R_K, which is minimized by the Epanechnikov kernel for which $C_K \simeq 1.05$ and is maximized (among single-peaked kernels) by the rectangular kernel for which $C_K \simeq 1.06$. Thus the constant C_K is essentially invariant to the kernel. Consequently, the Silverman rule (17.14) can be used by any kernel with unit variance.

The unknown standard deviation σ needs to be replaced with a sample estimator. Using the sample standard deviation s, we obtain a classical reference rule for the Gaussian kernel, sometimes referred to as the optimal bandwidth under the assumption of normality:

$$h_r = s 1.06 n^{-1/5}. \tag{17.15}$$

Silverman (1986, section 3.4.2) recommended the robust estimator $\widetilde{\sigma}$ of Section 11.14. This gives rise to a second form of the reference rule

$$h_r = \widetilde{\sigma} 1.06 n^{-1/5}. \tag{17.16}$$

Silverman (1986) observed that the constant $C_K = 1.06$ produces a bandwidth which is a bit too large when the density $f(x)$ is thick-tailed or bimodal. He therefore recommended using a slightly smaller bandwidth in practice, and based on simulation evidence, he specifically recommended $C_K = 0.9$. This leads to a third form of the reference rule:

$$h_r = 0.9\widetilde{\sigma} n^{-1/5}. \tag{17.17}$$

The rule (17.17) is popular in package implementations and is commonly known as **Silverman's Rule of Thumb**.

The kernel density estimator implemented with any of the above reference bandwidths is fully data dependent and thus a valid estimator. (That is, it does not depend on user-selected tuning parameters.) This is a good property.

Let us close this section by justifying the claim $R(\phi'') = 3/8\sqrt{\pi}$ with a more general calculation. Its proof is presented in Section 17.16.

Theorem 17.5 For any integer $m \geq 0$,

$$R\left(\phi^{(m)}\right) = \frac{\mu_{2m}}{2^{m+1}\sqrt{\pi}} \tag{17.18}$$

where $\mu_{2m} = (2m-1)!! = \mathbb{E}\left[Z^{2m}\right]$ is the $2m$th moment of the standard normal density.

17.10 SHEATHER-JONES BANDWIDTH*

In this section, we discuss a bandwidth selection rule derived by Sheather and Jones (1991), which has improved performance over the reference rule. The AIMSE-optimal bandwidth (17.11) depends on the unknown roughness $R(f'')$. An improvement on the reference rule may be obtained by replacing $R(f'')$ with a nonparametric estimator.

Consider the general problem of estimation of $S_m = \int_{-\infty}^{\infty}\left(f^{(m)}(x)\right)^2 dx$ for some integer $m \geq 0$. By m applications of integration by parts, we can calculate that

$$S_m = (-1)^m \int_{-\infty}^{\infty} f^{(2m)}(x) f(x) dx = (-1)^m \mathbb{E}\left[f^{(2m)}(X)\right]$$

where the second equality uses the fact that $f(x)$ is the density of X. Let $\widehat{f}(x) = (nb_m)^{-1}\sum_{i=1}^{n}\phi\left((X_i - x)/b_m\right)$ be a kernel density estimator using the Gaussian kernel and bandwidth b_m. An estimator of $f^{(2m)}(x)$ is

$$\widehat{f}^{(2m)}(x) = \frac{1}{nb_m^{2m+1}}\sum_{i=1}^{n}\phi^{(2m)}\left(\frac{X_i - x}{b_m}\right).$$

A nonparametric estimator of S_m is

$$\widehat{S}_m(b_m) = \frac{(-1)^m}{n}\sum_{i=1}^{n}\widehat{f}^{(2m)}(X_i) = \frac{(-1)^m}{n^2 b_m^{2m+1}}\sum_{i=1}^{n}\sum_{j=1}^{n}\phi^{(2m)}\left(\frac{X_i - X_j}{b_m}\right).$$

Jones and Sheather (1991) calculated that the MSE-optimal bandwith b_m for the estimator $\widehat{S}_m$ is

$$b_m = \left(\sqrt{\frac{2}{\pi}}\frac{\mu_{2m}}{S_{m+1}}\right)^{1/(3+2m)} n^{-1/(3+2m)} \tag{17.19}$$

where $\mu_{2m} = (2m-1)!!$. The bandwidth (17.19) depends on the unknown S_{m+1}. One solution is to replace S_{m+1} with a reference estimate. Given Theorem 17.5, this is $S_{m+1} = \sigma^{-3-2m}\mu_{2m+2}/2^{m+2}\sqrt{\pi}$. Substituted into (17.19) and simplifying, we obtain the reference bandwidth

$$\widetilde{b}_m = \sigma\left(\frac{2^{m+5/2}}{2m+1}\right)^{1/(3+2m)} n^{-1/(3+2m)}.$$

Used to estimate S_m, we obtain the feasible estimator $\widetilde{S}_m = \widehat{S}_m(\widetilde{b}_m)$. It turns out that two reference bandwidths of interest are

$$\widetilde{b}_2 = 1.24\sigma n^{-1/7}$$

and

$$\widetilde{b}_3 = 1.23\sigma n^{-1/9}$$

for $\widetilde{S}_2$ and $\widetilde{S}_3$.

A **plug-in bandwidth** h is obtained by replacing the unknown $S_2 = R\left(f''\right)$ in (17.11) with $\widetilde{S}_2$. Its performance, however, depends critically on the preliminary bandwidth b_2, which depends on the reference rule estimator $\widetilde{S}_3$.

Sheather and Jones (1991) improved on the plug-in bandwidth with the following algorithm, which takes into account the interactions between h and b_2. Take the two equations for optimal h and b_2 with S_2 and S_3 replaced with the reference estimates $\widetilde{S}_2$ and $\widetilde{S}_3$:

$$h = \left(\frac{R_K}{\widetilde{S}_2}\right)^{1/5} n^{-1/5}$$

$$b_2 = \left(\sqrt{\frac{2}{\pi}} \frac{3}{\widetilde{S}_3}\right)^{1/7} n^{-1/7}.$$

Solve the first equation for n and plug it into the second equation, viewing it as a function of h. We obtain

$$b_2(h) = \left(\sqrt{\frac{2}{\pi}} \frac{3}{R_K} \frac{\widetilde{S}_2}{\widetilde{S}_3}\right)^{1/7} h^{5/7}.$$

Now use $\widetilde{b}_2(h)$ to make the estimator $\widehat{S}_2(\widetilde{b}_2(h))$ a function of h. Find the h that is the solution to the equation

$$h = \left(\frac{R_K}{\widehat{S}_2(\widetilde{b}_2(h))}\right)^{1/5} n^{-1/5}. \tag{17.20}$$

The solution for h must be found numerically, but it is easy to solve by the Newton-Raphson method. Theoretical and simulation analysis have shown that the resulting bandwidth h and density estimator $\widehat{f}(x)$ perform quite well in a range of contexts.

When the kernel $K(u)$ is Gaussian, the relevant formulas are

$$b_2(h) = 1.357 \left(\frac{\widetilde{S}_2}{\widetilde{S}_3}\right)^{1/7} h^{5/7}$$

and

$$h = \frac{0.776}{\widehat{S}_2(\widetilde{b}_2(h))^{1/5}} n^{-1/5}.$$

17.11 RECOMMENDATIONS FOR BANDWIDTH SELECTION

In general it is advisable to try several bandwidths and use your judgment. Estimate the density function using each bandwidth. Plot the results and compare. Select your density estimator based on the evidence, your purpose for estimation, and your judgment.

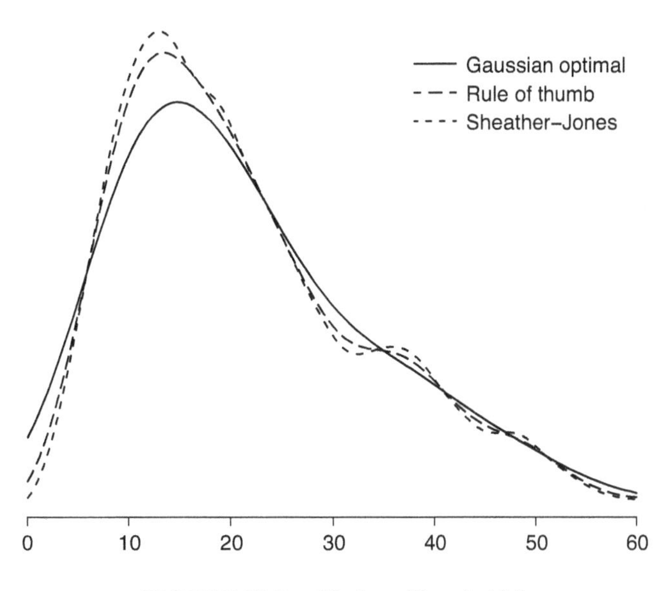

FIGURE 17.5 Choice of bandwidth

For example, take the empirical example presented in Section 17.3, which deals with wages for the subsample of Asian women. There are $n = 1149$ observations. Thus $n^{-1/5} = 0.24$. The sample standard deviation is $s = 20.6$. Thus the Gaussian optimal rule (17.15) is

$$h = s1.06n^{-1/5} = 5.34.$$

The interquartile range is $\widehat{R} = 18.8$. The robust estimate of standard deviation is $\widetilde{\sigma} = 14.0$. The rule-of-thumb (17.17) is

$$h = 0.9sn^{-1/5} = 3.08.$$

This value is smaller than the Gaussian optimal bandwidth, mostly because the robust standard deviation is much smaller than the sample standard deviation.

The Sheather-Jones bandwidth that solves equation (17.20) is $h = 2.14$, which is significantly smaller than the other two bandwidths. This is because the empirical roughness estimate $\widehat{S}_2$ is much larger than the normal reference value.

The density is estimated using these three bandwidths and the Gaussian kernel and is displayed in Figure 17.5. What we can see is that the estimate using the largest bandwidth (the Gaussian optimal) is the smoothest, and the estimate using the smallest bandwidth (Sheather-Jones) is the least smooth. The Gaussian optimal estimate understates the primary density mode and overstates the left tail, relative to the other two estimates. The Gaussian optimal estimate seems over-smoothed. The estimates using the rule-of-thumb and the Sheather-Jones bandwidth are reasonably similar, and the choice between the two may be made partly on aesthetics. The rule-of-thumb estimate produces a smoother estimate, which may be more appealing to the eye, while the Sheather-Jones estimate produces more detail. My preference leans toward detail, and hence the Sheather-Jones bandwidth. This is the justification for the choice $h = 2.14$ used for the density estimate displayed in Figure 17.3.

If you are working in a package that only produces one bandwith rule (such as Stata), then it is advisable to experiment by trying alternative bandwidths obtained by adding and subtracting modest deviations (e.g., 20–30%) and then assess the density plots obtained.

We can also assess the impact of the choice of kernel function. Figure 17.6 displays the density estimates calculated using the rectangular, Gaussian, and Epanechnikov kernel functions, and the Sheather-Jones

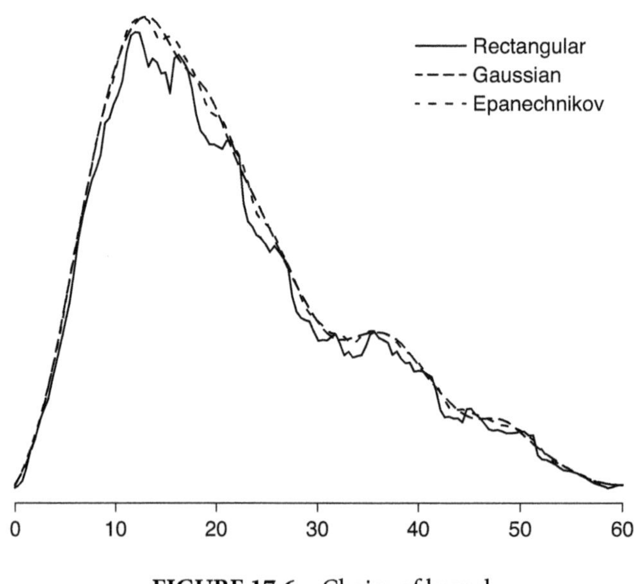

FIGURE 17.6 Choice of kernel

bandwidth (optimized for the Gaussian kernel). The shapes of the three density estimates are very similar, and the Gaussian and Epanechnikov estimates are nearly indistinguishable, with the Gaussian being slightly smoother. The estimate using the rectangular kernel, however, is noticably different. It is erratic and non-smooth. This figure illustrates how the rectangular kernel is a poor choice for density estimation, and the differences between the Gaussian and Epanechnikov kernels are typically minor.

17.12 PRACTICAL ISSUES IN DENSITY ESTIMATION

The most common purpose for a density estimator $\widehat{f}(x)$ is to produce a display such as Figure 17.3. In this case, the estimator $\widehat{f}(x)$ is calculated on a grid of values of x and then plotted. Typically the use of 100 gridpoints is sufficient for a reasonable density plot. However, if the density estimate has a section with a steep slope, it may be poorly displayed unless more gridpoints are used.

Sometimes it is questionable as to whether a density estimator can be used when the observations are somewhat in between continuous and discrete. For example, many variables are recorded as integers, even though the underlying model treats them as continuous. A practical suggestion is to refrain from applying a density estimator unless there are at least 50 distinct values in the dataset.

There is also a practical question about sample size. How large should the sample be to apply a kernel density estimator? The convergence rate is slow, so we should expect to require a larger number of observations than for parametric estimators. I suggest a minimal sample size of $n = 100$, and even then, estimation precision may be poor.

17.13 COMPUTATION

In Stata, the kernel density estimator (17.2) can be computed and displayed using the `kdensity` command. By default, it uses the Epanechnikov kernel and selects the bandwidth using the reference rule (17.17). One deficiency of the Stata `kdensity` command is that it incorrectly implements the reference rule for kernels with non-unit variances (which includes all kernel options in Stata other than the Epanechnikov and Gaussian). Consequently, the `kdensity` command should only be used with either the Epanechnikov or the Gaussian kernel.

R has several commands for density estimation, including the built-in command `density`. By default, the latter uses the Gaussian kernel and the reference rule (17.17). The reference rule can be explicitly specified using the option `nrd0`. Other kernels and bandwidth selection methods are available, including equation (17.16) as the option `nrd` and the Sheather-Jones method as the option `SJ`.

MATLAB has the built-in function `kdensity`. By default, it uses the Gaussian kernel and the reference rule (17.16).

17.14 ASYMPTOTIC DISTRIBUTION

In this section, we discuss asymptotic limit theory for the kernel density estimator (17.2). We first state a consistency result.

Theorem 17.6 If $f(x)$ is continuous in $\mathscr{N}$, then as $h \to 0$ and $nh \to \infty, \widehat{f}(x) \underset{p}{\longrightarrow} f(x)$.

This theorem shows that the nonparametric estimator $\widehat{f}(x)$ is consistent for $f(x)$ under quite minimal assumptions. Theorem 17.6 follows from (17.4) and (17.7).

I now provide an asymptotic distribution theory.

Theorem 17.7 If $f''(x)$ is continuous in $\mathscr{N}$, then as $nh \to \infty$ such that $h = O\left(n^{-1/5}\right)$,

$$\sqrt{nh}\left(\widehat{f}(x) - f(x) - \frac{1}{2}f''(x)h^2\right) \underset{d}{\longrightarrow} \mathrm{N}\left(0, f(x)R_K\right).$$

The proof is given in Section 17.16.

Theorems 17.1 and 17.2 characterize the asymptotic bias and variance. Theorem 17.7 extends this by applying the Lindeberg CLT to show that the asymptotic distribution is normal.

The convergence result in Theorem 17.7 is different from typical parametric results in two aspects. The first is that the convergence rate is $\sqrt{nh}$ rather than $\sqrt{n}$. This is because the estimator is based on local smoothing, and the effective number of observations for local estimation is nh rather than the full sample n. The second notable aspect of the theorem is that the statement includes an explicit adjustment for bias, as the estimator $\widehat{f}(x)$ is centered at $f(x) + \frac{1}{2}f''(x)h^2$. This is because the bias is not asymptotically negligible and needs to be acknowledged. The presence of a bias adjustment is typical in the asymptotic theory for kernel estimators.

Theorem 17.7 adds the extra technical condition that $h = O\left(n^{-1/5}\right)$, which strengthens the assumption $h \to 0$ by saying it must decline at least at the rate $n^{-1/5}$. This condition ensures that the remainder from the bias approximation is asymptotically negligible. It can be weakened somewhat if the smoothness assumptions on $f(x)$ are strengthened.

17.15 UNDERSMOOTHING

A technical way to eliminate the bias term in Theorem 17.7 is by using an **undersmoothing** bandwidth, which is a bandwidth h that converges to 0 faster than the optimal rate $n^{-1/5}$, so that $nh^5 = o(1)$. In practice this means that h is smaller than the optimal bandwidth, so the estimator $\widehat{f}(x)$ is AIMSE inefficient.

An undersmoothing bandwidth can be obtained by setting $h = n^{-\alpha} h_r$, where h_r is a reference or plug-in bandwidth and $\alpha > 0$.

With a smaller bandwidth, the estimator has reduced bias and increased variance. Consequently, the bias is asymptotically negligible.

Theorem 17.8 If $f''(x)$ is continuous in $\mathscr{N}$, then as $nh \to \infty$ such that $nh^5 = o(1)$,

$$\sqrt{nh}\left(\widehat{f}(x) - f(x)\right) \underset{d}{\longrightarrow} \mathrm{N}\left(0, f(x) R_K\right).$$

This theorem looks identical to Theorem 17.7 with the notable difference that the bias term is omitted. At first, this result appears to be a "better" distribution, as it is certainly preferred to have (asymptotically) unbiased estimators. However this rationale is incomplete. Theorem 17.7 (with the bias term included) is a better distribution result precisely because it captures the asymptotic bias. Theorem 17.8 is inferior precisely because it avoids characterizing the bias. Another way of thinking about it is that Theorem 17.7 is a more honest characterization of the distribution than is Theorem 17.8.

It is worth noting that the assumption $nh^5 = o(1)$ is the same as $h = o\left(n^{-1/5}\right)$. Some authors will state it one way, and some the other. The assumption means that the estimator $\widehat{f}(x)$ is converging at a slower rate than is optimal, and is thus AIMSE inefficient.

While the undersmoothing assumption $nh^5 = o(1)$ technically eliminates the bias from the asymptotic distribution, it does not actually eliminate the finite sample bias. Thus it is better in practice to view an undersmoothing bandwidth as producing an estimator with "low bias" rather than with "zero bias". It should also be remembered that an undersmoothing bandwidth has larger variance than the IMSE-optimal bandwidth.

17.16 TECHNICAL PROOFS*

For simplicity, all formal results assume that the kernel $K(u)$ has bounded support; that is, for some $a < \infty$, $K(u) = 0$ for $|u| > a$. This includes most kernels used in applications with the exception of the Gaussian kernel. The results apply as well to the Gaussian kernel but with a more detailed argument.

Proof of Theorem 17.1 We first show (17.4). Fix $\epsilon > 0$. Since $f(x)$ is continuous in some neighborhood $\mathscr{N}$, there exists a $\delta > 0$ such that $|v| \leq \delta$ implies $\left|f(x+v) - f(x)\right| \leq \epsilon$. Set $h \leq \delta/a$. Then $|u| \leq a$ implies $|hu| \leq \delta$ and $\left|f(x+hu) - f(x)\right| \leq \epsilon$. Then using (17.3),

$$\begin{aligned}
\left|\mathbb{E}\left[\widehat{f}(x) - f(x)\right]\right| &= \left|\int_{-a}^{a} K(u)\left(f(x+hu) - f(x)\right) du\right| \\
&\leq \int_{-a}^{a} K(u)\left|f(x+hu) - f(x)\right| du \\
&\leq \epsilon \int_{-a}^{a} K(u)\, du \\
&= \epsilon.
\end{aligned}$$

Since ϵ is arbitrary, this shows that $\left|\mathbb{E}\left[\widehat{f}(x) - f(x)\right]\right| = o(1)$ as $h \to 0$, as claimed.

We next prove equation (17.5). By the mean-value theorem,

$$\begin{aligned}f(x+hu) &= f(x)+f'(x)hu+\frac{1}{2}f''(x+hu^*)h^2u^2\\ &= f(x)+f'(x)hu+\frac{1}{2}f''(x)h^2u^2+\frac{1}{2}\left(f''(x+hu^*)-f''(x)\right)h^2u^2\end{aligned}$$

where u^* lies between 0 and u. Substituting into (17.3) and using $\int_{-\infty}^{\infty} K(u)\,u du=0$ and $\int_{-\infty}^{\infty} K(u)\,u^2 du=1$, we find

$$\mathbb{E}\left[\widehat{f}(x)\right]=f(x)+\frac{1}{2}f''(x)h^2+h^2R(h)$$

where

$$R(h)=\frac{1}{2}\int_{-\infty}^{\infty}\left(f''(x+hu^*)-f''(x)\right)u^2K(u)\,du.$$

It remains to show that $R(h)=o(1)$ as $h\to 0$. Fix $\epsilon>0$. Since $f''(x)$ is continuous in some neighborhood $\mathcal{N}$, there exists a $\delta>0$ such that $|v|\le\delta$ implies $\left|f''(x+v)-f''(x)\right|\le\epsilon$. Set $h\le\delta/a$. Then $|u|\le a$ implies $|hu^*|\le|hu|\le\delta$ and $\left|f''(x+hu^*)-f''(x)\right|\le\epsilon$. Then

$$|R(h)|\le\frac{1}{2}\int_{-\infty}^{\infty}\left|f''(x+hu^*)-f''(x)\right|u^2K(u)\,du\le\frac{\epsilon}{2}.$$

Since ϵ is arbitrary, this shows that $R(h)=o(1)$. This completes the proof. ■

Proof of Theorem 17.2 As mentioned at the beginning of this section, for simplicity, assume $K(u)=0$ for $|u|>a$.

Equation (17.6) has been established in Section 17.5. We now prove equation (17.7). By a derivation similar to that for Theorem 17.1, since $f(x)$ is continuous in $\mathcal{N}$, we have

$$\begin{aligned}\frac{1}{h}\mathbb{E}\left[K\left(\frac{X-x}{h}\right)^2\right] &= \int_{-\infty}^{\infty}\frac{1}{h}K\left(\frac{v-x}{h}\right)^2 f(v)dv\\ &= \int_{-\infty}^{\infty}K(u)^2 f(x+hu)du\\ &= \int_{-\infty}^{\infty}K(u)^2 f(x)du+o(1)\\ &= f(x)R_K+o(1).\end{aligned}$$

Then since the observations are i.i.d. and using (17.5), we have

$$\begin{aligned}nh\,\mathrm{var}\left[\widehat{f}(x)\right] &= \frac{1}{h}\mathrm{var}\left[K\left(\frac{X-x}{h}\right)\right]\\ &= \frac{1}{h}\mathbb{E}\left[K\left(\frac{X-x}{h}\right)^2\right]-h\left(\mathbb{E}\left[\frac{1}{h}K\left(\frac{X-x}{h}\right)\right]\right)^2\\ &= f(x)R_K+o(1).\end{aligned}$$

This is identical to the stated result. ■

Proof of Theorem 17.5 By m applications of integration by parts, the fact $\phi^{(2m)}(x) = He_{2m}(x)\phi(x)$ where $He_{2m}(x)$ is the $2m$th Hermite polynomial (see Section 5.10), the fact $\phi(x)^2 = \phi\left(\sqrt{2}x\right)/\sqrt{2\pi}$, the change-of-variables $u = x/\sqrt{2}$, an explicit expression for the Hermite polynomial, the normal moment $\int_{-\infty}^{\infty} u^{2mj}\phi(u)du = (2m-1)!! = (2m)!/\left(2^m m!\right)$ (for the second equality, see Section A.3), and the binomial theorem (Theorem 1.11), we find

$$\begin{aligned}
R\left(\phi^{(m)}\right) &= \int_{-\infty}^{\infty} \phi^{(m)}(x)\phi^{(m)}(x)dx \\
&= (-1)^m \int_{-\infty}^{\infty} He_{2m}(x)\phi(x)^2 dx \\
&= \frac{(-1)^m}{\sqrt{2\pi}} \int_{-\infty}^{\infty} He_{2m}(x)\phi\left(\sqrt{2}x\right) dx \\
&= \frac{(-1)^m}{2\sqrt{\pi}} \int_{-\infty}^{\infty} He_{2m}\left(u/\sqrt{2}\right)\phi(u)\, du \\
&= \frac{(-1)^m}{2\sqrt{\pi}} \int_{-\infty}^{\infty} \sum_{j=0}^{m} \frac{(2m)!}{j!\,(2m-2j)!2^m} (-1)^j u^{2m-2j}\phi(u)\, du \\
&= \frac{(-1)^m (2m)!}{2^{2m+1} m! \sqrt{\pi}} \sum_{j=0}^{m} \frac{m!}{j!\,(m-j)!} (-1)^j \\
&= \frac{(2m)!}{2^{2m+1} m! \sqrt{\pi}} \\
&= \frac{\mu_{2m}}{2^{m+1}\sqrt{\pi}}
\end{aligned}$$

as claimed. ■

Proof of Theorem 17.7 Define

$$Y_{ni} = h^{-1/2}\left(K\left(\frac{X_i - x}{h}\right) - \mathbb{E}\left[K\left(\frac{X_i - x}{h}\right)\right]\right)$$

so that

$$\sqrt{nh}\left(\widehat{f}(x) - \mathbb{E}\left[\widehat{f}(x)\right]\right) = \sqrt{n}\overline{Y}_n.$$

We must verify the conditions for the Lindeberg CLT (Theorem 9.1). It is necessary to verify the Lindeberg condition, because Lyapunov's condition fails.

In the notation of Theorem 9.1, $\overline{\sigma}_n^2 = \operatorname{var}\left[\sqrt{n}\overline{Y}_n\right] \to R_K f(x)$ as $h \to 0$. Notice that since the kernel function is positive and finite, $0 \le K(u) \le \overline{K}$, say, then $Y_{ni}^2 \le h^{-1}\overline{K}^2$. Fix $\epsilon > 0$. Then

$$\lim_{n\to\infty} \mathbb{E}\left[Y_{ni}^2 \mathbb{1}\left\{Y_{ni}^2 > \epsilon n\right\}\right] \le \lim_{n\to\infty} \mathbb{E}\left[Y_{ni}^2 \mathbb{1}\left\{\overline{K}^2/\epsilon > nh\right\}\right] = 0$$

the final equality holds since $nh > \overline{K}^2/\epsilon$ for sufficiently large n. This establishes the Lindeberg condition (9.1). The Lindeberg CLT (Theorem 9.1) shows that

$$\sqrt{nh}\left(\widehat{f}(x) - \mathbb{E}\left[\widehat{f}(x)\right]\right) = \sqrt{n}\,\overline{Y} \xrightarrow[d]{} \mathrm{N}\left(0, f(x)R_K\right).$$

Equation (17.5) establishes

$$\mathbb{E}\left[\widehat{f}(x)\right] = f(x) + \frac{1}{2}f''(x)h^2 + o(h^2).$$

Since $h = O\left(h^{-1/5}\right)$,

$$\sqrt{nh}\left(\widehat{f}(x) - f(x) - \frac{1}{2}f''(x)h^2\right) = \sqrt{nh}\left(\widehat{f}(x) - \mathbb{E}\left[\widehat{f}(x)\right]\right) + o(1)$$

$$\underset{d}{\longrightarrow} \mathrm{N}\left(0, f(x)R_K\right).$$

This completes the proof. ■

17.17 EXERCISES

Exercise 17.1 If X^* is a random variable with density $\widehat{f}(x)$ from (17.2), show that

(a) $\mathbb{E}[X^*] = \overline{X}_n$.
(b) $\mathrm{var}[X^*] = \widehat{\sigma}^2 + h^2$.

Exercise 17.2 Show that (17.11) minimizes (17.10).
Hint: Differentiate (17.10) with respect to h and set to 0. This is the first-order condition for optimization. Solve for h. Check the second-order condition to verify that this is a minimum.

Exercise 17.3 Suppose that $f(x)$ is the uniform density on $[0, 1]$. What does (17.11) suggest should be the optimal bandwidth h? How do you interpret this?

Exercise 17.4 You estimate a density for expenditures measured in dollars, and then re-estimate measuring in millions of dollars, but use the same bandwidth h. How do you expect the density plot to change? What bandwidth should you use so that the density plots have the same shape?

Exercise 17.5 You have a sample of wages for 1,000 men and 1,000 women. You estimate the density functions $\widehat{f}_m(x)$ and $\widehat{f}_w(x)$ for the two groups using the same bandwidth h. You then take the average $\widehat{f}(x) = \left(\widehat{f}_m(x) + \widehat{f}_w(x)\right)/2$. How does this compare to applying the density estimator to the combined sample?

Exercise 17.6 You increase your sample from $n = 1,000$ to $n = 2,000$. For univariate density estimation, how does the AIMSE-optimal bandwidth change? What happens if the sample increases from $n = 1,000$ to $n = 10,000$?

Exercise 17.7 Using the asymptotic formula (17.9) to calculate standard errors $s(x)$ for $\widehat{f}(x)$, find an expression that indicates when $\widehat{f}(x) - 2s(x) < 0$, which means that the asymptotic 95% confidence interval contains negative values. For what values of x is this likely (that is, around the mode or toward the tails)? If you generate a plot of $\widehat{f}(x)$ with confidence bands, and the latter include negative values, how should you interpret this?

CHAPTER 18

EMPIRICAL PROCESS THEORY

18.1 INTRODUCTION

An advanced and powerful branch of asymptotic theory is known as empirical process theory, which concerns the asymptotic distributions of random functions. Two of the primary results are the uniform law of large numbers and the functional central limit theorem.

This chapter is a brief introduction, but it is still the most advanced material presented in this textbook. For more detail, I recommend Pollard (1990), Andrews (1994), van der Vaart and Wellner (1996), and chapters 18–19 of van der Vaart (1998).

18.2 FRAMEWORK

The general setting for empirical process theory is the context of random functions—random elements of general function spaces. Practical applications include the empirical distribution function, partial sum processes, and log likelihood viewed as functions of the parameter space. For concreteness, most of the treatment here will focus on the setting where the random function of interest is a sample average or normalized sample average.

Let $X_i \in \mathscr{X} \subset \mathbb{R}^m$ be an i.i.d. set of random vectors, and $g(x, \theta)$ a vector function over $x \in \mathscr{X}$ and $\theta \in \mathbb{R}^k \subset \Theta$. Its average is

$$\overline{g}_n(\theta) = \frac{1}{n}\sum_{i=1}^{n} g(X_i, \theta). \tag{18.1}$$

We are interested in the properties of $\overline{g}_n(\theta)$ as a function over $\theta \in \Theta$. For example, we are interested in the conditions under which $\overline{g}_n(\theta)$ satisfies a uniform law of large numbers. One application arises in the theory of consistent estimation of nonlinear models (MLE and method of moments), in which cases the function $g(x, \theta)$ equals the log-density function (for MLE) or the moment function (for method of moments).

We are also interested in the properties of the normalized average

$$\nu_n(\theta) = \sqrt{n}\left(\overline{g}_n(\theta) - \mathbb{E}\left[\overline{g}_n(\theta)\right]\right) = \frac{1}{\sqrt{n}}\sum_{i=1}^{n}\left(g(X_i, \theta) - \mathbb{E}\left[g(X_i, \theta)\right]\right) \tag{18.2}$$

as a function over $\theta \in \Theta$. One application arises in advanced proofs of asymptotic normality of the MLE (e.g., Theorem 10.9) with $g(x, \theta)$ equaling the score function (the derivative of the log density).

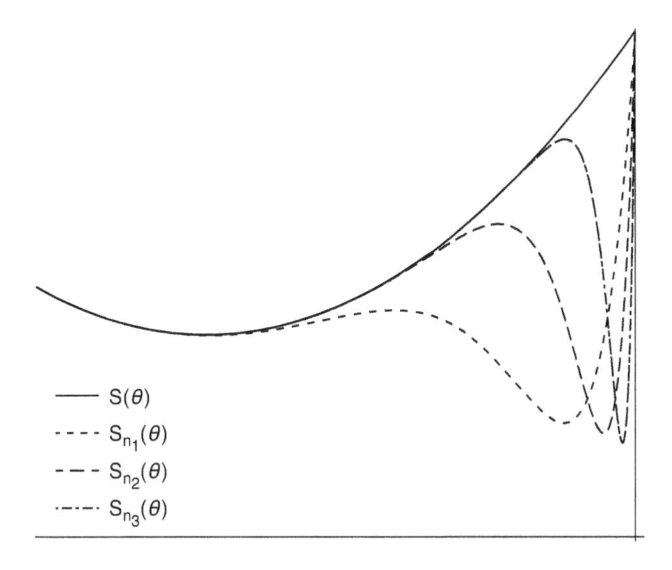

FIGURE 18.1 Non-uniform convergence

18.3 GLIVENKO-CANTELLI THEOREM

Empirical process theory arose from the study of the empirical distribution function (EDF)

$$F_n(\theta) = \frac{1}{n}\sum_{i=1}^{n} \mathbb{1}\{X_i \le \theta\}. \tag{18.3}$$

As discussed in Section 11.12, this is an estimator of the distribution function $F(\theta) = \mathbb{E}[\mathbb{1}\{X \le \theta\}]$. The EDF (18.3) is an example of equation (18.1) with $g(x,\theta) = \mathbb{1}\{x \le \theta\}$. Empirical process theory was developed to demonstrate that $F_n(\theta)$ is uniformly consistent for $F(\theta)$ and to provide its asymptotic distribution viewed as a function of θ.

For any $\theta \in \mathbb{R}$, $F_n(\theta)$ is the sample average of the i.i.d. random variables $\mathbb{1}\{X_i \le \theta\}$. From the WLLN, we deduce that $F_n(\theta) \underset{p}{\longrightarrow} F(\theta)$. This is true for any given θ. Since this convergence holds at a fixed value of θ, we call this **pointwise** convergence in probability. **Uniform** converence in probability is the following stronger property.

Definition 18.1 $S_n(\theta)$ **converges in probability** to $S(\theta)$ **uniformly** over $\theta \in \Theta$ if

$$\sup_{\theta\in\Theta} \|S_n(\theta) - S(\theta)\| \underset{p}{\longrightarrow} 0.$$

To see the issue, examine Figure 18.1. This displays a sequence of functions $S_n(\theta)$ (the dashed lines) for three values of n. The figure illustrates that for each θ, the function $S_n(\theta)$ converges toward the limit function $S(\theta)$. However for each n, the function $S_n(\theta)$ has a severe dip in the right-hand region. The result is that the global behavior of $S_n(\theta)$ is distinct from that of the limit function $S(\theta)$. In particular, the sequence of minimizers of $S_n(\theta)$ converge to the right-limit of the parameter space, while the minimizer of the limit criterion $S(\theta)$ is in the interior of the parameter space.

In contrast, Figure 18.2 displays an example of uniform convergence. The heavy solid line is the function $S(\theta)$. The dashed lines are $S(\theta) + \epsilon$ and $S(\theta) - \epsilon$. The thin solid line is $S_n(\theta)$. The figure illustrates a situation

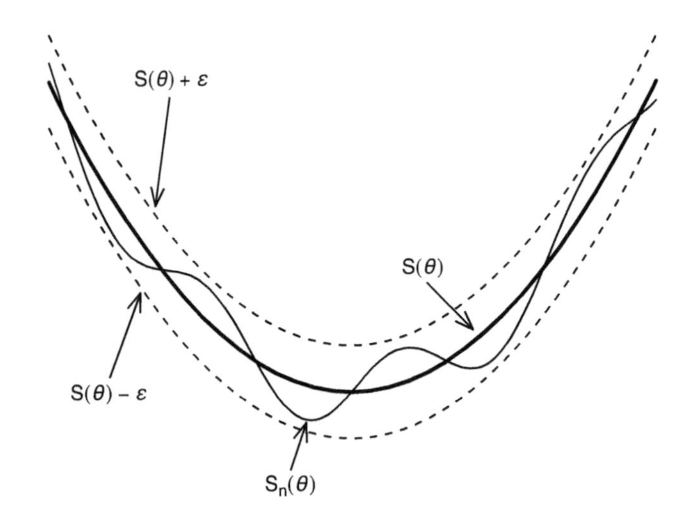

FIGURE 18.2 Uniform convergence

where $S_n(\theta)$ satisfies $\sup_{\theta\in\Theta}|S_n(\theta)-S(\theta)|<\epsilon$. The function $S_n(\theta)$ as displayed weaves up and down but stays within ϵ of $S(\theta)$. Uniform convergence holds if the event shown in this figure holds with high probability for n sufficiently large, for any arbitrarily small ϵ. While $S_n(\theta)$ as shown in Figure 18.2 is quite wiggly, the vertical magnitude of the wiggles is bounded by 2ϵ. Uniform convergence holds if for any $\epsilon>0$, we obtain this bound for sufficiently large n.

The following famous result establishes that the EDF (18.3) converges uniformly.

Theorem 18.1 Glivenko-Cantelli. If $X_i\in\mathbb{R}$ are i.i.d. with distribution $F(\theta)$, then

$$\sup_{\theta\in\mathbb{R}}|F_n(\theta)-F(\theta)|\underset{p}{\longrightarrow}0. \tag{18.4}$$

The Glivenko-Cantelli Theorem is a foundational result in empirical process theory. One of the wonderful features of the theorem is that it does not require any technical or regularity conditions. It holds for all distribution functions $F(\theta)$, including continuous, discrete, and mixed. The proof of Theorem 18.1 is not particularly challenging, but is deferred for now, as it emerges as a special case of the uniform law of large numbers under bracketing (see Theorem 18.2 in Section 18.5).

18.4 PACKING, COVERING, AND BRACKETING NUMBERS

To exclude the erratic behavior displayed in Figure 18.1, we want functions that are not too variable over Θ. In this section, we describe variability in terms of our ability to approximate the function space with a finite set of functions. There are three widely used concepts (packing, covering, and bracketing) based on somewhat different metaphors.

To start, we require a norm on the function space. The most common is the L_r norm for $r\geq 1$. For a function $h(x)$ the L_r **norm** is

$$\|h\|_r=\left(\mathbb{E}\left[\|h(X)\|^r\right]\right)^{1/r}.$$

This norm depends on the distribution F of X. It will be sometimes useful to make this dependence explicit. For a distribution Q, let $\mathbb{E}_Q[h(X)] = \int h(x)dQ(x)$, and let

$$\|h\|_{Q,r} = \left(\mathbb{E}_Q\left[\|h(X)\|^r\right]\right)^{1/r}.$$

This is the L_r norm when the expectation is calculated with the distribution Q.

We will use the L_r and $L_r(Q)$ norms to measure distances between values of θ. The $\boldsymbol{L_r}$ **distance** (or **metric**) between θ_1 and θ_2 is

$$d_r(\theta_1, \theta_2) = \left\|g(X, \theta_1) - g(X, \theta_2)\right\|_r$$

and the $L_r(Q)$ distance is

$$d_{Q,r}(\theta_1, \theta_2) = \left\|g(X, \theta_1) - g(X, \theta_2)\right\|_{Q,r}.$$

These metrics measure the gap between θ_1 and θ_2 by the rth moment of the difference between the random variables $g(X, \theta_1)$ and $g(X, \theta_2)$. When $g(\theta)$ is nonrandom, the L_r and $L_r(Q)$ distances simplify to $\left\|g(\theta_1) - g(\theta_2)\right\|$.

In addition to the above notation, the theory requires the the following construct.

Definition 18.2 A function $g(x, \theta)$ has an **envelope function** $G(x)$ if $\left\|g(x, \theta)\right\| \leq G(x) < \infty$ for every $x \in \mathscr{X}$ and $\theta \in \Theta$.

For example, take $g(x, \theta) = \mathbb{1}\{x \leq \theta\}$ for the EDF. This has envelope $G(x) = 1$. As another example, take $g(x, \theta) = x^\theta$ with $x \in [0, 1]$, which has envelope $G(x) = 1$. As a final example, take $g(x, \theta) = x\left(y - x'\theta\right)$ with $\|\theta\| \leq C$, which has envelope $G(x) = \left\|xy\right\| + \|x\|^2 C$. The last example shows that in some cases, a finite envelope can only be achieved on a bounded parameter space.

Let us now define the packing and covering numbers. The packing numbers are the number of points which you can "pack" into Θ while maintaining a minimum distance[1] between the points. The covering numbers are the number of balls required to cover Θ. An $L_r(Q)$-ball of radius ϵ centered at θ_j is the set $\left\{\theta : d_{Q,r}\left(\theta, \theta_j\right) < \epsilon\right\}$.

Definition 18.3 The **packing number** $D_r(\epsilon, Q)$ is the largest number of points θ_j that can be packed into Θ with each pair satisfying $d_{Q,r}(\theta_1, \theta_2) > \epsilon$. The **covering number** $N_r(\epsilon, Q)$ is the minimal number of $L_r(Q)$-balls of radius ϵ needed to cover Θ. The **uniform packing** and **uniform covering numbers** are

$$D_r(\epsilon) = \sup_Q D_r(\epsilon \|G\|_{Q,r}, Q)$$

$$N_r(\epsilon) = \sup_Q N_r(\epsilon \|G\|_{Q,r}, Q).$$

Packing and covering numbers are closely related, because they satisfy $N(\epsilon) \leq D(\epsilon) \leq N(\epsilon/2)$. Thus both have the same approximation properties. Therefore the choice of packing versus covering is a matter of convenience. We will focus on the covering numbers $N(\epsilon)$ without loss of generality. The role of the uniform packing/covering numbers is to eliminate dependence on the specific probability distribution Q. Notice that their definition effectively normalizes ϵ by the magnitude of the envelope function. This normalization is needed; otherwise, $D_r(\epsilon)$ and $N_r(\epsilon)$ could be infinite.

[1]Ironically, I am writing this at the time of the COVID-19 pandemic, during which individuals were instructed to maintain a "social distance" between individuals of 1.5 meters.

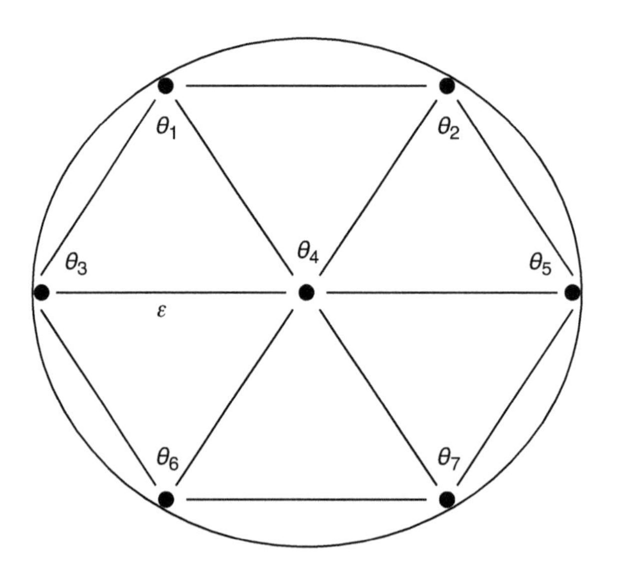

FIGURE 18.3 Packing numbers

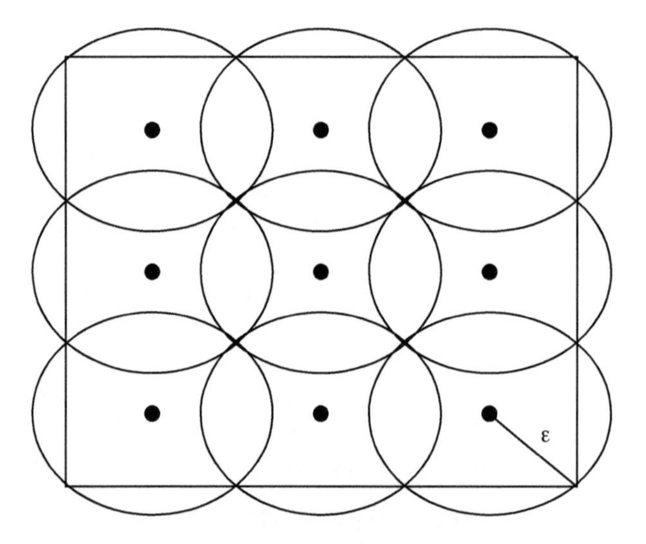

FIGURE 18.4 Covering numbers

Packing is illustrated in Figure 18.3. Displayed is a circle. Marked are seven points θ_j, which are each separated by ϵ. We can see that it would be impossible to place another point in the circle while maintaining the ϵ separation. Thus $D(\epsilon) = 7$.

Covering is illustrated in Figure 18.4. Displayed is a square. Marked are nine balls, each of radius ϵ. The balls precisely cover the square, and this could not be done with fewer balls. Thus, $N(\epsilon) = 9$.

The bracketing approach makes a different type of approximation. We say that a function $h(x) \in \mathbb{R}$ is **bracketed** by the functions $\ell(x)$ and $u(x)$ if $\ell(x) \leq h(x) \leq u(x)$ for all $x \in \mathscr{X}$. Given a pair of functions ℓ and u, the **bracket** $[\ell, u]$ is the set of all functions h which satisfy $\ell(x) \leq h(x) \leq u(x)$ for all $x \in \mathscr{X}$. It is an ϵ-$L_r(Q)$-**bracket** if $\|u(X) - \ell(X)\|_{Q,r} \leq \epsilon$. We say that a set of brackets $\left[\ell_j, u_j\right]$ **covers** Θ if for all $\theta \in \Theta$, $g(x, \theta)$

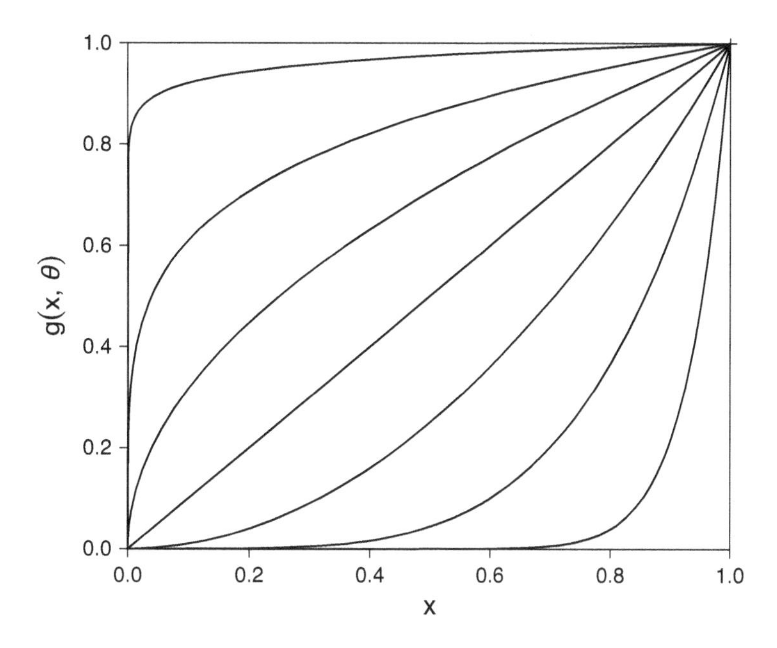

FIGURE 18.5 Bracketing numbers

is bracketed by $\left[\ell_j, u_j\right]$ for some j. The bracketing functions $\ell(x)$ and $u(x)$ do not need to be members of the function class $g(x, \theta)$.

Definition 18.4 The **bracketing number** $N_{[\ ]}(\epsilon, L_r(Q))$ is the minimum number of ϵ-$L_r(Q)$-brackets needed to cover Θ.

Bracketing is illustrated in Figure 18.5. Displayed is the region $[0,1]^2$. Consider the function $g(x,\theta)=x^{\theta}$ with $X \sim U[0,1]$. The function $g(x,\theta)$ is displayed in Figure 18.5 for seven values of θ. The values have been selected so that the L_2 distance is equal between each neighboring pair. Thus the region between the lowest and highest functions contains six brackets.

If $g(x,\theta)$ is in a 2ϵ-$L_r(Q)$-bracket $[\ell, u]$, then it is in a ball of radius ϵ about $(\ell+u)/2$. Thus the covering and bracketing numbers satisfy the relationship $N_r(\epsilon, Q) \leq N_{[\ ]}(2\epsilon, L_r(Q))$. Hence bounds on bracketing numbers imply the same bounds for covering numbers, suggesting that the former are more restrictive. However, our theorems (in the following sections) impose restrictions on the uniform covering numbers $N_r(\epsilon)$ which are more restrictive than the analogous restrictions on the bracketing numbers $N_{[\ ]}(\epsilon, L_r)$. Hence, neither the uniform covering number $N_r(\epsilon)$ nor the bracketing number $N_{[\ ]}(\epsilon, L_r)$ is more or less restrictive than the other.

What is useful about the packing, covering, and bracketing numbers is that they reveal the complexity of the function $g(x,\theta)$. The rate at which $D(\epsilon)$, $N(\epsilon)$, and $N_{[\ ]}(\epsilon)$ increase to infinity as ϵ decreases to 0 describes the the difficulty of approximating the function g with a finite number of elements.

To illustrate, take the empirical distribution function, for which $g(x,\theta)=\mathbb{1}\{x \leq \theta\}$. Recall the quantile function $q(\alpha)$ of a distribution $Q(x)$ (Definition 11.2). Given ϵ set $N=2\epsilon^{-r}$, $\theta_j = q\left(j/N\right)$, $\ell_j(x)=\mathbb{1}\left\{x \leq \theta_j\right\}$, and $u_j(x)=\mathbb{1}\left\{x \leq \theta_{j-1}\right\}$. The pair $\left[\ell_j, u_j\right]$ brackets $\mathbb{1}\{x \leq \theta\}$ for $\theta_{j-1} < \theta \leq \theta_j$. It is useful to observe that $(j/N) \leq Q(\theta_j) < (j+1)/N$. The bracket satisfies $\left\|u_j(X)-\ell_j(X)\right\|_{Q,r} = \left(Q(\theta_j)-Q(\theta_{j-1})\right)^{1/r} \leq (2/N)^{1/r} = \epsilon$.

Thus, $N_{[\,]}(\epsilon, L_r(Q)) \le 2\epsilon^{-r}$. From the relationship between bracketing and covering numbers, we deduce that $N_r(\epsilon, Q) \le 2^{1-r}\epsilon^{-r}$. Both are independent of Q and $\|G\|_{Q,r} \le 1$, so $N_{[\,]}(\epsilon, L_r) \le 2\epsilon^{-r}$ and $N_r(\epsilon) \le 2^{1-r}\epsilon^{-r}$.

18.5 UNIFORM LAW OF LARGE NUMBERS

Let us now generalize the uniform convergence (18.4) for the EDF to sample averages of general functions.

Definition 18.5 The sample average $\overline{g}_n(\theta)$ defined in (18.1) satisfies the **uniform law of large numbers (ULLN)** over Θ if

$$\sup_{\theta\in\Theta} \left\|\overline{g}_n(\theta) - g(\theta)\right\| \underset{p}{\longrightarrow} 0.$$

In the empirical process literature, an alternative label for the ULLN is that $\overline{g}_n(\theta)$ is **Glivenko-Cantelli.**

Theorem 18.2 **ULLN.** $\overline{g}_n(\theta)$ satisfies the ULLN over Θ if

1. X_i are i.i.d.
2. $\mathbb{E}[G(X)] < \infty$.
3. One of the following conditions holds:
   (a) For all $\epsilon > 0$, $N_{[\,]}(\epsilon, L_1) < \infty$,
   (b) For all $\epsilon > 0$, $N_1(\epsilon) < \infty$,
   or
   (c) With probability 1, $g(X,\theta)$ is continuous in θ, and Θ is compact.

Theorem 18.2 states that the sample mean $\overline{g}_n(\theta)$ satisfies the ULLN under one of three conditions: on the bracketing numbers, uniform covering numbers, or continuity of $g(x,\theta)$. Proofs for (a) and (c) are provided in Section 18.9. For case (b), see Theorem 9.13 of van der Vaart (1998).

After Theorem 18.1, I stated that its proof follows from Theorem 18.2. As shown in Section 18.4, for the EDF, the envelope function is $G(X) = 1$ which is bounded, and the bracketing numbers satisfy $N_{[\,]}(\epsilon, L_1) \le 2\epsilon^{-1} < \infty$. This satisfies the conditions for Theorem 18.2. Thus, $F_n(\theta)$ satisfies the ULLN. This completes the proof of Theorem 18.1.

Part (c) of Theorem 18.2 shows that the ULLN holds for sample averages of continuous functions. This condition is quite mild, as it imposes no smoothness requirements beyond continuity. The stated ULLN even allows for discontinuous functions as long as the discontinuities have zero probability. For example, in the case of the EDF, the function $g(x,\theta) = \mathbb{1}\{x \le \theta\}$ is discontinuous at $\theta = x$, but $g(X,\theta)$ is continuous with probability 1 if X has a continuous distribution. Similarly, functions such as $g(Y,X,\theta) = Y\mathbb{1}\{X \le \theta\}$ satisfy part (c) when X has a continuous distribution.

Most applications of the ULLN are to continuous functions (or continuous with probability 1), so part (c) of Theorem 18.2 can be applied. For functions that do not satisfy this condition, we can appeal to either part (a) or part (b). In many cases, the bracketing numbers $N_{[\,]}(\epsilon, L_1)$ can be explicitly calculated (as done for the EDF at the end of Section 18.4), in which case part (a) is convenient. In other cases, general technical conditions bound the uniform covering numbers $N_1(\epsilon)$; see van der Vaart and Wellner (1996) and van der Vaart (1998) for details.

The intuition for Theorem 18.2 is as follows. Consider part (a). For any $\epsilon > 0$, the number of balls needed to cover Θ is finite. Fix these balls. The WLLN can be applied at the center point of each ball, and since the number of balls is fixed, the error can be made uniformly smaller than ϵ. Thus the function is uniformly within 2ϵ of its expectation. Now consider part (b). For any $\epsilon > 0$, the number of brackets needed to cover Θ is finite. Fix these brackets. The WLLN can be applied at each lower bracket $\ell_j(X)$ and at each upper bracket $u_j(X)$, and the error can be made uniformly smaller than ϵ. Finally, consider part (c). If $g(x,\theta)$ is continuous in θ, then for any ϵ and θ, it has an ϵ-L_r-bracket. Since Θ is compact, it can be covered by a finite number of such brackets. Thus the conditions of part (b) are satisfied.

18.6 FUNCTIONAL CENTRAL LIMIT THEORY

Recall the normalized average $\nu_n(\theta)$ defined in (18.2). It is a random function over $\theta \in \Theta$ and is called a **stochastic process**. We sometimes write the process as $\nu_n(\theta)$ to indicate its dependence on the argument θ, and sometimes as ν_n when we want to indicate the entire function. Some authors use the notation $\nu_n(\cdot)$ for the latter.

We are interested in the asymptotic distribution of the process ν_n. It is scaled so that $\nu_n(\theta)$ satisfies the CLT pointwise in θ. We want to extend this to the entire joint process.

Asymptotic distribution theory for stochastic processes is known by a variety of labels including **functional central limit theory**, **invariance principles**, **empirical process limit theory**, **weak convergence**, and **Donsker's theorem**. Modern usage leans toward the label **convergence in distribution** to emphasize the connections with the finite-dimensional case.

As a leading example, consider the normalized empirical distribution function (18.3):

$$\nu_n(\theta) = \sqrt{n}\left(F_n(\theta) - F(\theta)\right). \tag{18.5}$$

By the CLT, $\nu_n(\theta)$ is asymptotically $\mathrm{N}(0, F(\theta)(1 - F(\theta)))$ for any individual θ. However, for some purposes, we require the distribution of the entire function ν_n. For example, to test the hypothesis that the distribution $F(\theta)$ is a specific distribution, we can reject for large values of the Kolmogorov-Smirnov statistic $\mathrm{KS}_n = \max_\theta |\nu_n(\theta)|$. This is the largest discrepancy between the EDF and a null distribution $F(\theta)$. The statistic KS_n is a function of the entire random function ν_n, so for the distribution of KS_n, we require the distribution of the entire function ν_n.

Let V denote the space of functions from Θ to $\mathbb{R}$. The random function ν_n is an element of V. To measure the distance between two elements in V, we use the uniform[2] metric

$$\rho(\nu_1, \nu_2) = \sup_{\theta\in\Theta} |\nu_1(\theta) - \nu_2(\theta)|. \tag{18.6}$$

The convergence in distribution for random functions is defined next.

Definition 18.6 The stochastic process $\nu_n(\theta)$ **converges in distribution** over $\theta \in \Theta$ to a limit random process $\nu(\theta)$ as $n \to \infty$, written $\nu_n \underset{d}{\longrightarrow} \nu$, if $\mathbb{E}\left[f(\nu_n)\right] \to \mathbb{E}\left[f(\nu)\right]$ for every bounded, continuous $f \in V$, where continuity is defined with respect to the uniform metric (18.6).

Some authors instead use the label "weak convergence". It is also common to see the notation $\nu_n \Rightarrow \nu$ (e.g., Billingsley, 1999) or $\nu_n \rightsquigarrow \nu$ (e.g., van der Vaart, 1998). For details, see chapters 1–2 of Billingsley (1999) and chapter 18 of van der Vaart (1998). Convergence in distribution for stochastic processes satisfies standard properties, including the CMT, where continuity is defined with respect to the uniform metric (18.6).

[2] This is an appropriate choice when ν_n is asymptotically continuous. For processes with jumps, an alternative metric is required.

For the normalized average (18.2), convergence in distribution is often called the **Donsker** property.

Just as for the ULLN, convergence of stochastic processes requires that we exclude functions that are "too erratic". The following technical condition is sufficient.

Definition 18.7 A random function $S_n(\theta)$ is **asymptotically equicontinuous** with respect to a metric $d(\theta_1, \theta_2)$ on $\theta \in \Theta$ if for all $\eta > 0$ and $\epsilon > 0$, there is some $\delta > 0$ such that

$$\limsup_{n\to\infty} \mathbb{P}\left[\sup_{d(\theta_1,\theta_2)\leq\delta} \|S_n(\theta_1) - S_n(\theta_2)\| > \eta\right] \leq \epsilon.$$

If no specific metric is mentioned, the default is Euclidean: $d(\theta_1, \theta_2) = \|\theta_1 - \theta_2\|$.

For background, a function $g(\theta)$ is **uniformly continuous** if a small distance between θ_1 and θ_2 implies that the distance between $g(\theta_1)$ and $g(\theta_2)$ is small. A family of functions is **equicontinuous** if this holds uniformly across all functions in the family. Asymptotic equicontinuity states this probabilistically: $g_n(\theta)$ is asymptotically equicontinuous if a small distance between θ_1 and θ_2 implies that the distance between $g_n(\theta_1)$ and $g_n(\theta_2)$ is small with high probability in large samples. This holds if $g_n(\theta)$ is equicontinuous but is broader. The function $g_n(\theta)$ can be discontinuous but the magnitude and probability of such discontinuities must be small. Asymptotically (as $n \to \infty$), an asymptotically equicontinuous function approaches continuity. Asymptotic equicontinuity is also called **stochastic equicontinuity**.

A deep result from probability theory gives conditions for convergence in distribution.

Theorem 18.3 Functional Central Limit Theorem. $\nu_n \underset{d}{\longrightarrow} \nu$ over $\theta \in \Theta$ as $n \to \infty$ if and only if the following conditions hold:

1. $(\nu_n(\theta_1), \ldots, \nu_n(\theta_m)) \underset{d}{\longrightarrow} (\nu(\theta_1), \ldots, \nu_n(\theta_m))$ as $n \to \infty$ for every finite set of points $\theta_1, \ldots, \theta_m$ in Θ.
2. There exists a finite partition $\Theta = \cup_{j=1}^{J} \Theta_j$ such that $\nu_n(\theta)$ is asymptotically equicontinuous over $\theta \in \Theta_j$ for $j = 1, \ldots, J$.

The proof is technically advanced. See theorems 7.1–7.2 of Billingsley (1999) or theorem 18.14 of van der Vaart (1998).

Condition 1 is sometimes called "finite-dimensional distributional convergence". This is typically verified by a multivariate CLT.

The assumption that $\nu_n(\theta)$ is asymptotically equicontinuous is high-level and difficult to directly verify in specific applications. The most common conditions are based on covering and bracketing numbers. Define the **bracketing integral** as

$$J_{[\,]}(\delta, L_2) = \int_0^\delta \sqrt{\log N_{[\,]}(x, L_2)}\,dx.$$

This integral is finite and decreases to 0 as $\delta \to 0$ if the bracketing numbers $N_{[\,]}(\epsilon, L_2)$ do not increase to infinity as $\epsilon \to 0$ too quickly. A sufficient condition for $J_{[\,]}(\delta, L_2) < \infty$ is that $N_{[\,]}(\epsilon, L_2) \leq O\left(\epsilon^{-\rho}\right)$ for some $0 < \rho < \infty$. Such rates are called **Euclidean** by Pollard (1990) and **polynomial** by van der Vaart (1998).

Similarly, the uniform covering integral is

$$J_2(\delta) = \int_0^\delta \sqrt{\log N_2(x)}\,dx = \int_0^\delta \sqrt{\log \sup_Q N_2(x\, \|G\|_{Q,2}, Q)}\,dx.$$

It is finite if the uniform covering numbers do not increase to infinity too quickly. A sufficient condition for this is a polynomial rate $N_2(\epsilon) \le O\left(\epsilon^{-\rho}\right)$.

Theorem 18.4 $\nu_n(\theta)$ is asymptotically equicontinuous over $\theta \in \Theta$ if $J_{[\,]}(\delta, L_2) < \infty$, or if $J_2(\delta) < \infty$ and $\mathbb{E}\left[G(X)^2\right] < \infty$.

See, for example, theorems 19.5 and 19.14 of van der Vaart (1998).

The fact that a finite bracketing integral or uniform covering integral implies asymptotic equicontinuity and thus the functional central limit theorem is not intuitive. A formal proof is highly advanced, technical, and also not particularly intuitive. Instead, I provide here a few comments to attempt a nonrigorous explanation.

Consider the event described in Definition 18.7 for the process $\nu_n(\theta)$. It involves the largest discrepancy between $\nu_n(\theta_1)$ and $\nu_n(\theta_2)$ for all parameters in the uncountable set $A = \{d(\theta_1, \theta_2) \le \delta\}$. This can be managed by what is known as a **chaining** argument. Imagine constructing a sequence of finite covers of the set A, where the jth cover consists of balls of radius $\delta_j = \delta/2^j$. By definition, there are at least $N_2(\delta_j)$ balls in the jth cover. Take any points $\theta_1, \theta_2 \in A$. From the sequence of covers, we can find a sequence of balls with center points θ_j that successively link θ_1 to θ_2, and the gaps satisfy $d\left(\theta_{j-1}, \theta_j\right) \le \delta_j$. This sequence is called a **chain**. The supremum over A can be written as the supremum over all such chains. This involves an infinite sum (over the covers $j = 1, \ldots, \infty$) and over the number $N_2(\delta_j)$ of balls in the cover. Using Markov's (Theorem 7.3) and Boole's (Theorem 1.2) inequalities, the probability in Definition 18.7 can be bounded by an expression of the form $\sum_{j=0}^{\infty} \mathbb{E}\left[\max_{\theta_j} \left\|\nu_n\left(\theta_{j-1}\right) - \nu_n\left(\theta_j\right)\right\|\right]$, where the maximum is over the $N_2(\delta_j)$ balls in the jth cover. The expectation of the maximum of N random variables with a finite pth moment satisfies $O\left(N^{1/p}\right)$ (similar to Theorem 9.7), a bound that becomes approximately logarithmic as $p \to \infty$. In fact, it can be shown that if the random variables are normal, then the expectation of this maximum satisfies $O\left(\sqrt{\log N}\right)$. While the normalized averages $\nu_n(\theta)$ are not finite-sample normal, they are approximately normal in large samples, and so asymptotically are bounded $O\left(\sqrt{\log N}\right)$. This results in a bound of the form $\sum_{j=0}^{\infty} \sqrt{\log N_2(\delta_j)}\delta_j$, which can be bounded by the integral $J_2(\delta)$. When the latter is finite, it can be made arbitrarily small by selecting δ sufficiently small. Thus Definition 18.7 is satisfied, so the process is asymptotically equicontinuous. This explanation is very brief (a complete proof requires considerable probabilistic machinery) and only provides a taste of the argument. What it is meant to reveal is the source of the square-root-log of the covering/bracketing numbers (due to a maximal inequality), and the integral (due to the infinite sum from the approximating chain).

18.7 CONDITIONS FOR ASYMPTOTIC EQUICONTINUITY

In Section 18.6, Theorem 18.3 stated that the functional central limit theorem holds for asymptotically equicontinuous functions, and Theorem 18.4 stated that this holds when either the bracketing integral or the uniform covering integral is finite. These conditions are abstract and nonintuitive. In this section, we consider simple and interpretable conditions under which these conditions hold.

The following result allows a broad class of functions that includes most parametric applications.

Theorem 18.5 Suppose that for all $\delta > 0$ and all $\theta_1 \in \Theta$,

$$\left(\mathbb{E}\left[\sup_{\|\theta-\theta_1\|<\delta} \left\|g(X,\theta) - g(X,\theta_1)\right\|^2\right]\right)^{1/2} \le C\delta^{\psi} \tag{18.7}$$

for some $C < \infty$ and $0 < \psi < \infty$. Then $J_{[\,]}(\delta, L_2) < \infty$, and $\nu_n(\theta)$ is asymptotically equicontinuous.

See theorem 5 of Andrews (1994).

An important special case of functions satisfying condition (18.7) are Lipschitz-continuous functions.

Definition 18.8 The function $g(x,\theta)$ is **Lipschitz-continuous** over $\theta \in \Theta$ if for some $B(x)$ such that $\mathbb{E}\left[B(X)^2\right] < \infty$,

$$\left\|g(x,\theta_1) - g(x,\theta_2)\right\| \leq B(x)\,\|\theta_1 - \theta_2\|$$

for all $x \in \mathscr{X}$ and all $\theta_1, \theta_2 \in \Theta$.

By direct substitution, if $g(x,\theta)$ is Lipschitz-continuous, it satisfies condition (18.7) with $C = \|B(X)\|_2$ and $\psi = 1$. This shows that Lipschitz-continuity implies asymptotic equicontinuity, as claimed. Condition (18.7) is broader, allowing for discontinuous (e.g., indicator) functions.

The following result allows the combination of different function classes.

Theorem 18.6 Suppose that $\mathbb{E}\left[G(X)^2\right] < \infty$, and

1. $g(x,\theta)$ is Lipschitz-continuous,
2. $g(x,\theta) = h(\theta'\psi(x))$, where $h(u)$ has finite total variation,

   or
3. $g(x,\theta)$ is a combination of functions of the form given in parts 1 and 2 obtained by addition, multiplication, minimum, maximum, and composition.

   Then $J_2(\delta) < \infty$, and $\nu_n(\theta)$ is asymptotically equicontinuous.

See theorems 2 and 3 of Andrews (1994). For a broader list of allowed combinations, see theorem 2.6.18 of van der Vaart (1998). The class of functions in part 2 may appear narrow, but it includes many relevant econometric applications. Allowable functions $h(u)$ include the indicator and sign functions, which are the most common sources of discontinuities.

A broad class of functions that satisfy finite uniform covering are called **Vapnik-Červonenkis (VC)**. The theory of VC classes is combinatorial. For details, see chapter 3 of Pollard (1990), chapter 2.6 of van der Vaart and Wellner (1996), or chapter 19 of van der Vaart (1998). Essentially, VC classes are sufficiently broad to include most applications of interest.

Theorem 18.7 If $\{g(\cdot,\theta) : \theta \in \Theta\}$ is a VC class, then $J_2(\delta) < \infty$. If, in addition, $\mathbb{E}\left[G(X)^2\right] < \infty$, then $\nu_n(\theta)$ is asymptotically equicontinuous over $\theta \in \Theta$.

See lemma 19.15 of van der Vaart (1998).

In this section, we have discussed Theorems 18.5, 18.6, and 18.7, which provide a variety of conditions that can be used to verify asymptotic equicontinuity in a specific application. Theorem 18.5 is useful for continuous or approximately continuous functions; Theorem 18.6 for functions that can be written as a combination of Lipschitz-continuous, finite-variation, and linear functions; and Theorem 18.7 for functions that are known to be members of a VC class.

18.8 DONSKER'S THEOREM

Let us illustrate the application of the functional CLT (Theorem 18.3) to the normalized empirical distribution function (18.5).

First, we verify condition 1 of Theorem 18.3. For any θ,

$$\nu_n(\theta) = \frac{1}{\sqrt{n}} \sum_{i=1}^{n} \left(\mathbb{1}\{X_i \le \theta\} - \mathbb{E}\left[\mathbb{1}\{X_i \le \theta\}\right] \right) = \frac{1}{\sqrt{n}} \sum_{i=1}^{n} U_i$$

where U_i is i.i.d., mean zero, and has variance $F(\theta)(1 - F(\theta))$. By the CLT, we have

$$\nu_n(\theta) \underset{d}{\longrightarrow} \mathrm{N}\left(0, F(\theta)(1 - F(\theta))\right).$$

For any pair of points (θ_1, θ_2),

$$\begin{pmatrix} \nu_n(\theta_1) \\ \nu_n(\theta_2) \end{pmatrix} = \frac{1}{\sqrt{n}} \sum_{i=1}^{n} \begin{pmatrix} (\mathbb{1}\{X_i \le \theta_1\} - F(\theta_1)) \\ (\mathbb{1}\{X_i \le \theta_2\} - F(\theta_2)) \end{pmatrix} = \frac{1}{\sqrt{n}} \sum_{i=1}^{n} U_i$$

where U_i is i.i.d., mean zero, and has covariance matrix

$$\mathbb{E}\left[U_i U_i'\right] = \begin{pmatrix} F(\theta_1)(1 - F(\theta_1)) & F(\theta_1 \wedge \theta_2) - F(\theta_1)F(\theta_2) \\ F(\theta_1 \wedge \theta_2) - F(\theta_1)F(\theta_2) & F(\theta_2)(1 - F(\theta_2)) \end{pmatrix}.$$

Thus by the multivariate CLT,

$$\begin{pmatrix} \nu_n(\theta_1) \\ \nu_n(\theta_2) \end{pmatrix} \underset{d}{\longrightarrow} \mathrm{N}\left(0, \begin{pmatrix} F(\theta_1)(1 - F(\theta_1)) & F(\theta_1 \wedge \theta_2) - F(\theta_1)F(\theta_2) \\ F(\theta_1 \wedge \theta_2) - F(\theta_1)F(\theta_2) & F(\theta_2)(1 - F(\theta_2)) \end{pmatrix}\right).$$

This extends to any set of points $(\theta_1, \theta_2, \ldots, \theta_m)$. Hence Condition 1 is satisfied.

Second, we verify Condition 2 of Theorem 18.3 (asymptotic equicontinuity). From the discussion at the end of Section 18.4, we know that for any distribution function $F(\theta)$, the bracketing numbers satisfy the polynomial rate $N_{[\,]}(\epsilon, L_2) \le 2\epsilon^{-2}$. Thus the bracketing integral is finite. By Theorem 18.4, $\nu_n(\theta)$ is asymptotically equicontinuous. This establishes Donsker's theorem.

Theorem 18.8 Donsker. If X_i are i.i.d. with distribution function $F(\theta)$, then $\nu_n \underset{d}{\longrightarrow} \nu$, where ν is a stochastic process whose marginal distributions are $\nu(\theta) \sim \mathrm{N}\left(0, F(\theta)(1 - F(\theta))\right)$ with covariance function

$$\mathbb{E}\left[\nu(\theta_1)\nu(\theta_2)\right] = F(\theta_1 \wedge \theta_2) - F(\theta_1)F(\theta_2).$$

This was the first functional limit theorem to be established, and it is due to Monroe Donsker. The result as stated is quite general, as it holds for all distributions $F(\theta)$, including both continuous and discrete distributions.

An important special case occurs when $F(\theta)$ is the uniform distribution $U[0,1]$. Then the limit stochastic process has the distribution $B(\theta) \sim \mathrm{N}\left(0, \theta(1 - \theta)\right)$ with covariance function

$$\mathbb{E}\left[B(\theta_1)B(\theta_2)\right] = \theta_1 \wedge \theta_2 - \theta_1\theta_2.$$

This stochastic process is known as a **Brownian bridge**. A corollary is the asymptotic distribution of the Kolmogorov-Smirnov statistic.

Theorem 18.9 When $X \sim U[0,1]$, then $\nu_n \underset{d}{\longrightarrow} B$, a standard Brownian bridge on $[0,1]$, and the null asymptotic distribution of the Kolmogorov-Smirnov statistic is

$$KS_n \underset{d}{\longrightarrow} \sup_{0 \le \theta \le 1} |B(\theta)|.$$

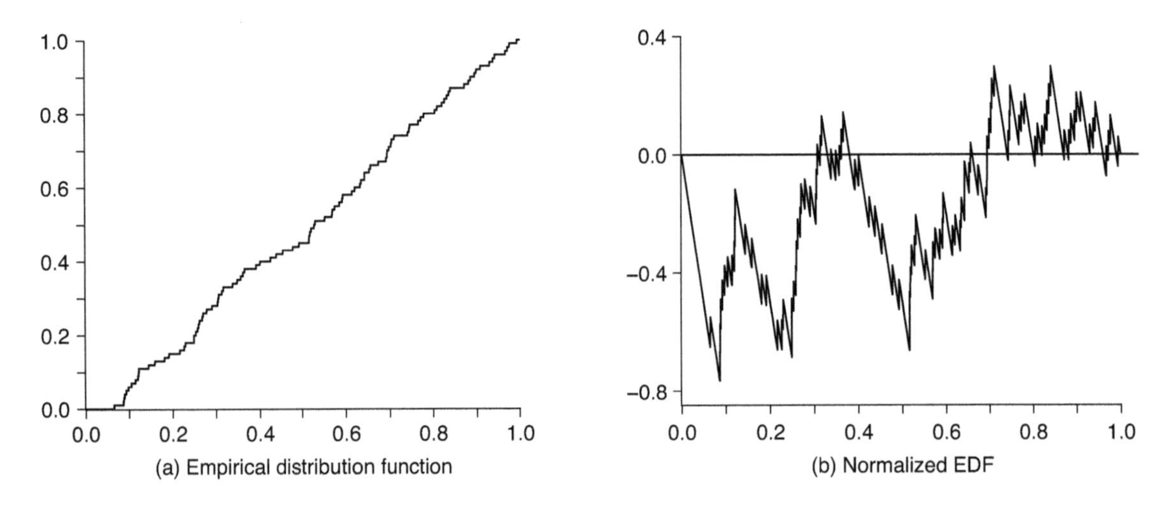

FIGURE 18.6 Empirical distribution function and process—uniform random variables

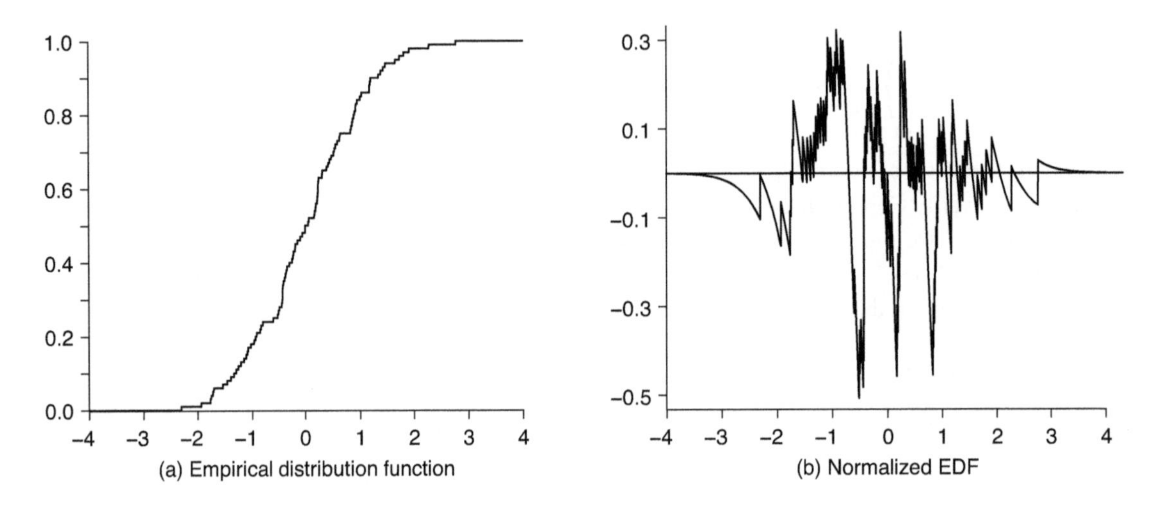

FIGURE 18.7 Empirical distribution function and process - normal random variables

By a change-of-variables argument, the assumption that $X \sim U[0,1]$ can be replaced by the assumption that $F(\theta)$ is continuous. While the limit distribution of ν_n is no longer a Brownian bridge, the asymptotic distribution of the Kolmogorov-Smirnov statistic is unaltered.

We illustrate in Figure 18.6 uniform random variables and in Figure 18.7 normal random variables. In each Figure, panel (a) displays the EDF calculated on a random sample of $n = 100$ observations. Panel (b) displays the normalized EDF (18.5). In Figure 18.6 (uniform random variables), the EDF is close to a straight line, as the true distribution function is linear. The normalized empirical distribution has domain $[0, 1]$. In Figure 18.7 (normal random variables), the empirical distribution function has the curved shape of the normal distribution. The normalized empirical distribution has domain $\mathbb{R}$, and the display is for the region $[-4, 4]$. The process has more variation in the center of the domain than at the edges, since the center has the greatest number of observations.

18.9 TECHNICAL PROOFS*

Proof of Theorem 18.2 under Bracketing. Fix $\epsilon > 0$ and $\eta > 0$. Let $\left[\ell_j, u_j\right], j = 1, \ldots, N = N_{[\,]}(\epsilon, L_1)$ be a set of ϵ-L_1-brackets which cover Θ. By construction, $\left|u_j(X)\right| \leq G(X)$ and $\left|\ell_j(X)\right| \leq G(X)$, so they have finite expectation. By assumption, $N < \infty$, and

$$\mathbb{E}\left[u_j(X)\right] - \mathbb{E}\left[\ell_j(X)\right] = \left\|u_j(X) - \ell_j(X)\right\|_1 \leq \epsilon. \tag{18.8}$$

The WLLN implies that their sample averages converge in probability to their expectations, which means that for n sufficiently large,

$$\mathbb{P}\left[\left|\frac{1}{n}\sum_{i=1}^{n}\left(u_j(X_i) - \mathbb{E}\left[u_j(X)\right]\right)\right| > \epsilon\right] \leq \frac{\eta}{2N} \tag{18.9}$$

and

$$\mathbb{P}\left[\left|\frac{1}{n}\sum_{i=1}^{n}\left(\ell_j(X_i) - \mathbb{E}\left[\ell_j(X)\right]\right)\right| > \epsilon\right] \leq \frac{\eta}{2N}.$$

Boole's Inequality (Theorem 1.2, property 6) and (18.9) imply

$$\begin{aligned}\mathbb{P}\left[\max_{1\leq j\leq n}\left|\frac{1}{n}\sum_{i=1}^{n}\left(u_j(X_i) - \mathbb{E}\left[u_j(X)\right]\right)\right| > \epsilon\right] &= \mathbb{P}\left[\bigcup_{j=1}^{N}\left\{\left|\frac{1}{n}\sum_{i=1}^{n}\left(u_j(X_i) - \mathbb{E}\left[u_j(X)\right]\right)\right| > \epsilon\right\}\right] \\ &\leq \sum_{j=1}^{N}\mathbb{P}\left[\left|\frac{1}{n}\sum_{i=1}^{n}\left(u_j(X_i) - \mathbb{E}\left[u_j(X)\right]\right)\right| > \epsilon\right] \leq \frac{\eta}{2}\end{aligned} \tag{18.10}$$

and similarly

$$\mathbb{P}\left[\max_{1\leq j\leq n}\left|\frac{1}{n}\sum_{i=1}^{n}\left(\ell_j(X_i) - \mathbb{E}\left[\ell_j(X)\right]\right)\right| > \epsilon\right] \leq \frac{\eta}{2}. \tag{18.11}$$

Take any $\theta \in \Theta$ and its bracket $\left[\ell_j, u_j\right]$. Since $\ell_j \leq g \leq u_j$,

$$\frac{1}{n}\sum_{i=1}^{n}\left(g(X_i,\theta) - \mathbb{E}\left[g(X,\theta)\right]\right) \leq \frac{1}{n}\sum_{i=1}^{n}\left(u_j(X_i) - \mathbb{E}\left[\ell_j(X)\right]\right) \leq \frac{1}{n}\sum_{i=1}^{n}\left(u_j(X_i) - \mathbb{E}\left[u_j(X)\right]\right) + \epsilon$$

where the second inequality is (18.8). Similarly,

$$\frac{1}{n}\sum_{i=1}^{n}\left(g(X_i,\theta) - \mathbb{E}\left[g(X,\theta)\right]\right) \geq \frac{1}{n}\sum_{i=1}^{n}\left(\ell_j(X_i) - \mathbb{E}\left[u_j(X)\right]\right) \leq \frac{1}{n}\sum_{i=1}^{n}\left(\ell_j(X_i) - \mathbb{E}\left[\ell_j(X)\right]\right) - \epsilon.$$

Together these results imply that

$$\sup_{\theta\in\Theta}\left\|\overline{g}_n(\theta) - g(\theta)\right\| \leq \max_{1\leq j\leq n}\left|\frac{1}{n}\sum_{i=1}^{n}\left(u_j(X_i) - \mathbb{E}\left[u_j(X)\right]\right)\right| + \max_{1\leq j\leq n}\left|\frac{1}{n}\sum_{i=1}^{n}\left(\ell_j(X_i) - \mathbb{E}\left[\ell_j(X)\right]\right)\right| + \epsilon.$$

Combined with (18.10) and (18.11), we deduce that

$$\mathbb{P}\left[\sup_{\theta\in\Theta}\left\|\overline{g}_n(\theta)-g(\theta)\right\|>3\epsilon\right]\leq\mathbb{P}\left[\max_{1\leq j\leq n}\left|\frac{1}{n}\sum_{i=1}^{n}\left(u_j(X_i)-\mathbb{E}\left[u_j(X)\right]\right)\right|>\epsilon\right]$$

$$+\mathbb{P}\left[\max_{1\leq j\leq n}\left|\frac{1}{n}\sum_{i=1}^{n}\left(\ell_j(X_i)-\mathbb{E}\left[\ell_j(X)\right]\right)\right|>\epsilon\right]\leq\eta.$$

Since ϵ and η are arbitrary, this completes the proof. ■

Proof of Theorem 18.2 under Continuity. Fix $\epsilon>0$ and $\eta>0$. For a given θ_j, let $B_j(\delta)$ be a ball of radius $\delta/2$ about θ_j. Set $u_j(x,\delta)=\sup_{\theta\in B_j(\delta)} g(x,\theta)$, and $\ell_j(x,\delta)=\inf_{\theta\in B_j(\delta)} g(x,\theta)$. By construction, $[\ell_j,u_j]$ is the bracket including all $g(x,\theta)$ for $\theta\in B_j(\delta)$. Since $g(X,\theta)$ is continuous in θ with probability 1, $u_j(X,\delta)-\ell_j(X,\delta)\to 0$ as $\epsilon\to 0$ with probability 1. Since

$$\mathbb{E}\left|u_j(X,\delta)-\ell_j(X,\delta)\right|\leq 2\mathbb{E}\left[G(X)\right]<\infty,$$

by dominated convergence (Theorem A.26), $\mathbb{E}\left|u_j(X,\delta)-\ell_j(X,\delta)\right|\to 0$ as $\delta\to 0$. Thus for δ sufficiently small, $\mathbb{E}\left|u_j(X,\delta)-\ell_j(X,\delta)\right|\leq\epsilon$, so $[\ell_j,u_j]$ is an ϵ-L_1-bracket. The collection of all such ϵ-L_1-brackets $[\ell_j,u_j]$ covers Θ. Since Θ is compact, there exists a finite cover. Thus $N_{[\ ]}(\epsilon,L_1)<\infty$. The conditions of Theorem 18.2 under bracketing are satisfied, which states that $\overline{g}_n(\theta)$ satisfies the ULLN over Θ. ■

18.10 EXERCISES

Exercise 18.1 Let $g(x,\theta)=\mathbb{1}\{x\leq\theta\}$ for $\theta\in[0,1]$ and assume $X\sim F=U[0,1]$. Let $N_1(\epsilon,F)$ be L_1 packing numbers.

(a) Show that $N_1(\epsilon,F)$ equal the packing numbers constructed with respect to the Euclidean metric $d(\theta_1,\theta_2)=|\theta_2-\theta_2|$.
(b) Verify that $N_1(\epsilon,F)\leq\lceil 1/\epsilon\rceil$.

Exercise 18.2 Find conditions under which sample averages of the following functions are stochastically equicontinuous.

(a) $g(X,\theta)=X\theta$ for $\theta\in[0,1]$.
(b) $g(X,\theta)=X\theta^2$ for $\theta\in[0,1]$.
(c) $g(X,\theta)=X/\theta$ for $\theta\in[a,1]$ and $a>0$. Can $a=0$?

Exercise 18.3 Define $\nu_n(\theta)=\frac{1}{\sqrt{n}}\sum_{i=1}^{n}X_i\mathbb{1}\{X_i\leq\theta\}$ for $\theta\in[0,1]$, where $\mathbb{E}[X]=0$ and $\mathbb{E}\left[X^2\right]=1$.

(a) Show that $\nu_n(\theta)$ is stochastically equicontinuous.
(b) Find the stochastic process $\nu(\theta)$ that has asymptotic finite-dimensional distributions of $\nu_n(\theta)$.
(c) Show that $\nu_n\underset{d}{\longrightarrow}\nu$.

APPENDIX

MATHEMATICS REFERENCE

A.1 LIMITS

Definition A.1 A sequence a_n has the **limit** a, written $a_n \to a$ as $n \to \infty$, or alternatively as $\lim_{n\to\infty} a_n = a$, if for all $\delta > 0$ there is some $n_\delta < \infty$ such that for all $n \geq n_\delta$, $|a_n - a| \leq \delta$.

Definition A.2 When a_n has a finite limit a, we say that a_n **converges** or is **convergent**.

Definition A.3 $\liminf_{n\to\infty} a_n = \lim_{n\to\infty} \inf_{m\geq n} a_m$.

Definition A.4 $\limsup_{n\to\infty} a_n = \lim_{n\to\infty} \sup_{m\geq n} a_m$.

Theorem A.1 If a_n has a limit a, then $\liminf_{n\to\infty} a_n = \lim_{n\to\infty} a_n = \limsup_{n\to\infty} a_n$.

Theorem A.2 **Cauchy Criterion**. The sequence a_n converges if for all $\epsilon > 0$

$$\inf_m \sup_{j>m} \left| a_j - a_m \right| \leq \epsilon.$$

A.2 SERIES

Definition A.5 **Summation Notation**. The sum of $a_1, \ldots, a_n$ is

$$S_n = a_1 + \cdots + a_n = \sum_{i=1}^{n} a_i = \sum\nolimits_{i=1}^{n} a_i = \sum\nolimits_{1}^{n} a_i = \sum a_i.$$

The sequence of sums S_n is called a **series**.

Definition A.6 The series S_n is **convergent** if it has a finite limit as $n \to \infty$, thus $S_n \to S < \infty$.

Definition A.7 The series S_n is **absolutely convergent** if $\sum_{i=1}^{n} |a_i|$ is convergent.

Theorem A.3 **Tests for Convergence**. The series S_n is absolutely convergent if any of the following hold:

1. **Comparison Test**. If $0 \leq a_i \leq b_i$ and $\sum_{i=1}^{n} b_i$ converges.

2. **Ratio Test**. If $a_i \geq 0$ and $\lim_{n\to\infty} \dfrac{a_{n+1}}{a_n} < 1$.

3. **Integral Test**. If $a_i = f(i) > 0$, where $f(x)$ is monotonically decreasing, and $\int_1^\infty f(x)dx < \infty$.

Theorem A.4 Theorem of Cesaro Means. If $a_i \to a$ as $i \to \infty$, then $n^{-1} \sum_{i=1}^{n} a_i \to a$ as $n \to \infty$.

Theorem A.5 Toeplitz Lemma. Suppose w_{ni} satisfies $w_{ni} \to 0$ as $n \to \infty$ for all i, $\sum_{i=1}^{n} w_{ni} = 1$, and $\sum_{i=1}^{n} |w_{ni}| < \infty$. If $a_n \to a$, then $\sum_{i=1}^{n} w_{ni}a_i \to a$ as $n \to \infty$.

Theorem A.6 Kronecker Lemma. If $\sum_{i=1}^{n} i^{-1}a_i \to a < \infty$ as $n \to \infty$, then $n^{-1} \sum_{i=1}^{n} a_i \to 0$ as $n \to \infty$.

A.3 FACTORIALS

Factorials are widely used in probability formulas.

Definition A.8 For a positive integer n, the **factorial** $n!$ is the product of all integers between 1 and n:

$$n! = n \times (n-1) \times \cdots \times 1 = \prod_{i=1}^{n} i.$$

Furthermore, $0! = 1$.

A simple recurrance property is $n! = n \times (n-1)!$

Definition A.9 For a positive integer n, the **double factorial** $n!!$ is the product of every second positive integer up to n:

$$n!! = n \times (n-2) \times (n-4) \times \cdots = \prod_{i=0}^{\lceil n/2 \rceil - 1} (n-2i)\,.$$

Furthermore, $0!! = 1$.

For even n

$$n!! = \prod_{i=1}^{n/2} 2i.$$

For odd n

$$n!! = \prod_{i=1}^{(n+1)/2} (2i-1)\,.$$

The double factorial satisfies the recurrance property $n!! = n \times (n-2)!!$

The double factorial can be written in terms of the factorial as follows. For even and odd integers, we have:

$$(2m)!! = 2^m m!$$

$$(2m-1)!! = \frac{(2m)!}{2^m m!}$$

A.4 EXPONENTIALS

An exponential is a function of the form a^x. We typically use the name "exponential function" to refer to the function $e^x = \exp(x)$, where e is the exponential constant $e \simeq 2.718 \ldots$.

Definition A.10 The **exponential function** is $e^x = \exp(x) = \sum_{i=0}^{\infty} \frac{x^i}{i!}$.

Theorem A.7 Properties of the exponential function

1. $e = e^0 = \exp(0) = \sum_{i=0}^{\infty} \frac{1}{i!}$.
2. $\exp(x) = \lim_{n\to\infty} \left(1 + \frac{x}{n}\right)^n$.
3. $(e^a)^b = e^{ab}$.
4. $e^{a+b} = e^a e^b$.
5. $\exp(x)$ is strictly increasing on $\mathbb{R}$, everywhere positive, and convex.

A.5 LOGARITHMS

In probability, statistics, and econometrics, the term "logarithm" always refers to the natural logarithm, which is the inverse of the exponential function. We use the notation "log" rather than the less description notation "ln".

Definition A.11 The **logarithm** is the function on $(0, \infty)$ which satisfies $\exp(\log(x)) = x$, or equivalently, $\log(\exp(x)) = x$.

Theorem A.8 Properties of the logarithm

1. $\log(ab) = \log(a) + \log(b)$.
2. $\log(a^b) = b \log(a)$.
3. $\log(1) = 0$.
4. $\log(e) = 1$.
5. $\log(x)$ is strictly increasing on $\mathbb{R}_+$ and concave.

A.6 DIFFERENTIATION

Definition A.12 A function $f(x)$ is **continuous** at $x = c$ if for all $\epsilon > 0$ there is some $\delta > 0$ such that $\|x - c\| \leq \delta$ implies $\|f(x) - f(c)\| \leq \epsilon$.

Definition A.13 The **derivative** of $f(x)$, denoted $f'(x)$ or $\frac{d}{dx}f(x)$, is

$$\frac{d}{dx}f(x) = \lim_{h\to 0} \frac{f(x+h) - f(x)}{h}.$$

A function $f(x)$ is **differentiable** if the derivative exists and is unique.

Definition A.14 The **partial derivative** of $f(x, y)$ with respect to x is

$$\frac{\partial}{\partial x}f(x, y) = \lim_{h\to 0} \frac{f(x+h, y) - f(x, y)}{h}.$$

Theorem A.9 **Chain Rule of Differentiation.** For real-valued functions $f(x)$ and $g(x)$,

$$\frac{d}{dx}f\left(g(x)\right) = f'\left(g(x)\right) g'(x).$$

Theorem A.10 **Derivative Rule of Differentiation.** For real-valued functions $u(x)$ and $v(x)$,

$$\frac{d}{dx}\frac{u(x)}{v(x)} = \frac{v(x)u'(x) - u(x)v'(x)}{v(x)^2}.$$

Theorem A.11 **Linearity of Differentiation**

$$\frac{d}{dx}\left(ag(x) + bf(x)\right) = ag'(x) + bf'(x).$$

Theorem A.12 **L'Hôpital's Rule.** For real-valued functions $f(x)$ and $g(x)$ such that $\lim_{x\to c} f(x) = 0$, $\lim_{x\to c} g(x) = 0$, and $\lim_{x\to c} \frac{f(x)}{g(x)}$ exists,

$$\lim_{x\to c} \frac{f(x)}{g(x)} = \lim_{x\to c} \frac{f'(x)}{g'(x)}.$$

Theorem A.13 **Common Derivatives**

1. $\frac{d}{dx}c = 0.$
2. $\frac{d}{dx}x^a = ax^{a-1}.$
3. $\frac{d}{dx}e^x = e^x.$
4. $\frac{d}{dx}\log(x) = \frac{1}{x}.$
5. $\frac{d}{dx}a^x = a^x \log(x).$

A.7 MEAN VALUE THEOREM

Theorem A.14 Mean Value Theorem. If $f(x)$ is continuous on $[a, b]$ and differentiable on (a, b), then there exists a point $c \in (a, b)$ such that

$$f'(c) = \frac{f(b) - f(a)}{b - a}.$$

The mean value theorem is frequently used to write $f(b)$ as the sum of $f(a)$ and the product of the slope times the difference:

$$f(b) = f(a) + f'(c)(b - a).$$

Theorem A.15 Taylor's Theorem. Let s be a positive integer. If $f(x)$ is s times differentiable at a, then there exists a function $r(x)$ such that

$$f(x) = f(a) + f'(a)(x - a) + \frac{f''(a)}{2}(x - a)^2 + \cdots + \frac{f^{(s)}(a)}{s!}(x - a)^s + r(x)$$

where

$$\lim_{x \to a} \frac{r(x)}{(x - a)^s} = 0.$$

The term $r(x)$ is called the **remainder**. The final equation of the theorem shows that the remainder is of smaller order than $(x - a)^s$.

Theorem A.16 Taylor's Theorem, Mean-Value Form. Let s be a positive integer. If $f^{(s-1)}(x)$ is continuous on $[a, b]$ and differentiable on (a, b), then there exists a point $c \in (a, b)$ such that

$$f(b) = f(a) + f'(a)(x - a) + \frac{f''(a)}{2}(b - a)^2 + \cdots + \frac{f^{(s)}(c)}{s!}(b - a)^s.$$

Taylor's theorem is local in nature, as it is an approximation at a specific point x. It shows that locally, $f(x)$ can be approximated by an s-order polynomial.

Definition A.15 The **Taylor series expansion** of $f(x)$ at a is

$$\sum_{k=0}^{\infty} \frac{f^{(k)}(a)}{k!}(x - a)^k.$$

Definition A.16 The **Maclaurin series expansion** of $f(x)$ is the Taylor series at $a = 0$, thus

$$\sum_{k=0}^{\infty} \frac{f^{(k)}(0)}{k!} x^k.$$

A necessary condition for a Taylor series expansion to exist is that $f(x)$ is infinitely differentiable at a, but this is not a sufficient condition. A function $f(x)$ which equals a convergent power series over an interval is called **analytic** in this interval.

A.8 INTEGRATION

Definition A.17 The **Riemann integral** of $f(x)$ over the interval $[a, b]$ is

$$\int_a^b f(x)dx = \lim_{N\to\infty} \frac{1}{N}\sum_{i=1}^{N} f\left(a + \frac{i}{N}(b-a)\right).$$

The sum on the right is the sum of the areas of the rectangles of width $(b-a)/N$ approximating $f(x)$.

Definition A.18 The **Riemann-Stieltijes** integral of $g(x)$ with respect to $f(x)$ over $[a, b]$ is

$$\int_a^b g(x)df(x) = \lim_{N\to\infty} \sum_{j=0}^{N-1} g\left(a + \frac{j}{N}(b-a)\right)\left(f\left(a + \frac{j+1}{N}(b-a)\right) - f\left(a + \frac{j}{N}(b-a)\right)\right).$$

The sum on the right is a weighted sum of the area of the rectangles weighted by the change in the function f.

Definition A.19 The function $f(x)$ is **integrable** on $\mathscr{X}$ if $\int_{\mathscr{X}} |f(x)|\, dx < \infty$.

Theorem A.17 Linearity of Integration

$$\int_a^b (cg(x) + df(x))\, dx = c\int_a^b g(x)dx + d\int_a^b f(x)dx.$$

Theorem A.18 Common Integrals

1. $\displaystyle\int x^a dx = \frac{1}{a+1}x^{a+1} + C.$
2. $\displaystyle\int e^x dx = e^x + C.$
3. $\displaystyle\int \frac{1}{x}dx = \log|x| + C.$
4. $\displaystyle\int \log x = x\log x - x + C.$

Theorem A.19 First Fundamental Theorem of Calculus. Let $f(x)$ be a continuous real-valued function on $[a, b]$, and define $F(x) = \int_a^x f(t)dt$. Then $F(x)$ has derivative $F'(x) = f(x)$ for all $x \in (a, b)$.

Theorem A.20 Second Fundamental Theorem of Calculus. Let $f(x)$ be a real-valued function on $[a, b]$ and $F(x)$ an antiderivative satisfying $F'(x) = f(x)$. Then

$$\int_a^b f(x)dx = F(b) - F(a).$$

Theorem A.21 Integration by Parts. For real-valued functions $u(x)$ and $v(x)$,

$$\int_a^b u(x)v'(x)dx = u(b)v(b) - u(a)v(a) - \int_a^b u'(x)v(x)dx.$$

This equation is often written compactly as

$$\int u dv = uv - \int v du.$$

Theorem A.22 Leibniz Rule. For real-valued functions $a(x)$, $b(x)$, and $f(x,t)$,

$$\frac{d}{dx}\int_{a(x)}^{b(x)} f(x,t)dt = f(x, b(x))\frac{d}{dx}b(x) - f(x, a(x))\frac{d}{dx}a(x) + \int_{a(x)}^{b(x)} \frac{\partial}{\partial x} f(x,t)dt.$$

When a and b are constants, the equation simplifies to

$$\frac{d}{dx}\int_a^b f(x,t)dt = \int_a^b \frac{\partial}{\partial x} f(x,t)dt.$$

Theorem A.23 Fubini's Theorem. If $f(x,y)$ is integrable, then

$$\int_{\mathcal{Y}}\int_{\mathcal{X}} f(x,y)dxdy = \int_{\mathcal{X}}\int_{\mathcal{Y}} f(x,y)dydx.$$

Theorem A.24 Fatou's Lemma. If f_n is a sequence of nonnegative functions, then

$$\int \liminf_{n\to\infty} f_n(x)dx \le \liminf_{n\to\infty} \int f_n(x)dx.$$

Theorem A.25 Monotone Convergence Theorem. If $f_n(x)$ is an increasing sequence of functions which converges pointwise to $f(x)$, then $\int f_n(x)dx \to \int f(x)dx$ as $n \to \infty$.

Theorem A.26 Dominated Convergence Theorem. If $f_n(x)$ is a sequence of functions which converges pointwise to $f(x)$, and $|f_n(x)| \le g(x)$, where $g(x)$ is integrable, then $f(x)$ is integrable and $\int f_n(x)dx \to \int f(x)dx$ as $n \to \infty$.

A.9 GAUSSIAN INTEGRAL

Theorem A.27 $\int_{-\infty}^{\infty} \exp\left(-x^2\right) dx = \sqrt{\pi}.$

Proof:

$$\begin{aligned}\left(\int_0^\infty \exp\left(-x^2\right) dx\right)^2 &= \int_0^\infty \exp\left(-x^2\right) dx \int_0^\infty \exp\left(-y^2\right) dy \\ &= \int_0^\infty \int_0^\infty \exp\left(-\left(x^2+y^2\right)\right) dxdy \\ &= \int_0^\infty \int_0^{\pi/2} r\exp\left(-r^2\right) d\theta\, dr \\ &= \frac{\pi}{2}\int_0^\infty r\exp\left(-r^2\right) dr \\ &= \frac{\pi}{4}.\end{aligned}$$

The third equality is the key. It makes the change of variables to polar coordinates $x = r\cos\theta$ and $y = r\sin\theta$, so that $x^2 + y^2 = r^2$. The Jacobian of this transformation is r. The region of integration in the (x, y) units is the positive quadrant (upper-right region), which corresponds to integrating θ from 0 to $\pi/2$ in polar coordinates. The final two equalities are simple integration. Taking the square root, we obtain

$$\int_0^\infty \exp\left(-x^2\right) dx = \frac{\sqrt{\pi}}{2}.$$

Since the integrals over the positive and negative real line are identical, we obtain the stated result.

A.10 GAMMA FUNCTION

Definition A.20 The **gamma function** for $x > 0$ is

$$\Gamma(x) = \int_0^\infty t^{x-1} e^{-t} dt.$$

The gamma function is not available in closed form. For computation, numerical algorithms are used.

Theorem A.28 Properties of the gamma function

1. For positive integer n, $\Gamma(n) = (n-1)!$
2. For $x > 1$, $\Gamma(x) = (x-1)\Gamma(x-1)$.
3. $\displaystyle\int_0^\infty t^{x-1} \exp(-\beta t)\, dt = \beta^{-\alpha}\Gamma(x)$.
4. $\Gamma(1) = 1$.
5. $\Gamma(1/2) = \sqrt{\pi}$.
6. $\displaystyle\lim_{n\to\infty} \frac{\Gamma(n+x)}{\Gamma(n)\, n^x} = 1$.
7. **Legendre's Duplication Formula**: $\Gamma(x)\Gamma\left(x+\frac{1}{2}\right) = 2^{1-2x}\sqrt{\pi}\,\Gamma(2x)$.
8. **Stirling's Approximation**: $\Gamma(x+1) = \sqrt{2\pi x}\left(\frac{x}{e}\right)^x \left(1 + O\left(\frac{1}{x}\right)\right)$ as $x \to \infty$.

Properties 1 and 2 can be shown by integration by parts. Property 3 can be shown by a change of variables. Property 4 is an exponential integral. Property 5 can be shown by applying a change of variables and the Gaussian integral. Proofs of the remaining properties are advanced and not provided.

A.11 MATRIX ALGEBRA

This is an abbreviated summary. For a more extensive review, see Appendix A of *Econometrics*.

A **scalar** a is a single number. A **vector** a or $\boldsymbol{a}$ is a $k \times 1$ list of numbers, typically arranged in a column, written as

$$\boldsymbol{a} = \begin{pmatrix} a_1 \\ a_2 \\ \vdots \\ a_k \end{pmatrix}$$

A **matrix** $\boldsymbol{A}$ is a $k \times r$ rectangular array of numbers, written as

$$\boldsymbol{A} = \begin{bmatrix} a_{11} & a_{12} & \cdots & a_{1r} \\ a_{21} & a_{22} & \cdots & a_{2r} \\ \vdots & \vdots & & \vdots \\ a_{k1} & a_{k2} & \cdots & a_{kr} \end{bmatrix}$$

By convention, a_{ij} refers to the element in the ith row and jth column of $\boldsymbol{A}$. If $r = 1$, then $\boldsymbol{A}$ is a column vector. If $k = 1$, then $\boldsymbol{A}$ is a row vector. If $r = k = 1$, then $\boldsymbol{A}$ is a scalar. Sometimes a matrix $\boldsymbol{A}$ is denoted by the symbol (a_{ij}).

The **transpose** of a matrix $\boldsymbol{A}$, denoted $\boldsymbol{A}'$, $\boldsymbol{A}^\top$, or $\boldsymbol{A}^t$, is obtained by flipping the matrix on its diagonal. In most of the econometrics literature and this textbook, we use $\boldsymbol{A}'$ to denote the transpose of $\boldsymbol{A}$. In the mathematics literature, $\boldsymbol{A}^\top$ is the convention. Thus

$$\boldsymbol{A}' = \begin{bmatrix} a_{11} & a_{21} & \cdots & a_{k1} \\ a_{12} & a_{22} & \cdots & a_{k2} \\ \vdots & \vdots & & \vdots \\ a_{1r} & a_{2r} & \cdots & a_{kr} \end{bmatrix}$$

Alternatively, letting $\boldsymbol{B} = \boldsymbol{A}'$, then $b_{ij} = a_{ji}$. Note that if $\boldsymbol{A}$ is $k \times r$, then $\boldsymbol{A}'$ is $r \times k$. If $\boldsymbol{a}$ is a $k \times 1$ vector, then $\boldsymbol{a}'$ is a $1 \times k$ row vector.

A matrix is **square** if $k = r$. A square matrix is **symmetric** if $\boldsymbol{A} = \boldsymbol{A}'$, which requires $a_{ij} = a_{ji}$. A square matrix is **diagonal** if the off-diagonal elements are all zero, so that $a_{ij} = 0$ if $i \neq j$.

An important diagonal matrix is the **identity matrix**, which has ones on the diagonal. The $k \times k$ identity matrix is denoted as

$$\boldsymbol{I}_k = \begin{bmatrix} 1 & 0 & \cdots & 0 \\ 0 & 1 & \cdots & 0 \\ \vdots & \vdots & & \vdots \\ 0 & 0 & \cdots & 1 \end{bmatrix}.$$

The **matrix sum** of two matrices of the same dimensions is

$$\boldsymbol{A} + \boldsymbol{B} = \left(a_{ij} + b_{ij}\right).$$

The product of a matrix $\boldsymbol{A}$ and scalar c is real is defined as

$$\boldsymbol{A}c = c\boldsymbol{A} = \left(a_{ij}c\right).$$

If $\boldsymbol{a}$ and $\boldsymbol{b}$ are both $k \times 1$, then their inner product is

$$\boldsymbol{a}'\boldsymbol{b} = a_1 b_1 + a_2 b_2 + \cdots + a_k b_k = \sum_{j=1}^{k} a_j b_j.$$

Note that $\boldsymbol{a}'\boldsymbol{b} = \boldsymbol{b}'\boldsymbol{a}$. We say that two vectors $\boldsymbol{a}$ and $\boldsymbol{b}$ are **orthogonal** if $\boldsymbol{a}'\boldsymbol{b} = 0$.

If $\boldsymbol{A}$ is $k \times r$ and $\boldsymbol{B}$ is $r \times s$, so that the number of columns of $\boldsymbol{A}$ equals the number of rows of $\boldsymbol{B}$, we say that $\boldsymbol{A}$ and $\boldsymbol{B}$ are **conformable**. In this case, the matrix product $\boldsymbol{AB}$ is defined. Writing $\boldsymbol{A}$ as a set of row vectors and $\boldsymbol{B}$ as a set of column vectors (each of length r), then the **matrix product** is defined as

$$\boldsymbol{AB} = \begin{bmatrix} \boldsymbol{a}_1'\boldsymbol{b}_1 & \boldsymbol{a}_1'\boldsymbol{b}_2 & \cdots & \boldsymbol{a}_1'\boldsymbol{b}_s \\ \boldsymbol{a}_2'\boldsymbol{b}_1 & \boldsymbol{a}_2'\boldsymbol{b}_2 & \cdots & \boldsymbol{a}_2'\boldsymbol{b}_s \\ \vdots & \vdots & & \vdots \\ \boldsymbol{a}_k'\boldsymbol{b}_1 & \boldsymbol{a}_k'\boldsymbol{b}_2 & \cdots & \boldsymbol{a}_k'\boldsymbol{b}_s \end{bmatrix}.$$

The **trace** of a $k \times k$ square matrix $\boldsymbol{A}$ is the sum of its diagonal elements:

$$\operatorname{tr}(\boldsymbol{A}) = \sum_{i=1}^{k} a_{ii}.$$

A useful property is

$$\operatorname{tr}(\boldsymbol{AB}) = \operatorname{tr}(\boldsymbol{BA}).$$

A square $k \times k$ matrix $\boldsymbol{A}$ is said to be **nonsingular** if there is no $k \times 1$ $\boldsymbol{c} \neq 0$ such that $\boldsymbol{Ac} = 0$.

If a square $k \times k$ matrix $\boldsymbol{A}$ is nonsingular, then there exists a unique $k \times k$ matrix $\boldsymbol{A}^{-1}$ called the **inverse** of $\boldsymbol{A}$ which satisfies

$$\boldsymbol{AA}^{-1} = \boldsymbol{A}^{-1}\boldsymbol{A} = \boldsymbol{I}_k.$$

For nonsingular $\boldsymbol{A}$ and $\boldsymbol{C}$, some useful properties include

$$\left(\boldsymbol{A}^{-1}\right)' = \left(\boldsymbol{A}'\right)^{-1}$$
$$(\boldsymbol{AC})^{-1} = \boldsymbol{C}^{-1}\boldsymbol{A}^{-1}.$$

A $k \times k$ real symmetric square matrix $\boldsymbol{A}$ is **positive semi-definite** if for all $\boldsymbol{c} \neq 0$, $\boldsymbol{c}'\boldsymbol{Ac} \geq 0$. This is written as $\boldsymbol{A} \geq 0$. A $k \times k$ real symmetric square matrix $\boldsymbol{A}$ is **positive definite** if for all $\boldsymbol{c} \neq 0$, $\boldsymbol{c}'\boldsymbol{Ac} > 0$. This is written as $\boldsymbol{A} > 0$. If $\boldsymbol{A}$ and $\boldsymbol{B}$ are each $k \times k$, we write $\boldsymbol{A} \geq \boldsymbol{B}$ if $\boldsymbol{A} - \boldsymbol{B} \geq 0$. This means that the difference between $\boldsymbol{A}$ and $\boldsymbol{B}$ is positive semi-definite. Similarly, we write $\boldsymbol{A} > \boldsymbol{B}$ if $\boldsymbol{A} - \boldsymbol{B} > 0$.

Many students misinterpret "$\boldsymbol{A} > 0$" to mean that $\boldsymbol{A} > 0$ has nonzero elements. This is incorrect. The inequality applied to a matrix means that it is positive definite.

Some properties include the following:

1. If $\boldsymbol{A} = \boldsymbol{G}'\boldsymbol{BG}$ with $\boldsymbol{B} \geq 0$, then $\boldsymbol{A} \geq 0$.
2. If $\boldsymbol{A} > 0$, then $\boldsymbol{A}$ is nonsingular, $\boldsymbol{A}^{-1}$ exists, and $\boldsymbol{A}^{-1} > 0$.
3. If $\boldsymbol{A} > 0$, then $\boldsymbol{A} = \boldsymbol{CC}'$, where $\boldsymbol{C}$ is nonsingular.

The **determinant**, written $\det \boldsymbol{A}$ or $|\boldsymbol{A}|$, of a square matrix $\boldsymbol{A}$ is a scalar measure of the transformation $\boldsymbol{Ax}$. The precise definition is technical. See Appendix A of *Econometrics* for details. Useful properties are:

1. $\det \boldsymbol{A} = 0$ if and only if $\boldsymbol{A}$ is singular.
2. $\det \boldsymbol{A}^{-1} = \dfrac{1}{\det \boldsymbol{A}}$.

REFERENCES

Amemiya, Takeshi (1994): *Introduction to Statistics and Econometrics*. Cambridge, MA: Harvard University Press.

Andrews, Donald W. K. (1994): "Empirical process methods in econometrics", in *Handbook of Econometrics*, Volume 4, chapter 37. Robert F. Engle and Daniel L. McFadden, eds., 2247–2294. Amsterdam: Elsevier.

Ash, Robert B. (1972): *Real Analysis and Probability*. Cambridge, MA: Academic Press.

Billingsley, Patrick (1995): *Probability and Measure*, Third Edition. New York: Wiley.

Billingsley, Patrick (1999): *Convergence of Probability Measure*, Second Edition. New York: Wiley.

Bock, Mary Ellen (1975): "Minimax estimators of the mean of a multivariate normal distribution," *Annals of Statistics* 3, 209–218.

Casella, George, and Roger L. Berger (2002): *Statistical Inference*, Second Edition. Pacific Grove, CA: Duxbury Press.

Efron, Bradley (2010): *Large-Scale Inference: Empirical Bayes Methods for Estimation, Testing and Prediction*. Cambridge: Cambridge University Press.

Epanechnikov, V. I. (1969): "Non-parametric estimation of a multivariate probability density," *Theory of Probability and Its Application* 14, 153–158.

Gallant, A. Ronald (1997): *An Introduction to Econometric Theory*. Princeton, NJ: Princeton University Press.

Gosset, William S. (a.k.a. "Student") (1908): "The probable error of a mean," *Biometrika* 6, 1–25.

Hall, Peter (1992): *The Bootstrap and Edgeworth Expansion*. New York: Springer-Verlag.

Hansen, Bruce E. (2022a): *Econometrics*. Princeton, NJ: Princeton University Press.

Hansen, Bruce E. (2022b): "A modern Gauss-Markov theorem," *Econometrica*.

Hodges Joseph L., and Erich L. Lehmann (1956): "The efficiency of some nonparametric competitors of the t-test," *Annals of Mathematical Statistics* 27, 324–335.

Hogg, Robert V., and Allen T. Craig (1995): *Introduction to Mathematical Statistics*, Fifth Edition. Hoboken, NJ: Prentice Hall.

Hogg, Robert V., and Elliot A. Tanis (1997): *Probability and Statistical Inference*, Fifth Edition. Hoboken, NJ: Prentice Hall.

James, W., and Charles M. Stein (1961): "Estimation with quadratic loss," *Proceedings of the Fourth Berkeley Symposium on Mathematical Statistics and Probability* 1, 361–380.

Jones, M. C., and S. J. Sheather (1991): "Using non-stochastic terms to advantage in kernel-based estimation of integrated squared density derivatives," *Statistics and Probability Letters* 11, 511–514.

Koop, Gary, Dale J. Poirier, and Justin L. Tobias (2007): *Bayesian Econometric Methods*. Cambridge: Cambridge University Press.

Lehmann, Erich L., and George Casella (1998): *Theory of Point Estimation*, Second Edition. New York: Springer.

Lehmann, Erich L., and Joseph P. Romano (2005): *Testing Statistical Hypotheses*, Third Edition. New York: Springer.

Li, Qi, and Jeffrey Racine (2007): *Nonparametric Econometrics*. Princeton, NJ: Princeton University Press.

Linton, Oliver (2017): *Probability, Statistics, and Econometrics*. Cambridge, MA: Academic Press.

Mann, Henry B., and Abraham Wald (1943): "On stochastic limit and order relationships," *Annals of Mathematical Statistics* 14, 217–226.

Marron, James S., and Matt P. Wand (1992): "Exact mean integrated squared error," *Annals of Statistics* 20, 712–736.

Pagan, Adrian, and Aman Ullah (1999): *Nonparametric Econometrics*. Cambridge: Cambridge University Press.

Parzen, Emanuel (1962): "On estimation of a probability density function and mode," *Annals of Mathematical Statistics* 33, 1065–1076.

Pollard, David (1990): *Empirical Processes: Theory and Applications*. Hayward, CA: Institute of Mathematical Statistics.

Ramanathan, Ramu (1993): *Statistical Methods in Econometrics*. Cambridge, MA: Academic Press.

Rosenblatt, Murrey (1956): "Remarks on some non-parametric estimates of a density function," *Annals of Mathematical Statistics* 27, 832–837.

Rudin, Walter (1976): *Principles of Mathematical Analysis*, Third Edition. New York: McGraw Hill.

Rudin, Walter (1987): *Real and Complex Analysis*, Third Edition. New York: McGraw Hill.

Scott, David W. (1992): *Multivariate Density Estimation*. New York: Wiley-Interscience.

Shao, Jun (2003): *Mathematical Statistics*, Second Edition. New York: Springer.

Sheather, Simon J., and M. C. Jones (1991): "A reliable data-based bandwidth selection method for kernel density estimation," *Journal of the Royal Statistical Society, Series B* 53, 683–690.

Silverman, Bernard W. (1986): *Density Estimation for Statistics and Data Analysis*. London: Chapman and Hall.

van der Vaart, Aad W. (1998): *Asymptotic Statistics*. Cambridge: Cambridge University Press.

van der Vaart, Aad W., and Jon A. Wellner (1996): *Weak Convergence and Empirical Processes*. New York: Springer.

Wasserman, Larry (2006): *All of Nonparametric Statistics*. New York: Springer.

White, Halbert (1982): "Instrumental variables regression with independent observations," *Econometrica* 50, 483–499.

White, Halbert (1984): *Asymptotic Theory for Econometricians*. Cambridge MA: Academic Press.

INDEX

www.ingramcontent.com/pod-product-compliance
Ingram Content Group UK Ltd.
Pitfield, Milton Keynes, MK11 3LW, UK
UKHW010047010325
455720UK00003B/4